MARINE FUNGI

Biodiversity, Genomics and Applications

Series

Progress in Mycological Research

- Fungi from Different Environments (2009)
- Systematics and Evolution of fungi (2011)
- Fungi from Different Substrates (2015)
- Fungi: Applications and Management Strategies (2016)
- Advances in Macrofungi: Diversity, Ecology and Biotechnology (2019)
- Advances in Macrofungi: Industrial Avenues and Prospects (2021)
- Advances in Macrofungi: Pharmaceuticals and Cosmeceuticals (2021)
- Fungal Biotechnology: Prospects and Avenues (2022)
- Bioprospects of Macrofungi: Recent Developments (2023)
- Ecology of Macrofungi: An Overview (2024)
- Advances in Cordyceps Research: Prospects and Avenues (2025)
- The Role of Entomopathogenic Fungi in Agriculture (2025)
- Truffles: Diversity, Ecology and Biotechnology (2025)
- Marine Fungi: Biodiversity, Genomics and Applications (2027)

Series: Progress in Mycological Research

MARINE FUNGI

Biodiversity, Genomics and Applications

Editors

Kandikere R. Sridhar
Central Research Laboratory, NITTE (Deemed to be University)
KS Hegde Medical Academy (KSHEMA), Mangalore, India

Sunil K. Deshmukh
Scientific Advisor
Greenvention Biotech Pvt. Ltd. Pune, India

Patricia Velez
Institute of Biology, National Autonomous University of Mexico
Mexico City, Mexico

CRC Press
Taylor & Francis Group
Boca Raton London New York

CRC Press is an imprint of the
Taylor & Francis Group, an **informa** business
A SCIENCE PUBLISHERS BOOK

First edition published 2027
by CRC Press
2385 NW Executive Center Drive, Suite 320, Boca Raton FL 33431

and by CRC Press
4 Park Square, Milton Park, Abingdon, Oxon, OX14 4RN

CRC Press is an imprint of Taylor & Francis Group, LLC

Library of Congress Cataloging-in-Publication Data (applied for)

ISBN: 978-1-041-29214-2 (hbk)
ISBN: 978-1-041-29256-2 (pbk)
ISBN: 978-1-003-76870-8 (ebk)

DOI: 10.1201/9781003768708

Typeset in Times New Roman by
Prime Publishing Services

Preface

Mycology has advanced as a prime area of contemporary research. Besides basic aspects, mycological research has expanded its arena into several applied fields in agriculture, biofuels, biotherapeutics, environment, human health, industries, and medicine. They become a major source of value-added products like antibiotics, beverages, biofuels, biopesticides, drugs, enzymes, natural products, nutraceuticals, organic acids, prebiotics, pigments, polysaccharides, and proteins (Niego et al., 2023; Hyde et al., 2024). The diversity of the fungal kingdom expanded from tiny unicellular organisms (e.g., chytrids) to large multicellular organisms (e.g., mushrooms) with varied morphological and ecological dimensions, having different strategies of life cycle and lifestyle. However, the enormous global diversity of fungi allowed us to explore only 5% of the expected conventional global estimate of 2.2 to 3.8 million (Hawksworth and Lucking, 2017).

Marine and marine-influenced ecosystems are the major biodiversity hotspots for an innumerable number of fungi, which serve as a rich source for biotechnological and bioprospecting applications. There are two main types of marine fungi: the obligate marine fungi survive and sporulate exclusively in the marine environment, while the facultative marine fungi mainly originate from freshwater or terrestrial environments and can grow and sporulate in the marine environments (Kohlmeyer and Kohlmeyer, 1979). These fungi have been isolated from a wide range of substrates (e.g., algae, driftwood, foam, mangrove leaves, pneumatophores, roots, seagrasses, seawater, sediments, seedlings, invertebrates, and wood) (Jones et al., 2019). Currently, 2,254 fungi are considered marine (in 882 genera, representing 12 phyla and 36 classes) (marinefungi.org: 7th December, 2025). The first marine fungus was described from the shores of North Africa about two centuries ago (Amend et al., 2025). However, until today, the diversity and functions of marine fungi remain a 'dark matter' in the marine ecosystems.

The book, *Marine Fungi: Biodiversity, Genomics and Applications*, emphasizes several fundamental aspects (diversity, ecology, genomics, and systematics) as well as applied aspects (bioactive metabolites, human health-promoting bioactives, medicinal potential, and ecosystem health). The current volume comprises ten chapters contributed by 34 authors from seven countries divided into three subdivisions: (1) Diversity and Ecology; (2) Genomics and Phylogeny; (3) Human and Ecosystem Health. The first subdivision, in three chapters, contemplates global diversity, diversity in oxygen minimum zones, and diversity in sandy beaches. The second subdivision discusses genomics, biosynthetic gene clusters, biodiversity, and systematics in Southeast Asia in two chapters. The last subdivision, in five chapters, deals with bioactive potential, bioactive metabolites in health promotion, medicinal potential, therapeutic potential, symbiosis, ecosystem stability, climate resilience, and the importance of yeasts and endophytes.

We anticipate that the expectations of the contributors and editors in following different aspects of marine mycological research and applications will be valuable for readers with a wide interest in marine mycology and allied areas (applied biology, biochemistry, biodiversity, biotechnology, ecology, genetics, marine biology, medicine, metabolites, mutualism, and systematics).

The editors appreciate the punctuality of all the contributors and reviewers who delivered the chapters and scrupulous evaluation on time, respectively. We are indebted to the CRC Press, Boca Raton, USA, for its conscientiousness in fulfilling many official formalities on short notice to bring out this book on time.

Mangalore, Karnataka, India — Kandikere R. Sridhar
Pune, Maharashtra, India — Sunil K. Deshmukh
Mexico City, Mexico — Patricia Velez

References

Amend, A.S., Gunde-Cimerman, N., Coelho, M.A., Durkin, C.A. et al. (2025). Ten pressing questions (and answers) about marine fungi and opportunities for collaborations in the ocean sciences. Oceanography, 38(4): 40–49. https://doi.org/10.5670/oceanog.2025.e402.

Hawksworth, D.L. and Lücking, R. (2017). Fungal diversity revisited: 2.2 to 3.8 million species. Microbiology Spectrum, 5: FUNK-0052-2016. https://doi.org/10.1128/microbiolspec.funk-0052-2016.

Hyde, K.D., Baldrin, P., Chen, Y., Chethana, K.W.T., De Hong, S. et al. (2024). Current trends, limitations, and future research in the fungi? Fungal Divers., 125: 1–71. https://doi.org/10.1007/s13225-023-00532-5.

Jones, E.B.G., Pang, K.L., Abdel-Wahab, M.A., Scholz, B., Hyde, K.D. et al. (2019). An online resource for marine fungi. Fungal Diver., 96(1): 347–433. https://doi.org/10.1007/s13225-019-00426-5.

Kohlmeyer, J. and Kohlmeyer, E. (1979). Marine Mycology: The Higher Fungi. Elsevier-Academic Press, New York and London, 705 pp.

Neigo, A.G.T., Lambert, C., Mortimer, P., Thongklang, N., Rapior, S. et al. (2023). The contribution of fungi to the global economy. Fungal Divers., 121: 95–137. https://doi.org/10.1007/s13225-023-00520-9.

Contents

List of Contributors

Adamantidi, Theodora
Hephaestus Laboratory, School of Chemistry, Faculty of Sciences, Democritus University of Thrace, Kavala University Campus, St. Lucas, 65404 Kavala, Greece.

Alex, Nisha
Department of Marine Biosciences, Faculty of Ocean Science and Technology, Kerala University of Fisheries and Ocean Studies, Kochi, Kerala, India.

Anastasiadou, Chryssa
Fisheries Research Institute, Hellenic Agricultural Organization-DIMITRA (ELGO-DIMITRA), Nea Peramos, 64007 Kavala, Greece.

Augustine, Ashly
Department of Marine Biosciences, Faculty of Ocean Science and Technology, Kerala University of Fisheries and Ocean Studies, Kochi, Kerala, India.

Bagacay, Jan Felnesh Exe
Division of Biological Sciences, College of Arts and Sciences, University of the Philippines Visayas, Miagao, Iloilo, Philippines; Graduate School, University of the Philippines Visayas, Miagao, Iloilo, Philippines.

Bhaskar, Sneha
Department of Studies in Biotechnology, University of Mysore, Manasagangotri, Mysore 570 006, India.

Calabon, Mark S.
Division of Biological Sciences, College of Arts and Sciences, University of the Philippines Visayas, Miagao, Iloilo, Philippines; Graduate School, University of the Philippines Visayas, Miagao, Iloilo, Philippines.

Carmona-Almeraya, G.
Posgrado en Ciencias del Mar y Limnología, Universidad Nacional Autónoma de México, Mexico City, Mexico; Instituto de Biología, Universidad Nacional Autónoma de México, Mexico City, Mexico.

Devadatha, B.
Microbial Type Culture Collection and Gene Bank (MTCC), CSIR-Institute of Microbial Technology, Chandigarh, 160036, India; Academy of Scientific and Innovative Research (AcSIR), Ghaziabad 201002, India.

Dinesh, B.
Department of Biochemistry, Mangalore University, Mangalore 574 199, Karnataka, India.

Gonçalves, Micael F.M.
Centre for Environmental and Marine Studies, Department of Biology, University of Aveiro, 3810-193 Aveiro, Portugal.

Gonçalves, Tiago M.
Centre for Environmental and Marine Studies, Department of Biology, University of Aveiro, 3810-193 Aveiro, Portugal.

Jones E.B. Gareth
Department of Botany and Microbiology, King Saudi University, Riyadh, Saudi Arabia.

Joshi, Chandrashekhar G.
Department of Biochemistry, Mangalore University, Mangalore 574 199, Karnataka, India.

Kini, K. Ramachandra
Department of Studies in Biotechnology, University of Mysore, Manasagangotri, Mysore 570 006, India.

Kumar, Vinit
State Key Laboratory Breeding Base of Green Pesticide and Agricultural Bioengineering, Guizhou University, Guizhou, China.

Manjusha, K.
Department of Marine Biosciences, Faculty of Ocean Science and Technology, Kerala University of Fisheries and Ocean Studies, Kochi, Kerala, India.

Nimsi, K.A.
Department of Biosciences, MES College, Marampally, Aluva, Kerala, India.

Ofrydopoulou, Anna
Fisheries Research Institute, Hellenic Agricultural Organization-DIMITRA (ELGO-DIMITRA), Nea Peramos, 64007 Kavala, Greece.

Opiña, Lilcah Angelique D.
Division of Biological Sciences, College of Arts and Sciences, University of the Philippines Visayas, Miagao, Iloilo, Philippines; Graduate School, University of the Philippines Visayas, Miagao, Iloilo, Philippines..

Papikinou, Mria Aliki
Hephaestus Laboratory, School of Chemistry, Faculty of Sciences, Democritus University of Thrace, Kavala University Campus, St. Lucas, 65404 Kavala, Greece

Pavlidis, Konstantinos
Hephaestus Laboratory, School of Chemistry, Faculty of Sciences, Democritus University of Thrace, Kavala University Campus, St. Lucas, 65404 Kavala, Greece.

Posadas, J.
Posgrado en Ciencias del Mar y Limnología, Universidad Nacional Autónoma de México, Mexico City, Mexico; Instituto de Biología, Universidad Nacional Autónoma de México, Mexico City, Mcxico.

Rapuru, Rushendran
Department of Pharmacology, SRM College of Pharmacy, Faculty of Medicine and Health Sciences, SRM Institute of Science and Technology, Kattankulathur, Chengalpet 603203, Tamil Nadu, India.

Reisinho, Nuno
Centre for Environmental and Marine Studies, Department of Biology, University of Aveiro, 3810-193 Aveiro, Portugal.

Rivera-Zizumbo, Aurora
Instituto de Biología, Universidad Nacional Autónoma de México, Mexico City, Mexico; Posgrado en Ciencias Biológicas, Universidad Nacional Autónoma de México, Mexico City, Mexico.

Sara, Fred Ric C.
Division of Biological Sciences, College of Arts and Sciences, University of the Philippines Visayas, Miagao, Iloilo, Philippines.

Sekhar, Shailasree
Division of Biochemistry, School of Life Science, JSS Academy of Higher Education and Research, Mysore, India.

Shenoy, Priyanka
SBRR Mahajana's first grade college, Mysore, India.

Siccio, Majoerly Augustine B.
Division of Biological Sciences, College of Arts and Sciences, University of the Philippines Visayas, Miagao, Iloilo, Philippines.

Sridhar, Kandikere R.
Central Research Laboratory, NITTE (Deemed to be University), KS Hegde Medical Academy (KSHEMA), Mangalore 575 018, Karnataka, India.

Tsoupras, Alexandros
Hephaestus Laboratory, School of Chemistry, Faculty of Sciences, Democritus University of Thrace, Kavala University Campus, St. Lucas, 65404 Kavala, Greece.

Varma, Konderu Niteesh
Microbial Type Culture Collection and Gene Bank (MTCC), CSIR-Institute of Microbial Technology, Chandigarh, 160036, India.

Velez, Patricia
Instituto de Biología, Universidad Nacional Autónoma de México, Mexico City, Mexico.

Biodiversity and Ecology

1

Global Diversity of Marine Fungi in Underexplored Regions and their Environmental Determinants

Konderu Niteesh Varma,[1] *B. Devadatha*[1,2,*] and *E.B. Gareth Jones*[3,*]

Marine fungi comprise a taxonomically diverse ecological group and are reported from various marine environments on different substrates and habitats. In contrast to other microbes, they remain ignored and underrated worldwide. Marine fungal research is limited to certain regions of the Arctic and Antarctic, Africa, and Australasia, indicating substantial research gaps due to lack of expertise, logistical, environmental factors, and funding challenges. The present study documents the marine fungi reported from these underexplored territories based on already published literature. Our compiled list includes 207 fungal species spreading over 102 different genera of marine fungi reported from the Arctic; 112 fungal species belonging to 60 genera reported from Antarctica; 176 fungal species from 125 distinct genera in African countries, and 149 fungi in 108 genera of marine fungi recorded in Australasia across different locations and various marine habitats. This chapter reviews the impact of different environmental factors on the

[1] Microbial Type Culture Collection (MTCC) and Gene bank, CSIR-Institute of Microbial Technology, Sector 39A, Chandigarh, India.
[2] Academy of Scientific and Innovative Research (AcSIR), Ghaziabad, 201002, India.
[3] Department of Botany and Microbiology, King Saudi University, Riyadh, Saudi Arabia.
* Corresponding authors: devadatha.b@csir.res.in/torperadgj@gmail.com

DOI: 10.1201/9781003768708-2

distribution and composition of various marine fungal communities across varied marine habitats.

1. Introduction

Marine microbial research chiefly focused on bacterial diversity by employing both culturable and metagenomics studies. The studies on marine fungi remained limited globally as compared to other microbes despite the fact that these fungi help in nutrient recycling and play a major role in overall functioning of marine ecosystems. Marine fungi constitute a taxonomically distinct ecological group and are reported from a wide array of substrates and the habitats within the marine environment. They are mainly classified into two types: "Obligate" marine fungi can survive and sporulate solely in the marine environment, while "facultative" marine fungi mostly come from freshwater or terrestrial environments but can grow and sporulate in the marine environment (Kohlmeyer and Kohlmeyer, 1979). They have been isolated from a wide range of substrates, including algae, driftwood, mangrove leaves, roots, pneumatophores, seedlings, wood, seawater, sediments, marine invertebrates, seagrasses, and sea foam (Jones et al., 2015; 2019). Recent advances in marine fungal classification based on DNA barcoding, multi locus phylogenetic analysis, divergence time estimation, and metagenomics have significantly enhanced our understanding of marine fungi. These approaches have expanded molecular databases by incorporating data from previously unclassified and unculturable taxa, thereby advancing our comprehension of their diversity, evolutionary associations and ecological functions within marine milieu (Jones et al., 2019). Few studies on the global diversity of marine fungi have been undertaken, although their occurrence have been well documented over 100 years, with marine mycologists content to survey different localities and produce long lists of species (Hughes, 1986). This was also true for the biogeography of terrestrial fungi. Hughes (1974, 1975, 1986) has been the leading light in underscoring the principles of biogeography and the distribution of marine fungi, and producing many world distribution maps. Subsequent marine mycologists (Boyd and Kohlmeyer, 1982; Jones, 1968) have widely used these maps, but they convey little ecological information. These studies were primarily concerned with global distribution, with fungi characterised as temperate, tropical or cosmopolitan, with specific focus on those species growing on wood (Brooks, 1972). Others classified marine fungi as Arctic-Antarctic, temperate, sub tropical, and tropical (Booth, 1979; Vrijmoed et al., 1982).

Booth and Kenkel (1986) employed various statistical methods to analyse the distribution of 68 marine lignicolous fungi from 31 worldwide collecting sites along with data on temperature, salinity, and proposed the model of cool (eurythermic thermic homeohaline; homeohalothermic; homeothermic euryhaline; euryhalothermic), Mixed (eryhaliothermic), and warm (homeothermic euryhaline; euryhaliothermic; eurythermic homeohyaline; homeohaliothermic) groupings. They acknowledged this was not a definitive assessment, but was intended to encourage discussion. Hughes (1974, 1975) divided the major littoral biogeographical regions based on isothermal lines: Arctic, North Temperate, North sub tropical, Tropical, South sub tropical, South temperate, and Antarctic, and this has been widely accepted. There are limited research studies available on marine fungal communities in various marine habitats, markedly in few developing countries and Polar Regions, which contributes to their poor representation in marine environments. Therefore, this study aims to compile a list of marine fungi documented in the unexplored regions of the world and summarise the factors affecting their diversity and distribution in these marine habitats.

2. Current Knowledge of Marine Fungi

Marine fungi exhibit a global distribution, inhabiting diverse environments ranging from the Arctic to Antarctic, from estuaries to oceans, from shallow waters to the deep-sea, and all substrates so far surveyed (Fig. 1). Currently 2,254 fungi are regarded as marine, in 882 genera, representing 12 phyla and 36 classes (marinefungi.org website as of 7th December, 2025). Thus, all major fungal taxa are to be found in the world's oceans, although some are poorly represented: the Aphelidiomycota, Blastocladiomycota, Chytridiomycota, Mortierellomycota, and Mucoromycota (Hassett et al., 2020). The reason why they are so poorly recovered remains obscure, but different study methods, such as baiting different materials (snake skin, baby hair, pollen) in liquid culture and water sampling for algal blooms, may account for this in part (Breyer and Baltar, 2023; Peng et al., 2024). Ascomycota dominate most marine habitats studied, with the Basidiomycota less well documented (Jones et al., 2019; Calabon et al., 2023; Devadatha et al., 2021).

The predicted number of marine fungi ranges from 10,000 to 380,000, based on the algorithm of 5% of all documented fungal species, which totals between 143,000–623,915 species (Fryar, personal communication, Jones et al., 2025). Documented marine fungi have dramatically increased in the decades 1980–1989, 1990–1999 and 2010–2019 as the result of the widening of substrates/environments undergoing investigation (Calabon, 2023; Jones et al., 2019). These include studies on macroalgae (Ogaki et al., 2019; Poli et al., 2020), corals (Marchese et al., 2021), sponges (Bovio et al., 2018; Pérez-Llano et al., 2023), plastics (Ekanayaka et al., 2022), seawater (Wijayawardene et al., 2022), planktonic marine fungi (Sen et al., 2022; Peng et al., 2024), sandy beaches (Velez et al., 2022; Pang et al., 2023), sediments:193 species in 103 genera in mangrove sediments (Devadatha et al., 2021), ten new Eurotiales species from mud flats in China (Liu et al., 2023), and 38 species from mangrove sediments from Zhanjiang, Guangdong province, China (Li et al., 2023). Substrates with the greatest diversity of marine fungi include mangroves, which account for 850 taxa (Devadatha et al., 2021), and salt marshes with 486 taxa (Calabon, 2021).

However, many potential new marine fungi will be found by exploring the vast array of substrates in the oceans that have yet to be investigated: 24,000 algae occur in marine habitats, 12,000 are "larger" seaweeds and the remainder include unicellular, colonial, and planktonic species (Phillips, 2023). Despite their prevalence in salt marshes and mangroves, very few marine fungi have been reported from fruits and seeds, marine rocks, sea mud and sediments, especially deep-sea sediments (below 1,000 m) which are likely to host marine fungi. Taylor and Cunliffe (2016) highlighted that mycoplankton is a much-neglected and understudied component of

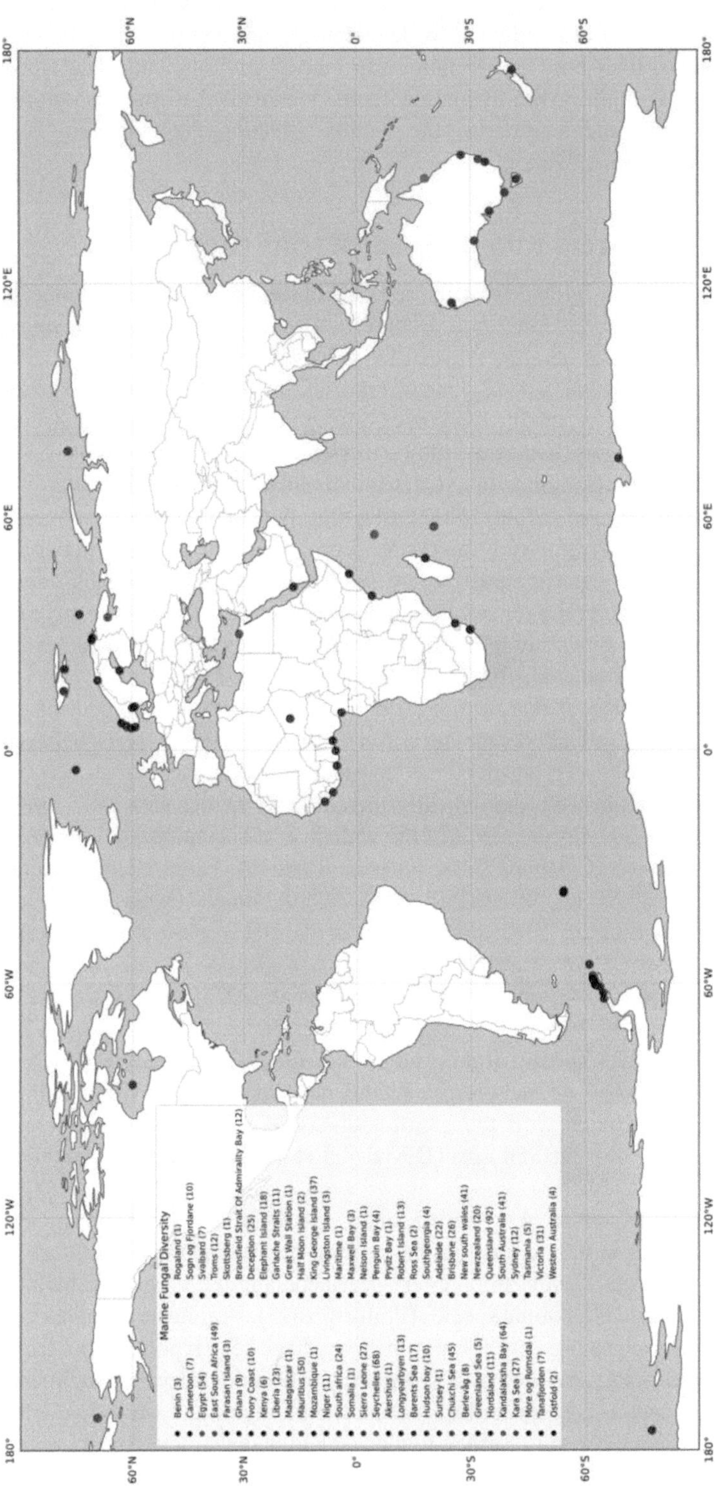

Fig. 1. Global diversity of marine fungi from unexplored regions.

pelagic marine habitat. It is estimated there are 2.2 million marine animals and a potential source of marine fungi: 18,000 marine fish (it is estimated that approximately 3.5 trillion fish live in the oceans); 75,000 crustacean species worldwide, of which 17,500 are decapods; over 3,000 formally described marine sponge species; and 110,000 known marine molluscs (Costello et al., 2017). This huge marine diversity remains untapped in the search for marine fungi. Fungi expected in marine habitats include endophytes of seaweeds, microsporidia and chytrid fungi on various animals (Yarden, 2014; Amend et al., 2019).

With the advances in molecular techniques and sequencing platforms, fungal genomics and metagenomics progressed through high-throughput sequencing (HTS) technologies, including 454-Pyrosequencing, Illumina, Ion Torrent, PacBio, and Oxford Nanopore sequencing platforms (Nilsson et al., 2019). The application of these technologies has resulted in the discovery of unculturable fungal diversity and numerous unidentified lineages of marine fungi, particularly in neglected marine sediments and habitats, highlighting the importance of further exploration and DNA sequencing studies (Amend et al., 2019). Nagano and Nagahama (2012) noted a marine Chytridiomycota lineage in deep-sea sediments; Le Calvez et al. (2009) noted unknown phyllotypes branching within the basidiomycete radiation in a deep-sea hydrothermal vent, 36 novel marine lineages (Richards et al., 2012), and seven deep-branching lineages from the seagrass *Halophila stipulacea* (Abdel-Wahab et al., 2021), and 44 new lineages.

While surveys of different habitats for marine fungi, along with their classification, have dominated their study, other aspects have also been ongoing. Investigating their growth in response to salinity and temperature, salinity is the main limiting parameter of pelagic fungal community composition, while the benthic fungal diversity is shaped by water depth, salinity and sediment C and N availability (Jones and Jennings, 1964; Jennings, 1986; Yang et al., 2023; Mazur-Marzec et al., 2024), their ability to decay wood (Jones and Irvine, 1971; Mouzouros, 1986; Pointing et al., 1998); secondary metabolite production (Ebel, 2012; Yurchenko et al., 2025); and experimental studies into the colonisation of wood, succession and interactions between different species (Alias and Jones, 2000). More recently, studies explored fungal community structure, their role in the breakdown of plastics in the oceans and their potential commercial applications (Wang et al., 2022).

3. Global Marine Fungal Diversity

Pang et al. (2023) reviewed the progress in marine mycology research conducted on a diverse range of substrata and marine habitats across 20 countries. This study provided insights into the global occurrence of marine fungi and compiled a comprehensive number of fungal taxa reported, the current research status, and potential research gaps. Marine fungi are distributed worldwide; however, not all geographic regions have been equally studied (Fig. 1). Many biodiversity-rich coastal countries known for high marine fungal diversity, such as China, India, Thailand, Japan, Malaysia, Brunei, and Florida (USA), contribute substantially to describing novel fungal taxa annually. Nevertheless, marine fungi remain relatively underreported in most of these countries despite their contributions to terrestrial fungal taxonomy. Most of these countries currently have well-established funding and expertise dedicated specifically to marine mycology.

In contrast, marine fungal research is limited to the Arctic and Antarctic polar regions, Africa and parts of Australasia, reflecting significant research gaps due to logistical, environmental and funding challenges. This study documents marine fungi reported from these underexplored regions by referring to published studies up to the present. The marine fungal taxa that have been identified to the species level were considered, whereas those identified only to the genus level were excluded, particularly from the polar, African, and Australasian regions. Individual regional lists were compiled and the total number of documented marine fungal taxa from these regions was tabulated and represented in the maps.

3.1 Diversity and Distribution of Marine Fungi in Underexplored Habitats and Regions

3.1.1 Polar Regions (Arctic and Antarctic)

Marine fungi in Polar regions have developed highly complex physiological and biological mechanisms to resist the freezing temperatures, higher salinity, and lack of sunlight during the winter. The marine fungal diversity composition and distribution is considerably influenced by the global climate change, which is incited by unfavourable weather conditions, increased temperatures and the melting of glaciers in the Polar regions (Gonçalves et al., 2017; Newsham et al., 2022). Both facultative and obligate marine fungi have been reported from a wide range of substrates, including animals, driftwood, invertebrates, macroalgae, plants, and sediments found in polar regions (Godinho et al., 2013; Duarte et al., 2016; Hagestad et al., 2019). Recent studies have revealed that Arctic and Antarctic oceans are rich reservoirs of marine fungal diversity (Hassett and Gradinger, 2016; Jiya et al., 2023). Few marine fungal genera, *viz*., *Cadophora, Geomyces, Thelebolus*, and *Penicillium* found to be present in Antarctic and Arctic environments are known to possess many psychrophilic or psychrotolerant properties. The diversity of marine fungi reported from these polar regions to date is summarised below.

3.1.2 Arctic

Arctic waters are acclaimed for hosting a diverse array of microorganisms inhabiting various substrates and habitats. Nevertheless, in recent years, they have been progressively confronted by fluctuations in rising water temperatures, salinity levels, changes in sunlight permeability, contaminant accumulation, and build-up of sediments, which are mainly attributed to climate change (Wang et al., 2022). According to Grossart et al. (2016), these environmental modifications are having an impact on fungal diversity, species richness, and their ecological functions. The majority of research on marine fungi has relied on the conventional isolation and phenotypic characterization of cultivated fungi from different Arctic marine habitats (Shearer et al., 2007; Pang et al., 2011). Rämä et al. (2017) provided a list of 100 marine fungi characterised morphologically from diverse substrates of the Arctic region. Recent investigations on fungal diversity in Arctic marine sediments, from Barents Sea, Chukchi Sea, Kara Sea, Kandalaksha Bay, Finnmark, Troms and Svalbard revealed diverse and rich fungal communities in the Arctic Ocean (Zhang et al., 2015; Bubnova and Nikitin, 2017; Bubnova and Konovalova, 2019) (Fig. 2).

Studies on diversity, distribution, and composition of marine fungal communities from sediments of Kongsfjorden, Arctic, revealed that *Ascomycota* and *Basidiomycota* are dominant phyla, with *Hypocreales* and *Saccharomycetales* as predominant orders (Zhang et al., 2015). Chen et al. (2023) elucidated the diversity and functions of mycobiomes across nine unique habitats in the High Arctic, using a high-throughput sequencing approach determining a total of 10,419 ASVs of which 2,884 ASVs (amplicon sequence variants) were categorized into 11 phyla and 261 fungal species whereas 7,535 ASVs were classified as unidentified phyla, indicating the prevalence of undescribed fungal taxa. Several studies revealed that fungal diversity distribution and composition is mostly influenced by salinity gradients, with chytrids (*Chytridiomycota*) often dominant among the fungal communities in Arctic benthic habitats (Comeau et al., 2016; Hassett et al., 2017; 2019). Ilicic et al. (2024) opine that Arctic aquatic ecosystems are subject to localised changes in salinity levels due to intensified glacial melting and subsequent terrestrial runoff. High-throughput sequencing analyses of ice cores and sediments from the Arctic marine ecosystem revealed that *Chytridiomycota* predominate among fungal phyla by actively parasitising diatoms and also contribute to the carbon cycle (Hassett and Gradinger, 2016). Few marine fungi have evolved as potential fungal pathogens, such as *Aureobasidium melanogenum*, *Naganishia albida*, and *Rhodotorula mucilaginosa*, due to climate change in Arctic habitats. These fungi can proliferate at 37°C, demonstrate β-haemolytic activity and display resistance to azoles and echinocandins (Perini et al., 2019). Furthermore, Chen et al. (2022) identified 31 fungal pathogens in these environments, comprising seven species classified as Biosafety Level-2 from the High Arctic region. Our list of marine fungi reported from the Arctic shows that 209 fungal species spread over 102 genera (Table 1). The genus *Penicillium* dominates with the highest number of species, that is 37, followed by *Aspergillus*, *Trichoderma*, and *Cladosporium* with 12, 11, and 9, respectively.

3.1.3 Antarctic

The Antarctic ecosystem is one of the pristine regions on Earth, characterized by severe environmental conditions, including arid, frigid temperatures, oligotrophic extremes, extended periods of darkness and intense ultraviolet radiation. However, this ecosystem exhibits a surprisingly very high biodiversity, with an increasing number of species documented annually (Griffiths, 2010; Chown and Brooks, 2019). Antarctica is one of the major reservoirs of microbial diversity that stayed concealed for several decades or longer and is well suited to extreme environmental conditions (Vincent, 2000). These adaptations result from variation in the regulation of genes and metabolic processes, resulting in the production of novel bioactive compounds and metabolites that facilitate effective survival of these microbes in Antarctica (Varrella et al., 2021). The marine fungal diversity of Antarctica is mainly undescribed, rendering it a classical model for understanding fungal

Table 1. Table showing the diversity of marine fungi in the Arctic region (Substrate/Host: A, Algae; C, Coral; F, Seawater/Froth; I, Invertebrate; S, Sediment; U, Unknown substrate; W, Wood/Plant).

Taxon	Family	Substrate/ Host	Distributution	References
Absidia caerulea	*Cunninghamellaceae*	S	Kandalaksha Bay	Khusnullina et al. (2018)
Acremonium murorum	*Bionectriaceae*	S	Barents Sea and Kara Sea, Chukchi Sea	Bubnova and Nikitin 2017)
Acremonium potronii	*Bionectriaceae*	S	Kandalaksha Bay	Khusnullina et al. (2019)
Akanthomyces muscarium	*Cordycipitaceae*	S	Kandalaksha Bay	Schaumann (1975), Khusnullina et al. (2018)
Algochytrops polysiphoniae	*Chytridiaceae*	A	Greenland	ACIA (2004)
Alternaria alternata	*Pleosporaceae*	S	Barents Sea and Kara Sea, Chukchi Sea	Kochkina et al. (2024)
Alternaria botrytis	*Pleosporaceae*	A	Barents Sea, Kara Sea, Chukchi Sea	Blanchette et al. (2010)
Alternaria chartarum	*Pleosporaceae*	S	Barents Sea, Kara Sea, Chukchi Sea	Gams et al. (2003)
Alternaria raphani	*Pleosporaceae*	W, S	Barents Sea, Kara Sea, Chukchi Sea	Björdal (2021)
Amylocarpus encephaloides	*Heliotales*	W	Troms	Pang et al. (2011)
Appendichordella amicta	*Halosphaeriaceae*	W	Hudson Bay	Buchan et al. (2002)
Aspergillus brasiliensis	*Aspergillaceae*	S	Kandalaksha Bay	Varga et al. (2011)
Aspergillus candidus	*Aspergillaceae*	W, S	Chukchi Sea	Björdal (2021)
Aspergillus carneus	*Aspergillaceae*	W, S	Chukchi Sea	Björdal (2021)
Aspergillus cibarius	*Aspergillaceae*	S	Greenland Sea	Kochkina et al. (2024)
Aspergillus flavus	*Aspergillaceae*	S	Barents Sea, Kara Sea, Kandalaksha Bay	Khusnullina et al. (2019)
Aspergillus fumigatus	*Aspergillaceae*	S	Surtsey	Chen et al. (2023)
Aspergillus glaucus	*Aspergillaceae*	A	Chukchi Sea	Blanchette et al. (2010)
Aspergillus niger	*Aspergillaceae*	A, S, W	Chukchi Sea	Björdal (2021); Blanchette et al. (2010)
Aspergillus pulverulentus	*Aspergillaceae*	A	Chukchi Sea	Blanchette et al. (2010)

Table 1 contd. ...

...Table 1 contd.

Taxon	Family	Substrate/ Host	Distributution	References
Aspergillus sydowii	*Aspergillaceae*	S	Kandalaksha Bay	Khusnullina et al. (2019)
Aspergillus tamarii	*Aspergillaceae*	S	Kandalaksha Bay	Khusnullina et al. (2019)
Aspergillus wentii	*Aspergillaceae*	S	Chukchi Sea	Blanchette et al. (2004)
Asteromyces cruciatus	*Pleosporaceae*	W	Tanafjorden	Rämä et al. (2014)
Aureobasidium melanogenum	*Saccotheciaceae*	F	Greenland	Perini et al. (2019)
Aureobasidium pullulans	*Saccotheciaceae*	A	Chukchi Sea	Kuznetsova et al. (2019)
Beauveria bassiana	*Cordycipitaceae*	A, S	Chukchi Sea	Blanchette et al. (2004); Booth (1981)
Cadophora fastigiata	*Ploettnerulaceae*	S	Barents Sea, Kara Sea	Lagerberg et al. (1927)
Cadophora malorum	*Ploettnerulaceae*	F	Chukchi Sea	Bubnova (2005)
Calycina marina	*Pezizellaceae*	A, I	Finnmark, Troms	Alexander et al. (2009); Arenz and Blanchette (2009, 2011); Pang et al. (2011)
Capronia ciliomaris	*Herpotrichiellaceae*	W	Troms	Rämä et al. (2014)
Ceriosporopsis halima	*Halosphaeriaceae*	W	Sogn og Fjordane	Jones et al. (1972)
Chaetomium difforme	*Chaetomiaceae*	A	Chukchi Sea	Blanchette et al. (2010)
Chaetomium globosum	*Chaetomiaceae*	A	Chukchi Sea	Blanchette et al. (2010)
Cirrenalia macrocephala	*Halosphaeriaceae*	W	Sogn og Fjordane	Jones et al. (1972)
Cladosporium cladosporioides	*Cladosporiaceae*	S	Kandalaksha Bay, Barents Sea, Kara Sea	Rämä et al. (2017)
Cladosporium fusiforme	*Cladosporiaceae*	S	Barents Sea, Kara Sea	Booth (1981), Rämä et al. (2017)
Cladosporium herbarum	*Cladosporiaceae*	F	Chukchi Sea	Björdal (2021); Blanchette et al. (2004); Blanchette et al. (2010); Bubnova (2005)
Cladosporium iridis	*Cladosporiaceae*	S	Greenland Sea	Kochkina et al. (2024)
Cladosporium macrocarpum	*Cladosporiaceae*	S	Chukchi Sea	Bubnova (2010)
Cladosporium perangustum	*Cladosporiaceae*	S	Chukchi Sea	Luo et al. (2020)

Table 1 contd. ...

...Table 1 contd.

Taxon	**Family**	**Substrate/ Host**	**Distributution**	**References**
Cladosporium psychrotolerans	*Cladosporiaceae*	S	Barents Sea, Kara Sea	Booth (1981), Rämä et al. (2017)
Cladosporium sphaerospermum	*Cladosporiaceae*	S	Barents Sea, Kara Sea	Rämä et al. (2017)
Cladosporium tenuissimum	*Cladosporiaceae*	S	Chukchi Sea	Luo et al. (2020)
Coprinellus disseminatus	*Psathyrellaceae*	S	Kandalaksha Bay	Khusnullina et al. (2019)
Corollospora maritima	*Halosphaeriaceae*	W	Østfold	Eriksson (1974)
Cremasteria cymatilis	*Incertae sedis*	W	Sogn og Fjordane	Jones et al. (1972)
Debaryomyces hansenii	*Saccharomycetaceae*	A	Chukchi Sea	Booth (1981); Bubnova (2005)
Dictyosporium pelagicum	*Dictyosporiaceae*	W	Hudson Bay	Buchan et al. (2002)
Didymella fucicola	*Didymellaceae*	W	Hordaland	Rämä et al. (2014)
Digitatispora marina	*Niaceae*	W	Troms, Finnmark	Rämä et al. (2014)
Eiona tunicata	*Incertae sedis*	W	Tanafjorden	Rämä et al. (2014)
Emericella versicolor	*Aspergillaceae*	A	Chukchi Sea	Booth (1981)
Emericellopsis fuci	*Bionectriaceae*	A, S	Chukchi Sea	Blanchette et al. (2004); Blanchette et al. (2010); Booth (1981)
Emericellopsis sphaerospora	*Bionectriaceae*	S	Kandalaksha Bay	Rämä et al. (2017)
Emericellopsis tubakii	*Bionectriaceae*	S	Surtsey	Rämä et al. (2014)
Exophiala angulospora	*Herpotrichiellaceae*	I	West-Finnmark	Gjessing et al. (2011)
Exophiala psychrophila	*Herpotrichiellaceae*	I	More og Romsdal	Pederson and Langvad (1989)
Furcasterigmium furcatum	*Plectospharellaceae*	S	Chukchi Sea	Blanchette et al. (2004)
Fusarium culmorum	*Nectriaceae*	S	Kandalaksha Bay	Saccardo (1892); Khusnullina et al. (2019)
Fusarium graminearum	*Nectriaceae*	S	Kandalaksha Bay	Petch (1936); Khusnullina et al. (2019)
Fusarium oxysporum	*Nectriaceae*	S	Kandalaksha Bay	Schlechtendal (1824); Khusnullina et al. (2019)
Fusarium solani	*Nectriaceae*	U	Surtsey	Chen et al. (2023)
Gymnostellatospora bhattii	*Pseudeurotiaceae*	S	Kara Sea	Kochkina et al. (2024)

Table 1 contd. ...

...Table 1 contd.

Taxon	**Family**	**Substrate/ Host**	**Distributution**	**References**
Halenospora varia	*Tricladiaceae*	W	Svalbard	Jones et al. (1972)
Halobyssothecium obiones	*Lentitheciaceae*	W	Svalbard	Jones et al. (1972)
Halokirschsteiniothelia maritima	*Mytilinidiaceae*	W	Troms/Sogn og Fjordane	Rämä et al. (2014)
Halosphaeriopsis mediosetigera	*Halosphaeriaceae*	W	Hudson Bay	Schaumann (1975)
Havispora longyearbyenensis	*Halosphaeriaceae*	W	Svalbard/ Surtsey	Jones et al. (1972)
Humicola alopallonella	*Halosphaeriaceae*	W	Sogn og Fjordane	Rämä et al. (2014)
Lautisporopsis circumvestita	*Halosphaeriaceae*	W	Troms	Pang et al. (2011)
Lentescospora submarina	*Halosphaeriaceae*	W	Tanafjorden	Jones et al. (1972)
Lindra inflata	*Lulworthiaceae*	W	Sogn og Fjordane	Jones et al. (1972)
Lulwoana uniseptata	*Lulworthiaceae*	W	Sogn og Fjordane/ Tanafjorden	Jones et al. (1972)
Lulworthia floridana	*Lulworthiaceae*	W	Akershus	Rämä et al. (2014)
Lulworthia halima	*Lulworthiaceae*	W	Hordaland	Jones et al. (1972)
Lulworthia purpurea	*Lulworthiaceae*	W	Sogn og Fjordane	Rämä et al. (2014)
Malassezia globosa	*Malasseziaceae*		Surtsey	Chen et al. (2023)
Malassezia restricta	*Malasseziaceae*	W	Surtsey	Chen et al. (2023)
Meyerozyma guilliermondii	*Debaryomycetaceae*	S	Kandalaksha Bay	Schaumann (1975)
Microascus trigonosporus	*Microascaceae*	S	Hordaland	Rämä et al. (2014)
Monodictys surtseya	*Incertae sedis*	U	Surtsey	Schaumann (1975)
Mrakia blollopis	*Mrakiaceae*	W	Surtsey	Wehmer (1903)
Mucor hiemalis	*Mucoraceae*	S	Kandalaksha Bay	Khusnullina et al. (2018)
Mucor zonatus	*Mucoraceae*	S	Kara Sea	Kochkina et al. (2024)
Naganishia albida	*Filobasidiaceae*	F	Surtsey	Chen et al. (2023)
Naïs inornata	*Halosphaeriaceae*	W	Troms	Eriksson (1974)
Nectriella laminariae	*Bionectriaceae*	W	Rogaland	Arenz and Blanchette (2011); Buchan et al. (2002)
Neosetophoma cerealis	*Phaeosphaeriaceae*	A	Chukchi Sea	Booth (1981)
Nereiospora comata	*Halosphaeriaceae*	W	Hudson Bay, Finnmark	Barr (1980); Buchan et al. (2002)

Table 1 contd. ...

...Table 1 contd.

Taxon	Family	Substrate/ Host	Distributution	References
Nereiospora pelagica	*Halosphaeriaceae*	W	Tanafjorden	Rämä et al. (2014)
Nereiospora cristata	*Halosphaeriaceae*	W	Tanafjorden, Hudson Bay, Finnmark, Chukchi Sea	Rämä et al. (2014); Kuznetsova et al. (2019)
Neurospora tetraspora	*Sordariaceae*	A	Chukchi Sea	Kuznetsova et al. (2019)
Oidiodendron periconioides	*Myxotrichaceae*	S	Kandalaksha Bay, Kara Sea	Kochkina et al. (2024)
Oidiodendron tenuissimum	*Myxotrichaceae*	S	Kara Sea	Kochkina et al. (2024)
Olpidium maritimum	*Olpidiaceae*	W, S, F	Chukchi Sea	Blanchette et al. (2004); Blanchette et al. (2010)
Paracorollospora luteola	*Halosphaeriaceae*	W	Tanafjorden	Rämä et al. (2014)
Paradendryphiella arenariae	*Pleosporaceae*	A, S	Chukchi Sea	Woudenberg et al. (2013); Jones et al. (1972)
Paradendryphiella salina	*Pleosporaceae*	S, W	Kandalaksha Bay, Sogn og Fjordane	Khusnullina et al. (2019)
Parengyodontium album	*Cordycipitaceae*	S	Kandalaksha Bay	Hocking and Mcrae (1999)
Penicillium antarcticum	*Aspergillaceae*	S	Kandalaksha Bay	Khusnullina et al. (2019)
Penicillium aurantiogriseum	*Aspergillaceae*	S	Kandalaksha Bay	Dierckx (1901); Khusnullina et al. (2019)
Penicillium brevicompactum	*Aspergillaceae*	S	Kandalaksha Bay	Björdal (2021); Blanchette et al. (2010); Blanchette et al. (2016)
Penicillium canescens	*Aspergillaceae*	S	Chukchi Sea	Woudenberg et al. (2013)
Penicillium chrysogenum	*Aspergillaceae*	S	Chukchi Sea, Kandalaksha Bay	Thom (1910), Khusnullina et al. (2019)
Penicillium citrinum	*Aspergillaceae*	S	Kandalaksha Bay	Blanchette et al. (2016)
Penicillium commune	*Aspergillaceae*	S	Chukchi Sea	Woudenberg et al. (2013)
Penicillium corylophilum	*Aspergillaceae*	S	Kara Sea	Banke et al. (1997)
Penicillium dipodomyis	*Aspergillaceae*	S	Kandalaksha Bay	Khusnullina et al. (2018)

Table 1 contd. ...

...Table 1 contd.

Taxon	**Family**	**Substrate/ Host**	**Distributution**	**References**
Penicillium echinulatum	*Aspergillaceae*	S	Kandalaksha Bay	Khusnullina et al. (2019)
Penicillium expansum	*Aspergillaceae*	S	Kandalaksha Bay	Khusnullina et al. (2018)
Penicillium glabrum	*Aspergillaceae*	S	Barents Sea	Zaleski (1928); Khusnullina et al. (2019)
Penicillium janczewskii	*Aspergillaceae*	S	Kandalaksha Bay	Zaleski (1927); Khusnullina et al. (2019)
Penicillium jensenii	*Aspergillaceae*	S	Kandalaksha Bay	Westling (1911); Khusnullina et al. (2019)
Penicillium lagena	*Aspergillaceae*	S	Kara Sea	Peterson et al. (2015)
Penicillium lanosum	*Aspergillaceae*	S	Kandalaksha Bay	Khusnullina et al. (2018)
Penicillium lemhiflumine	*Aspergillaceae*	S	Kandalaksha Bay	Khusnullina et al. (2018)
Penicillium madriti	*Aspergillaceae*	S	Barents Sea, Kara Sea	Orpurt and Fennell (1955); Blanchette et al. (2016)
Penicillium megasporum	*Aspergillaceae*	S	Barents Sea, Kara Sea	Blanchette et al. (2016)
Penicillium miczynskii	*Aspergillaceae*	S	Chukchi Sea	Novak and Backus (1963); Kuznetsova et al. (2019)
Penicillium montanense	*Aspergillaceae*	S	Kandalaksha Bay	Currie and Thom (1915); Khusnullina et al. (2019)
Penicillium oxalicum	*Aspergillaceae*	S	Kandalaksha Bay	Khusnullina et al. (2018)
Penicillium paneum	*Aspergillaceae*	W	Chukchi Sea	Blanchette et al. (2004); Kuznetsova et al. (2019)
Penicillium paxilli	*Aspergillaceae*	S	Chukchi Sea	Woudenberg et al. (2013)
Penicillium purpurogenum	*Aspergillaceae*	W	Chukchi Sea	Blanchette et al. (2004); Blanchette et al. (2010); Booth (1981)
Penicillium raistrickii	*Aspergillaceae*	A, S	Chukchi Sea	Blanchette et al. (2004); Kuznetsova et al. (2019)

Table 1 contd. ...

...Table 1 contd.

Taxon	Family	Substrate/ Host	Distributution	References
Penicillium restrictum	*Aspergillaceae*	S	Chukchi Sea	Thom (1906); Kuznetsova et al. (2019)
Penicillium roqueforti	*Aspergillaceae*	S	Kandalaksha Bay	Dierckx (1901); Khusnullina et al. (2018)
Penicillium roseopurpureum	*Aspergillaceae*	S	Kandalaksha Bay	Biourge (1923); Khusnullina et al. (2018)
Penicillium sanguifluum	*Aspergillaceae*	S	Kandalaksha Bay	Thom (1930); Khusnullina et al. (2018)
Penicillium simplicissimum	*Aspergillaceae*	S	Kandalaksha Bay	Westling (1911); Khusnullina et al. (2018)
Penicillium solitum	*Aspergillaceae*	S	Kandalaksha Bay	Thom (1910); Khusnullina et al. (2018)
Penicillium spinulosum	*Aspergillaceae*	S	Kandalaksha Bay	Khusnullina et al. (2018)
Penicillium swiecickii	*Aspergillaceae*	S	Kara Sea	Maire (1917); Khusnullina et al. (2018)
Penicillium thomii	*Aspergillaceae*	S	Kandalaksha Bay	Dierckx (1901); Khusnullina et al. (2018)
Penicillium verrucosum	*Aspergillaceae*	S	Kandalaksha Bay	Zaleski (1927); Khusnullina et al. (2018)
Penicillium waksmanii	*Aspergillaceae*	S	Kandalaksha Bay	Khusnullina et al. (2018)
Peroneutypa scoparia	*Diatrypaceae*	S	Chukchi Sea	Luo et al. (2020)
Phaeosphaeria neomaritima	*Phaeosphaeriaceae*	W	Chukchi Sea	Kohlmeyer (1973); Kuznetsova et al. (2019)
Phanerodontia chrysosporium	*Aspergillaceae*	S	Kara Sea	Brewer (1958); Khusnullina et al. (2018)
Pharcidia laminaricola	*Mycosphaerellaceae*	W	Hordaland	Kohlmeyer (1973)
Pharcidia rhachiana	*Mycosphaerellaceae*	W	Hordaland	Kohlmeyer (1973)
Phellinidium ferrugineofuscum	*Hymenochaetaceae*	S	Greenland Sea	Kochkina et al. (2024)
Phenoliferia glacialis	*Sporidiobolaceae*	W	Surtsey	Pang et al. (2011)
Phenoliferia psychrophenolica	*Sporidiobolaceae*	W	Surtsey	Pang et al. (2011)
Phialophorophoma litoralis	*Didymellaceae*	W	Troms	Langvad (1989), Rämä et al. (2014)

Table 1 contd. ...

...Table 1 contd.

Taxon	**Family**	**Substrate/ Host**	**Distributution**	**References**
Phoma herbarum	*Didymellaceae*	I	Unknown	Rämä et al. (2014)
Phycomelaina laminariae	*Phyllachoraceae*	-	Hordaland	Kohlmeyer (1973)
Pleospora pelvetiae	*Pleosporaceae*	I	Hordaland	Rehm (1896); Kohlmeyer (1973)
Pseudeurotium hygrophilum	*Pseudeurotiaceae*	S	Kandalaksha Bay	Khusnullina et al. (2018)
Pseudogymnoascus appendiculatus	*Pseudeurotiaceae*	S	Kara Sea	Minnis and lindner (2013)
Pseudogymnoascus pannorum	*Pseudeurotiaceae*	S	Kandalaksha Bay	Arenz and Blanchette (2011); Buchan et al. (2002)
Remispora maritima	*Halosphaeriaceae*	W	Hordaland/ Troms	Arenz and Blanchette (2011)
Remispora pileata	*Halosphaeriaceae*	W	Svalbard, Finnmark	Pang et al. (2011)
Remispora quadri-remis	*Halosphaeriaceae*	W	Hudson Bay, Finnmark, Troms	Schaumann (1975); Rämä et al. (2014); Pang et al. (2011)
Remispora spitsbergensis	*Halosphaeriaceae*	W	Svalbard	Rämä et al. (2014)
Remispora stellata	*Halosphaeriaceae*	W	Svalbard	Buchan et al. (2002)
Rhizophydium angulosum	*Rhizophydiaceae*	W, S	Hudson Bay	Buchan et al. (2002)
Rhizophydium sphaerotheca	*Rhizophydiaceae*	W, S	Hudson Bay	Fischer (1892); Buchan et al. (2002)
Rhizopus arrhizus	*Mucoraceae*	S	Kandalaksha Bay	Khusnullina et al. (2018)
Rhodotorula mucilaginosa	*Sporidiobolaceae*	F	Surtsey	Chen et al. (2023)
Robbauera albescens	*Robbaueraceae*	A	Chukchi Sea	Gams (1971); Kuznetsova et al. (2019)
Sablicola chinensis	*Halosphaeriaceae*	W	Troms	Rämä et al. (2014)
Sarocladium kiliense	*Sarocladiaceae*	S	Surtsey	Summerbell et al. (2011)
Sarocladium strictum	*Sarocladiaceae*	S	Kandalaksha Bay, Barents Sea and Kara Sea	Hughes (1958); Khusnullina et al. (2018)
Scopulariopsis asperula	*Microascaceae*	S	Kandalaksha Bay	Duval (1935); Khusnullina et al. (2018)

Table 1 contd. ...

...Table 1 contd.

Taxon	**Family**	**Substrate/ Host**	**Distributution**	**References**
Scopulariopsis brumptii	*Microascaceae*	S	Barents Sea and Kara Sea	Zare and gams (2001); Khusnullina et al. (2018)
Simplicillium lamellicola	*Cordycipitaceae*	S	Kandalaksha Bay	Khusnullina et al. (2018)
Sphaerulina orae-maris	*Mycosphaerellaceae*	W	Troms	Buchan et al. (2002)
Spizellomyces palustris	*Spizellomycetaceae*	S	Hudson Bay	Hughes (1958); Khusnullina et al. (2018)
Stachybotrys chartarum	*Stachybotryaceae*	S	Kandalaksha Bay	Rämä et al. (2014)
Stagonospora abundata	*Phaeosphaeriaceae*	I	Hordaland	Kohlmeyer (1973)
Stagonospora haliclysta	*Phaeosphaeriaceae*	I	Hordaland	Rämä et al. (2014)
Stemphylium triglochinicola	*Pleosporaceae*	I	Tanafjorden	Minnis and lindner (2013)
Stigmidium ascophylli	*Mycosphaerellaceae*	I	Hordaland	Samson et al. (2011)
Talaromyces diversus	*Aspergillaceae*	S	Kandalaksha Bay	Samson et al. (2011)
Talaromyces funiculosus	*Aspergillaceae*	S	Kandalaksha Bay	Samson et al. (2011)
Talaromyces minioluteus	*Aspergillaceae*	S	Kandalaksha Bay	Khusnullina et al. (2018)
Talaromyces proteolyticus	*Aspergillaceae*	S	Kara Sea	Samson et al. (2011)
Talaromyces variabilis	*Aspergillaceae*	S	Kandalaksha Bay	Samson et al. (2011)
Talaromyces verruculosus	*Aspergillaceae*	S	Kandalaksha Bay	Khusnullina et al. (2018)
Thelebolus ellipsoideus	*Thelebolaceae*	S	Chukchi Sea	Booth (1981); Kuznetsova et al. (2019)
Tilletiopsis albescens		A	Chukchi Sea	Luo et al. (2020)
Tolypocladium cylindrosporum	*Ophiocordycipitaceae*	S	Kandalaksha Bay	Gams (1971); Khusnullina et al. (2018)
Tolypocladium inflatum	*Ophiocordycipitaceae*	S	Kandalaksha Bay	Bissett (1983); Khusnullina et al. (2018)
Tolypocladium tundrense	*Ophiocordycipitaceae*	S	Kandalaksha Bay	Artemchuk (1972); Barr (1980); Buchan et al. (2002)

Table 1 contd. ...

...Table 1 contd.

Taxon	Family	Substrate/ Host	Distributution	References
Toriella tubulifera	*Halosphaeriaceae*	W	Hudson bay, Finmark	Jones et al. (1972); Buchan et al. (2002)
Torpedospora radiata	*Torpedosporaceae*	W	Sogn og Fjordane	Karsten (1892); Jones et al. (1972)
Trichoderma aureoviride	*Hypocreaceae*	S	Kandalaksha Bay	Blanchette et al. (2004); Blanchette et al. (2016)
Trichoderma citrinoviride	*Hypocreaceae*	W	Ostfold	Eriksson (1974)
Trichoderma hamatum	*Hypocreaceae*	S	Chukchi Sea	Kuznetsova et al. (2019)
Trichoderma harzianum	*Hypocreaceae*	S	Kandalaksha Bay	Björdal (2021); Blanchette et al. (2004); Blanchette et al. (2016)
Trichoderma koningii	*Hypocreaceae*	S	Chukchi Sea	Bissett (1992); Kuznetsova et al. (2019)
Trichoderma minutisporum	*Hypocreaceae*	S	Kandalaksha Bay	Webster et al. (1969)
Trichoderma paraviridescens	*Hypocreaceae*	S	Kara Sea	Bjørdal (2019, 2021); Blanchette et al. (2010); Bubnova (2005); Bubnova (2010)
Trichoderma parceramosum	*Hypocreaceae*	S	Kandalaksha Bay	Rifai (1969); Khusnullina et al. (2018)
Trichoderma piluliferum	*Hypocreaceae*	S	Kandalaksha Bay	Rifai (1969); Khusnullina et al. (2018)
Trichoderma polysporum	*Hypocreaceae*	S	Kandalaksha Bay	Khusnullina et al. (2018)
Trichoderma viride	*Hypocreaceae*	A, W, S	Chukchi Sea	Bissett (1992); Kuznetsova et al. (2019)
Trichosporon asahii	*Trichosporonaceae*		Surtsey	Chen et al. (2023)
Tubakiella galerita	*Halosphaeriaceae*	W	Svalbard	Buchan et al. (2002)
Uebelmesseromyces harderi	*Uebelmesseromycetaceae*	W, S	Hudson Bay	Simmons (1967); Buchan et al. (2002)
Umbelopsis isabellina	*Umbelopsidaceae*	S	Barents Sea, Kara Sea	Gams et al. (2003)

Table 1 contd. ...

...Table 1 contd.

Taxon	Family	Substrate/ Host	Distributution	References
Umbelopsis ramanniana	*Umbelopsidaceae*	S	Kandalaksha Bay	Arx (1982); Khusnullina et al. (2018)
Umbelopsis vinacea	*Umbelopsidaceae*	S	Kandalaksha Bay	Khusnullina et al. (2018)
Verruciconidia persicina	*Bionectriaceae*	S	Greenland Sea	Kochkina et al. (2024)
Wickerhamomyces anomalus	*Wickerhamomycetaceae*	U	Surtsey	Chen et al. (2023)

evolution. Da Silva et al. (2022) studies revealed high diversity and fungal richness, with 133 distinct fungal amplicon sequence variants (ASVs) in various phyla, with *Agaricomycetes* sp., *Chaetoceros* sp., and *Ciliophora* sp. as the predominant taxa of marine fungi in sediments of the South Shetland Islands archipelago of Antarctica (Fig. 3). Zhang et al. (2022) investigated marine fungal diversity across eleven diverse habitats in an ice-free region of maritime Antarctica, which detected a total 14,814 fungal ASVs with 313 recognised genera and 320 identified species, indicating the significant diversity of fungal communities.

The DNA metagenomic analysis from the Larsemann Hills, Antarctic ecosystem, covering four diverse habitats, including the deep sea, exposed lands, lakes and snow-covered areas, resulted in *Ascomycota* as the dominant phylum, followed by *Basidiomycota* and others. Among these habitats, lake sediments revealed the highest amplicon sequence variants in contrast to others (Jiya et al., 2023). Our documented list includes 114 fungal species in 60 genera reported from Antarctica (Table 2). The genus *Penicillium* (15 species) hosts the highest number of species, followed by *Aspergillus* (7 species) and other fungal species. Despite several studies investigating marine fungal diversity through metagenomics and culture-based approaches in algae, seawater, and sediments, a significant portion of this fungal diversity remains molecularly unexamined.

3.1.4 Africa

Limited studies on marine fungi have been conducted in several African countries, including Egypt (North Africa), Kenya, Mauritius and Seychelles (East Africa), Ghana, Nigeria, and Niger (West Africa), as well as South Africa (Southern Africa), as illustrated in Fig. 4. Even though it has a vast coastal area with significant mangroves, most of these regions are poorly studied despite being recognised as marine biodiversity hotspots. The earliest account of marine fungi in Kenya was provided by Erneholm (1972). Kohlmeyer and Kohlmeyer (1971) and Jones et al. (1972) documented three and nine marine fungal species from the Durban coast along the Indian Ocean coast, while Gorter (1978) collected ten species from the Atlantic coast. Steinke and Jones (1993) documented 93 marine fungi across three distinct habitats: mangroves, intertidal rocky coasts, and sandy beaches on the east coast of South Africa, with mangroves hosting the highest number of species (55).

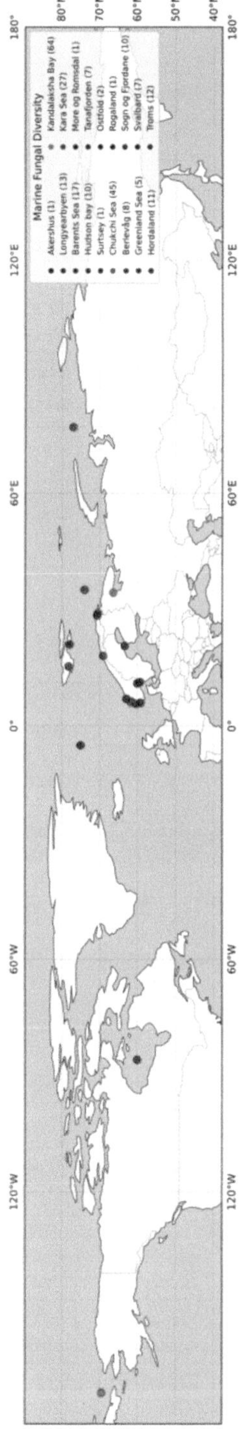

Fig. 2. Diversity of marine fungi reported in various regions within the Arctic.

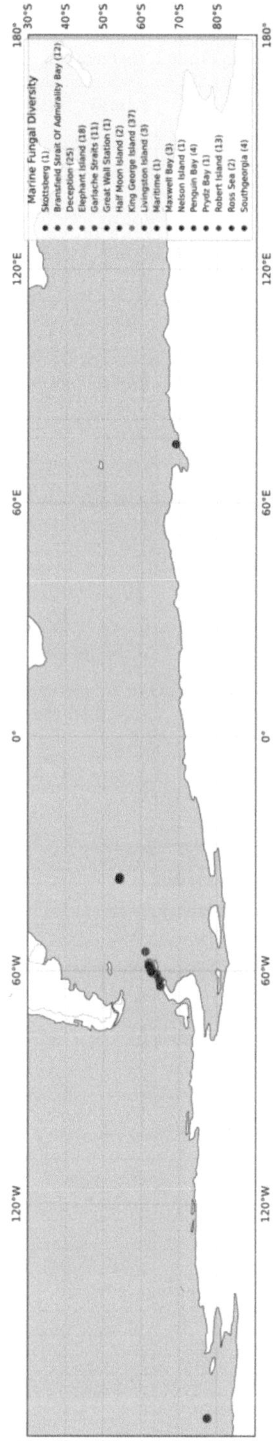

Fig. 3. Diversity of marine fungi reported across various regions within the Antarctic.

Table 2. Table showing the diversity of marine fungi in the Antarctic region (Substrate/Host: A, Algae; C, Coral; F, Seawater/Froth; I, Invertebrate; S, Sediment; U, Unknown substrate; W, Wood/Plant).

Taxon	Family	Substrate/ Host	Distribution	References
Akanthomyces attenuatum	*Cordycipitaceae*	F	Brainsfield and Garlache Straits	Gonçalves et al. (2017)
Akanthomyces muscarium	*Cordycipitaceae*	F	Ross Sea	Varella et al. (2021)
Antarctomyces pellizariae	*Thelebolaceae*	A	King George Island	Furbino et al. (2018)
Antarctomyces psychrotrophicus	*Thelebolaceae*	A	Elephant Island, King George Island	Loque et al. (2010); Godinho et al. (2013); Furbino et al. (2014)
Anthracocystis penniseti	*Ustilaginaceae*	A	King George Island	Duarte et al. (2016)
Aspergillus conicus	*Aspergillaceae*	A	Deception	Godinho et al. (2013)
Aspergillus glaucus	*Aspergillaceae*	A	Elephant Island	Godinho et al. (2013)
Aspergillus protuberus	*Aspergillaceae*	A	Elephant Island	Furbino et al. (2014)
Aspergillus pseudoglaucus	*Aspergillaceae*	F	Brainsfield and Garlache Straits	Gonçalves et al. (2017)
Aspergillus repens	*Aspergillaceae*	A	Elephant Island	Godinho et al. (2013)
Aspergillus tabacinus	*Aspergillaceae*	A	Deception	Furbino et al. (2014)
Aspergillus terreus	*Aspergillaceae*	A	Elephant Island	Godinho et al. (2013)
Aureobasidium pullulans	*Saccotheciaceae*	A	King George Island	Loque et al. (2010)
Beauveria bassiana	*Cordycipitaceae*	A	King George Island	Furbino et al. (2018)
Cadophora dextrinospora	*Mollisiaceae*	I	King George Island, Deception	Godinho et al. (2019)
Cadophora luteo-olivacea	*Ploettnerulaceae*	S	King George Island	Varella et al. (2021)
Cadophora malorum	*Ploettnerulaceae*	A	Elephant Island	Furbino et al. (2014)
Candida sake	*Debaryomycetaceae*	I	King George Island	Godinho et al. (2013); Duarte et al. (2016)

Table 2 contd. ...

...*Table 2 contd.*

Taxon	**Family**	**Substrate/ Host**	**Distribution**	**References**
Candida spencermartinsiae	*Debaryomycetaceae*	F	King George Island	Vaz et al. (2011)
Candida zeylanoides	*Debaryomycetaceae*	F	Brainsfield and Garlache Straits	Vaz et al. (2011)
Ceriosporopsis halima	*Halosphaeriaceae*	W	Brainsfield and Garlache Straits	Pugh and Jones (1986)
Cladosporium lignicola	*Cladosporiaceae*	A	Elephant Island	Furbino et al. (2014)
Cladosporium sphaerospermum	*Cladosporiaceae*	F	Brainsfield and Garlache Straits	Gonçalves et al. (2017)
Cladosporium tenuissimum	*Cladosporiaceae*	A	Robert Island	Godinho et al. (2013)
Cladosporium verrucocladosporioides	*Cladosporiaceae*	I	King George Island, Robert Island, Elephant Island	Godinho et al. (2013, 2019); Furbino et al. (2014, 2018)
Clavispora lusitaniae	*Metschnikowiaceae*	I	Deception	Godinho et al. (2019)
Coprinellus radians	*Psathyrellaceae*	A	Robert Island	Furbino et al. (2018)
Curcurbitinus constrictum	*Halosphaeriaceae*	W	Penguin Bay, South Georgia	Grasso et al. (1997)
Cystobasidium laryngis	*Cystobasidiaceae*	A	King George Island	Furbino et al. (2014)
Cystobasidium minutum	*Sporidiobolaceae*	A	King George Island	Furbino et al. (2014)
Cystobasidium pinicola	*Sporidiobolaceae*	A	King George Island	Duarte et al. (2016)
Cystobasidium slooffiae	*Cystobasidiaceae*	I	Maxwell Bay, King George Island, Brainsfield and Garlache Straits	Gonçalves et al. (2017)
Cystofilobasidium infirmominiatum	*Cystofilobasidiaceae*	A	King George Island	Furbino et al. (2014)

Table 2 contd. ...

...Table 2 contd.

Taxon	**Family**	**Substrate/ Host**	**Distribution**	**References**
Debaryomyces hansenii	*Saccharomycetaceae*	A	Deception, Robert Island	Godinho et al. (2013)
Didymella longicolla	*Didymellaceae*	I	Maritime	Godinho et al. (2019)
Dioszegia athyri	*Cryptococcaceae*	A	Deception	Duarte et al. (2016)
Dioszegia hungarica	*Cryptococcaceae*	A	King George Island	Rovati et al. (2013)
Dioszegia xingshanensis	*Cryptococcaceae*	A	King George Island	Duarte et al. (2016)
Exophiala xenobiotica	*Herpotrichiellaceae*	F	Brainsfield and Garlache Straits	Gonçalves et al. (2017)
Filobasidium magnum	*Filobasidiaceae*	A	Deception	Duarte et al. (2016)
Geomyces laevis	*Pseudeurotiaceae*	A	Elephant Island	Godinho et al. (2013)
Geotrichum australiense	*Dipodascaceae*	A	King George Island	Furbino et al. (2014)
Geotrichum silvicola	*Dipodascaceae*	I	Deception	Godinho et al. (2019)
Glaciozyma antarctica	*Camptobasidiaceae*	F	Brainsfield and Garlache Straits	Gonçalves et al. (2017); Varella et al. (2021)
Glaciozyma litorale	*Camptobasidiaceae*	A	King George Island, Robert Island	Duarte et al. (2016)
Glaciozyma martinii	*Camptobasidiaceae*	A	King George Island	Duarte et al. (2016)
Goffeauzyma gilvescens	*Filobasidiaceae*	A	King George Island	Rovati et al. (2013)
Graphium rubrum	*Graphiaceae*	F	Brainsfield and Garlache Straits	Gonçalves et al. (2017)
Guehomyces pullulans	*Cystofilobasidiaceae*	A	King George Island	Furbino et al. (2014)
Halosphaeriopsis mediosetigera	*Halosphaeriaceae*	W	Penguin Bay, South Georgia	Grasso et al. (1997)
Holtermanniella festucosa	*Holtermanniellaceae*	A	King George Island	Duarte et al. (2016)
Holtermanniella nyarrowii	*Holtermanniellaceae*	A	King George Island	Duarte et al. (2016)

Table 2 contd. ...

...*Table 2 contd.*

Taxon	**Family**	**Substrate/ Host**	**Distribution**	**References**
Leucosporidiella creatinivorum	*Leucosporidiaceae*	A, F	King George Island	Vaz et al. (2011); Vaca et al. (2013)
Leucosporidiella fragarium	*Leucosporidiaceae*	A	King George Island	Duarte et al. (2016); Furbino et al. (2018)
Leucosporidiella muscorum	*Leucosporidiaceae*	A, S	King George Island, Robert Island	Vaz et al. (2011); Duarte et al. (2016); Furbino et al. (2018)
Leucosporidium scottii	*Leucosporidiaceae*	S, F	King George Island	Vaz et al. (2011)
Macalpinomyces spermophorus	*Ustilaginaceae*	A	Robert Island	Duarte et al. (2016)
Metschnikowia bicuspidata	*Metschnikowiaceae*	A, F	Livingston Island, Robert Island, Nelson Island, Half-moon Island, King George Island, Deception	Loque et al. (2010); Vaz et al. (2011); Godinho et al. (2013); Furbino et al. (2014, 2018); Duarte et al. (2016)
Meyerozyma caribbica	*Debaryomycetaceae*	A	Deception	Godinho et al. (2013)
Meyerozyma guilliermondii	*Debaryomycetaceae*	A	King George Island, Deception	Furbino et al. (2014); Godinho et al. (2013)
Mortierella antarctica	*Mortierellaceae*	A	Elephant Island	Furbino et al. (2014)
Mortierella elongatula	*Mortierellaceae*	I	Deception	Godinho et al. (2019)
Mortierella fimbricystis	*Mortierellaceae*	I	Deception	Godinho et al. (2019)
Mrakia blollopis	*Mrakiaceae*	F	Brainsfield and Garlache Straits	Godinho et al. (2019)
Mrakia frigida	*Mrakiaceae*	A	Brainsfield and Garlache Straits	Godinho et al. (2019)
Mycoarthris cf. corallina	*Hyaloscyphaceae*	A	Elephant Island	Godinho et al. (2013)
Naganishia adeliensis	*Filobasidiaceae*	A	King George Island	Furbino et al. (2014)

Table 2 contd. ...

...Table 2 contd.

Taxon	Family	Substrate/ Host	Distribution	References
Naganishia albidosimilis	*Filobasidiaceae*	A	King George Island	Furbino et al. (2014)
Oidiodendron truncatum	*Myxotrichaceae*	A	Elephant Island	Furbino et al. (2014)
Papiliotrema laurentii	*Cryptococcaceae*	A	Deception	Godinho et al. (2013)
Penicillium camerunense	*Aspergillaceae*	I	Deception	Godinho et al. (2019)
Penicillium cavernicola	*Aspergillaceae*	I	Deception	Godinho et al. (2019)
Penicillium chrysogenum	*Aspergillaceae*	A, F	King George Island, Brainsfield and Garlache Straits	Gonçalves et al. (2017)
Penicillium citreonigrum	*Aspergillaceae*	S	Great wall Station	Varella et al. (2021)
Penicillium citreosulfuratum	*Aspergillaceae*	F	Brainsfield and Garlache Straits	Gonçalves et al. (2017)
Penicillium citrinum	*Aspergillaceae*	A	Elephant Island, Deception	Godinho et al. (2013); Furbino et al. (2014)
Penicillium discolor	*Aspergillaceae*	A	Elephant Island	Godinho et al. (2013)
Penicillium glabrum	*Aspergillaceae*	S	Ross Sea	Ha et al. (2020)
Penicillium goetzii	*Aspergillaceae*	I	Deception	Godinho et al. (2019)
Penicillium granulatum	*Aspergillaceae*	S	Prydz Bay	Varella et al. (2021)
Penicillium herpotrichoides	*Aspergillaceae*	A	King George Island	Godinho et al. (2019)
Penicillium rubens	*Aspergillaceae*	F	King George Island	Gonçalves et al. (2017)
Penicillium solitum	*Aspergillaceae*	S, A	Bransfield Strait of Admirality Bay, King George Island, Elephant Island	Gonçalves et al. (2013); Furbino et al. (2014)

Table 2 contd. ...

...Table 2 contd.

Taxon	Family	Substrate/ Host	Distribution	References
Penicillium spinulosum	*Aspergillaceae*	A	Elephant Island	Godinho et al. (2013)
Penicillium tardochrysogenum	*Aspergillaceae*	I	Deception	Godinho et al. (2019)
Phaeosphaeria herpotrichoides	*Phaeosphaeriaceae*	A	King George Island	Loque et al. (2010)
Phenoliferia glacialis	*Sporidiobolaceae*	A	King George Island	Duarte et al. (2016)
Piricauda pelagica	*Incertae sedis*	W	King George Island	Pugh and Jones (1986)
Polypaecilum botryoides	*Aspergillaceae*	I	Deception	Godinho et al. (2019)
Pseudogymnoascus appendiculatus	*Pseudeurotiaceae*	I	Robert Island, Livingston, King George Island	Godinho et al. (2019)
Pseudogymnoascus destructans	*Pseudeurotiaceae*	A	Deception	Godinho et al. (2013)
Pseudogymnoascus pannorum	*Pseudeurotiaceae*	A	King George Island	Loque et al. (2010)
Pseudogymnoascus verrucosus	*Pseudeurotiaceae*	I	Deception	Godinho et al. (2019)
Purpureocillium lilacinum	*Ophiocordycipitaceae*	F	Bransfield and Garlache Straits	Gonçalves et al. (2017)
Rhodotorula glutinis	*Sporidiobolaceae*	I	Maxwell Bay, King George Island	Herrera et al. (2017)
Rhodotorula mucilaginosa	*Sporidiobolaceae*	A	Maxwell Bay, King George Island, Robert Island	Loque et al. (2010); Furbino et al. (2014); Herrera et al. (2017)
Septoria chromolaenae	*Mycosphaerellaceae*	I	Robert Island, Livingston, King George Island	Godinho et al. (2019)
Simplicillium lamellicola	*Cordycipitaceae*		King George Island	Gonçalves et al. (2015)

Table 2 contd. ...

...Table 2 contd.

Taxon	Family	Substrate/ Host	Distribution	References
Simplicillium aogashimaense	*Cordycipitaceae*	F	Bransfield and Garlache Straits	Gonçalves et al. (2017)
Sporobolomyces shibatanus	*Sporidiobolaceae*	A	Robert Island, Halfmoon Island	Duarte et al. (2016)
Symmetrospora marina	*Symmetrosporaceae*	A	Deception	Duarte et al. (2016)
Tausonia pullulans	*Mrakiaceae*		King George Island	Erich et al. (2015); Ugidos-Rodríguez et al. (2018)
Thelebolus balaustiformis	*Thelebolaceae*	I	Deception	Godinho et al. (2019)
Thelebolus globosus	*Thelebolaceae*	A	Elephant Island, Robert Island	Godinho et al. (2013, 2019); Furbino et al. (2014)
Tilletiopsis washingtonensis	*Tilletiopsidaceae*	A	Deception	Duarte et al. (2016)
Tolypocladium tundrense	*Ophiocordycipitaceae*		King George Island	Godinho et al. (2019)
Toriella tubulifera	*Halosphaeriaceae*	W	King George Island	Pugh and Jones (1986)
Trichocladium lignicola	*Chaetomiaceae*	W	Penguin Bay, South Georgia	Grasso et al. (1997)
Trichoderma asperellum	*Hypocreaceae*	S	Penguin Bay, South Georgia	Varella et al. (2021)
Vishniacozyma carnescens	*Bulleribasidiaceae*	A	King George Island	Loque et al. (2010); Duarte et al. (2016)
Vishniacozyma victoriae	*Bulleribasidiaceae*	S	King George Island	Vaz et al. (2011); Godinho et al. (2013); Furbino et al. (2014); Duarte et al. (2016)
Yamadazyma mexicana	*Debaryomycetaceae*	A	Elephant Island	Godinho et al. (2013)

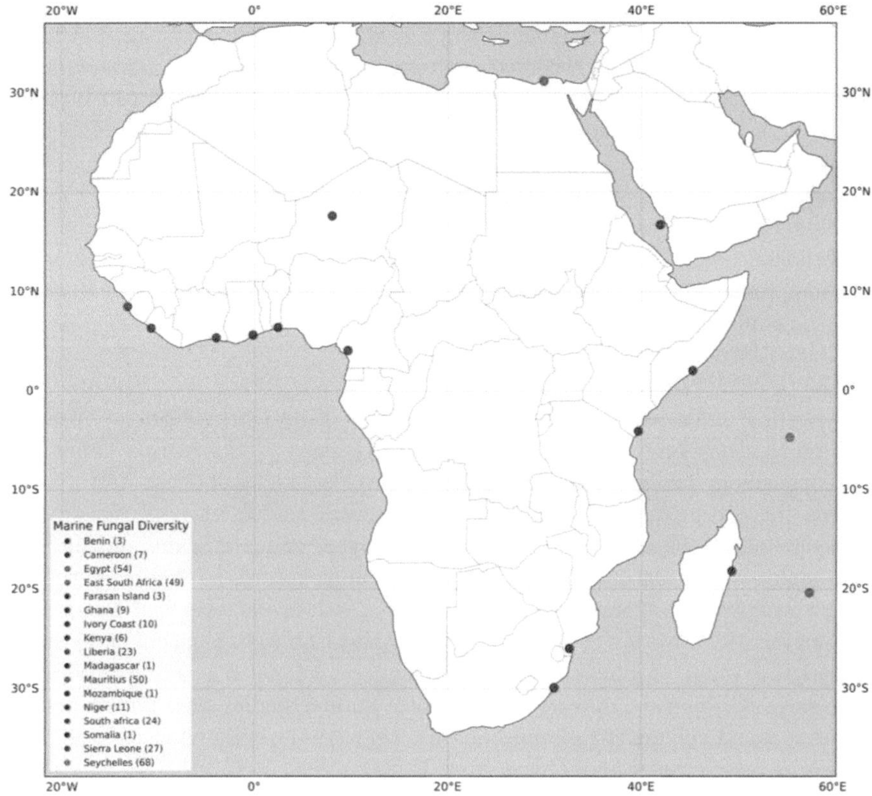

Fig. 4. Diversity of marine fungi reported across various countries in Africa.

In a subsequent study, Steinke (2000) recorded 38 marine fungi on decaying prop roots of *Rhizophora mucronata* in the Kosi Durban Bayhead and Mtata River, South Africa.

Studies on marine fungi on rhizosphere sediments from different mangrove plants of Kenya resulted in 33 fungal taxa in 11 genera based on ITS sequence data; however, none of these taxa have been identified to species level. In more recent times extensive studies on marine fungi have been conducted along the Egyptian and Saudi Arabian coastlines of the Red Sea (Abdel-Wahab, 2005; Abdel-Aziz, 2010; Abdel-Wahab et al., 2021; Pang et al., 2023). The majority of research on African marine fungi has mainly relied on conventional traditional morphological identification of culturable fungi (Steinke, 2000; Ukoima and Umechuruba, 2009), with a limited number of investigations incorporating molecular characterization. Muwawa et al. (2024) performed amplicon-based studies on the rhizospheres of four mangrove plant species at Gazi Bay and Mida Creek in Kenya, revealing that the predominant ASVs were *Ascomycota* (96%), followed by *Basidiomycota* (3%).

Human activities may impact fungal diversity, community structure and their ecological roles in these mangrove ecosystems. Our current compiled list includes 176 fungi in 125 different genera across various marine habitats in African countries, with Seychelles recorded with the highest number of marine fungal species (68), followed by Mauritius (49), Egypt (54), East South Africa (49) with Madagascar and Mozambique having the least (1) documented species (Table 3). The genus *Corollospora* (10) has the highest number of fungal species, followed by *Leptosphaeria* (4). Furthermore, there is presently limited information on unculturable marine fungi using HTS methods from this continent. This highlights a substantial research gap in marine fungal research across Africa, emphasising the necessity for advanced molecular analyses and culture-independent studies to better understand the diversity and ecology of marine fungi on the continent.

3.1.5 Australasia

The coastline of Australia amounts to more than 47,000 km (including many islands), with broad sandy beaches to rocky cliffs, salt marshes and mangrove swamps and its shores adjoin the Pacific, Indian, and Southern Oceans. Most of the Australasian marine fungi have been documented on substrates associated with mangroves. Although numerous other substrates such as algae, animals, corals, plants, and sand grains have been investigated for marine fungi, records of their presence on these substrates are sporadic (Kendrick et al., 1982; Kohlmeyer and Volkmann-Kohlmeyer, 1991; Norton et al., 1994). Cribb and Cribb were pioneers in the study of mangrove fungi, documenting and introducing several novel taxa from Australian mangroves (Cribb and Cribb, 1955a, b, 1960). Further studies were undertaken by Kohlmeyer and Volkmann-Kohlmeyer (1987, 1989, 1991), who conducted extensive studies on marine fungi from Queensland. Hyde's (1990, 1992, 1993) research was mainly focused on fungi on mangrove wood and driftwood in Queensland, New South Wales, and Victoria and resulted in the addition of numerous new marine fungal taxa and records to Australian marine fungi. Additionally, Morrison-Gardiner (2002) isolated 617 strains from coral reefs (algae and invertebrates) around Australia, but identified only the genus with *Alternaria*, *Aspergillus*, and *Penicillium* spp. as the dominant taxa. McCarthy and Kantvilas (2015) have extensively studied marine lichens.

Fryar et al. (2019, 2020) reported a novel genus, *Annabella australiensis* and 43 marine fungal species, including 12 taxa of unknown status that are likely new species, collected from driftwood and mangrove wood in Australia. A total of 132 marine fungi has been recorded for Australia, mainly on drift or mangrove wood (Pang et al., 2023). Marine fungi of New Zealand have been documented by Lintott and Lintott (1982, 2002) when 38 marine and three mangrove sites were sampled, resulting in 31 Ascomycota and 2 Basidiomycota (Fig. 5). The dominant species were *Halosphaeriopsis medisetigera* (36 sites) and *Corollospora maritima* (33 sites). Our composed list of marine fungi recorded at different locations in Australasia includes 149 fungi belonging to 108 diverse genera of marine fungi (Table 4). The genus *Halosarpheia* (6) has the highest number of fungal species, followed by *Haloguignardia, Lulworthia, Saagaromyces*, and *Savoryella* (4 species each). This study sampled a wide range of shores around both the North and South Islands. However, substantial locations in Australasia remain unexplored for marine fungi,

Table 3. Table showing the diversity of marine fungi in the Africa region (Substrate/Host: A, Algae; C, Coral; F, Seawater/Froth; I, Invertebrate; S, Sediment; U, Unknown substrate; W, Wood/Plant) (Distribution: B, Benin; Cm, Cameroon; E, Egypt; ESA, East South Africa; FA, Farasan Island; GH, Ghana; IC, Ivory Coast; KN, Kenya; LB, Liberia; MG, Madagascar; MR, Mauritius; MZ, Mozambique; N, Niger; SA, South Africa; SM, Somalia; SR, Sierra Leone; SY, Seychelles).

Taxon	Family	Substrate/ Host	Distribution	References
Acrocordiopsis patilii	*Melanommataceae*	W	ESA, MR, E, KN	Abdel-Wahab (2005); Abdel-Aziz (2010)
Adomia avicenniae	*Incertae sedis*	W	E, LB	Abdel-Wahab (2005); Abdel-Aziz (2010)
Aigialus grandis	*Aigialaceae*	W	SY	Devadatha et al. (2021)
Aigialus mangrovis	*Aigialaceae*	W	MR, SY	Devadatha et al. (2021)
Aigialus parvus	*Aigialaceae*	W, D	E, FA, GH, LB	Abdel-Wahab (2005); Abdel-Aziz (2010)
Ajigaurospora pseudopulchella	*Halosphaeriaceae*	S	SA	Devadatha et al. (2021)
Alternaria alternata	*Halosphaeriaceae*	I, W	MR, SY	Devadatha et al. (2021)
Aniptodera chesapeakensis	*Halosphaeriaceae*	I, W	MR, SA, SY	Steinke and Lubke (2003)
Aniptodera limnetica	*Halosphaeriaceae*	W	E, B, Cm, LB	Abdel-Wahab (2005); Abdel-Aziz (2010)
Aniptodera mangrovei	*Halosphaeriaceae*	I, W	MR, SY, LB, SR, E	Abdel-Wahab (2005); Abdel-Aziz (2010)
Aniptosporopsis lignatilis	*Halosphaeriaceae*	W	MR, SY	Devadatha et al. (2021)
Antennospora quadricornuta	*Halosphaeriaceae*	W	MR, MZ, SY, GH, FA, LB, GH, E	Abdel-Wahab (2005); Abdel-Aziz (2010)
Arenariomyces majusculus	*Halosphaeriaceae*	W	ESA, SY, E	Abdel-Wahab (2005); Abdel-Aziz (2010)
Arenariomyces parvulus	*Halosphaeriaceae*	S	SA	Steinke and Lubke (2003)
Arenariomyces trifurcatus	*Halosphaeriaceae*	S, W	ESA, SY, E, SR	Abdel-Wahab (2005); Abdel-Aziz (2010)
Ascocratera manglicola	*Aigialaceae*	W	E	Abdel-Wahab (2005); Abdel-Aziz (2010)
Aspergillus flavus	*Aspergillaceae*	S	KN	Devadatha et al. (2021)
Aspergillus luchuensis	*Aspergillaceae*	S	N	Devadatha et al. (2021)
Aspergillus niger	*Aspergillaceae*	S	N	Devadatha et al. (2021)
Aspergillus solani	*Aspergillaceae*	S	N	Devadatha et al. (2021)
Aspergillus sydowii	*Aspergillaceae*	S	N	Devadatha et al. (2021)
Asteromyces cruciatus	*Pleosporaceae*	W	E, ESA	Abdel-Wahab (2005); Abdel-Aziz (2010)
Bactrodesmium linderi	*Chaetomiaceae*	S, W	ESA, SY, E, SA	Abdel-Wahab (2005); Abdel-Aziz (2010); Steinke and Lubke (2003)

Table 3 contd. ...

...Table 3 contd.

Taxon	**Family**	**Substrate/ Host**	**Distribution**	**References**
Bathyascus grandisporus	*Halosphaeriaceae*	W	SY	Devadatha et al. (2021)
Belizeana tuberculata	*Incertaesedis*	W	SY, E	Abdel-Wahab (2005); Abdel-Aziz (2010)
Biatriospora marina	*Biatriosporaceae*	W	MR, SY	Devadatha et al. (2021)
Boeremia hedericola	*Didymellaceae*	W	E	Abdel-Wahab (2005); Abdel-Aziz (2010)
Camarosporium roumeguerei	*Camarosporiaceae*	W	ESA, Cm, IC	Steinke and Lubke (2003)
Candida albicans	*Debaryomycetaceae*	U	N	Devadatha et al. (2021)
Carbosphaerella leptosphaerioides	*Halosphaeriaceae*	S	SA	Steinke and Lubke (2003)
Carbosphaerella pleosporoides	*Halosphaeriaceae*	S	SA	Steinke and Lubke (2003)
Caryosporella rhizophorae	*Melanommataceae*	W	ESA, MR, SY, Cm, IC	Steinke and Jones (1993)
Ceriosporopsis halima	*Halosphaeriaceae*	W	ESA, SY, IC, LB	Devadatha et al. (2021)
Cirrenalia basiminuta	*Halosphaeriaceae*	W	MR	Devadatha et al. (2021)
Cirrenalia macrocephala	*Halosphaeriaceae*	W, D	ESA, Cm, GH, IC, LB, SR, SA, SY	Steinke and Lubke (2003)
Cirrenalia pseudomacrocephala	*Halosphaeriaceae*	D, W	MR, SY	Devadatha et al. (2021)
Colletotrichum gloeosporioides	*Glomerellaceae*	U	N	Devadatha et al. (2021)
Corollospora anglusa	*Halosphaeriaceae*	W, S	E, SA	Abdel-Wahab (2005); Abdel-Aziz (2010); Steinke and Lubke (2003)
Corollospora angusta	*Halosphaeriaceae*	S	SA	Steinke and Lubke (2003)
Corollospora armoricana	*Halosphaeriaceae*	U	SR	Aleem (1979)
Corollospora cinnamomea	*Halosphaeriaceae*	U	SR	Aleem (1979)
Corollospora colossus	*Halosphaeriaceae*	S	SR	Aleem (1979)
Corollospora cristata	*Halosphaeriaceae*	U	SR	Aleem (1979)
Corollospora filiformis	*Halosphaeriaceae*	U	SR	Aleem (1979)
Corollospora fusca	*Halosphaeriaceae*	U	SR	Aleem (1979)
Corollospora graccilis	*Halosphaeriaceae*	W	E	Abdel-Wahab (2005); Abdel-Aziz (2010)
Corollospora intermedia	*Halosphaeriaceae*	U	SR	Aleem (1979)
Corollospora lacera	*Halosphaeriaceae*	U	SR	Aleem (1979)
Corollospora luteola	*Halosphaeriaceae*	S	SA	Steinke and Lubke (2003)

Table 3 contd. ...

...Table 3 contd.

Taxon	Family	Substrate/ Host	Distribution	References
Corollospora maritima	*Halosphaeriaceae*	W	ESA, SY, Cm, IC, E	Steinke and Lubke (2003)
Corollospora prolifera	*Halosphaeriaceae*	S	SA	Steinke and Lubke (2003)
Corollospora pulchella	*Halosphaeriaceae*	W	ESA, SY, IC, SR	Aleem (1979)
Corollospora quinqueseptata	*Halosphaeriaceae*	S	SA	Steinke and Lubke (2003)
Corollospora ramulosa	*Halosphaeriaceae*	D	SY	Devadatha et al. (2021)
Corollosporopsis portsaidica	*Halosphaeriaceae*	W	E	Abdel-Wahab (2005); Abdel-Aziz (2010)
Coronopapilla mangrovei	*Zopfiaceae*	W	MR	Devadatha et al. (2021)
Crinigera maritima	*Incertae sedis*	W	E	Abdel-Wahab (2005); Abdel-Aziz (2010)
Cryptosphaeria bathurstensis	*Diatrypaceae*	U	SR	Aleem (1979)
Cucullosporella mangrovei	*Halosphaeriaceae*	A, W	MR, SY	Devadatha et al. (2021)
Cucurbitinus constrictus	*Halosphaeriaceae*	W	ESA	Steinke and Lubke (2003)
Cumulospora marina	*Lulworthiaceae*	W	E	Abdel-Wahab (2005); Abdel-Aziz (2010)
Cytospora rhizophorae	*Cytosporaceae*	S	MR, SY, LB, SR	Aleem (1979)
Dictyosporium pelagicum	*Dictyosporiaceae*	W	SY	Devadatha et al. (2021)
Dryosphaera navigans	*Incertae sedis*	S	SA	Steinke and Lubke (2003)
Epicoccum nigrum	*Didymellaceae*	W	E	Abdel-Wahab (2005); Abdel-Aziz (2010)
Etheirophora blepharospora	*Etheirophoraceae*	W	MR, SY	Devadatha et al. (2021)
Exserohilum rostratum	*Pleosporaceae*	W	SY	Devadatha et al. (2021)
Fulvocentrum aegyptiacum	*Juncigenaceae*	W	E	Abdel-Wahab (2005); Abdel-Aziz (2010)
Fulvocentrum clavatisporum	*Juncigenaceae*	W	E, SB	Abdel-Wahab (2005); Abdel-Aziz (2010)
Fusarium moniliforme	*Nectriaceae*	U	N	Devadatha et al. (2021)
Gloniella clavatispora	*Hysteriaceae*	W	ESA	Steinke and Lubke (2003)
Haiyanga salina	*Halosphaeriaceae*	W	ESA, SY, LB, SR	Aleem (1979)
Halenospora varia	*Leotiaceae*	W	SY, SR	Aleem (1979)
Halobyssothecium obiones	*Lentitheciaceae*	U	SR	Steinke and Jones (1993); Aleem (1979)

Table 3 contd. ...

...Table 3 contd.

Taxon	Family	Substrate/ Host	Distribution	References
Halocryptosphaeria bathurstensis	*Diatrypaceae*	W	E	Abdel-Wahab (2005): Abdel-Aziz (2010)
Halocyphina villosa	*Niaceae*	W	GH, LB, ESA, MR, SY, LB, MR, N, SY, FA	Steinke and Lubke (2003)
Halojulella avicenniae	*Halojulellaceae*	W	E	Abdel-Wahab (2005); Abdel-Aziz (2010)
Halokirschsteiniothelia maritima	*Mytilinidiaceae*	W	E	Abdel-Wahab (2005); Abdel-Aziz (2010)
Halorosellinia oceanica	*Xylariaceae*	S	SA	Steinke and Lubke (2003)
Halosarpheia fibrosa	*Halosphaeriaceae*	W	ESA, MR, SY, E	Abdel-Wahab (2005); Abdel-Aziz (2010)
Halosarpheia marina	*Halosphaeriaceae*	I, W	ESA, MR, SY, E	Abdel-Wahab (2005); Abdel-Aziz (2010)
Halosarpheia trullifera	*Halosphaeriaceae*	W	ESA	Steinke and Lubke (2003)
Halosarpheia viscosa	*Halosphaeriaceae*	W	ESA, MR, SY	Steinke and Jones (1993)
Halosphaeria appendiculata	*Halosphaeriaceae*	W	IC, LB	Steinke and Jones (1993)
Halosphaeriopsis mediosetigera	*Halosphaeriaceae*	I, W	ESA, SY, LB, SR, E	Abdel-Wahab (2005); Abdel-Aziz (2010)
Helicascus kanaloanus	*Morosphaeriaceae*	S, W	ESA	Steinke and Lubke (2003)
Humicola alopallonella	*Halosphaeriaceae*	W	ESA, SY, Cm, GH, IC, LB	Steinke and Lubke (2003)
Hydea pygmea	*Lulworthiaceae*	S	LB, SR, GH	Aleem (1979)
Hypophloeda rhizospora	*Melanconidaceae*	W	SY	Hyde (1990)
Hypoxylon lenormandii	*Hypoxylaceae*	S	SA	Steinke and Lubke (2003)
Hypoxylon ochraceum	*Hypoxylaceae*	W	ESA	Devadatha et al. (2021)
Kallichroma glabrum	*Nectriaceae*	U	SA	Steinke and Jones (1993)
Kallichroma tethys	*Ijuhyaceae*	W	E, ESA, MR, SY	Abdel-Wahab (2005); Abdel-Aziz (2010)
Kochiella crispa	*Halosphaeriaceae*	W	SY, LB, SR	Aleem (1979)
Lanspora coronata	*Phomatosporaceae*	W	SY	Devadatha et al. (2021)
Lasiodiplodia theobromae	*Botryosphaeriaceae*	U	N	Devadatha et al. (2021)
Lautospora gigantea	*Lautosporaceae*	W	MR	Devadatha et al. (2021)
Leptosphaeria australiensis	*Leptosphaeriaceae*	W	ESA, MR, SY, LB, SR,	Steinke and Lubke (2003)

Table 3 contd. ...

...Table 3 contd.

Taxon	Family	Substrate/ Host	Distribution	References
Leptosphaeria avicenniae	*Leptosphaeriaceae*	W	LB, IC	Devadatha et al. (2021)
Leptosphaeria nypicola	*Leptosphaeriaceae*	W	SY	Devadatha et al. (2021)
Leptosphaeria orae-maris	*Leptosphaeriaceae*	W	E	Abdel-Wahab (2005); Abdel-Aziz (2010)
Leptosphaeria pelagica	*Leptosphaeriaceae*	S	MR	Devadatha et al. (2021)
Leptosphaeria peruviana	*Leptosphaeriaceae*	W	E	Abdel-Wahab (2005); Abdel-Aziz (2010)
Lignincola laevis	*Halosphaeriaceae*	I, W	ESA, MR, SY	Steinke and Lubke (2003)
Lignincola tropica	*Halosphaeriaceae*	W	MR	Devadatha et al. (2021)
Limacospora sundica	*Halosphaeriaceae*	S	SA	Steinke and Lubke (2003)
Lindra obtusa	*Lulworthiaceae*	W	E	Abdel-Wahab (2005); Abdel-Aziz (2010)
Lineolata rhizophorae	*Lineolataceae*	W	ESA, KN, MR, SY, E	Abdel-Wahab (2005); Abdel-Aziz (2010)
Longicorpus striataspora	*Striatiguttulaceae*	F, W	SY	Devadatha et al. (2021)
Lulwoana uniseptata	*Lulworthiaceae*	W	E	Abdel-Wahab (2005); Abdel-Aziz (2010)
Lulworthia medusa	*Lulworthiaceae*	F, W	ESA, MR, SY	Steinke and Lubke (2003)
Manglicola guatemalensis	*Manglicolaceae*	W	ESA, MR	Devadatha et al. (2021)
Marinosphaera mangrovei	*Halosphaeriaceae*	W, S	B, ESA, MG, MR, SA, SY, E	Steinke and Lubke (2003)
Marinospora calyptrata	*Halosphaeriaceae*	S	SA	Steinke and Lubke (2003)
Massarina lacertensis	*Massarinaceae*	W	MR	Devadatha et al. (2021)
Matsusporium tropicale	*Lulworthiaceae*	S, W	LB, SY, SR	Devadatha et al. (2021)
Melomastia marinospora	*Dyfrolomycetaceae*	W	MR	Devadatha et al. (2021)
Moromyces varius	*Halosphaeriaceae*	W	E	Abdel-Wahab (2005); Abdel-Aziz (2010)
Morosphaeria ramunculicola	*Morosphaeriaceae*	W	MR	Devadatha et al. (2021)
Morosphaeria velatispora	*Morosphaeriaceae*	W	MR, SY	Devadatha et al. (2021)
Mycosphaerella pneumatophorae	*Mycosphaerellaceae*	W	E	Abdel-Wahab (2005); Abdel-Aziz (2010)
Nais aquatica	*Halosphaeriaceae*	W	E	Abdel-Wahab (2005); Abdel-Aziz (2010)
Natantispora retorquens	*Halosphaeriaceae*	I, W	ESA	Steinke and Lubke (2003)

Table 3 contd. ...

...Table 3 contd.

Taxon	Family	Substrate/ Host	Distribution	References
Neomassariosphaeria typhicola	*Amniculicolaceae*	S	SA	Steinke and Lubke (2003)
Neptunella longirostris	*Halosphaeriaceae*	S	SA	Steinke and Lubke (2003)
Nereiospora cristata	*Halosphaeriaceae*	S	SA	Steinke and Lubke (2003)
Nia globispora	*Niaceae*	W	E	Abdel-Wahab (2005); Abdel-Aziz (2010)
Nia vibrissa	*Niaceae*	U	SR	Aleem (1979)
Nigrospora oryzae	*Apiosporaceae*	W	SY	Devadatha et al. (2021)
Oceanitis cincinnatula	*Halosphaeriaceae*	I, W	SY	Devadatha et al. (2021)
Oceanitis viscidula	*Halosphaeriaceae*	U	SR	Aleem (1979)
Okeanomyces cucullatus	*Halosphaeriaceae*	I, W	ESA, MR, SY, E	Abdel-Wahab (2005); Abdel-Aziz (2010)
Papulaspora halima	*Lulworthiaceae*	W	SY	Devadatha et al. (2021)
Paraaniptodera longispora	*Halosphaeriaceae*	W	MR	Devadatha et al. (2021)
Pedumispora rhizophorae	*Diatrypaceae*	W	SY	Devadatha et al. (2021)
Penicillium herquei	*Aspergillaceae*	U	GH	Devadatha et al. (2021)
Penicillium sclerotiorum	*Aspergillaceae*	U	KN	Devadatha et al. (2021)
Pestalotiopsis microspora	*Pestalotiopsidaceae*	U	N	Devadatha et al. (2021)
Phaeosphaeria capensis	*Phaeosphaeriaceae*	W	E	Abdel-Wahab (2005); Abdel-Aziz (2010)
Phaeosphaeria halima	*Phaeosphaeriaceae*	W	ESA	Devadatha et al. (2021)
Piricauda elliptica	*incertae sedis*	W	SY	Devadatha et al. (2021)
Pseudocercospora avicenniicola	*Mycosphaerellaceae*	U	B	Devadatha et al. (2021)
Quintaria lignatilis	*Lophiostomataceae*	W	MR, SY	Devadatha et al. (2021)
Rhabdospora avicenniae	*Mycosphaerellaceae*	S	Cm, GH, IC, LB, ESA	Devadatha et al. (2021)
Rhizophila marina	*Diatrypaceae*	W	ESA, MR, SY	Steinke and Lubke (2003)
Rhizopus stolonifer	*Rhizopodaceae*	W	N	Devadatha et al. (2021
Rhytidenglerula landolphiae	*Englerulaceae*	W	E	Abdel-Wahab (2005); Abdel-Aziz (2010)
Rimora mangrovei	*Aigialaceae*	W	ESA, MR	Steinke and Lubke (2003)
Saagaromyces abonnis	*Halosphaeriaceae*	S, W	E, SA	Abdel-Wahab (2005); Abdel-Aziz (2010); Steinke and Lubke (2003)
Saagaromyces glitra	*Halosphaeriaceae*	W	ESA, MR, SY	Steinke and Lubke (2003)

Table 3 contd. ...

...Table 3 contd.

Taxon	Family	Substrate/ Host	Distribution	References
Saagaromyces fusca	*Halosphaeriaceae*	U	SR	Aleem (1979)
Saagaromyces ratnagirenis	*Halosphaeriaceae*	W	ESA, MR, SY, E	Abdel-Wahab (2005); Abdel-Aziz (2010)
Saagaromyces unicellularis	*Halosphaeriaceae*	W	E	Abdel-Wahab (2005); Abdel-Aziz (2010)
Sammeyersia grandispora	*Lulworthiaceae*	A, W	GH, LB, ESA, MR, SR, SY, E	Abdel-Wahab (2005); Abdel-Aziz (2010); Aleem (1979)
Savoryella appendiculata	*Savoryellaceae*	S	SR	Aleem (1979)
Savoryella lignicola	*Savoryellaceae*	I, W	ESA, MR, SY, E	Abdel-Wahab (2005); Abdel-Aziz (2010)
Savoryella paucispora	*Savoryellaceae*	W	ESA, MR, SY	Devadatha et al. (2021)
Sclerococcum haliotrephum	*Dactylosporaceae*	W, S	E, ESA, MR, SY, ESA	Abdel-Wahab (2005); Abdel-Aziz (2010)
Swampomyces armeniacus	*Etheirophoraceae*	W	E, MR, SY	Abdel-Wahab (2005); Abdel-Aziz (2010)
Swampomyces clavatispora	*Etheirophoraceae*	W	E	Abdel-Wahab (2005); Abdel-Aziz (2010)
Swampomyces triseptatus	*Etheirophoraceae*	W	MR	Devadatha et al. (2021)
Thalassogena sphaerica	*Halosphaeriaceae*	W	ESA, MR	Steinke and Lubke (2003)
Torpedospora radiata	*Torpedosporaceae*	W	E	Abdel-Wahab (2005); Abdel-Aziz (2010)
Trailia ascophylli	*Halosphaeriaceae*	S	SA	Steinke and Lubke (2003)
Trematosphaeria mangrovis	*Trematosphaeriaceae*	W	LB, MR	Devadatha et al. (2021)
Trichocladium lignicola	*Chaetomiaceae*	W	ESA	Steinke and Lubke (2003)
Trichoderma inhamatum	*Hypocreaceae*	I	KN	Devadatha et al. (2021)
Trichoderma reesei	*Hypocreaceae*	D	ESA	Steinke and Lubke (2003)
Tubakiella galerita	*Halosphaeriaceae*	W	SY	Devadatha et al. (2021)
Ulkenomyces aestuarii	*Halomycetaceae*	S	ESA	Steinke and Lubke (2003)
Verruculina enalia	*Didymosphaeriaceae*	W	LB, SR, ESA, KN, MR, SY	Aleem (1979)
Zasmidium mangrovei	*Mycosphaerellaceae*	S	SA	Steinke and Lubke (2003)

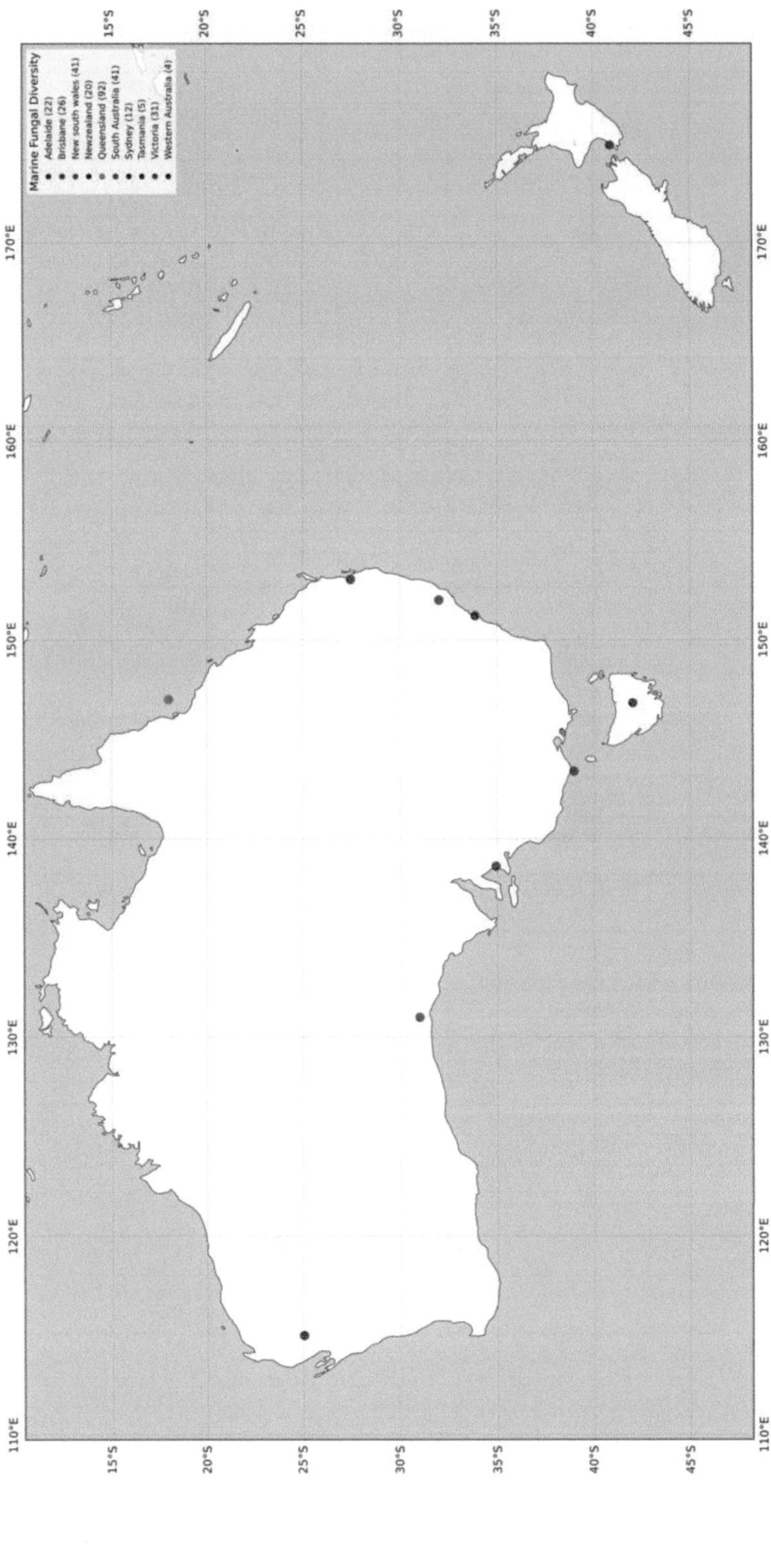

Fig. 5. Diversity of marine fungi reported in various regions of Australasia.

Table 4. Table showing the diversity of marine fungi in the Australasia region. (Substrate/Host: A. Algae; C, Coral; F, Seawater/Froth; I, Invertebrate; S, Sediment; U, Unknown substrate; W, Wood/Plant) (Distribution: A, Adelaide; Br, Brisbane; NSW, New South Wales; NZ, New Zealand; Q, Queensland; SA, South Australia; Sy, Sydney; TAS, Tasmania; VIC, Victoria; WA, Western Australia).

Taxon	Family	Substrate/ Host	Distribution	References
Acrocordiopsis patilii	*Melanommataceae*	W	Q	Kohlmeyer and Volkmann-Kohlmeyer (1991)
Adomia avicenniae	*Incertae sedis*	W	Br, NSW, NZ	Schatz (1985)
Aigialus grandis	*Aigialaceae*	W	Q	Kohlmeyer and Volkmann-Kohlmeyer (1991); Hyde (1992)
Aigialus parvus	*Aigialaceae*	W	Q	Kohlmeyer and Volkmann-Kohlmeyer (1991)
Aniptodera chesapeakensis	*Halosphaeriaceae*	W	Br, NSW, SY	Fryar et al. (2020); Hyde (1990)
Aniptodera juncicola	*Halosphaeriaceae*	W	SA	Fryar et al. (2020)
Aniptodera mangrovei	*Halosphaeriaceae*	W	Br, NSW	Hyde (1990)
Aniptodera salsuginosa	*Halosphaeriaceae*	W	SA	Fryar et al. (2020)
Aniptosporopsis lignatilis	*Halosphaeriaceae*	W	Sy, SA	Fryar et al. (2020)
Anisostagma rotundatum	*Halosphaeriaceae*	W	SA	Fryar et al. (2020)
Annabella australiensis	*Cordieritidaceae*	W	A, SA	Fryar et al. (2019)
Antennospora quadricornuta	*Halosphaeriaceae*	W	A, Q	Cribb and Cribb (1956, 1969); Kohlmeyer (1984); Kohlmeyer and Volkmann-Kohlmeyer (1991)
Arenariomyces majusculus	*Halosphaeriaceae*	W	Br, NSW	Kohlmeyer and Volkmann-Kohlmeyer (1989)
Arenariomyces parvulus	*Halosphaeriaceae*	A, W	Q	Kohlmeyer and Volkmann-Kohlmeyer (1989); Kohlmeyer and Volkmann-Kohlmeyer (1991)
Arenariomyces trifurcatus	*Halosphaeriaceae*	S, W, F	Q	Kendrick et al. (1982); Cribb and Cribb (1956); Kohlmeyer (1984)
Ascocratera manglicola	*Aigialaceae*	W	A, Br, NSW, NZ, Q	Kohlmeyer (1986); Hyde (1990); Kohlmeyer and Volkmann-Kohlmeyer (1991); Hyde (1992)

Table 4 contd. ...

...Table 4 contd.

Taxon	Family	Substrate/ Host	Distribution	References
Aspergillus restrictus	*Aspergillaceae*	C	Q	Kendrick et al. (1982)
Asteromyces cruciatus	*Pleosporaceae*	F	SA, VIC	Boyd and Kohlmeyer (1982)
Belizeana tuberculata	*Incertae sedis*	W	Br, NSW, Q	Kohlmeyer and Volkmann-Kohlmeyer (1987, 1991)
Biatriospora marina	*Biatriosporaceae*	W	Br, NSW, Q	Kohlmeyer and Volkmann-Kohlmeyer (1991)
Blodgettia bornetii	*Incertae sedis*	A, W	NSW, Q	Cribb and Cribb (1969)
Caryospora australiensis	*Melanommataceae*	W	VIC	Abdel-Wahab and Jones 2000)
Caryosporella rhizophorae	*Melanommataceae*	W	Q	Kendrick et al. (1982); Kohlmeyer and Volkmann-Kohlmeyer (1991)
Ceriosporopsis cambrensis	*Halosphaeriaceae*	W	SA	Fryar et al. (2020)
Ceriosporopsis halima	*Halosphaeriaceae*	W	Br, NSW, Q	Fryar et al. (2020); Cribb and Cribb (1956); Jones et al. (1972)
Chadefaudia balliae	*Halosphaeriaceae*	A	VIC, WA	Kohlmeyer (1973)
Chadefaudia corallinarum	*Halosphaeriaceae*	A	Q	Kendrick et al. (1982); Cribb and Cribb (1969)
Chadefaudia gymnogongri	*Halosphaeriaceae*	A	NSW, TAS, Q, SA	Cribb and Herbert (1954); Cribb and Cribb (1955, 1960); Kohlmeyer and Kohlmeyer (1979)
Cirrenalia macrocephala	*Halosphaeriaceae*	W	SA. NSW	Fryar et al. (2020); Jones et al. (1972)
Corollospora gracilis	*Halosphaeriaceae*	S	Q	Kendrick et al. (1982); Kohlmeyer and Volkmann-Kohlmeyer (1991)
Corollospora maritima	*Halosphaeriaceae*	W	Q, NZ, NSW, SA, VIC	Fryar et al. (2020); Kohlmeyer (1984); Bebout et al. (1987); Kohlmeyer and Volkmann-Kohlmeyer (1991)
Corollospora pulchella	*Halosphaeriaceae*	W	Q, NZ, NSW, SA	Kohlmeyer (1984); Kohlmeyer and Volkmann-Kohlmeyer (1991)
Corollospora quinqueseptata	*Halosphaeriaceae*	W	SA	Fryar et al. (2020)

Table 4 contd. ...

...Table 4 contd.

Taxon	Family	Substrate/ Host	Distribution	References
Cremasteria cymatilis	*Incertia sedis*	W	NSW	Jones et al. (1972)
Cryptosphaeria bathurstensis	*Diatrypaceae*	W	Q	Hyde and Rappaz (1993)
Cryptovalsa halosarceicola	*Diatrypaceae*	W	Q	Hyde (1993)
Cytospora rhizophorae	*Cytosporaceae*	W, S	Br, Q	Shaw (1989); Kohlmeyer and Volkmann-Kohlmeyer (1991)
Decaisnella formosa	*Massariaceae*	S	VIC	Abdel-Wahab and Jones (2003)
Diatrypasimilis australiensis	*Diatrypaceae*	W	Q	Chalkley et al. (2010)
Dictyosporium pelagicum	*Dictyosporiaceae*	W	NZ	Tennakoon et al. (2023)
Emericella versicolor	*Aspergillaceae*	C	Q	Kendrick et al. (1982)
Etheirophora blepharospora	*Etheirophoraceae*	W	Q	Kohlmeyer and Volkmann-Kohlmeyer (1991); Hyde (1992)
Eutypella naqsii	*Diatrypaceae*	W	Q	Hyde (1995)
Exophiala pisciphila	*Herpotrichiellaceae*	W	Q	Langdon and McDonald (1987)
Exserohilum rostratum	*Pleosporaceae*	W	NZ	Kendrick et al. (1982); Norton et al. (1994)
Fulvocentrum aegyptiaca	*Juncigenaceae*	W	SA	Fryar et al. (2020)
Haiyanga salina	*Halosphaeriaceae*	W	Q	Kendrick et al. (1982)
Halobyssothecium obiones	*Lentitheciaceae*	W	Q	Kendrick et al. (1982); Cribb and Cribb (1960)
Halocryptosphaeria bathurstensis	*Diatrypaceae*	W	SA	Fryar et al. (2020)
Halocyphina villosa	*Niaceae*	W	VIC	Nakagiri et al. (1991)
Halographis runica	*Roccellaceae*	S	Q	Kohlmeyer and Volkmann-Kohlmeyer (1992)
Haloguignardia decidua	*Lulworthiaceae*	A, W	Q	Kendrick et al. (1982)
Haloguignardia irritans	*Lulworthiaceae*	U	Q	Kendrick et al. (1982)
Haloguignardia longispora	*Lulworthiaceae*	A	Q, NSW	Cribb and Cribb (1956)
Haloguignardia tumefaciens	*Lulworthiaceae*	A	NSW, Q, SA, TAS	Cribb and Herbert (1954); Cribb and Cribb (1956, 1960), Kohlmeyer (1971)

Table 4 contd. ...

...Table 4 contd.

Taxon	Family	Substrate/ Host	Distribution	References
Halojulella avicenniae	*Halojulellaceae*	W	A	Kohlmeyer and Volkmann-Kohlmeyer (1991); Hyde (1992)
Halomassarina thalassiae	*Trematosphaeriaceae*	W	Q	Kohlmeyer and Volkmann-Kohlmeyer (1991); Hyde (1992a)
Halorosellinia oceanica	*Xylariaceae*	S, W	Br, NSW, Q	Hyde (1990)
Halosarpheia fibrosa	*Halosphaeriaceae*	W	SA, SY, A, Br, NSW, NZ, Q, VIC	Fryar et al. (2020)
Halosarpheia marina	*Halosphaeriaceae*	W	A, Br, NSW, Q, VIC, SA, SY	Fryar et al. (2020)
Halosarpheia minuta	*Halosphaeriaceae*	W	Br, NSW, Q, VIC	Fryar et al. (2020)
Halosarpheia trullifera	*Halosphaeriaceae*	W	NZ, SA	Fryar et al. (2020)
Halosarpheia unicellularis	*Halosphaeriaceae*	W	SA	Fryar et al. (2020)
Halosarpheia viscosa	*Halosphaeriaceae*	W	Q, VIC	Campbell et al. (2003)
Halosphaeria appendiculata	*Halosphaeriaceae*	W	A, Q, VIC, SA	Fryar et al. (2020)
Halosphaeriospsis alopallonella	*Halosphaeriaceae*	W	A, Q, VIC	Calabon et al. (2023)
Hydea pygmea	*Lulworthiaceae*	W	Q	Kendrick et al. (1982)
Hypoxylon ochraceum	*Hypoxylaceae*	W	A	Schmit and Shearer 2003)
Kallichroma glabrum	*Ijuhyaceae*	W	SA, Q	Kohlmeyer and Volkmann-Kohlmeyer 1991)
Kallichroma tethys	*Ijuhyaceae*	W	A, Br, NSW, Q	Hyde (1990)
Koralionastes ellipticus	*Koralionastestaceae*	W, C	Q	Volkmann-Kohlmeyer and Kohlmeyer (1992)
Koralionastes ovalis	*Koralionastestaceae*	C, W	Q	Volkmann-Kohlmeyer and Kohlmeyer (1992)
Koralionastes violaceus	*Koralionastestaceae*	U	Q	Volkmann-Kohlmeyer and Kohlmeyer (1990)
Leptosphaeria australiensis	*Leptosphaeriaceae*	W	A, Br, Q, VIC, SY, NSW	Hyde (1992)
Leptosphaeria avicenniae	*Leptosphaeriaceae*	W	NZ	Fryar et al. (2020)
Lignincola laevis	*Halosphaeriaceae*	W	SA, SY, NZ, NSW Q	Fryar et al. (2020)
Lindrella thalassiae	*Lulworthiaceae*	S	Q	Kendrick et al. (1982)

Table 4 contd. ...

...Table 4 contd.

Taxon	Family	Substrate/ Host	Distribution	References
Lineolata rhizophorae	*Lineolactaceae*	W, R	Q	Kendrick et al. (1982)
Linocarpon livistonae	*Linocarpaceae*	P	Q	Kendrick et al. (1982)
Longicorpus striataspora	*Striatiguttulaceae*	F, W	Br, NSW, Q	Fryar et al. (2020)
Lophiostoma acrostichi	*Lophiostomataceae*	U	Br	Fryar et al. (2020)
Lulworthia floridana	*Lulworthiaceae*	W	NZ, NSW	Jones et al. (1972)
Lulworthia longispora	*Lulworthiaceae*	W	Q	Cribb and Cribb (1955)
Lulworthia medusa	*Lulworthiaceae*	A, F, W	A, Q, Br, VIC, NSW, SA, SY, NZ	Fryar et al. (2020)
Lulworthia purpurea	*Lulworthiaceae*	W	NSW	Jones et al. (1972)
Lulwoana uniseptata	*Lulworthiaceae*	W	NSW, NZ	Fryar et al. (2020), Jones et al. (1972)
Manglicola guatemalensis	*Manglicolaceae*	W	Q	Kendrick et al. (1982)
Mangrovispora pemphii	*incertae sedis*	W	Q	Kendrick et al. (1982)
Marinosphaera mangrovei	*Halosphaeriaceae*	W	Q, SA	Fryar et al. (2020)
Massarina cystophorae	*Massarinaceae*	A	TAS	Cribb and Herbert (1954)
Massarina lacertensis	*Massarinaceae*	W	Q	Kohlmeyer and Volkmann-Kohlmeyer (1991)
Melanopsamma balani	*Chaetosphaeriaceae*	A	NSW, TAS	Kohlmeyer and Kohlmeyer (1979)
Melaspilea mangrovei	*Melaspileaceae*	W	Q	Vrijmoed et al. (1996)
Melomastia mangrovei	*Dyfrolomycetaceae*	W	Q	Hyde (1992)
Melomastia marinospora	*Dyfrolomycetaceae*	W	A	Hyde (1992)
Moromyces varius	*Lulworthiaceae*	W	SA	Fryar et al. (2020)
Morosphaeria velatispora	*Morosphaeriaceae*	W	Q	Kendrick et al. (1982)
Natantispora retorquens	*Halosphaeriaceae*	W	SA, A, Br, VIC	Fryar et al. (2020)
Neocosmospora solani	*Nectriaceae*	S	WA	Glazebrook et al. (1986)

Table 4 contd. ...

...Table 4 contd.

Taxon	Family	Substrate/ Host	Distribution	References
Neptunella longirostris	*Halosphaeriaceae*	W	Br, NSW, Q,	Cribb and Cribb (1956); Hyde (1990); Kohlmeyer and Volkmann-Kohlmeyer (1991)
Nia vibrissa	*Niaceae*	W	NSW, Q, SA, VIC	Fryar et al. (2020)
Nigrospora oryzae	*Apiosporaceae*	P	NZ	Kendrick et al. (1982)
Oceanitis unicaudata	*Halosphaeriaceae*	W	NSW	Kohlmeyer (1983)
Oceanitis viscidula	*Halosphaeriaceae*	W	Q	Kohlmeyer and Volkmann-Kohlmeyer (1991)
Ocostaspora apilongissima	*Halosphaeriaceae*	W	Q	Kohlmeyer and Volkmann-Kohlmeyer (1991)
Okeanomyces cucullatus	*Halosphaeriaceae*	W	NSW, A, Q, VIC	Kohlmeyer and Kohlmeyer (1979); Kohlmeyer (1984, Hyde (1990, 1992); Kohlmeyer and Volkmann-Kohlmeyer (1991)
Ophiobolus australiensis	*Phaeosphaeriaceae*	W	Q	Cribb and Cribb (1956)
Panorbis viscosus	*Halosphaeriaceae*	W	Q, SA	Kohlmeyer and Volkmann-Kohlmeyer (1991)
Papulaspora halima	*Lulworthiaceae*	W	NZ	Kendrick et al. (1982)
Paraaniptodera longispora	*Halosphaeriaceae*	W	Br, NSW	Kohlmeyer and Volkmann-Kohlmeyer (1991)
Penicillium bilaiae	*Aspergillaceae*	S	TAS	Capon et al. (2007)
Penicillium citrinum	*Aspergillaceae*	S, W	NZ	Kendrick et al. (1982)
Penicillium restrictum	*Aspergillaceae*	S	Q	Kendrick et al. (1982)
Phialophora bubaki	*Herpotrichiellaceae*	S	Q	Kendrick et al. (1982)
Phialophorophoma litoralis	*Incertae sedis*	W	A, Br, NSW, VIC, Q	Hyde (1990)
Pontogeneia valoniopsidis	*Lulworthiaceae*	A	Q	Cribb and Cribb (1960)
Pseudopithomyces chartarum	*Didymosphaeriaceae*	S	Q	Kendrick et al. (1982)
Pseudoplatystomum scabridisporum	*Lophiostomataceae*	W	VIC	Abdel-Wahab and Jones 2000)
Purpureocillium lilacinum	*Ophiocordycipitaceae*	S	Q	Kendrick et al. (1982)
Pyrenographa xylographoides	*Pyrenulaceae*	W	Q	Aptroot (1991); Alias et al. (1996)

Table 4 contd. ...

...Table 4 contd.

Taxon	Family	Substrate/ Host	Distribution	References
Quintaria lignatilis	*Lophiostomataceae*	W	Q	Kohlmeyer and Volkmann-Kohlmeyer (1991); Hyde (1992)
Remispora hamata	*Halosphaeriaceae*	W	NZ, SA	Fryar et al. (2020)
Remispora quadri-remis	*Halosphaeriaceae*	W	NZ	Kendrick et al. (1982)
Remispora stellata	*Halosphaeriaceae*	W	SA, NZ	Kendrick et al. (1982)
Rhizophila marina	*Incertae sedis*	W	Q	Kohlmeyer and Volkmann-Kohlmeyer (1991)
Rimora mangrovei	*Aigialaceae*	W	Q	Kohlmeyer and Volkmann-Kohlmeyer (1991)
Saagaromyces abonnis	*Halosphaeriaceae*	S, W	A, Br, NSW, Q, VIC, SA	Hyde (1990)
Saagaromyces glitra	*Halosphaeriaceae*	W	A, SA, SY	Hyde (1990); Hyde (1992)
Saagaromyces mangrovei	*Halosphaeriaceae*	W	SA	Fryar et al. (2020)
Saagaromyces ratnagiriensis	*Halosphaeriaceae*	W	Br, NSW, Q, VIC	Fryar et al. (2020)
Sammeyersia grandispora	*Lulworthiaceae*	A, F, W	A, Br, NSW, VIC, SA, SY, Q	Leightly (1980); Hyde (1990)
Savoryella longispora	*Savoryellaceae*	W	Q	Kendrick et al. (1982)
Savoryella melanospora	*Savoryellaceae*	W	VIC	Abdel-Wahab and Jones 2000)
Savoryella paucispora	*Savoryellaceae*	W	Q	Kendrick et al. (1982)
Savoryella verrucosa	*Savoryellaceae*	W	VIC	Fryar et al. (2020
Sclerococcum haliotrephum	*Dactylosporaceae*	W	A, Br, NSW, Q, SY, VIC, SA	Kohlmeyer (1983); Hyde (1990); Kohlmeyer and Volkmann-Kohlmeyer (1991); Hyde (1992)
Spathulospora adelpha	*Spathulosporaceae*	A	SA, VIC, WA	Inderbitzin et al. (2004)
Spathulospora calva	*Spathulosporaceae*	A	SA, VIC, WA	Kohlmeyer (1983)
Spathulospora phycophila	*Spathulosporaceae*	A	SA, VIC	Kohlmeyer (1983)
Swampomyces armeniacus	*Etheirophoraceae*	W	Q, SA, SY	Hyde and Nakagiri (1992)
Swampomyces triseptatus	*Etheirophoraceae*	W	Q	Hyde and Nakagiri (1992)
Sydowia polyspora	*Dothioraceae*	S	Q	Kendrick et al. (1982)

Table 4 contd. ...

...Table 4 contd.

Taxon	Family	Substrate/ Host	Distribution	References
Torpedospora radiata	*Torpedosporaceae*	S, W	Q, SA, SY	Kohlmeyer and Volkmann-Kohlmeyer (1991)
Tunicatispora australiensis	*Halosphaeriaceae*	W	A, NSW, VIC, SA	Hyde (1990)
Ulkenomyces aestuarii	*Incertae sedis*	F	NZ	Kendrick et al. (1982)
Verruculina enalia	*Didymosphaeriaceae*	W	A, Q, VIC	Hyde (1990)

which are vulnerable to anthropogenic activities. Further studies are crucial to explore the hidden marine fungal diversity, conservation and ecological roles in Australasia.

3.1.6 Deep Sea

Deep sea is a unique ecosystem that has captivated marine mycologists and gained lot of attention in recent years, as it hosts numerous unidentified lineages of marine fungi (Lai et al., 2007; Burgaud et al., 2009; Le Calvez et al., 2009; Nagano et al., 2010; Xu et al., 2014). This is an inhospitable habitat that lacks sunlight, has extremely cold temperatures (< 4°C), has high hydrostatic pressures (up to 110 MPa), and is devoid of substrates for growth (Nagano and Nagahama, 2012). Xu et al. (2018) investigated fungal communities in sediments from the Southwest Indian Ridge, identifying 14 fungal species by a culture-dependent approach, whereas HTS determined 79 fungal taxa. A culture-dependent approach examined deep-sea sediments at diverse depths in Antarctica, revealing fungi from the genera *Acremonium*, *Penicillium* and *Pseudogymnoascus* (Ascomycota). It was found that *Penicillium* represents an Antarctic polyextremophile that can withstand these extreme conditions (Ogaki et al., 2020).

Metabarcoding studies on sediments of Antarctica found *Ascomycota*, *Mortierellomycota*, *Basidiomycota*, *Chytridiomycota*, *Mucoromycota* and *Rozellomycota* as major phyla (Ogaki et al., 2021; De Souza et al., 2022). Many marine fungi have been reported from the deep sea, including the yeasts *Cryptococcus surugaensis* and *Rhodotorula lamellibrachi* (Nagahama et al., 2001, 2003) and filamentous fungi that may be regarded as obligate to this environment such as *Oceanitis abyssalis* and *O. scuticella*. These fungi were recovered from depths of 5707 m of the abyssal plain in the Northwest Pacific Ocean, east of the Japan Islands (Nagano et al., 2024).

4. Factors Affecting Global Diversity of Marine Fungi

The marine environment is highly multifaceted, determined by ecological factors such as high salinity, high pressure, pH, low water activity, light, oxygen level and low temperatures (Raghukumar et al., 2010; Tisthammer et al., 2016; Wijayawardene et al., 2022). Marine fungal distribution in marine ecosystems cannot be attributed to a single factor, which are more vulnerable to recurrent changes and are influenced by a multitude of parameters. However, salinity and temperature are the key factors that affect the distribution and diversity of marine fungi in marine environments (Booth and Kenkel, 1986; Hughes, 1986). Marine fungi possess different metabolic mechanisms that differ from those of terrestrial fungi that enable them to survive in harsh conditions of the marine environment (Bonugli-Santos et al., 2015).

The marine fungi are globally distributed in diverse marine environments influenced by multiple region-specific environmental conditions, which plays a crucial role in determining the diversity and distribution of these fungi (Jones et al., 2019; Kumar et al., 2021). Anthropogenic activities, biological interactions, dissolved oxygen, pH, nutrient availability and methodological factors determine the taxonomic structure and functional role of marine fungi (Chen et al., 2019). The effect of environmental variables on the distribution and composition of marine fungal communities across various maritime environments is discussed in this chapter.

4.1 *Effect of Temperature on Marine Fungal Diversity*

Temperature is the key factor that directly affects the marine fungal species composition and diversity in marine habitat (Kohlmeyer, 1983). Temperature fluctuations can influence the marine fungal composition in specific regions; for instance, some fungi are cosmopolitan, like *Lignincola laevis*, temperate, such as *Spathulospora antartica*, while some are tropical, like *Antennospora quadricornuta* (Jones et al., 2019; Kumar et al., 2021). Higher temperatures result in the eutrophication in marine waters, which promotes the growth and reproduction of marine fungi thus changing the composition of marine fungal communities (Jones et al., 2022). Temperature was the main factor that influenced the pelagic fungal CAZyme activity and plays a vital role in the degradation of organic material and biological functions in the ocean (Guo et al., 2025). According to Tisthammer et al. (2016) temperature and chlorophyll *a* are the chief environmental factors affecting the marine fungal diversity in the surface seawater of Helgoland Roads. Byrne and Jones (1974) discovered that the marine basidiomycete *Digitatispora marina* develops fruit bodies at temperatures below 10°C but fails to sporulate at or above 10°C. This temperature-dependent behaviour demonstrates the fungus' adaptation to cold marine conditions. Few studies have validated that the upsurge in temperature and salinity influences the growth of marine fungi such as *Phoma* sp. and *Pestalotia* sp., resulting in a phenomenon known as the "*Phoma pattern*" (Davidson, 1974).

A similar phenomenon has been reported in other marine fungi, including *Asteromyces cruciatus*, *Curvularia* sp., *Lignincola laevis*, *Nia vibrissa* and *Paradendryphiella salina* (Ritchie, 1957; Lorenz and Molitoris, 1992). Boyd and Kohlmeyer (1982) established that temperature profoundly influences the growth of *Asteromyces cruciatus*, *Sigmoidea marina*, and *Varicosporina ramulosa in vitro* conditions, comparable to their typical temperature levels. Certain marine fungal species are temperature dependent for growth and development for example

Fusarium fujikuroi, *Penicillium decumbens*, and *P. chrysogenum*, showed negligible growth at 9°C, increased growth at 13°C (except for *P. chrysogenum*) and optimum growth rates at 20°C (Fuentes et al., 2015). *Talaromyces albobiverticillius* exhibited maximum production of pigments without salinity at a culture temperature of 27°C (Venkatachalam et al., 2020). Studies by Asseri et al. (2025) showed that Chytridomycetes revealed a negative connection with temperature and a positive correlation with salinity. Conversely, Neocallimastigomycete species demonstrated increased abundance with rising temperatures and reduced abundance with higher salinity.

Sen et al. (2021) characterized the trophic modes within marine mycoplankton, highlighting saprotrophs and pathotrophs as the most prevalent across multiple guilds, indicating that chlorophyll-a, pH and temperature impact mycoplankton diversity. The shifts in the composition, distribution, and diversity of mycoplanktonic communities are caused by temperature, NO_2^- and total organic carbon (TOC) as demonstrated by Li et al. (2023). Abdel-Gawad et al. (2014) on their studies on marine fungi in the Red Sea of Egypt suggested that both temperature and salinity chiefly influence the metabolic activities and growth rate of marine fungi. The variation in temperature impacts the ability to degrade organic matter on marine fungal species which in turn affect the dynamics of nutrient cycling in marine habitats. These studies revealed that temperature is one of the most critical factors that influence the diversity and seasonal occurrence of marine fungi in marine environments. Jones et al. (2022) suggested that marine fungi can acclimatise efficiently to the challenges posed by climate change.

4.2 *Effect of Salinity on Marine Fungal Diversity*

Salinity is a major factor that affects the diversity, distribution, and composition of soil and water microbiomes, with the changes in salinity posing an enormous challenge for microbial metabolism and survival (Lozupone and Knight, 2007; Shearer et al., 2007; Herlemann et al., 2011). It accounts for the distribution of some marine fungi with oceanic, brackish, or elevated salinity (Jones et al., 2022). Marine fungi can survive at osmotic and ionic stress resulted by higher salinity *via* the salt-in mechanism by maintaining the balance between the internal and external salt concentrations and appropriate solutes such as polyols and high-osmolarity glycerol (HOG) (Luard et al., 1985; Dihazi et al., 2003; Gostinčar et al., 2011). The fungal cell wall of marine fungi is involved in sustaining the integrity and activity of the associated cells and guarding against environmental changes. It is a key factor that determines the fungal diversity and community composition in marine environments (Mohamed and Martiny, 2011; Sen et al., 2021; Asseri et al., 2025). Genomic and transcriptome studies of *Eurotium rubrum* and *Wallemia ichthyophaga* showed that these marine fungi thrive best at high salinity by accumulating various organic osmolytes that modulate the osmotic pressure (Zajc et al., 2013; Kis-Papo et al., 2014). A black yeast, *Hortaea werneckii*, is commonly found in hypersaline environments and can survive at higher salinity levels, as demonstrated by a few studies (Kogej et al., 2005; Gostinčar et al., 2011). However, the impact of salinity variations on marine fungi is yet unclear (Plemenitaš, 2021).

High salinity gradients negatively impact marine fungal diversity, while moderate salinity in tidal marshes supports enhanced fungal richness and diversity. (Mohamed and Martiny, 2011). Several studies confirmed that marine fungi can tolerate a wide range of salinity fluctuations in contrast to the freshwater and terrestrial fungi (Jones, 1971; Davidson, 1974; Jones et al., 2022). A few findings indicate that marine ascomycetes demonstrate a broad tolerance to both low and high salinities, whereas terrestrial ascomycetes exhibit restricted tolerance to elevated salinity levels (Davidson, 1974; Byrne and Jones, 1975). Numerous studies demonstrate that Chytridiomycota have higher susceptibility to salinity in contrast to Ascomycota and Basidiomycota. In contrast to saprolegniaceous species, Labyrinthulomycota and marine Oomycota are capable of surviving and reproducing in higher salinities (Amend, 2019; Jones et al., 2022). Gao et al. (2024) identified salinity as the principal factor affecting the distribution of phylum Ascomycota. While the distribution of Basidiomycota and lower fungal taxa was significantly influenced by ecological parameters.

Salinity is a major limiting factor influencing the composition of pelagic fungal communities, as evidenced by research conducted in the Baltic Sea, which indicates notable variations in fungal communities and diversity above and below a threshold of 8 PSU (Rojas-Jimenez et al., 2019; Mazur-Marzec et al., 2024). Yang et al. (2023) observed that *Candolleomyces brunneovagabundus* has obligately adapted to the marine environment by having a significantly increased rate of basidiospore germination in seawater in comparison to freshwater. Oceanic marine fungi include those that release their ascospores passively with deliquescing asci: *Antennospora quadricornuta, Remispora maritima*, and other species such as *Candida atlantica* and *Clavariopsis bulbosa.* A wide range of marine fungi are reported from brackish water, especially from mangroves, such as those host-specific on the palm *Nypa fruticans* (*Aniptodera nypae, Arecophila nypae*, and *Tirisporella beccariana*) and many yeasts (*Candida suwanaritii, Kwoniella mangrovensis*, and *Saturnispora sekii*) (Loilong et al., 2012). Few studies on the oligotrophic, highly saline coastal waters of the Arabian Gulf have demonstrated a greater abundance of marine yeasts, possibly due to their tendency to survive at various levels of salinity, temperature and other environmental conditions (Fotedar et al., 2018, 2019, 2022). The Torpedosporales are a unique group of marine fungi often restricted in their ecological niche and frequently reported from the Red Sea and neighbouring coasts, including *Adomia avicenniae*, *Fulvocentrum aegyptiacum*, *F. clavatisporum*, *Kallichroma asperum*, *K. ellipsoideum*, and *Torpedospora yanbuensis* (Abdel-Wahab et al., 2019, 2022).

4.3 *Effect of Physical Factors on Marine Fungal Diversity*

In addition to temperature and salinity, parameters like pH, light transmission, and hydrostatic pressure can also impact the composition of marine fungal communities. The distribution and diversity of fungal communities are influenced by the pH of the environment, which can also result in various physiological constraints on the establishment of marine fungal communities (Reich et al., 2017).

Aspergillus terreus NTOU4989 displayed an extensive range of physiological modifications when growing in aquatic and marine environments at pH 3 (highly acidic) to pH 9 (alkaline), and temperatures range between 15°C and 45°C. Later transcriptome analysis revealed that, in contrast to temperature of 25°C and pH 7, most stress response genes were detected and upregulated at a temperature of 45°C and pH 3 (Pang et al., 2020). The composition and distribution of marine fungal communities in Arctic and Antarctic locations can be affected by light transmission, which also impacts aquatic life, particularly at coastal polynyas and under low snow levels on sea ice (Rämä et al., 2017). Growth, development, and distribution of mycoplankton can be influenced by light and other factors, which can change the dynamics of the marine fungal species (Li et al., 2023). Fotedar et al. (2022) demonstrated that the diversity and yeast communities' structure in coastal waters are affected by total organic matter, organic carbon in the water and sulphur. Ascomycota, Basidiomycota, and Chytridiomycota are the dominant fungal phyla in hydrothermal vents with saprotrophic, symbiotic, and pathogenic lifestyles (Schultz et al., 2025).

4.4 *Effect of Nutrient Availability on Marine Fungal Diversity*

Marine fungi are crucial for nutrient recycling and biogeochemical cycles; they enable the degradation of organic matter and biomineralisation, affecting the productivity of marine ecosystems and availability of nutrients (Peng et al., 2024; Britto Martins de Oliveira et al., 2025). Nutrient availability has a significant role in determining marine fungal diversity and distribution, with different nutrients influencing the community composition across marine ecosystems (Wu et al., 2023). Few studies describe that the chemical oxygen demand, dissolved oxygen, carbon and nitrogen availability in the sediments, along with water depth, salinity, oxygen levels, and temperature decide the diversity dynamics of mycoplankton and marine fungi in the marine environment (Orsi et al., 2013). The regulation of an optimum intracellular potassium concentration by marine fungi is essential for their survival in a marine environment. Calcium is crucial for the absorption of potassium and biological solutes in the hyphae (Jennings, 1983). Organic matter depletion in the euphotic zone of ocean ecosystems is caused by fungal diversity and abundance of major phyla such as Ascomycota and Basidiomycota (Wang et al., 2014). It is evident from the few studies that periodic occurrence, distribution, and metabolic activity of marine fungi in seawater can affect the degradation of particulate organic matter (POM) into dissolved organic matter (DOM) in marine environments (Orsi et al., 2022; Sen et al., 2022).

The availability of nitrogen and particulate organic matter in marine environments demonstrate the existence and distribution of the mycoplankton in coastal and oceanic water (Taylor and Cunliffe, 2016). Ammonia, phosphate, silicates and nitrate in addition to C and N also affect the composition and distribution of marine fungi in particular environments (Breyer and Baltar, 2023). A few studies revealed that benthic fungal communities in marine environments may involve anaerobic processes for organic matter degradation, nitrogen cycling and fungal denitrification (Jebaraj et al., 2010; Guo et al., 2015). Eutrophication of marine environments significantly influences the composition of the planktonic fungal communities, affecting the predominant taxa of *Dothideomycetes* more than the less frequently occurring taxa of *Agaricomycetes*. The dominant fungal taxa were mostly influenced by carbon compounds, while the rare taxa were concurrently affected by nitrogen, phosphorus, and carbon compounds in the Beibu Gulf (Zhao et al., 2023).

4.5 *Effect of Marine Substrata on Marine Fungal Diversity*

The marine ecosystems comprise of a wide range of substrata that include algae, animal-associated substrata, driftwood, submerged wood debris, mangrove wood and leaves, seagrasses, particulate organic matter, salt marsh plants, and snow help in colonisation and reproduction of marine fungi (Jones et al., 2019; Devadatha et al., 2021). Compared to other maritime ecosystems, mangroves support a large number of marine fungi due to availability of rich and diverse substrates (Jones et al., 2019). The diversity of marine fungi is significantly influenced by the characteristics and availability of these substrata (Luo and Pang, 2014; Sridhar, 2017; Sen et al., 2022). Among them, lignocellulosic substrates, particularly wood, support the highest diversity of marine fungal taxa, whereas calcareous surfaces and sand grains tend to host comparatively fewer species (Jones, 2011; Luo and Pang, 2014; Raghukumar, 2017; Sridhar, 2017). Plant-derived habitats, such as mangrove ecosystems, have been extensively investigated and are known to harbour distinct fungal assemblages that vary with plant species and tidal elevation (Calado et al., 2015; Simões et al., 2015; Devadatha et al., 2021).

Others include host specificity of fungi on seaweeds and the distribution of host algae: *Mycaureola dilsea* on *Dilsea carnosa*, *Lautitia danica* on *Chondrus crispus*, *Mycophycias ascophylli* on *Ascophyllum nodosum* in temperate waters, and *Sapthulospora* species on *Ballia carnosa* in Arctic waters. Host-specific fungi on seagrasses *Halotthia posidonniae, Paralulworthia posidoniae*, and *Pontoporeia biturbinata* on *Posidonia oceanica* in temperate water seagrasses (Poli et al., 2020); *Lindriella thalassiae* on *Thalassia testudinum*; *Rambellisea gigliensis*, and *R. halocynthiae* on the tunicate *Halocynthia papillosa, Koralionastes ellipticus* and *K. giganteus* on corals in tropical habitats (Kohlmeyer and Volkm.-Kohlmeyer, 1987; Pasqualetti et al., 2024). Animal hosts, including corals, sponges, invertebrates, and marine mammals also serve as reservoirs for diverse fungal communities. Limited studies have documented the marine fungi collected on animal substrates such as calcareous shells and feathers from mangrove ecosystems of India (Ananda and Sridhar, 2001; Nambiar and Raveendran, 2015). Fungi allied with phytoplankton are recognised as commensals and pathogens of many marine species, such as corals and sponges (Amend et al., 2019). However, the diversity and ecological roles of fungal parasites are inadequately investigated, highlighting existing research gaps (Ilicic et al., 2022). Fuentes et al. (2015) studied the growth and respiration rates of five marine fungi, suggesting that growth and respiration of these fungi are enhanced with increases in the glucose and temperature levels. The search for host-specific fungi has increased recently due to their inherent ability to produce bioactive metabolites. Fan et al. (2019), identified 55 fungal species from *Fucus vesiculosus* and evaluated their anticancer potential using metabolomics and One-Strain-Many-Compounds approaches.

4.6 Effect of Anthropogenic Factors

Anthropogenic activities have gradually altered the maritime ecosystem in recent years, resulting in remarkable alteration to marine environment, which significantly affects the fungi residing in the marine waters (Fig. 6). Marine fungi are essential in degrading anthropogenic contaminants such as plastics and petroleum-derived compounds, as well as in sequestering heavy metals, therefore facilitating the natural detoxification of the marine environment through enzyme-mediated degradation (Velmurugan et al., 2017; Behera and Das, 2023). Human activities impact marine fungal composition by altering nutrient availability, causing chemical pollution and altering habitat (Nogales et al., 2011). Kettner et al. (2017) found that marine fungi from the Ascomycota, Chytridiomycota, and Cryptomycota phyla mainly inhabit plastic substrates in fresh to brackish waters of northeast Germany.

A study of the taxonomic composition of bacterial and fungal communities on submerged polyethylene in sediments demonstrated the dominance of Ascomycota and minor fractions of Basidiomycota and Mucoromycota (De Tender et al., 2017). Few marine fungi, such as *Scedosporium apiospermum*, *Scopulariopsis brevicaulis*, and *Stachybotrys chartarum* are rich in polluted ecosystems and may act as indicators of anthropogenic pollution, whereas fungi like *Mucor circinelloides* suggest the pristine, undisturbed nature of the marine environment. Furno et al. (2022) highlighted that changes in fungal community composition and species richness may result from

Fig. 6. Deposition of anthropogenic waste materials, including plastic and tire residues, along coastal shorelines.

the type of substrate and influence of human activities at sampling sites. Marine fungi are essential in the decomposition of oil and total petroleum hydrocarbons, functioning as major decomposers of high-molecular-weight hydrocarbons by the production of extracellular enzymes. Some research indicates that marine fungi from Ascomycota, particularly *Penicillium* spp., isolated from marine sediments, possess the capability to degrade n-alkanes, heavy metals, and polycyclic aromatic hydrocarbons (Simister et al., 2015; Maamar et al., 2020).

4.7 *Effect of Sampling, Seasonal Variations, and Methodological Approaches*

The sampling strategies, seasonal variations, time and location of sample collection can influence the reported fungal diversity, as several species may not be sporulating when collected from the marine habitat. This further necessitates additional incubation in humid chambers with periodic spraying of sterile seawater to induce sporulation (Jones, 2011). The understanding of marine fungal diversity mainly relies on methodological approaches employed in its study (Kumar et al., 2021). Traditional culture-based methods often underestimate fungal diversity because several marine fungi are unculturable or necessitate specific conditions for growth (Amend, 2019). Conventional marine fungal classification may result in taxonomic ambiguity due to difficulties in delineating closely related and cryptic fungal species. In contrast, multi locus-based phylogenetic analyses revealed several novel taxa, novel lineages, and lead to the taxonomic re-evaluation of certain orders, families, genera, and species of marine fungi (Calabon et al., 2023; Dayarathne et al., 2025). The HTS of environmental DNA from marine habitats has discovered diverse and phylogenetically rich fungal communities, along with the uncovering of dark taxa (Orsi et al., 2013, Burgaud et al., 2022). Although there are advantages, molecular approaches confront challenges, such as the incorrect amplification of other types of eukaryotes, which may affect accurate identification (Tsui et al., 2011). Studies on this area are currently limited, affecting our ability to determine whether specific fungi are truly endemic and hindering our understanding of the diversity of fungal communities in different marine environments (Tisthammer et al., 2016). Estuarine habitats comprise a significant portion of the fungal taxa commonly seen in terrestrial environments, whereas open ocean sites were predominantly characterised by unidentified taxa (Jeffries et al., 2016). Sarma and Hyde (2001) proposed detailed criteria for the collection of mangrove samples for examining marine fungal diversity. It includes the collection of 1,000 to 2,000 samples during both wet and dry seasons and the collection of various substrates and mangrove plant species from diverse locations with a uniform sample size considering the local environmental conditions. Few studies showed that the diversity and richness of mangrove fungi are highest during the monsoon in contrast to the summer season, due to increased moisture, nutrient availability, and favourable environmental conditions that promote fungal growth and sporulation during the monsoon (Sarma and Hyde, 2001). Further studies are essential to comprehend how sampling, seasonal differences, and methodological approaches affect the marine fungal distribution and analysis (Fig. 7).

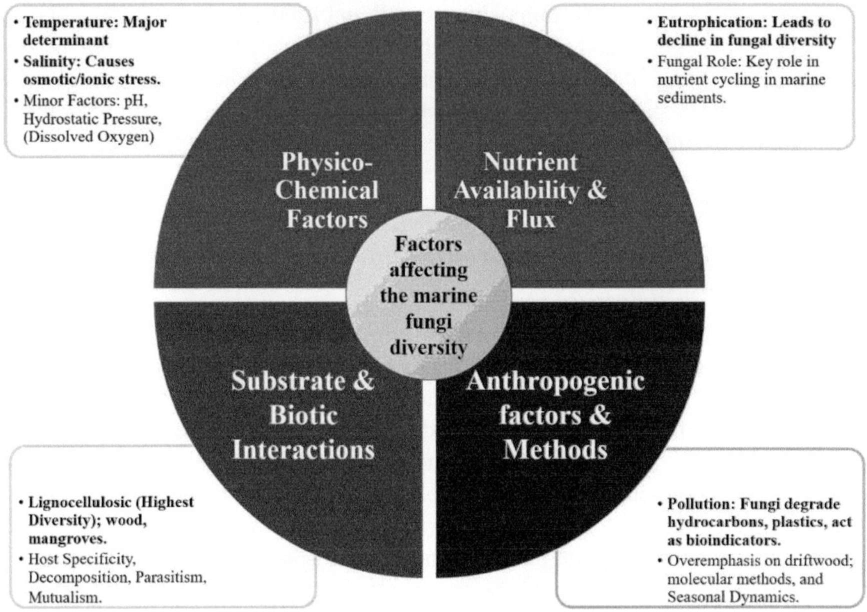

Fig. 7. Factors affecting global diversity and distribution of marine fungi.

5. Conclusion and Future Prospects

Marine fungi have the unique ability to reproduce both asexually and sexually in marine environments, which is essential for their perpetuation, as evidenced by the failure of terrestrial Ascomycota and saprolegniaceous species to sporulate in saline environments. Many marine-derived asexual fungal species from genera such as *Aspergillus*, *Cladosporium*, and *Penicillium* have extensive tolerance to salinity and environmental conditions, enabling their survival in marine environments. Most of the marine fungi cannot be characterised into global zones, as they are known only from their original collection or lack ecological data. Others can be classified as Arctic/Antarctic: *Sapthulospora antarctica*, *Leucosporidium antarcticum*, and *Monodictys austrina*; cold-water species: *Lautosporopsis* and *Toriella tubulifera*; temperate: *Ceriosporopsis halima, Haligena elaterophora*, and *Remispora stellata*; and tropical: *Antennospora quadricornuta, Cystospora rhizophorae*, and *Helicascus kanaloanus. Corollospora* and *Lulworthia* are the most specious genera, globally widely distributed, but some are ancient lineages, e.g., *C. angusta* and *C. parvula* (310 MYA) and others are more recent *C. cinnamomea* and *C. luteola* (12–31 MYA). Genera in the Lulworthiales received a stem age of 158.53 MYA for *Hydea pymea* and 62.32 for *Cumulospora marina* and *Linderilella thalssaiae*. Gunde-Cimerman et al. (2009) observed that many species within the phylum Ascomycota exhibited halotolerant and halophilic characteristics that play a key role in the decomposition of organic matter, including plant debris and animal remains (Hyde et al., 1998). Mangroves host a diverse range of marine fungal communities in contrast to other marine habitats, mainly due to the availability of abundant and diverse substrata.

Morphological, multi locus phylogenetic analysis and metagenomic studies of marine fungi advanced our knowledge on marine fungi. Very limited knowledge exists for unculturable marine fungal communities in the established databases, such as FUNGuild, UNITE, and SILVA. For accurate identification and evolutionary study of unculturable groups, we critically need dedicated databases and curated marine fungal genome resources. There is a dire need to enrich molecular data (nuclear ribosomal genes and protein-coding genes) for marine fungal taxa established based on morphology, culture-based methods or single marker genes to understand their taxonomic affiliations and ecological roles. Conventional and novel high-throughput culturomics methods must be employed for the efficient isolation and conservation of viable marine fungal cultures. This is crucial to comprehend their physiological characteristics, ecological functions and metabolic processes in marine habitats. Current global marine fungal resources, like the https://marinefungi.org/ database hosted by Prof. Gareth Jones and collaborators, represent a major advance, offering up-to-date classifications, species descriptions, sequence data information, and a comprehensive detail of marine fungal taxa. However, there is a strong need for broader global engagement to systematically upgrade marine fungal taxonomies, sequence reference databases and genome repositories, particularly regarding whole genome data and ecological metadata.

The lack of marine fungal herbarium and culture collections in most of the biodiversity-rich countries further hinders reference data and thus national and international repositories that are specific for marine fungi should be developed. The integration of *ex situ* collections, voucher specimens, living marine fungal cultures, related molecular data, and *in situ* conservation of endangered marine ecosystems should be given importance for progressing the fungal biodiversity and genomic research. Future research should extend to underexplored marine substrata (such as animals, corals, sediments, deep-sea habitats, and novel host associations) and geographic regions to reveal hidden diversity of marine fungi, combining high-throughput metagenomics, long-read sequencing and multigene phylogenetic analysis to overcome current taxonomic limitations. The majority of biodiversity-rich regions and countries continue to lack dedicated scientific funding and local taxonomic expertise to fully document marine fungal diversity, creating a critical bottleneck for global marine mycological research, database development, and conservation actions.

Most of the African, Australasian, and Polar Regions have rich biodiversity and include abundant habitats and substrata that are known or presumed to host marine fungi. Many of these habitats and substrates are unexplored, with certain areas, such as mangroves (Africa, Australasia), algae, animal substrata, and coral reefs being minimally investigated from a mycological perspective. Many of these neglected environments and substrates presumably harbour significant novel marine fungal diversity. Therefore, further studies focusing on omics-based innovative approaches such as metagenomics, genomics, and culturomics to understand fungal diversity and the functional role of mycobiomes from underexplored marine habitats and novel substrata are critically needed. This research is crucial for understanding the impacts of climate change on these largely unexplored marine ecosystems.

Acknowledgments

Devadatha acknowledges the financial support from CSIR, the Department of Biotechnology (DBT) and the Government of India. Gareth Jones is supported under the Distinguished Scientist Fellowship Program (DSFP), King Saud University, Kingdom of Saudi Arabia.

References

Abdel-Aziz, F.A. (2010). Marine fungi from two sandy Mediterranean beaches on the Egyptian north coast. Bot. Mar., 53: 283–289. https://doi.org/10.1515/bot.2010.023.

Abdel-Gawad, K.M., Hifney, A.F., Issa, A.A. and Gomaa, M. (2014). Spatio-temporal, environmental factors, and host identity shape culturable-epibiotic fungi of seaweeds in the Red Sea, Egypt. Hydrobiologia, 740(1): 37–49. https://doi.org/10.1007/s10750-014-1935-0.

Abdel-Wahab, B.A., Abd El-Kareem, H.F., Alzamami, A., Fahmy, C.A., Elesawy, B.H. et al. (2022). Novel exopolysaccharide from marine *Bacillus subtilis* with broad potential biological activities: Insights into antioxidant, anti-inflammatory, cytotoxicity, and anti-Alzheimer activity. Metabolites, 12(8): 715. https://doi.org/10.3390/metabo12080715.

Abdel-Wahab, M.A. (2005). Diversity of marine fungi from Egyptian Red Sea mangroves. Bot. Mar., 48: 348–355. https://doi.org/10.1515/bot.2005.047.

Abdel-Wahab, M.A. and Jones, E.B. (2003). *Decaisnella formosa* sp. nov. (Ascomycota, Massariaceae) from an Australian sandy beach. Can. J. Bot., 81(6): 598–600. https://doi.org/10.1139/b03-059.

Abdel-Wahab, M.A. and Jones, E.B.G. (2000). Three new marine ascomycetes from driftwood in Australia sand dunes. Mycoscience, 41(4): 379–388. https://doi.org/10.1007/BF02463951.

Abdel-Wahab, M.A., Bahkali, A.H., Elgorban, A.M. and Jones, E.B.G. (2021). High-throughput amplicon sequencing of fungi and microbial eukaryotes associated with the seagrass *Halophila stipulacea* (Forssk.) Asch. from Al-Leith mangroves, Saudi Arabia. Mycol. Prog., 20: 1365–1381. https://doi.org/10.1007/s11557-021-01744-2.

Abdel-Wahab, M.A., Jones, E.B.G., Bahkali, A.H. and El-Gorban, A.M. (2019). Marine fungi from Red Sea mangroves in Saudi Arabia with *Fulvocentrum rubrum* sp. nov. (Torpedosporales, Ascomycota). Nova Hedwigia, 108(3): 1–13. https://doi.org/10.1127/nova_hedwigia/2018/0511.

ACIA. (2004). Impacts of a Warming Arctic: Arctic Climate Impact Assessment. Cambridge University Press, Cambridge, 140 pp.

Aleem, A.A. (1979). Distribution and ecology of marine fungi in Sierra Leone (tropical west Africa). In: 3rd International Marine Mycology Symposium. Bot. Mar., 22: 403–412.

Alexander, E., Stock, A., Breiner, H.-W., Behnke, A., Bunge, J. et al. (2009). Microbial eukaryotes in the hypersaline anoxic L'Atalante deep-sea basin. Environ. Microbiol., 11: 360–381. https://doi.org/10.1111/j.1462-2920.2008.01777.x.

Alias, S.A. (1996). Ecological and Taxonomic Studies of Lignicolous Marine Fungi in Malaysian Mangroves. Ph.D. Thesis, University of Portsmouth, UK, 164 pp.

Alias, S.A. and Jones, E.B.G. (2000). Colonization of mangrove wood by marine fungi at Kuala Selangor mangrove stand, Malaysia. pp. 9–21. *In*: Hyde, K.D., Ho, W.H. and Pointing, S.B. (eds.). Aquatic Mycology Across the Millennium Fungal Diversity Research Series # 5.

Amend, A., Burgaud, G., Cunliffe, M., Edgcomb, V.P., Ettinger, C.L. et al. (2019). Fungi in the marine environment: Open questions and unsolved problems. MBio, 10(2): 10–1128. https://doi.org/10.1128/mBio.01189-18.

Ananda, K. and Sridhar, K.R. (2001). *Aniptodera indica*, a new species of mangrove-inhabiting ascomycete from the west coast of India. J. Environ. Biol., 22(4): 283–286.

Arenz, B. and Blanchette, R. (2011). Distribution and abundance of soil fungi in Antarctica at sites on the Peninsula, Ross Sea Region and McMurdo Dry Valleys. Soil Biol. Biochem., 43(2): 308–315. https://doi.org/10.1016/j.soilbio.2010.10.016.

Artemchuk, N. (1972). The fungi of the White Sea. New phycomycetes, discovered in the Velikaya Salma Strait of the Kandalakshal Bay. Veroeff. Inst. Meeresforsch. Bremerhav., 13: 213–237.

Asseri, A.A., Coello-Camba, A. and Agustí, S. (2025). Fungal planktonic community related to salinity and temperature in an oligotrophic sea. Front. Microbiol., 16: 1435925. https://doi.org/10.3389/fmicb.2025.1435925.

Banke, S., Frisvad, J.C. and Rosendahl, S. (1997). Taxonomy of *Penicillium chrysogenum* and related xerophilic species, based on isozyme analysis. Mycol. Res., 101(5): 617–624. https://doi.org/10.1017/S0953756296003048.

Barr, D.J. (1980). An outline for the reclassification of the Chytridiales, and for a new order, the Spizellomycetales. Can. J. Bot., 58(22): 2380–2394. https://doi.org/10.1139/b80-276.

Behera, A.D. and Das, S. (2023). Ecological insights and potential application of marine filamentous fungi in environmental restoration. Rev. Environ. Sci. Biotechnol., 22(2): 281–318. https://doi.org/10.1007/s11157-023-09655-2.

Bjørdal, Y. (2021). Rapid Microbial Responses to Temperature Changes in Arctic Anoxic Peat Soil. Master's Thesis, UiT Norges Arktiske Universitet, Norway, 60 pp. https://hdl.handle.net/10037/21757.

Blanchette, R.A., Held, B.W., Arenz, B.E., Jurgens, J.A., Baltes, N.J. et al. (2010). An Antarctic hot spot for fungi at Shackleton's historic hut on Cape Royds. Microb. Ecol., 60: 29–38. https://doi.org/10.1007/s00248-010-9664-z.

Blanchette, R.A., Held, B.W., Hellmann, L., Millman, L. and Büntgen, U. (2016). Arctic driftwood reveals unexpectedly rich fungal diversity. Fungal Ecol., 23: 58–65. https://doi.org/10.1016/j.funeco.2016.06.001.

Blanchette, R.A., Held, B.W., Jurgens, J.A., McNew, D.L., Harrington, T.C. et al. (2004). Wood-destroying soft rot fungi in the historic expedition huts of Antarctica. Appl. and Env. Microbio., 70(3): 1328–1335. https://doi.org/10.1128/aem.70.3.1328-1335.2004.

Bonugli-Santos, R.C., dos Santos Vasconcelos, M.R., Passarini, M.R., Vieira, G.A., Lopes, V.C. et al. (2015). Marine-derived fungi: Diversity of enzymes and biotechnological applications. Front. Microbiol., 6: 269. https://doi.org/10.3389/fmicb.2015.00269.

Booth, T. (1979). Strategies for study of fungi in marine and marine-influenced ecosystems. Rev. Microbiol. (S. Paulo), 10: 123–138.

Booth, T. (1981). Lignicolous and zoosporic fungi in marine environments of Hudson Bay. Can. J. Bot., 59: 1867–1881. https://doi.org/10.1139/b81-247.

Booth, T. and Kenkel, N. (1986). Ecological studies of lignicolous marine fungi: A distribution model based on ordination and classification. pp. 297–310. *In*: Moss, S.T. (ed.). The Biology of Marine Fungi. Cambridge Univ. Press.

Bovio, E., Garzoli, L., Poli, A., Prigione, V., Firsova, D. et al. (2018). The culturable mycobiota associated with three Atlantic sponges, including two new species: *Thelebolus balaustiformis* and *T. spongiae*. Fungal Syst. Evol., 1(1): 141–167. https://doi.org/10.3114/fuse.2018.01.07.

Boyd, P.E. and Kohlmeyer, K. (1982). The influence of temperature on the seasonal and geographic distribution of three marine fungi. Mycologia, 74(6): 894–902. https://doi.org/10.1080/00275514.1982.12021608.

Breyer, E. and Baltar, F. (2023). The largely neglected ecological role of oceanic pelagic fungi. Tr. Ecol. Evol., 38(9): 870–888. https://doi.org/10.1016/j.tree.2023.05.002.

Britto Martins de Oliveira, J., Corrêa Junior, D., Parente, C.E.T. and Frases, S. (2025). Fungi in mangrove: Ecological importance, climate change impacts, and the role in environmental remediation. Microorganisms, 13(4): 878. https://doi.org/10.3390/microorganisms13040878.

Brooks, R.D. (1972). The Occurrence and Distribution of Wood-inhabiting Marine Fungi in Point Judith Pond. Doctoral dissertation, University of Rhode Island, Kingston, USA.

Bubnova, E.N. (2005). Mycobiota Composition Changes along the Land: Marine Landscape gradient (a Case Study of the Kandalaksha Bay Shore of the White Sea). Ph.D. Thesis. Moscow State University, Moscow, 160 pp.

Bubnova, E.N. (2010). Fungal diversity in bottom sediments of the Kara Sea. Bot. Mar., 53(6): 595–600. https://doi.org/10.1515/bot.2010.063?urlappend=%3Futm_source%3Dresearchgate.net%26utm_medium%3Darticle.

Bubnova, E.N. and Konovalova, O.P. (2019). Fungi in bottom sediments of the Chukchi Sea. Rus. J. Mar. Biol., 45(2): 86–95. https://doi.org/10.1134/S1063074019020020.

Bubnova, E.N. and Nikitin, D.A. (2017). Fungi in bottom sediments of the Barents and Kara seas. Rus. J. Mar. Biol., 43(5): 400–406. https://doi.org/10.1134/S1063074017050029.

Buchan, A., Newell, S.Y., Moreta, J.L.L. and Moran, M.A. (2002). Analysis in Internal Transcribed Spacer (ITS) regions of rRNA genes fungal communities in a southeastern U.S. salt marsh. Microb. Ecol., 43(3): 329–340. https://doi.org/10.1007/s00248-001-1062-0.

Burgaud, G., Edgcomb, V., Hassett, B.T., Kumar, A., Li, W. et al. (2022). Marine fungi. pp. 243–295. *In*: The Marine Microbiome. Springer Cham. https://doi.org/10.1007/978-3-030-90383-1_5.

Burgaud, G., Le Calvez, T., Arzur, D., Vandenkoornhuyse, P. and Barbier, G. (2009). Diversity of culturable marine filamentous fungi from deep-sea hydrothermal vents. Environ. Microbiol., 11(6): 1588–1600. https://doi.org/10.1111/j.1462-2920.2009.01886.x.

Byrne, P.J. and Jones, E.B.G. (1974). Lignicolous marine fungi. Veröffentlichungen des Instituts für Meeresforschung Bremerhaven Supplement, 5: 301–320.

Byrne, P.J. and Jones, E.B.G. (1975). Effect of salinity on the reproduction of terrestrial and marine fungi. Trans. Br. Mycol. Soc., 65(2): 185–200.

Calabon, M., Jones, E.B.G., Promputtha, I. and Hyde, K.D. (2021). Fungal biodiversity in salt marsh ecosystems. J. Fungi, 7: 1–72. https://doi.org/10.3390/jof7080648.

Calabon, M.S., Jones, E.B.G., Pang, K.L., Abdel-Wahab, M.A., Jin, J. et al. (2023). Updates on the classification and numbers of marine fungi. Bot. Mar., 66(4): 213–238. https://doi.org/10.1515/bot-2023-0032.

Calado, M.D.L., Carvalho, L., Pang, K.L. and Barata, M. (2015). Diversity and ecological characterization of sporulating higher filamentous marine fungi associated with *Spartina maritima* (Curtis) fernald in two portuguese salt marshes. Microb. Ecol., 70: 612–633. https://doi.org/10.1007/s00248-015-0600-0.

Campbell, J., Shearer, C.A., Crane, J.L. and Fallah, P.M. (2003). A reassessment of two freshwater ascomycetes, *Ceriospora caudae-suis* and *Submersisphaeria aquatica*. Mycologia, 95(1): 41–53. https://doi.org/10.2307/3761960.

Capon, R.J., Stewart, M., Ratnayake, R., Lacey, E. and Gill, J.H. (2007). Citromycetins and bilains A–C: New aromatic polyketides and diketopiperazines from Australian marine-derived and terrestrial *Penicillium* spp. J. Nat. Prod., 70(11): 1746–1752. https://doi.org/10.1021/np0702483.

Chalkley, D.B., Suh, S.O., Volkmann-Kohlmeyer, B., Kohlmeyer, J. and Zhou, J.J. (2010). *Diatrypasimilis australiensis*, a novel xylarialean fungus from mangrove. Mycologia, 102(2): 430–437. https://doi.org/10.3852/09-142.

Chen, J., Wang, B., Lu, Y., Guo, Y., Sun, J. et al. (2019). Quorum sensing inhibitors from marine microorganisms and their synthetic derivatives. Mar. Drugs, 17(2): 80. https://doi.org/10.3390/md17020080.

Chen, J., Zang, Y., Yang, Z., Qu, T., Sun, T. et al. (2022). Composition and functional diversity of epiphytic bacterial and fungal communities on marine macrophytes in an intertidal zone. Front. Microbiol., 13: 839465. https://doi.org/10.3389/fmicb.2022.839465.

Chen, X., Yan, D., Yu, L. and Zhang, T. (2023). An integrative study of mycobiome in different habitats from a high Arctic region: Diversity, distribution, and functional role. J. Fungi, 9(4): 437. https://doi.org/10.3390/jof9040437.

Chown, S.L. and Brooks, C.M. (2019). The state and future of Antarctic environments in a global context. Annu. Rev. Environ. Resour., 44: 1–30. https://papers.ssrn.com/sol3/papers.cfm?abstract_id=3472228#.

Comeau, A.M., Vincent, W.F., Bernier, L. and Lovejoy, C. (2016). Novel chytrid lineages dominate fungal sequences in diverse marine and freshwater habitats. Sci. Rep., 6: 30120. https://doi.org/10.1038/srep30120.

Costello, M.J., Tsai, P., Wong, P.S., Cheung, A.K.L., Basher, Z. et al. (2017). Marine biogeographic realms and species endemicity. Nat. Comm., 8(1): 1057. https://doi.org/10.1038/s41467-017-01121-2.

Cribb, A.B. and Cribb, J.W. (1955a). Marine Fungi from Queensland. University of Queensland Press, Brisbane, 81 pp.

Cribb, A.B. and Cribb, J.W. (1955b). Marine Fungi from Queensland. 1. Pap. Dept. Bot. Univ. Qd., Australia, 3: 77–81.

Cribb, A.B. and Cribb, J.W. (1960). Some Marine Fungi on Algae in European Herbaria. University of Queensland Press, Brisbane, 48 pp.

Currie, J.N. and Thom, C. (1915). An oxalic acid producing *Penicillium*. Journal of Biological Chemistry, 22(2): 287–293. https://doi.org/10.1016/S0021-9258(18)87646-3.

Da Silva, T.H., Câmara, P.E., Pinto, O.H.B., Carvalho-Silva, M., Oliveira, F.S. et al. (2022). Diversity of fungi present in permafrost in the South Shetland Islands, Maritime Antarctic. Microbial. Ecol., 83(1): 58–67. https://doi.org/10.1007/s00248-021-01735-6.

Davidson, D.E. (1974). The effect of salinity on a marine and a freshwater Ascomycete. Can. J. Bot., 52(3): 553–563. https://doi.org/10.1139/b74-071.

Dayarathne, M.C., Jones, E.B.G., Rämä, T., Hagestad, O.C., Abdel-Wahab, M.A. et al. (2025). Towards an integrative morpho-molecular classification of the Lulworthiomycetidae. Mycosphere, 16(1): 79–168. https://doi.org/10.5943/mycosphere/16/1/2.

De Souza, L.M.D., Lirio, J.M., Coria, S.H., Lopes, F.A.C., Convey, P. et al. (2022). Diversity, distribution, and ecology of fungal communities present in Antarctic lake sediments uncovered by DNA metabarcoding. Sci. Rep., 12(1): 8407. https://doi.org/10.1038/s41598-022-12290-6.

De Tender, C., Devriese, L.I., Haegeman, A., Maes, S., Vangeyte, J. et al. (2017). Temporal dynamics of bacterial and fungal colonization on plastic debris in the North Sea. Environ. Sci. Technol., 51(13): 7350–7360. https://doi.org/10.1021/acs.est.7b00697.

Devadatha, B., Jones, E.B.G., Pang, K.L., Abdel-Wahab, M.A., Hyde, K.D. et al. (2021). Occurrence and geographical distribution of mangrove fungi. Fungal Divers., 106: 137–227. https://doi.org/10.1007/s13225-020-00468-0.

Dierckx, R.P. (1901). Un essai de revision du genre *Penicillium* Link. Annales de la Société Scientifique de Bruxelles, 25: 83–89.

Dihazi, A., Jaiti, F., El Hadrami, I., El Hassni, M. and Zouine, J. (2003). Effect of salicylic acid on phenolic compounds related to date palm resistance to *Fusarium oxysporum* f. sp. *albedinis*. Phytopathol. Mediterr., 42(1): 9–16.

Duarte, S., Bärlocher, F., Pascoal, C. and Cássio, F. (2016). Biogeography of aquatic hyphomycetes: Current knowledge and future perspectives. Fungal Ecol., 19: 169–181. https://doi.org/10.1016/j.funeco.2015.06.002.

Ebel, R. (2012). Natural products from marine-derived fungi. pp. 411–440. *In*: Jones, E.B.G. and Pang, K.L. (eds.). Marine Fungi and Fungal-like Organisms. De Gruyter, Berlin.

Ekanayaka, A.H., Tibpromma, S., Dai, D., Xu, R., Suwannarach, N. et al. (2022). A review of the fungi that degrade plastic. J. Fungi, 8: 772. https://doi.org/10.3390/jof808077.

Erich, S., Kuschel, B., Schwarz, T., Ewert, J., Böhmer, N. et al. (2015). Novel high-performance metagenome β-galactosidases for lactose hydrolysis in the dairy industry. J. Biotech., 210: 27–37. https://doi.org/10.1016/j.jbiotec.2015.06.411.

Eriksson, K.E. and Goodell, E.W. (1974). Pleiotropic mutants of the wood-rotting fungus *Polyporus adustus* lacking cellulase, mannanase, and xylanase. Can. J. Microbiol., 20(3): 371–378.

Erneholm, N. (1972). Marina svampar. Studier i den tempererade och tropiska marina svampfloran. Thesis, Department of Marine Botany, Göteborgs Universitet, Germany. http://hdl.handle.net/2077/10912.

Fan, B., Parrot, D., Blümel, M., Labes, A. and Tasdemir, D. (2019). Influence of OSMAC-based cultivation on metabolome and anticancer activity of fungi associated with the brown alga *Fucus vesiculosus*. Mar. Drugs, 17(1): 67. https://doi.org/10.3390/md17010067.

Fotedar, R., Chatting, M., Kolecka, A., Zeyara, A., Al Malki, A. et al. (2022). Communities of culturable yeasts and yeast-like fungi in oligotrophic hypersaline coastal waters of the Arabian Gulf surrounding Qatar. Ant. van Leeuwen., 115(5): 609–633. https://doi.org/10.1007/s10482-022-01722-y.

Fotedar, R., Fell, J.W., Boekhout, T., Kolecka, A., Zeyara, A. et al. (2019). *Cystobasidium halotolerans* sp. nov., a novel basidiomycetous yeast species isolated from the Arabian Gulf. Int. J. Syst. Evol. Microbiol., 69(3): 839–845. https://doi.org/10.1099/ijsem.0.003250.

Fotedar, R., Kolecka, A., Boekhout, T., Fell, J.W., Al-Malki, A. et al. (2018). Fungal diversity of the hypersaline Inland Sea in Qatar. Bot. Mar., 61(6): 595–609. https://doi.org/10.1515/bot-2018-0048?urlappend=%3Futm_source%3Dresearchgate.net%26utm_medium%3Darticle.

Fryar, S.C., Haelewaters, D. and Catcheside, D.E. (2019). *Annabella australiensis* gen. and sp. nov. (Helotiales, Cordieritidaceae) from South Australian mangroves. Mycol. Prog., 18(7): 973–981. https://doi.org/10.1007/s11557-019-01499-x.

Fryar, S.C., Hyde, K.D. and Catcheside, D.E. (2020). A survey of marine fungi on wood in South Australia. Bot. Mar., 63(5): 469–478. https://doi.org/10.1515/bot-2020-0005.

Fuentes, M.E., Quinones, R.A., Gutierrez, M.H. and Pantoja, S. (2015). Effects of temperature and glucose concentration on the growth and respiration of fungal species isolated from a highly productive coastal upwelling ecosystem. Fungal Ecol., 13: 135–149. https://doi.org/10.1016/j.funeco.2014.09.006.

Furbino, L.E., Godinho, V.M., Santiago, I.F., Pellizari, F.M., Alves, T.M. et al. (2014). Diversity patterns, ecology and biological activities of fungal communities associated with the endemic macroalgae across the Antarctic Peninsula. Microb. Ecol., 67(4): 775–787. https://doi.org/10.1007/s00248-014-0374-9.

Furbino, L.E., Pellizzari, F.M., Neto, P.C., Rosa, C.A. and Rosa, L.H. (2018). Isolation of fungi associated with macroalgae from maritime Antarctica and their production of agarolytic and carrageenolytic activities. Polar Biol., 41(3): 527–535. https://doi.org/10.1007/s00300-017-2213-1.

Furno, M.F., Poli, A., Ferrero, D., Tardelli, F., Manzini, C. et al. (2022). The culturable mycobiota of sediments and associated microplastics: From a harbour to a marine protected area, a comparative study. J. Fungi, 8(9): 927. https://doi.org/10.3390/jof8090927.

Gams, W. (1971). *Tolypocladium*, eine Hyphomyceten Gattung mit geschwollenen Phialiden. Persoonia-Mol. Phylog. Evol. Fungi., 6(2): 185–191.

Gams, W., Korf, R.P., Pitt, J.I., Hawksworth, D.L., Berbee, M.L. et al. (2003). Has dual nomenclature for fungi run its course? The Article 59 debate. Mycotaxon, 88: 493–508.

Gao, M., Liu, B., Li, J., Deng, Y., Zhang, Y. et al. (2024). Diversity and distribution of fungi in the marine sediments of Zhanjiang Bay, China. J. Fungi, 10(12): 867. https://doi.org/10.3390/jof10120867.

Gjessing, M.C., Davey, M., Kvellestad, A. and Vrålstad, T. (2011). *Exophiala angulospora* causes systemic inflammation in Atlantic cod Gadus morhua. Dis. of Aqua. Org., 96(3): 209–219. https://doi.org/10.3354/dao02381.

Godinho, V.M., de Paula, M.T.R., Silva, D.A.S., Pareswue, K., Martins, A.P. et al. (2019). Diversity and distribution of hidden cultivable fungi associated with marine animals of Antarctica. Fungal Biol., 123(7): 507–516. https://doi.org/10.1016/j.funbio.2019.05.001.

Godinho, V.M., Furbino, L.E., Santiago, I.F., Pellizzari, F.M., Yokoya, N.S. et al. (2013). Diversity and bioprospecting of fungal communities associated with endemic and cold-adapted macroalgae in Antarctica. The ISME J., 7(7): 1434–1451. https://doi.org/10.1038/ismej.2013.77.

Gonçalves, V.N., Campos, L.S., Melo, I.S., Pellizari, V.H., Rosa, C.A. et al. (2013). *Penicillium solitum*: A mesophilic, psychrotolerant fungus present in marine sediments from Antarctica. Polar Biol., 36(12): 1823–1831. https://doi.org/10.1007/s00300-013-1403-8.

Gonçalves, V.N., Carvalho, C.R., Johann, S., Mendes, G., Alves, T.M. et al. (2015). Antibacterial, antifungal, and antiprotozoal activities of fungal communities present in different substrates from Antarctica. Polar Biol., 38(8): 1143–1152. https://doi.org/10.1007/s00300-015-1672-5.

Gonçalves, V.N., Vitoreli, G.A., de Menezes, G.C., Mendes, C.R., Secchi, E.R. et al. (2017). Taxonomy, phylogeny, and ecology of cultivable fungi present in seawater gradients across the northern Antarctica Peninsula. Extremophiles, 21(6): 1005–1015. https://doi.org/10.1007/s00792-017-0959-6.

Gorter, G.J.M.R. (1978). Lignicolous marine fungi on submerged wood from the Atlantic coast of South Africa. Nova Hedwigia, 29: 895–903.

Gostinčar, C., Lenassi, M., Gunde-Cimerman, N. and Plemenitaš, A. (2011). Fungal adaptation to extremely high salt concentrations. Adv. Appl. Microbiol., 77: 71–96. https://doi.org/10.1016/b978-0-12-387044-5.00003-0.

Grasso, R., Schena, F., Gulli, G. and Cevese, A. (1997). Does low-frequency variability of heart period reflect a specific parasympathetic mechanism? J. Auton. Nerv. Syst., 63(1-2): 30–38. https://doi.org/10.1016/s0165-1838(96)00128-2.

Grasso, S., Bruni, V. and Maio, G. (1997). Marine fungi in Terra Nova Bay (Ross Sea, Antarctica). New Microbiol., 20(4): 371–376.

Griffiths, H.J. (2010). Antarctic marine biodiversity: What do we know about the distribution of life in the Southern Ocean? PLoS One, 5(8): e11683. https://doi.org/10.1371/journal.pone.0011683.

Grossart, H.P., Wurzbacher, C., James, T.Y. and Kagami, M. (2016). Discovery of dark matter fungi in aquatic ecosystems demands a reappraisal of the phylogeny and ecology of zoosporic fungi. Fungal Ecol., 19: 28–38. https://doi.org/10.1016/j.funeco.2015.06.004.

Gunde-Cimerman, N., Ramos, J. and Plemenitaš, A. (2009). Halotolerant and halophilic fungi. Mycol. Res., 113(11): 1231–1241. https://doi.org/10.1016/j.mycres.2009.09.002.

Guo, R., Wolf, K.K.E., Hoppe, C.J.M., Wu, L. and Lohmann, G. (2025). The changing nature of future Arctic marine heatwaves and its potential impacts on the ecosystem. Nat. Clim. Change, 15: 162–170. https://doi.org/10.1038/s41558-024-02224-7.

Guo, X., Zhang, Q., Zhang, X., Zhang, J. and Gong, J. (2015). Marine fungal communities in water and surface sediment of a sea cucumber farming system: Habitat-differentiated distribution and nutrients driving succession. Fungal Ecol., 14: 87–98. https://doi.org/10.1016/j.funeco.2014.12.001.

Ha, T.M., Kim, D.C., Sohn, J.H., Yim, J.H. and Oh, H. (2020). Anti-inflammatory and protein tyrosine phosphatase 1b inhibitory metabolites from the antarctic marine-derived fungal strain *Penicillium glabrum* SF-7123. Mar. Drugs, 18(5): 247. https://doi.org/10.3390/md18050247.

Hagestad, O.C., Andersen, J.H., Altermark, B., Hansen, E. and Rämä, T. (2019). Cultivable marine fungi from the Arctic Archipelago of Svalbard and their antibacterial activity. Mycology, 11(3): 230–242. https://doi.org/10.1080/21501203.2019.

Hassett, B.T. and Gradinger, R. (2016). Chytrids dominate Arctic marine fungal communities. Environ. Microbiol., 18(6): 2001–2009. https://doi.org/10.1111/1462-2920.13216.

Hassett, B.T., Borrego, E.J., Vonnahme, T.R., Rämä, T., Kolomiets, M.V. and Gradinger, R. (2019). Arctic marine fungi: Biomass, functional genes, and putative ecological roles. The ISME J., 13(6): 1484–1496. https://doi.org/10.1038/s41396-019-0368-1.

Hassett, B.T., Ducluzeau, A.-L.L., Collins, R.E. and Gradinger, R. (2017). Spatial distribution of aquatic marine fungi across the western Arctic and subarctic. Environ. Microbiol., 19(2): 475–484. https://doi.org/10.1111/1462-2920.13371.

Hassett, B.T., Vonnahme, T.R., Peng, X., Jones, E.B.G. and Heuze, C. (2020). Global diversity and geography of planktonic marine fungi. Bot. Mar., 63(2): 121–139. https://doi.org/10.1515/bot-2018-0113.

Herlemann, D.P., Labrenz, M., Jürgens, K., Bertilsson, S., Waniek, J.J. and Andersson, A.F. (2011). Transitions in bacterial communities along the 2000 km salinity gradient of the Baltic Sea. The ISME J., 5(10): 1571–1579. https://doi.org/10.1038/ismej.2011.41.

Herrera, L.M., García-Laviña, C.X., Marizcurrena, J.J., Volonterio, O., de León, R.P. et al. (2017). Hydrolytic enzyme-producing microbes in the Antarctic oligochaete *Grania* sp. (Annelida). Pol. Biol., 40(4): 947–953. https://doi.org/10.1007/s00300-016-2012-0.

Hughes, G.C. (1974). Geographical distribution of the higher marine fungi. Veröff. Inst. Meeresforsch. Bremerh. Suppl., 5: 419–441.

Hughes, G.C. (1975). Studies of fungi in oceans and estuaries since 1961, 1. Lignicolous, caulicolous and foliicolous species. Oceanogr. Mar. Biol. Ann. Rev., 13: 69–180.

Hughes, G.C. (1986). Biogeography and the marine fungi. pp. 275–295. *In*: Moss, S.T. (ed.). The Biology of Marine Fungi. Cambridge University Press.

Hyde, K.D. (1990). A comparison of the intertidal mycota of five mangrove tree species. Asian Mar. Biol., 7: 93–107.

Hyde, K.D. (1990). Intertidal fungi from warm temperate mangroves of Australia, including *Tunicatispora australiensis* gen. et sp. nov. Aust. Syst. Bot., 3(4): 711–718. https://doi.org/10.1071/SB9900711.

Hyde, K.D. (1992). Fungi from decaying intertidal fronds of *Nypa fruticans*, including three new genera and four new species. Bot. J. Linn. Soc., 110(2): 95–110. https://doi.org/10.1111/j.1095-8339.1992.tb00284.x.

Hyde, K.D. (1992). The effect of incubation on the occurrence of marine fungi on randomly collected lignocellulose samples. Sydowia, 44: 131–136.

Hyde, K.D. (1993). Fungi from palms V. *Phomatospora nypae* sp. nov. and notes on marine fungi from *Nypa fruticans* in Malaysia. Sydowia, 45(2): 199–203.

Hyde, K.D. (1995). Fungi from palms XXII. A new species of *Ascotaiwania*. Sydowia, 47: 213–216.

Hyde, K.D. and Nakagiri, A. (1992). Intertidal fungi from Australia. The genus Swampomyces including *S. triseptatus* sp. nov. Sydowia, 44: 122–130.

Hyde, K.D. and Rappaz, F. (1993). *Eutypa bathurstensis* sp. nov. from intertidal *Avicennia*. Mycol. Res., 97(7): 861–864. https://doi.org/10.1016/S0953-7562(09)81163-5.

Hyde, K.D., Jones, E.G., Leaño, E., Pointing, S.B., Poonyth, A.D. et al. (1998). Role of fungi in marine ecosystems. Biodiver. Conser., 7(9): 1147–1161. https://doi.org/10.1023/A:1008823515157.

Ilicic, D., Woodhouse, J., Karsten, U., Schimani, K., Zimmermann, J. et al. (2024). Chytrid fungi infecting Arctic microphytobenthic communities under varying salinity conditions. Sci. Rep., 14: 25821. https://doi.org/10.1038/s41598-024-77202-2.

Ilicic, D., Woodhouse, J., Karsten, U., Zimmermann, J., Wichard, T. et al. (2022). Antarctic glacial meltwater impacts the diversity of fungal parasites associated with benthic diatoms in shallow coastal zones. Front. Microbiol., 13: 805694. https://doi.org/10.3389/fmicb.2022.805694.

Inderbitzin, P., Lim, S.R., Volkmann-Kohlmeyer, B., Kohlmeyer, J. and Berbee, M.L. (2004). The phylogenetic position of *Spathulospora* based on DNA sequences from dried herbarium material. Mycol. Res., 108(7): 737–748. https://doi.org/10.1017/S0953756204000206.

Jebaraj, C.S., Raghukumar, C., Behnke, A. and Stoeck, T. (2010). Fungal diversity in oxygen-depleted regions of the Arabian Sea revealed by targeted environmental sequencing combined with cultivation. FEMS Microbiol. Ecol., 71(3): 399–412. https://doi.org/10.1111/j.1574-6941.2009.00804.x.

Jeffries, T.C., Curlevski, N.J., Brown, M.V., Harrison, D.P., Doblin, M.A. et al. (2016). Partitioning of fungal assemblages across different marine habitats. Environ. Microbiol. Rep., 8(2): 235–238. https://doi.org/10.1111/1758-2229.12373.

Jennings, D.H. (1983). Some aspects of the physiology and biochemistry of marine fungi. Biol. Rev., 58(3): 423–459. https://doi.org/10.1111/j.1469-185X.1983.tb00395.x.

Jennings, D.H. (1986). Fungal growth in the sea. pp. 1–10. *In*: Moss, S.T. (ed.). The Biology of Marine Fungi. Cambridge University Press.

Jiya, N., Shede, P. and Sharma, A. (2023). Diversity and composition of fungal communities across diverse environmental niches in Antarctica. Polar Sci., 38: 100973. https://doi.org/10.1016/j.polar.2023.100973.

Jones, E.B.G. (1968). Marine fungi. Curr. Sci., 37: 378–379.

Jones, E.B.G. (1971). Aquatic fungi. pp. 335–365. *In*: Booth, C. (ed.). Methods in Microbiology, 4. Academic Press.

Jones, E.B.G. (2011). Are there more marine fungi to be described? Bot. Mar., 54(4): 343–354. https://doi.org/10.1515/bot.2011.043.

Jones, E.B.G. and Irvine, J. (1971). The role fungi play in the deterioration of wood in the sea. J. Inst. Wood Sci., 5: 31–40.

Jones, E.B.G. and Jennings, D.H. (1964). The effect of salinity on the growth of marine fungi in comparison with non-marine species. Trans. Br. Mycol. Soc., 47: 619–625.

Jones, E.B.G., Kohne, H., Trussell, P.C. and Turner, R.D. (1972). Results of an international co-operative research programme on the biodeterioration of timber submerged in the sea. Mater. Org., 7: 93–118.

Jones, E.B.G., Pang, K.L., Abdel-Wahab, M.A., Scholz, B., Hyde, K.D. et al. (2019). An online resource for marine fungi. Fungal Diver., 96(1): 347–433. https://doi.org/10.1007/s13225-019-00426-5.

Jones, E.B.G., Ramakrishna, S., Vikineswary, S., Das, D., Bahkali, A.H. et al. (2022). How do fungi survive in the sea and respond to climate change? J. Fungi, 8(3): 291. https://doi.org/10.3390/jof8030291.

Jones, E.B.G., Suetrong, S., Sakayaroj, J., Bahkali, A.H., Abdel-Wahab, M.A. et al. (2015). Classification of marine Ascomycota, Basidiomycota, Blastocladiomycota, and Chytridiomycota. Fungal Diver., 73: 1–72. https://doi.org/10.1007/s13225-015-0339-4.

Jones, S.L., King, D.H., Cummings, V.J., Robinson, S.A. and Waterman, M.J. (2025). Research bias in long-term monitoring of Antarctic nearshore marine and terrestrial biota. Global Change Biol., 31(8): e70392. https://doi.org/10.1111/gcb.70392.

Kendrick, B., Risk, M.J., Michaelides, J. and Bergman, K. (1982). Amphibious microborers: Bioeroding fungi isolated from live corals. Bull. Mar. Sci., 32(4): 862–867.

Kettner, M.T., Rojas-Jimenez, K., Oberbeckmann, S., Labrenz, M. and Grossart, H.P. (2017). Microplastics alter composition of fungal communities in aquatic ecosystems. Environ. Microbiol., 19(11): 4447–4459. https://doi.org/10.1111/1462-2920.13891.

Khusnullina, A.I., Bilanenko, E.N. and Kurakov, A.V. (2018). Microscopic fungi of White Sea sediments. Contemp. Prob. of Ecol., 11(5): 503–513. https://doi.org/10.1134/S1995425518050062.

Kis-Papo, T., Weig, A.R., Riley, R., Peršoh, D., Salamov, A. et al. (2014). Genomic adaptations of the halophilic Dead Sea filamentous fungus *Eurotium rubrum*. Nat. Comm., 5: 3745. https://doi.org/10.1038/ncomms4745.

Kochkina, G.A., Pinchuk, I.P., Ivanushkina, N.E., Avtukh, A.N. and Pimenov, N.V. (2024). Fungi of the Arctic seas. Microbiology, 93(3): 282–292. https://doi.org/10.1134/S002626172360502X.

Kogej, T., Ramos, J., Plemenitaš, A. and Gunde-Cimerman, N. (2005). The halophilic fungus *Hortaea werneckii* and the halotolerant fungus *Aureobasidium pullulans* maintain low intracellular cation concentrations in hypersaline environments. Appl. Environ. Microbiol., 71(11): 6600–6605. https://doi.org/10.1128/aem.71.11.6600-6605.2005.

Kohlmeyer, J. (1983). Geography of marine fungi. Aust. J. Bot. (Supplementary Series), 13(10): 67–76.

Kohlmeyer, J. (1984). Tropical marine fungi. Mar. Ecol., 5(4): 329–378. https://doi.org/10.1111/j.1439-0485.1984.tb00130.x.

Kohlmeyer, J. (1986). Taxonomic studies of the Marine Ascomycotina. pp. 199–210. *In*: Moss, S.T. (ed.). The Biology of Marine Fungi. Cambridge University Press, Cambridge, UK.

Kohlmeyer, J. and Kohlmeyer, E. (1971). Marine fungi from tropical America and Africa. Mycologia, 63: 831–861. https://doi.org/10.1080/00275514.1971.12019172.

Kohlmeyer, J. and Kohlmeyer, E. (1979). Marine Mycology: The Higher Fungi. Elsevier-Academic Press, New York and London, 705 pp.

Kohlmeyer, J. and Volkmann-Kohlmeyer, B. (1987). Koralionastetaceae fam. nov. (Ascomycetes) from coral rock. Mycologia, 79(5): 764–778. https://doi.org/10.1080/00275514.1987.12025457.

Kohlmeyer, J. and Volkmann-Kohlmeyer, B. (1989). Hawaiian marine fungi, including two new genera of Ascomycotina. Mycol. Res., 92(4): 410–421. https://doi.org/10.1016/S0953-7562%2889%2980185-6.

Kohlmeyer, J. and Volkmann-Kohlmeyer, B. (1991). Marine fungi of Queensland, Australia. Mar. Freshwat. Res., 42(1): 91–99.

Kumar, V., Sarma, V.V., Thambugala, K.M., Huang, J.J., Li, X.Y. et al. (2021). Ecology and evolution of marine fungi with their adaptation to climate change. Front. Microbiol., 12: 719000. https://doi.org/10.3389/fmicb.2021.719000.

Kuznetsova, T.A., Kirichuk, N.N. and Buzoleva, L.S. (2019). Fungi in bottom sediments of the chukchi sea. *In*: Species Diversity and Biology of Fungi Isolated from the Dead Wood, Soils, and Sediments. Springer, Cham, (pp. 95–108). https://doi.org/10.1134/S1063074019020020.

Lagerberg, T., Lundberg, G. and Melin, E. (1927). Biological and practical researches into blueing in pine and spruce. Centraltryckeriet., 25: 145–272.

Lai, X., Cao, L., Tan, H., Fang, S., Huang, Y. and Zhou, S. (2007). Fungal communities from methane hydrate-bearing deep-sea marine sediments in the South China Sea. The ISME J., 1(8): 756–762. https://doi.org/10.1038/ismej.2007.51.

Langdon, J.S. and McDonald, W.L. (1987). Cranial *Exophiala pisciphila* infection in *Salmo salar* in Australia. Bull. Eur. Ass. Fish Pathol., 7: 35–37.

Le Calvez, T., Burgaud, G., Mahé, S., Barbier, G. and Vandenkoornhuyse, P. (2009). Fungal diversity in deep-sea hydrothermal ecosystems. Appl. Environ. Microbiol., 75: 6415–6421. https://doi.org/10.1128/AEM.00653-09.

Li, X., Huang, M., Li, N., Zhao, H., Pu, Y. et al. (2023). Effects of environmental factors on mycoplankton diversity and trophic modes in coastal surface water. Ecol. Ind., 146: 109778. https://doi.org/10.1016/j.ecolind.2022.109778.

Lintott, W.H. and Lintott, E.A. (1982). Lignicolous marine fungi from New Zealand. Mauri Ora, 10: 73–76.

Lintott, W.H. and Lintott, E.A. (2002). Marine fungi from New Zealand. pp. 285–292. *In*: Hyde, K.D. (ed.). Fungi in Marine Environments. Fungal Diversity Press. Hong Kong.

Liu, C., Wang, X.-C., Yu, Z.-H., Zhuang, W.-Y. and Zen, Z.-Q. (2023). Seven new species of *Eurotiales* (Ascomycota) isolated from tidal flat sediments in China. J. Fungi, 9: 960. https://doi.org/10.3390/jof9100960.

Loilong, A., Sakayaroj, J., Rungjindamai, N., Choeyklin, R. and Jones, E. (2012). Biodiversity of fungi on the palm *Nypa fruticans*. pp. 273–290. *In*: Jones, E. and Pang, K. (eds.). Marine Fungi. Berlin, Boston: De Gruyter. https://doi.org/10.1515/9783110264067.273.

Loque, C.P., Medeiros, A.O., Pellizzari, F.M., Oliveira, E.C., Rosa, C.A. et al. (2010). Fungal community associated with marine macroalgae from Antarctica. Polar Biol., 33(5): 641–648. https://doi.org/10.1007/s00300-009-0740-0.

Lorenz, R. and Molitoris, H.P. (1992). Combined influence of salinity and temperature (Phoma-pattern) on growth of marine fungi. Can. J. Bot., 70(10): 2111–2115. https://doi.org/10.1139/b92-262.

Lozupone, C.A. and Knight, R. (2007). Global patterns in bacterial diversity. Proc. Natl. Acad. Sci. USA, 104(27): 11436–11440. https://doi.org/10.1073/pnas.0611525104.

Luard, E.J. (1985). Respiration of *Penicillium chrysogenum* in relation to the osmotic potential of the growth medium. Exp. Mycol., 9(2): 99–107. https://doi.org/10.1016/0147-5975(85)90029-5.

Luo, Y., Wei, X., Yang, S., Gao, Y.H. and Luo, Z.H. (2020). Fungal diversity in deep-sea sediments from the Magellan seamounts as revealed by a metabarcoding approach targeting the ITS2 regions. Mycology, 11(3): 214–229. https://doi.org/10.1080/21501203.2020.1799878.

Luo, Y., Xu, W., Luo, Z.H. and Pang, K.L. (2020). Diversity and temperature adaptability of cultivable fungi in marine sediments from the Chukchi Sea. Bot. Mar., 63(2): 197–207. https://doi.org/10.1515/bot-2018-0119.

Luo, Z.H. and Pang, K.L. (2014). Fungi from substrates in marine environment. pp. 97–114. *In*: Misra, J.K., Tewari, J.P., Deshmukh, S.K. and Vágvölgyi, C. (eds.). Fungi from Different Substrates. CRC Press, Boca Raton, USA. https://doi.org/10.1201/b17646-5.

Maamar, A., Lucchesi, M.E., Debaets, S., Nguyen van Long, N., Quemener, M. et al. (2020). Highlighting the crude oil bioremediation potential of marine fungi isolated from the Port of Oran (Algeria). Diversity, 12(5): 196. https://doi.org/10.3390/d12050196.

Marchese, P., Garzoli, L., Young, R., Allcock, L., Barry, F. et al. (2021). Fungi populate deep-sea coral gardens as well as marine sediments in the Irish Atlantic Ocean. Environ. Microbiol., 23(8): 4168–4184. https://doi.org/10.1111/1462-2920.15560.

Mazur-Marzec, H., Andersson, A.F., Błaszczyk, A., Dąbek, P., Górecka, E. et al. (2024). Biodiversity of microorganisms in the Baltic Sea: The power of novel methods in the identification of marine microbes. FEMS Microbiol. Rev., 48: fuae024. https://doi.org/10.1093/femsre/fuae024.

McCarthy, P.M. and Kantvilas, G. (2015). *Thelidium robustum* sp. nov. (lichenized Ascomycota, Verrucariaceae) from Kangaroo Island, South Australia. J. Adelaide Bot. Gardens, 29(2016): 37–40. https://biostor.org/reference/281785.

McKenzie, E.H. and McKenzie, E.H. (2001). Fungi anamorphic in Australasia. Aus. Syst. Bot., 14(3): 485–500. https://doi.org/10.1071/SB99032.

McRae, C.F., Hocking, A.D. and Seppelt, R.D. (1999). *Penicillium* species from terrestrial habitats in the Windmill Islands, East Antarctica, including a new species, *Penicillium antarcticum*. Pol. Biol., 21(2): 97–111. https://doi.org/10.1007/s003000050340.

Mohamed, D.J. and Martiny, J.B. (2011). Patterns of fungal diversity and composition along a salinity gradient. ISME J., 5(3): 379–388. https://doi.org/10.1038/ismej.2010.137.

Morrison-Gardiner, S. (2002). Dominant fungi from Australian coral reefs. Fungal Divers, 9: 105–121.

Mouzouros, R. (1986). Patterns of timber decay caused by marine fungi. pp. 341–353. *In*: Moss, S.T. (ed.). The Biology of Marine Fungi. Cambridge University Press.

Muwawa, E.M., Makonde, H.M., Obieze, C.C., de Oliveira, I.G., Jefwa, J.M. et al. (2024). Diversity and assembly patterns of mangrove rhizosphere mycobiome along the coast of Gazi Bay and Mida Creek in Kenya. PLoS One, 19(4): e0298237. https://doi.org/10.1371/journal.pone.0298237.

Nagahama, T., Hamamoto, M., Nakase, T. and Horikoshi, K. (2001). *Rhodotorula lamellibrachii* sp. nov., a new yeast species from a tubeworm collected at the deep-sea floor in Sagami Bay. Ant. van Leeuwen., 80(3-4): 317–323. https://doi.org/10.1023/a:1013043301388.

Nagahama, T., Hamamoto, M., Nakase, T., Takaki, Y. and Horikoshi, K. (2003). *Cryptococcus surugaensis* sp. nov., a novel yeast species from sediment collected at the deep-sea floor of Suruga Bay. Int. J. Syst. Evol. Microbiol., 53(6): 2095–2098. https://doi.org/10.1099/ijs.0.02712-0.

Nagano, Y. and Nagahama, T. (2012). Fungal diversity in deep-sea extreme environments. Fungal Ecol., 5(4): 463–471. https://doi.org/10.1016/j.funeco.2012.01.004.

Nagano, Y., Abdel-Wahab, M.A., Nakajima, R. and Yabuki, A. (2024). *Oceanitis abyssalis* sp. nov., a new deep-sea fungus from sunken wood collected at 5707 m depth in the Northwest Pacific Ocean. Phytotaxa, 663(4): 171–183. https://doi.org/10.11646/phytotaxa.663.4.1.

Nagano, Y., Nagahama, T., Hatada, Y., Nunoura, T., Takami, H. et al. (2010). Fungal diversity in deep-sea sediments: The presence of novel fungal groups. Fungal Ecol., 3(4): 316–325. https://doi.org/10.1016/j.funeco.2010.01.002.

Nakagiri, A. and Ito, T. (1991). Basidiocarp development of the cyphelloid gasteroid aquatic basidiomycetes *Halocyphina villosa* and *Limnoperdon incarnatum*. Can. J. Bot., 69(10): 2320–2327. https://doi.org/10.1139/b91-292.

Nalim, F.A., Samuels, G.J., Wijesundera, R.L. and Geiser, D.M. (2011). New species from the *Fusarium solani* species complex derived from perithecia and soil in the Old World tropics. Mycologia, 103(6): 1302–1330. https://doi.org/10.3852/10-307.

Nambiar, R. and Raveendran, K. (2015). Identification of marine mycoflora in Amini Dweep of Lakshadweep Islands. Bioinfolet, 12(2B): 518–520.

Newsham, K.K., Misiak, M., Goodall-Copestake, W.P., Dahl, M.S., Boddy, L. et al. (2022). Experimental warming increases fungal alpha diversity in an oligotrophic maritime Antarctic soil. Front. Microbiol., 13: 1050372. https://doi.org/10.3389/fmicb.2022.1050372.

Nilsson, R.H., Anslan, S., Bahram, M., Wurzbacher, C., Baldrian, P. et al. (2019). Mycobiome diversity: High-throughput sequencing and identification of fungi. Nat. Rev. Microbiol., 17(2): 95–109. https://doi.org/10.1038/s41579-018-0116-y.

Nogales, B., Lanfranconi, M.P., Piña-Villalonga, J.M. and Bosch, R. (2011). Anthropogenic perturbations in marine microbial communities. FEMS Microbiol. Rev., 35(2): 275–298. https://doi.org/10.1111/j.1574-6976.2010.00248.x.

Norton, R.A. (1994). Evolutionary aspects of oribatid mite life histories and consequences for the origin of the Astigmata. pp. 99–135. *In*: Houck, M.A. (ed.). Mites. Springer, Boston, MA. https://doi.org/10.1007/978-1-4615-2389-5_5.

Novak, R.O. and Backus, M.P. (1963). A new species of *Mycotypha* with a zygosporic stage. Mycologia, 55(6): 790–798. https://doi.org/10.1080/00275514.1963.12018071.

Ogaki, M.B., Coelho, L.C., Vieira, R., Neto, A.A., Zani, C.L. et al. (2020). Cultivable fungi present in deep-sea sediments of Antarctica: Taxonomy, diversity, and bioprospecting of bioactive compounds. Extremophiles, 24(2): 227–238. https://doi.org/10.1007/s00792-019-01148-x.

Ogaki, M.B., de Paula, M.T., Ruas, D., Pellizzari, F.M., García-Laviña, C.X. et al. (2019). Marine fungi associated with Antarctic macroalgae. pp. 239–255. *In*: Castro-Sowinski, S. (ed.). The Ecological Role of Micro-organisms in the Antarctic Environment. Springer Cham. https://doi.org/10.1007/978-3-030-02786-5_11.

Ogaki, M.B., Pinto, O.H.B., Vieira, R., Neto, A.A., Convey, P. et al. (2021). Fungi present in Antarctic deep-sea sediments assessed using DNA metabarcoding. Microb. Ecol., 82(1): 157–164. https://doi.org/10.1007/s00248-020-01658-8.

Orpurt, P.A. and Fennell, D.I. (1955). A new species of *Penicillium* from soil. Mycologia, 47(2): 233–237. https://doi.org/10.1080/00275514.1955.12024447.

Orsi, W., Biddle, J.F. and Edgcomb, V. (2013). Deep sequencing of subseafloor eukaryotic rRNA reveals active fungi across marine subsurface provinces. PLOS One, 8(2): e56335. https://doi.org/10.1371/journal.pone.0056335.

Orsi, W.D., Vuillemin, A., Coskun, Ö.K., Rodriguez, P., Oertel, Y. et al. (2022). Carbon assimilating fungi from surface ocean to subseafloor revealed by coupled phylogenetic and stable isotope analysis. ISME J., 16(5): 1245–1261. https://doi.org/10.1038/s41396-021-01169-5.

Pang, K.L., Chow, R.K., Chan, C.W. and Vrijmoed, L.P. (2011). Diversity and physiology of marine lignicolous fungi in Arctic waters: A preliminary account. Polar Res., 30(1): 5859. https://doi.org/10.3402/polar.v30i0.5859.

Pang, K.L., Jones, E.B.G., Abdel-Wahab, M.A., Adams, S.J., Alves, A. et al. (2023). Recent progress in marine mycological research in different countries, and prospects for future developments worldwide. Bot. Mar., 66(4): 239–269. https://doi.org/10.1515/bot-2023-0015.

Pang, Z., Zhao, Y., Xu, P. and Yu, D. (2020). Microbial diversity of upland rice roots and their influence on rice growth and drought tolerance. Microorganisms, 8(9): 1473269. https://doi.org/10.3390/microorganisms8091329.

Pasqualetti, M., Braconcini, M., Barghini, P., Gorrasi, S., Schillaci, D. et al. (2024). From marine neglected substrata new fungal taxa of potential biotechnological interest: The case of *Pelagia noctiluca*. Front. Microbiol., 15: 1473269. https://doi.org/10.3389/fmicb.2024.1473269.

Pedersen, O.A. and Langvad, F. (1989). *Exophiala psychrophila* sp. nov., a pathogenic species of the black yeasts isolated from farmed Atlantic salmon. Mycol. Res., 92(2): 153–156. https://doi.org/10.1016/S0953-7562(89)80004-8.

Peng, X., Amend, A.S., Baltar, F., Blanco-Bercial, L., Breyer, E. et. al. (2024). Planktonic marine fungi: A review. J. Geophys. Res. Biogeosci., 129(3): e2023JG007887. https://doi.org/10.1029/2023JG007887.

Pérez-Llano, Y., Yarzábal Rodríguez, L.A., Martínez-Romero, E., Dobson, A.D., Gunde-Cimerman, N. et al. (2023). From friends to foes: Fungi could be emerging marine sponge pathogens under global change scenarios. Front. Microbiol., 14: 1213340. https://doi.org/10.3389/fmicb.2023.1213340.

Perini, L., Mogrovejo, D.C., Tomazin, R., Gostinčar, C., Brill, F.H. et al. (2019). Phenotypes associated with pathogenicity: Their expression in arctic fungal isolates. Microorganisms, 7(12): 600. https://doi.org/10.3390/microorganisms7120600.

Petch, T. (1936). Cordyceps militaris and Isaria farinosa. Trans. Br. Mycol. Soc., 20(3-4): 216–224. https://doi.org/10.1016/S0007-1536(36)80014-X.

Phillips, J.A. (2023). The Lives of Seaweeds: A Natural History of Our Planet's Seaweeds and Other Algae. Kate Shanahan (ed.). Princeton University Press, USA, 289 pp.

Plemenitaš, K. (2021). Quality of Life of Patients with Amyotrophic Lateral Sclerosis. Diploma Thesis. Digital library University of Maribor, Slovenia. https://doi.org/10.1080/13548506.2011.651149.

Pointing, S.B., Vrijmoed, L.L.P. and Jones, E.B.G. (1998). A qualitative assessment of lignocellulose-degrading enzyme activity in marine fungi. Bot. Mar., 41: 293–298. https://api.semanticscholar.org/CorpusID:84941770.

Poli, A., Bovio, E., Ranieri, L., Varese, G.C. and Prigione, V. (2020). Fungal diversity in the Neptune Forest: Comparison of the mycobiota of *Posidonia oceanica*, *Flabellia petiolata* and *Padina pavonica*. Front. Microbiol., 11: 933. https://doi.org/10.3389/fmicb.2020.00933.

Pugh, G.J.F. and Jones, E.B.G. (1986). Antarctic marine fungi: A preliminary account. pp. 323–330. *In*: Moss, S.T. (ed.). The Biology of Marine Fungi. Cambridge University Press.

Raghukumar, C., Damare, S.R. and Singh, P. (2010). A review on deep-sea fungi: Occurrence, diversity, and adaptations. Bot. Mar., 53(6): 479–492. https://doi.org/10.1515/bot.2010.076.

Raghukumar, S. (2017). Fungi in Coastal and Oceanic Marine Ecosystems: Marine Fungi. Springer Cham, 378 pp. https://doi.org/10.1007/978-3-319-54304-8.

Rämä, T., Hassett, B.T. and Bubnova, E. (2017). Arctic marine fungi: From filaments and flagella to operational taxonomic units and beyond. Bot. Mar., 60(4): 433–452. https://doi.org/10.1515/bot-2016-0104.

Rämä, T., Mathiassen, G.H. and Kauserud, H. (2014). Marine fungi new to Norway, with an outlook to the overall diversity. Agarica, 35: 35−47. https://oceanrep.geomar.de/id/eprint/29967.

Reich, M., Wichels, A., Panzer, K., Krause, E., Giménez, L. and Gerdts, G. (2017). Impacts of a reduction in seawater pH mimicking ocean acidification on the structure and diversity of mycoplankton communities. Aq. Microb. Ecol., 79(3): 221–233. https://doi.org/10.3354/ame01831.

Richards, T.A., Jones, M.D.M., Leonard, G. and Bass, D. (2012). Marine fungi: Their ecology and molecular diversity. Ann. Rev. Mar. Sci., 4: 495–522. https://doi.org/10.1146/annurev-marine-120710-100802.

Ritchie, D. (1957). Salinity optima for marine fungi affected by temperature. Am. J. Bot., 44(10): 870–874. https://doi.org/10.2307/2438907.

Rojas-Jimenez, K., Rieck, A., Wurzbacher, C., Jürgens, K., Labrenz, M. et al. (2019). A salinity threshold separating fungal communities in the Baltic Sea. Front. Microbiol., 10: 680. https://doi.org/10.3389/fmicb.2019.00680.

Rovati, J.I., Pajot, H.F., Ruberto, L., Mac Cormack, W. and Figueroa, L.I. (2013). Polyphenolic substrates and dyes degradation by yeasts from 25 de Mayo/King George Island (Antarctica). Yeast, 30(11): 459–470. https://doi.org/10.1002/yea.2982.

Saccardo, P.A. (1892). Sylloge Fungorum omnium hucusque cognitorum (Volume 10). Patavii, Sumptibus Auctoris, Pitavii, Italy. https://doi.org/10.5962/bhl.title.5371.

Sarma, V.V. and Hyde, K.D. (2001). A review on frequently occurring fungi in mangroves. Fungal Diver., 8: 1–34.

Schatz, S. (1985). *Adomia avicenniae*: A new ascomycetous genus from Red Sea and Australian mangroves. Tr. Brit. Mycol. Soc., 84(3): 555–559. https://doi.org/10.1016/S0007-1536(85)80025-5.

Schaumann, K. (1975). Ökologische Untersuchungen über höhere Pilze im Meer-und Brackwasser der Deutschen Bucht unter besonderer Berücksichtigung der holzbesiedelnden. Arten, 15: 75–182.

Schlechtendahl, D.F.L. (1824). Flora Berolinensis: Pars secunda: Cryptogamia (Volume 2). Ferdinandi Dümmler, Germany, 284 pp.

Schmit, J.P. and Shearer, C.A. (2003). A checklist of mangrove-associated fungi, their geographical distribution, and known host plants. Mycotaxon, 85: 423–477. https://doi.org/10.5962/p.414675.

Schultz, J., Altalhi, S. and Rosado, A.S. (2025). Diving into the deep: Fungal diversity in newly discovered hydrothermal vents of Hatiba Mons, Red Sea. Front. Mar. Sci., 12: 1649339. https://doi.org/10.3389/fmars.2025.1649339.

Sen, K., Bai, M., Li, J., Ding, X., Sen, B. and Wang, G. (2023). Spatial patterns of planktonic fungi indicate their potential contributions to biological carbon pump and organic matter remineralization in the water column of South China Sea. J. Fungi, 9(6): 640. https://doi.org/10.3390/jof9060640.

Sen, K., Bai, M., Sen, B. and Wang, G. (2021). Disentangling the structure and function of mycoplankton communities in the context of marine environmental heterogeneity. Sci. Tot. Environ., 766: 142635. https://doi.org/10.1016/j.scitotenv.2020.142635.

Sen, K., Sen, B. and Wang, G. (2022). Diversity, abundance, and ecological roles of planktonic fungi in marine environments. J. Fungi, 8(5): 491. https://doi.org/10.3390/jof8050491.

Shearer, C.A., Descals, E., Kohlmeyer, B., Kohlmeyer, J., Marvanová, L. et al. (2007). Fungal biodiversity in aquatic habitats. Biodiver. Conser., 16(1): 49–67. https://doi.org/10.1007/s10531-006-9120-z.

Simister, R.L., Poutasse, C.M., Thurston, A.M., Reeve, J.L., Baker, M.C. and White, H.K. (2015). Degradation of oil by fungi isolated from Gulf of Mexico beaches. Mar. Poll. Bull., 100(1): 327–333. https://doi.org/10.1016/j.marpolbul.2015.08.029.

Simões, M.F., Antunes, A., Ottoni, C.A., Amini, M.S., Alam, I. et al. (2015). Soil and rhizosphere-associated fungi in gray mangroves (*Avicennia marina*) from the Red Sea: A metagenomic approach. Genom. Proteom. Bioinform., 13(5): 310–320. https://doi.org/10.1016/j.gpb.2015.07.002.

Sridhar, K.R. (2017). Marine filamentous fungi: Diversity, distribution and bioprospecting. pp. 59–73. *In*: Satyanarayana, T., Deshmukh, S.K. and Johri, B. (eds.). Developments in Fungal Biology and Applied Mycology. Springer Singapore. https://doi.org/10.1007/978-981-10-4768-8_5.

Steinke, T.D. (2000). Mangrove fungi on dead prop roots of *Rhizophora mucronata* at three localities in South Africa. Afr. J. Bot., 66(2): 95–95. https://doi.org/10.1016/S0254-6299(15)31067-X.

Steinke, T.D. and Jones, E.B.G. (1993). Marine and mangrove fungi from the Indian Ocean coast of South Africa. South Afr. J. Bot., 59(4): 385–390. https://doi.org/10.1016/S0254-6299(16)30711-6.

Steinke, T.D. and Lubke, R.A. (2003). Arenicolous marine fungi from southern Africa. South Afr. J. Bot., 69(4): 540–545. https://doi.org/10.1016/S0254-6299(15)30292-1.

Taylor, J.D. and Cunliffe, M. (2016). Multi-year assessment of coastal planktonic fungi reveals environmental drivers of diversity and abundance. ISME J., 10: 2118–2128. https://doi.org/10.1038/ismej.2016.24.

Tennakoon, D.S., de Silva, N.I., Maharachchikumbura, S.S., Bhat, D.J., Kumla, J. et al. (2023). Exploring more on Dictyosporiaceae: The species geographical distribution and intriguing novel additions from plant litter. Diversity, 15(3): 410. https://doi.org/10.3390/d15030410.

Thom, C. (1906). Fungi in cheese ripening; Camembert and Roquefort. U.S. Department of Agriculture, Bureau of Animal Industry Bulletin, 82: 1–39.

Thom, C. (1910). Cultural studies of species of *Penicillium*. Bull. of the U.S. Dep. of Agri., Bureau Animal Industry, 118: 1–109.

Thom, C. (1930). *The Penicillia*. Williams & Wilkins Co., Baltimore, USA, 644 pp.

Tisthammer, K.H., Cobian, G.M. and Amend, A.S. (2016). Global biogeography of marine fungi is shaped by the environment. Fungal Ecol., 19: 39–46. https://doi.org/10.1016/j.funeco.2015.09.003.

Tsui, C.K., Woodhall, J., Chen, W., Lévesque, C.A., Lau, A. et al. (2011). Molecular techniques for pathogen identification and fungus detection in the environment. IMA Fungus, 2(2): 177–189. https://doi.org/10.5598/imafungus.2011.02.02.09.

Ugidos-Rodríguez, S., Matallana-González, M.C. and Sánchez-Mata, M.C. (2018). Lactose malabsorption and intolerance: A review. Food and Function, 9(8): 4056–4068. https://doi.org/10.1039/c8fo00555a.

Ukoima, H.N. and Umechuruba, C.I. (2009). Effect of relative humidity and temperature on fungi isolated from marine habitat in Port Harcourt. Int. J. Biosci., 4(2): 104–107.

Vaca, I., Faúndez, C., Maza, F., Paillavil, B., Hernández, V., Acosta, F. et al. (2013). Cultivable psychrotolerant yeasts associated with Antarctic marine sponges. World J. Microbiol. Biotech., 29(1): 183–189. https://doi.org/10.1007/s11274-012-1159-2.

Varella, S., Brunati, M., Tosi, S. and Marino, I.A.M. (2021). Diversity, ecological role and biotechnological potential of Antarctic marine fungi. J. of Fungi, 7(5): 391. https://doi.org/10.3390/jof7050391.

Varga, J., Frisvad, J.C. and Samson, R. (2011). Two new aflatoxin producing species, and an overview of *Aspergillus* section Flavi. Stud. Mycol., 69(1): 57–80. https://doi.org/10.3114/sim.2011.69.05.

Varrella, S., Barone, G., Tangherlini, M., Rastelli, E., Dell'Anno, A. et al. (2021). Diversity, ecological role and biotechnological potential of antarctic marine fungi. J. Fungi, 7(5): 391. https://doi.org/10.3390/jof7050391.

Vaz, J.A., Barros, L., Martins, A., Santos-Buelga, C., Vasconcelos, M.H. et al. (2011). Chemical composition of wild edible mushrooms and antioxidant properties of their water soluble polysaccharidic and ethanolic fractions. Food Chem., 126(2): 610–616. https://doi.org/10.1016/j.foodchem.2010.11.063.

Velez, P., Salcedo, D.L., Espinosa-Asuar, L., Gasca-Pineda, J., Hernandez-Monroy, A. et al. (2022). Fungal diversity in sediments from deep-sea extreme ecosystems: Insights into low-and high-temperature hydrothermal vents, and an oxygen minimum zone in the southern Gulf of California, Mexico. Front. Mar. Sci., 9: 802634. https://doi.org/10.3389/fmars.2022.802634.

Velmurugan, N., Lee, H.M., Cha, H.J. and Lee, Y.S. (2017). Proteomic analysis of the marine-derived fungus *Paecilomyces* sp. strain SF-8 in response to polycyclic aromatic hydrocarbons. Bot. Mar., 60(4): 381–392. https://doi.org/10.1515/bot-2016-0101.

Venkatachalam, M., Shum-Chéong-Sing, A., Dufossé, L. and Fouillaud, M. (2020). Statistical optimization of the physico-chemical parameters for pigment production in submerged fermentation of *Talaromyces albobiverticillius* 30548. Microorganisms, 8(5): 711. https://doi.org/10.3390/microorganisms8050711

Vincent, W.F. (2000). Evolutionary origins of Antarctic microbiota: Invasion, selection, and endemism. Antarct. Sci., 12(3): 374–385. https://doi.org/10.1017/S0954102000000420.

Vrijmoed, L.L.P., Hodgkiss, I.J. and Thrower, L.B. (1982). Factors affecting the distribution of lignicolous marine fungi in Hong Kong. Hydrobiologia, 87(2): 143–160. https://doi.org/10.1007/BF00015195.

Vrijmoed, L.L.P., Hyde, K.D. and Jones, E.B.G. (1996). *Melaspilea mangrovei* sp. nov., from Australian and Hong Kong mangroves. Mycol. Res., 100(3): 291–294. https://doi.org/10.1016/S0953-7562(96)80156-0.

Wang, H.N., Sun, S.S., Liu, M.Z., Yan, M.C., Liu, Y.F. et al. (2022). Natural bioactive compounds from marine fungi (2017–2020). J. Asian Nat. Prod. Res., 24(3): 203–230. https://doi.org/10.1080/10286020.2021.1947254.

Wang, X., Singh, P., Gao, Z., Johnson, Z.I. and Wang, G. 2014. Distribution and diversity of planktonic fungi in the west pacific warm pool. PLOS One, 9(7): e101523. https://doi.org/10.1371/journal.pone.0101523.

Wehmer, C. (1903). Der *Mucor* der Hanfrötte, *M. hiemalis* nov. spec. Ann. Mycol., 1(1): 37–41.

Westling, R. (1911). Über die grünen Spezies der Gattung *Penicillium*. Arkiv för Botanik utgifvet af K. Svenska Vetenskaps-Akademien i Stockholm, 11(1): 1–156.

Wijayawardene, N.N., Dai, D.Q., Jayasinghe, P.K., Gunasekara, S.S., Nagano, Y. et al. (2022). Ecological and oceanographic perspectives in future marine fungal taxonomy. J. Fungi, 8(11): 1141. https://doi.org/10.3390/jof8111141.

Woudenberg, J.H.C., Groenewald, J.Z., Binder, M. and Crous, P.W. (2013). *Alternaria* redefined. Stud. Mycol., 75(1): 171–212. https://doi.org/10.3114/sim0015.

Wu, K., Liu, Y., Liao, X., Yang, X., Chen, Z. et al. (2023). Fungal diversity and its relationship with environmental factors in coastal sediments from Guangdong, China. J. Fungi, 9(1): 101. https://doi.org/10.3390/jof9010101.

Xu, W., Gong, L.F., Pang, K.L. and Luo, Z.H. (2018). Fungal diversity in deep-sea sediments of a hydrothermal vent system in the Southwest Indian Ridge. Deep-Sea Res. Part I: Oceanographic Research Papers, 131: 16–26. https://doi.org/10.1016/j.dsr.2017.11.001.

Xu, W., Pang, K.L. and Luo, Z.H. (2014). High fungal diversity and abundance in deep-sea sediments of the Pacific Ocean. Microb. Ecol., 68(4): 688–698. https://doi.org/10.1007/s00248-014-0448-8.

Yang, K.L., Lin, J.Y., Li, G.M. and Yang, Z.L. (2023). Mushrooms adapted to seawater: Two new species of Candolleomyces (Basidiomycota, Agaricales) from China. J. Fungi, 9(12): 1204. https://doi.org/10.3390/jof9121204.

Yarden, O. (2014). Fungal association with sessile marine invertebrates. Front. Microbiol., 5: 228. https://doi.org/10.3389/fmicb.2014.00228.

Yurchenko, E.A., Chingizova, E.A., Aminin, D.L. and Yurchenko, A.N. (2025). Marine fungi: In search of new antibacterial drugs. Mol. Biol., 59(1): 35–50. https://doi.org/10.1134/S0026893324700705.

Zajc, J., Liu, Y., Dai, W., Yang, Z., Hu, J. et al. (2013). Genome and transcriptome sequencing of the halophilic fungus *Wallemia ichthyophaga*: Haloadaptations present and absent. BMC Genomics, 14(1): 617. https://doi.org/10.1186/1471-2164-14-617.

Zaleski, K.W. (1928). Über die in Polen gefundenen Arten der Gruppe Penicillium Link. I, II, III. Bulletin International de l'Academie Polonaise des Sciences et des Lettres Classe des sciences mathématiques et naturelles. Série B, Sci. Nat., 1927: 417–563.

Zare, R. and Gams, W. (2001). The genera *Lecanicillium* and *Simplicillium* gen. nov. Nova Hedwigia, 73: 1–50.

Zhang, L., Delworth, T.L., Yang, X., Zeng, F., Lu, F. et al. (2022). The relative role of the subsurface Southern Ocean in driving negative Antarctic Sea ice extent anomalies in 2016–2021. Comm. Earth Environ., 3(1): 302. https://doi.org/10.1038/s43247-022-00624-1.

Zhang, T., Fei Wang, N., Qin Zhang, Y., Yu Liu, H. and Yan Yu, L. (2015). Diversity and distribution of fungal communities in the marine sediments of Kongsfjorden, Svalbard (High Arctic). Sci. Rep., 5(1): 14524. https://doi.org/10.1038/srep14524.

Zhao, H., Brearley, F.Q., Huang, L., Tang, J., Xu, Q. et al. (2023). Abundant and rare taxa of planktonic fungal community exhibit distinct assembly patterns along a coastal eutrophication gradient. Microb. Ecol., 85(2): 495–507. https://doi.org/10.1007/s00248-022-01976-z.

2

Fungal Diversity Patterns in the Oxygen Minimum Zones

Posadas, J.[1,2,*] and *Carmona-Almeraya, G.*[1,2]

Oxygen minimum zones (OMZ) are mid-water oceanic regions characterized by seasonal or permanent low dissolved oxygen concentrations, typically occurring between 50–1000 meters depth and covering about 8% of the ocean's surface. Prominent OMZ are in the Eastern Tropical Pacific, the northern Indian Ocean, especially in the Arabian Sea and Bay of Bengal, and the Eastern Tropical Atlantic. Additionally, seasonal hypoxic zones may occur along continental margins, such as in the Gulf of Mexico, where they are mainly driven by anthropogenic nutrient inputs. These regions represent dynamic biogeochemical environments that host specialized microbial communities. While most studies have traditionally focused on prokaryotes, recent efforts have begun to uncover a rich and diverse mycobiota. Fungal communities in OMZ, detected through culture-based and molecular techniques, include a wide taxonomic range of fungi, particularly within Ascomycota and Basidiomycota. Genera such as *Aspergillus*, *Aureobasidium, Cladosporium*, and *Rhodotorula* have been found across multiple OMZ regions, whereas others appear to be region-specific, reflecting both biogeographic patterns and methodological biases. Notably, a significant proportion of environmental sequences remain taxonomically unidentified, indicating the presence of undescribed fungal lineages adapted to hypoxic environments.

[1] Posgrado en Ciencias del Mar y Limnología, Universidad Nacional Autónoma de México, Mexico City, Mexico.
[2] Instituto de Biología, Universidad Nacional Autónoma de México, Mexico City, Mexico.
* Corresponding author: judith.posadas@st.ib.unam.mx

DOI: 10.1201/9781003768708-3

Studies have revealed spatial gradients in fungal diversity, with higher richness in nearshore OMZ, likely influenced by nutrient inputs and terrestrial runoff. Furthermore, several fungal taxa exhibit functional adaptations to low-oxygen conditions, including nitrate reduction and denitrification, suggesting potential roles in marine nitrogen cycling. Early-diverging fungal lineages and novel environmental clades have also been detected, offering insights into fungal evolution under hypoxic conditions. Current knowledge about fungal relevance in OMZ remains geographically and taxonomically incomplete. Broader integrative research combining culture-based and molecular approaches is essential to fully understand their ecological roles, metabolic capabilities, evolutionary adaptations, and potential biotechnological applications.

1. Introduction

The oxygen minimum zones (OMZ) are oceanic regions characterized by oxygen (O_2) concentrations < 22 μmol kg^{-1} (Levin, 2018) regarding the normal average in the sea. They are usually distributed at intermediate depths around 50–1000 m water depth, covering approximately 8% of the oceanic surface and 1% of its volume (Paulmier and Ruiz-Pino, 2009). The Eastern Tropical Pacific Ocean contains one of the largest and one of the most extensively studied oxygen minimum zones, off the coasts of Peru, Ecuador, and Mexico, where elevated biological productivity coupled with restricted oceanic circulation leads to oxygen depletion (Paulmier and Ruiz-Pino, 2009; Ulloa et al., 2012).

The Indian Ocean is one of the world's major oceans, and the Arabian Sea is a large, marginal sea that is located within and forms the northwestern part of the Indian Ocean. The Northern Indian Ocean, in the Arabian Sea hosts another prominent OMZ, characterized by its exceptional intensity and persistence, making it one of the most pronounced low-oxygen regions in the global ocean (Naqvi et al., 2006). The Eastern Tropical Atlantic Ocean also hosts a significant OMZ off the West African coast, particularly near Angola and Namibia, where coastal upwelling and water mass stratification contribute to persistent deoxygenation (Brandt et al., 2015). In contrast, the hypoxic zone in the Gulf of Mexico, though coastal and seasonal in nature, represents an important example of anthropogenically induced hypoxia. This zone is primarily driven by nutrient-rich runoff from the Mississippi River, which fuels phytoplankton blooms and subsequent oxygen consumption during organic matter decomposition (Rabalais et al., 2010) (Fig. 1). Collectively, these OMZ represent unique biogeochemical environments that influence the composition and functional dynamics of marine microbial communities.

Although OMZ play a role in global biogeochemistry, they are a critical source of nitrous oxide, approximately 20% of total global emissions and are the cause of the loss of nitrogen fixed in the oceans, close to 50%, releasing nitrous oxide (an ozone-destroying potent greenhouse gas) into the atmosphere and limiting global productivity with the consequent feedback on the carbon cycle (Codispoti et al., 2010). The formation of these regions is multifactorial driven by physical,

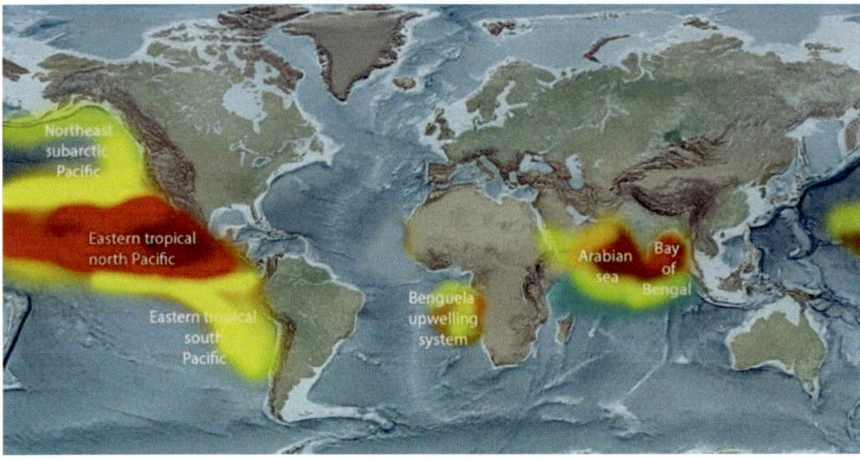

Fig. 1. Global map showing the distribution and intensity of oxygen minimum zones. Regions shaded in red indicate areas with more severe and persistent oxygen depletion, while yellow areas represent zones with less intense or more variable oxygen deficiency.

***Source*:** Adapted from Long et al., 2021; Moffit et al., 2015.

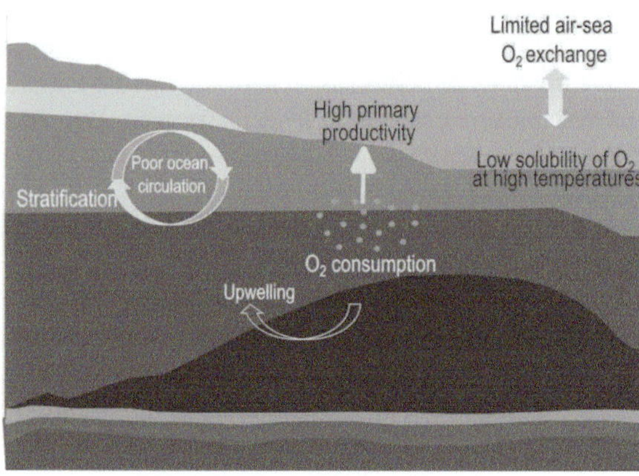

Fig. 2. Main physical, chemical, and biological factors contributing to the formation and expansion of oxygen minimum zones in the ocean.

chemical, and biological processes (Fig. 2). Key factors include thermal stratification, weak ocean circulation, limited air-sea O_2 exchange, reduce O_2 solubility at high temperatures, and the upwelling of nutrient-rich waters to the surface that enhance biological productivity and O_2 consumption (Paulmier and Ruiz-Pino, 2009; Helly and Levin, 2004). Over the past 30–50 years, anthropogenic activity has driven climate change, leading to a decline in oceanic oxygen availability (Papiol et al., 2017; Bertagnolli et al., 2018). The expansion of OMZ is associated with adverse impacts on marine ecosystems, including fungal communities in both the water column and sediments, as OMZ waters intersect approximately 0.3% of the global seafloor (Møller et al., 2022; Papiol et al., 2017). The aim of this review is to provide a synthesis of current knowledge on fungal diversity, ecology, and functional roles in OMZ, highlighting their taxonomic composition and biogeographic patterns within these extreme, yet globally, significant marine ecosystems.

2. Fungal Diversity

Biotic communities in OMZ are dominated by microorganisms, as oxygen deficiency exerts strong selective pressures that restrict multicellular life (Long et al., 2021). However, the diversity of fungi in these extreme environments has been little explored until recently, as several studies have focused on the assessment of the prokaryotic community composition.

Research on fungal communities within OMZ has been unevenly distributed across oceanic regions, with a notable concentration of studies in the Indian and Pacific Oceans. In the Indian Ocean, particularly in the Arabian Sea, investigations on mycology have focused on both seasonal and permanent OMZ, emphasizing taxonomic identification, phylogenetic affiliations, patterns of distribution and biomass in sediments, and the functional potential of fungal isolates for nitrate reduction (Cathrine and Raghukumar, 2009; Jebaraj et al., 2010; Manohar et al., 2014; Manohar et al., 2015; Stief et al., 2014). In contrast, studies conducted in the Pacific Ocean have covered its northern, eastern, and southern OMZ regions, using primarily culture-independent approaches to explore fungal taxonomic and functional diversity. These efforts include morphological assessments via epifluorescence microscopy (Gutiérrez et al., 2010), analyses of biochemical composition (Gutiérrez et al., 2020), and evaluations of nitrous oxide production potential (Peng and Valentine, 2021). In comparison, the Atlantic Ocean OMZ region remains largely unexplored in terms of mycobiota, with available information limited to broader surveys of microbial eukaryotes (Morgan-Smith et al., 2013). This geographical disparity highlights the need for expanded and integrative mycological research across all OMZ regions to better understand the ecological roles and global biogeochemical relevance of marine fungi.

Fungal diversity in OMZ has been increasingly recognized through both culture-dependent (Table 1) and culture-independent approaches (Table 2), each offering unique insights into the taxonomic and functional diversity of fungi in these extreme marine habitats. Culture-based studies have historically provided critical baseline information, though they capture only a fraction of the fungal community. Notably, around 95 isolates recovered through traditional cultivation techniques have belonged predominantly to the phylum Ascomycota, including species such as *Tritirachium candoliense* sp. nov. or *Aspergillus terreus*, as well as species closely related (> 93% sequence similarity) to *A. cervinus*, *A. versicolor*, *Eupenicillium javanicum*, *Penicillium* sp., *P. brevicompactum*, *P. citreonigrum*, *P. roseopurpureum*, *Cladosporium cladosporioides*, *Cladosporium* sp., *Zasmidium cellare*, *Davidiella tassiana*, *Aureobasidium pullulans*, *Hortaea werneckii*, *Nectria cinnabarina*, *Paecilomyces* sp., *Engyodontium album*, *Acremonium* sp., *Cochliobolus lunatus*, *Fusarium oxysporum*, *Chaetomium globosum*, *Chaetomium* sp., *Phaeosphaeria* sp., *Didymocrea sadasivanii*, *Apiospora montagnei*, *Pichia guilliermondii*, and *Sarocladium strictum*. Representatives of *Basidiomycota* have also recovered, showing close phylogenetic relationships to species such as *Rhodotorula minuta*,

Table 1. Key studies on fungal diversity in oxygen minimum zones based on culture-dependent approaches (NA, data not specified).

Isolate	Functional Trait	References
This study recovered 26 fungal cultures from the sediment sites, mostly closely related (> 97% sequence similarity) to known Dikarya, with two isolates likely representing novel fungal lineages.	NA	Jebaraj et al. (2010)
Tritirachium candoliense sp. nov.	NA	Manohar et al. (2014)
Aspergillus terreus	Dissimilatory nitrate reduction	Stief et al. (2014)
This study recovered 28 fungal cultures from the sediment sites, mostly closely related (> 93% sequence similarity) to known Dikarya, and revealed several novel or environmentally significant taxa.	Potential nitrate-reducing activity	Manohar et al. (2015)
This study documents filamentous fungi in both the water column and sediments.	NA	Gutiérrez et al. (2010)
From seawater samples, the study recovered 99 filamentous Ascomycete isolates grouped into 39 OTUs, predominantly *Penicillium* spp., with ~11 representing potentially novel taxa.	NA	Vera et al. (2017)

Rhodosporidium diobovatum, *Melanopsichium pennsylvanicum*, *Tilletiopsis albescens*, *Cerrena unicolor*, and *Coriolopsis byrsina* (Manohar et al., 2014; Manohar et al., 2015; Stief et al., 2014; Vera et al., 2017).

These findings reveal a wide taxonomic range of filamentous fungi and yeasts capable of thriving under low-oxygen conditions. Taxonomic diversity was observed in OMZ from both the Indian and Pacific Oceans, indicating that these ecosystems host a rich and varied fungal assemblage (Fig. 3). To date, around 27 genera have been documented in the Indian Ocean OMZ, whereas 170 genera have been reported from the Pacific Ocean. Interestingly, several genera, such as *Acremonium*, *Alternaria*, *Aspergillus*, *Aureobasidium*, *Chaetomium*, *Cladosporium*, *Fusarium*, *Hortaea*, *Phaeosphaeria*, and *Rhodotorula* have been detected in both regions, suggesting a core group of fungi potentially well-adapted to hypoxic marine environments. In contrast, some taxa appeared to be exclusive to a particular ocean basin, pointing to possible regional endemism or localized environmental selection pressures.

The fungal communities identified through molecular methods in both sediments, and the water column are predominantly composed of Ascomycota and Basidiomycota, along with a minor presence of early-diverging fungal lineages and uncultured taxa (Peng and Valentine, 2021; Posadas et al., 2024; Velez et al., 2022). In sediments from the Eastern Tropical North Pacific, the main Ascomycota sequences belong to the classes *Dothideomycetes*, *Sordariomycetes*, and *Eurotiomycetes*, whereas Basidiomycota sequences belong to *Agaricomycetes* and a minor proportion of *Malasseziomycetes* and *Ustilaginomycetes* (Posadas et al., 2024; Velez et al., 2022). In contrast, water column samples from the Pacific, analyzed through amplicon-based sequencing, consistently show a higher relative

Table 2. Key studies on fungal diversity in oxygen minimum zones based on culture-independent approaches (NA, data not specified; ASV, amplicon sequence variants; OTU, operational taxonomic units).

Region	Geographic Location	Sample Type	Molecular Methodology	NCBI BioProject	ASV/ OTU Count	References
Indian Ocean	Arabian Sea	Water column and sediment	Restriction fragment length polymorphism (RFLP)	GenBank accession ranges: GU071985 through GU072533, GU072549 through GU072590, and GQ120105 through GQ120153	48	Jebaraj et al. (2010)
	Bay of Bengal	Water column	The ITS2 region of the fungal rRNA gene was amplified using fungal-specific primers ITS1-F/ITS2 and paired-end sequenced using 2 × 250 nt on an Illumina HiSeq 2500 platform.	PRJNA794046	NA	Li et al. (2024)
Pacific Ocean	Eastern Tropical North Pacific	Water column	ITS2 region of the fungal rRNA gene using primers ITS3tagmix/ ITS4tag001 and paired-end sequenced using 2 × 250 nt on an Illumina MiSeq platform. Additionally, a metagenomic dataset was mined to complement the amplicon-based analyses.	PRJNA623945	237	Peng and Valentine (2021)
	Eastern Tropical North Pacific	Sediment	Illumina MiSeq paired-end (2 × 300) sequencing, targeting the ITS1 region of the ribosomal RNA gene cluster by using the primers set ITS1F/ITS2	PRJNA793088	43	Velez et al. (2022)
	Eastern Tropical North Pacific	Sediment	Illumina MiSeq paired-end (2 × 300) sequencing, targeting the ITS1 of the ribosomal RNA gene cluster with the primer set ITS1F/ITS2.	PRJNA822656	353	Posadas et al. (2024)
NA	Various	NA	Phylogenetic analyses of fungal sequences were conducted using open-access SSU rRNA genes datasets obtained from 12 environmental diversity studies across 14 oxygen-depleted sampling sites.	NA	NA	Jebaraj et al. (2012)
NA	Various	NA	A comprehensive culture-independent review of fungal diversity across marine habitats, including oxygen-depleted environments, emphasizing the vast and largely unexplored diversity of marine fungi.	NA	NA	Manohar and Raghukumar (2013)

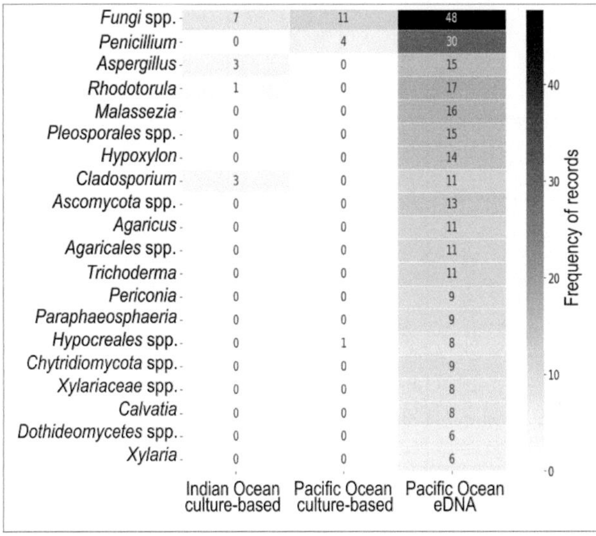

Fig. 3. Heatmap showing the frequency of records for the top 20 fungal genera (out of 186 in total) across oceanic regions (Indian and Pacific oceans) and study techniques (isolation-based and molecular-based approaches). Frequencies are grouped by genus (rows) and by region-technique combinations (columns). Grayscale shading indicates the number of records, with dark gray corresponding to higher frequencies.

abundance of Basidiomycota compared to Ascomycota, with the most prevalent and abundant basidiomycetous yeast class being Microbotryomycetes, primarily consisting of the genera *Rhodotorula*, *Rhodosporidiobolus*, and *Sporobolomyces* (Peng and Valentine, 2021). Thus, community compositions between different habitat types and the need to complement them with various molecular methods, because using the metagenomic assessment, they record higher relative abundance of Ascomycota in the same column water samples (Peng and Valentine, 2021).

Generally, a large fraction of the genetic sequences recovered from OMZ studies do not match known organisms. For instance, in sediments of the Gulf of California and Mazatlan, ~20.58% and ~10.2% of ASV (amplicon sequence variants), respectively, were identified only to the kingdom level as *Fungi* spp. (Posadas et al., 2024; Velez et al., 2022), which evidences the copious occurrence of uncultured fungal phylotypes in these environments, including water column research (Li et al., 2024; Peng and Valentine, 2021). Besides, a negligible proportion of early-diverging phyla have been detected, including *Mucoromycota*, *Zoopagomycota*, *Chytridiomycota*, *Blastocladiomycota*, *Microsporidia*, and *Rozellomycota* (Peng and Valentine, 2021; Posadas et al., 2024; Velez et al., 2022).

Environmental sequences from formally undescribed and uncultured fungi, examined through metagenomic phylogenies, have revealed novel environmental clades. The Pezizomycotina clone group (PCG), within the Ascomycota, shows phylogenetic affiliation with *Aspergillus* sp., *Penicillium* sp., and *Eupenicillium* sp. (Dawson and Pace, 2002; Jebaraj et al., 2012; Manohar and Raghukumar, 2013; Nagano et al., 2010). Likewise, environmental sequences from the deep-sea fungal group I (DSF-Group 1) clustered with yeast *Metschnikowia bicuspidata* within the

Saccharomycotina (Nagano et al., 2010; Thaler et al., 2012). In the Basidiomycota, new environmental sequences have been detected in Ustilaginomycotina, included in the hydrothermal and/or anaerobic fungal group (Hy-An group), which is taxonomically close to *Malassezia* sp. Basal fungal lineages have also been identified, such as the basal clone group I (BCG I) and the LKM11 group, closely associated with the Cryptomycota fungal clade (Nagahama et al., 2011; Thaler et al., 2012). Mainly occurring in oxygen-depleted habitats, these basal lineages could offer important perspectives on their fungal evolutionary trajectories due to their early divergence.

Interestingly, in the Eastern Tropical North Pacific (ETNP) OMZ several of the most widely distributed and dominant ASV showed affinity to the previously mentioned DSF-group1 (Ascomycota; AB507843, AB507839, KT758162.1, and KT758149.1). This uncultured fungal phylotype is recognized as a globally distributed deep-sea endemic group, probably adapted to low-oxygen conditions, as evidence by its continuous records and high abundance in subsurface sediments and oxygen-depleted deep-sea habitats (Bass et al., 2007; Takishita et al., 2007; Nagano et al., 2010; Nagano and Nagahama, 2012; Velez et al., 2022; Posadas et al., 2024). This finding suggests that fungi are far more diverse and abundant in suboxic habitats than previously acknowledged.

Regarding spatial patterns of fungal diversity within OMZ, studies consistently indicate higher taxonomic richness in nearshore areas compared to offshore sites (Li et al., 2024; Posadas et al., 2024; Thompson et al., 2024; Peng and Valentine, 2025). This trend has been documented across both water column and sediment samples, suggesting that coastal proximity exerts a significant influence on the composition and structure of fungal communities. Patterns in fungal taxonomic composition with depth differ. For instance, the relative abundance of molecular signals from groups such as *Chytridiomycota* tends to be higher in samples collected at depths shallower than 90 meters compared to those from deeper layers (Peng and Valentine, 2025). Several studies have characterized the spatial dynamics of OMZ; minimum oxygen concentrations vary with latitude; a pattern often associated with differences in water column depth (Paulmier and Pino, 2009). The position of the oxycline also shifts with depth, typically becoming shallower in regions where OMZ are more intense. In such environments, characterized by greater depth and a lower oxycline, reduced microbial taxonomic richness has been observed (Thompson et al., 2024). These patterns are related with the fluctuations in organic matter inputs caused by decreases in evenness towards the open ocean (Posadas et al., 2024). Gutiérrez et al. (2010) reported a greater fungal diversity in coastal OMZ sediments and overlying waters relative to oceanic zones, while Vera et al. (2017) observed a comparable pattern in pelagic fungal assemblages. More recently, Posadas et al. (2024) further supported these observations, suggesting that the elevated fungal diversity detected in coastal zones may result from increased inputs of organic matter, elevated nutrient availability, and the influence of terrestrial runoff. These studies support the existence of a spatial gradient in fungal community composition within OMZ ecosystems, likely driven by fine-scale environmental heterogeneity associated with coastal dynamics. In addition, these findings underscore the urgent need for targeted research to better understand the ecological drivers shaping fungal diversity in these transitional environments.

Furthermore, previous studies have documented clear structural differences in fungal communities depending on oxygen availability, distinguishing across environments with normal, seasonal or permanent oxygen depletion (Jebaraj et al., 2010). However, more recent investigations reveal a patchy distribution of fungal diversity without consistent correlations to depth or oxygen concentration (Li et al., 2024; Posadas et al., 2024). These observations imply that localized biogeochemical conditions play a significant role in driving fine-scale heterogeneity in fungal assemblage structure (Li et al., 2024; Posadas et al., 2024; Feng et al., 2021). The ecological flexibility of fungi to diverse and extreme environments underpins their success as one of the most widespread eukaryotic kingdoms. To advance our understanding of their distribution and ecological functions within marine OMZ, future research must prioritize their interactions with specific environmental variables.

3. Ecological Role of Fungi

Although the full scope of fungal diversity and its ecological roles in low-oxygen environments remains largely unknown, culture-based methods have revealed the occurrence of fungi with facultative parasitic or opportunistic lifestyles, such as members of *Chytridiomycota*, *Aspergillus sydowii*, *Alternaria alternata*, *Exobasidium gracile*, and *Aureobasidium pullulans* (Vera et al., 2017), as well as *Cryptococcus* sp., *Cystofilobasidium* sp., *Mortierella wolfii*, *Rhizophydium* sp., and *Malassezia* spp., including *M. restricta* (Manohar et al., 2013). In lower proportions, saprotrophic fungi such as *Globulicium hiemale*, *Neurospora crassa*, members of the genus *Armillaria*, *Penicillium* spp., and various *Agaricomycetes* have also been recorded. Finally, the least represented group includes putative symbiotrophs, such as *Geosmithia namyslowskii*. In parallel, culture-independent analyses using bioinformatic tools (e.g., FUNGuild) based on taxonomic affiliation have allowed functional inferences of fungal lifestyles across OMZ regions. These analyses revealed distinct trophic guild patterns: facultative parasitic or opportunistic fungi dominated nearshore environments, whereas saprotrophs were more prevalent offshore. Notably, at offshore sites such as the Alfonso Basin, fungal guilds with potential symbiotic roles dominated the community, suggesting associations with animal hosts. It is important to note that only approximately 19.62% of total fungal reads could be assigned to functional guilds, underscoring the need for further ecological and functional studies to clarify fungal contributions in OMZ and their potential environmental and biotechnological relevance.

In oxygen-limited environments, several fungi have been observed to exhibit remarkable flexibility in coping with fluctuations in oxygen availability by employing alternative respiratory strategies. *Fusarium oxysporum* has been widely documented to perform incomplete nitrate reduction under low-oxygen conditions (Shoun et al., 1992; Jebaraj et al., 2010; Manohar et al., 2015; Rédou et al., 2015). Similarly, *Aspergillus terreus* has demonstrated dissimilatory nitrate reduction under anaerobic conditions (Manohar et al., 2014), and *Mucor* spp. have also revealed metabolic flexibility in low-oxygen environments (Manohar et al., 2015). In addition, two ASV detected in the water column of the ETNP OMZ exhibited approximately 97% sequence similarity to the ITS2 region of *Penicillium melinii* and *Chaetomium* sp., both of which are known to produce nitrous oxide as part of their nitrogen metabolism (Maeda et al., 2015; Peng and Valentine, 2021). Other fungal taxa such as *Alternaria* spp., *Aspergillus* spp., *Aureobasidium* spp., *Cladosporium* spp., *Fusarium* spp., *Penicillium* spp., *Rhodotorula* spp., and *Trichoderma* spp. show evidence of denitrifying activity even under simulated deep-sea, anoxic, or suboxic conditions (Cathrine and Raghukumar, 2009; Manohar et al., 2015). These observations indicate that certain fungal groups possess functional traits enabling them to contribute to nitrogen cycling processes in OMZ ecosystems, potentially playing roles analogous to prokaryotic denitrifiers.

Additionally, while numerous fungal taxa have been detected in different OMZ samples using both culture-dependent and -independent approaches (indicating their potential tolerance to low-oxygen environments), their specific role in denitrification remains largely uncharacterized. For instance, *Phaeosphaeria* sp. (Manohar et al., 2015; Posadas et al., 2024), *Tritirachium* sp., *Byssochlamys* sp., and *Paecilomyces* sp. (Jebaraj et al., 2010), as well as *A. cervinus*, *A. versicolor*, *Parengyodontium album*, *Alternaria alternata*, *Aureobasidium pullulans*, *Rhodosporidium diobovatum*, *Melanopsichium pennsylvanicum*, *Rhodotorula minuta* (Manohar et al., 2015), and *Roussoella solani* (Posadas et al., 2024) have been reported from OMZ sediments and water columns; however, their functional roles in nitrogen cycling remain poorly understood. This gap is critical, as recent investigations in the ETNP OMZ have demonstrated that fungi can contribute significantly to nitrous oxide, a potent greenhouse gas. In fact, researchers estimated that fungal processes may account for approximately 18–22% of total N_2O emissions in specific water column layers (Peng and Valentine, 2021). These findings suggest that fungal denitrification may constitute a significant and previously underestimated source of N_2O in marine OMZ ecosystems. They also suggest the potential existence of additional fungal mechanisms for coping with fluctuating oxygen levels, highlighting the need for targeted functional studies.

4. Future Prospects

Future research on fungal diversity in OMZ should adopt integrative strategies that combine both culture-dependent and culture-independent methodologies to achieve a polyphasic understanding of fungal taxonomic and functional composition. Culture-based techniques are essential for the physiological and metabolic characterization of isolates, enabling the identification of traits such as oxygen tolerance and denitrification capacity. In parallel, culture-independent approaches, including high-throughput amplicon sequencing, metagenomic analysis, and bioinformatic tools for functional guild assignment (e.g., FUNGuild), are indispensable for uncovering the vast, often hidden diversity of fungal taxa, including those that are rare or unculturable. The ecological relevance of fungi in OMZ is becoming increasingly apparent, as evidence suggests that these organisms not only survive in hypoxic and anoxic conditions but also contribute meaningfully to biogeochemical processes, hinting at their participation in nutrient turnover, and nitrogen cycling. Recent findings indicate that several fungal taxa may perform denitrification or produce nitrous oxide, implicating them as underrecognized but potentially significant contributors to nitrogen transformations within these unique marine environments. Moreover, the roles of fungi in inorganic compound metabolism or symbiotic interactions with other organisms at depth remain practically uncharacterized. A substantial proportion of environmental fungal sequences also remains unclassified at both taxonomic and functional levels, underscoring the urgent need for enhanced reference databases, targeted functional studies, and curated repositories of type species. Although metabarcoding offers a powerful tool for detecting greater hidden diversity, it remains subject to methodological limitations such as primer bias, particularly against basal fungal lineages when using ITS-based markers. Overall, advancing the study of fungal communities in OMZ is essential to fully elucidate their ecological roles, assess their contributions to marine nitrogen fluxes and greenhouse gas emissions, and predict their responses to ongoing oceanic changes.

5. Conclusion

Fungal diversity in OMZ represents an important yet still underexplored component of marine microbial ecosystem. Current evidence indicates that fungi not only persist but have also evolved under hypoxic conditions, encompassing a wide taxonomic spectrum dominated by Ascomycota and Basidiomycota, as well as early-diverging lineages and novel environmental clades. Although knowledge about biogeographic patterns remains limited, available studies indicate that often the distribution of fungal communities in OMZ is shaped by site-specific environmental heterogeneity, with enhanced diversity in nearshore areas possibly influenced by nutrient inputs. Functionally, several taxa exhibit adaptations to oxygen depletion, including nitrate reduction and denitrification, pointing to meaningful roles in nitrogen cycling and greenhouse gas emissions. Despite these advances, major knowledge gaps persist. The potential involvement of fungi in nitrogen fixation, inorganic compound metabolism, or symbiotic associations at depth remains largely unknown, while a substantial fraction of environmental sequences cannot yet be taxonomically or functionally assigned. Addressing these gaps requires polyphasic strategies that integrate culture-dependent and culture-independent techniques to clarify taxonomic identities, characterize metabolic traits, and uncover the ecological contributions of rare or unculturable taxa. Strengthening genomic databases, expanding functional studies, and reducing methodological biases will be key to advancing this field. Finally, such efforts will enhance our understanding of fungal diversity, function, and evolution under low-oxygen conditions, while illuminating their broader roles in marine biogeochemical processes, nutrient fluxes, and ecosystem responses to climate-driven ocean change.

Acknowledgements

The authors are doctoral students from the Posgrado en Ciencias del Mar y Limnología at the Universidad Nacional Autónoma de México and acknowledge financial support from the Programa de Apoyo a Proyectos de Investigación e Innovación Tecnológica (DGAPA-PAPIIT, number IN203524) as well as fellowships from the Secretaría de Ciencia, Humanidad, Tecnología e Innovación (SECIHTI; numbers 1082039 and 1082184).

References

Bass, D., Howe, A., Brown, N., Barton, H., Demidova, M. et al. (2007). Yeast forms dominate fungal diversity in the deep oceans. Proc. R. Soc. B Biol. Sci., 274(1629): 3069–3077. https://doi.org/10.1098/rspb.2007.1067.

Bertagnolli, A.D. and Stewart, F.J. (2018). Microbial niches in marine oxygen minimum zones. Nat. Rev. Microbiol., 16(12): 723–729. https://doi.org/10.1038/s41579-018-0087-z.

Brant, P., Bange, H.W., Banyte, D., Dengler, M., Didwischus, S.H. et al. (2015). On the role of circulation and mixing in the ventilation of oxygen minimum zones with a focus on the eastern tropical North Atlantic. Biogeosciences, 12: 489–512. https://doi.org/10.5194/bg-12-489-2015.

Cathrine, S.J. and Raghukumar, C. (2009). Anaerobic denitrification in fungi from the coastal marine sediments off Goa, India. Mycol. Res., 113: 100–109. https://doi.org/10.1016/j.mycres.2008.08.009.

Codispoti, L.A. (2010). Interesting times for marine N_2O. Sci., 327(5971): 1339–1340.

Dawson, S.C. and Pace, N.R. (2002). Novel kingdom-level eukaryotic diversity in anoxic environments. Proc. Natl. Acad. Sci., 99(12): 8324–8329. https://doi.org/10.1073/pnas.062169599.

Feng, L., Song, Q., Jiang, Q. and Li, Z. (2021). The horizontal and vertical distribution of deep-sea sediments fungal community in the South China Sea. Front. Mar. Sci., 8: 592784. https://doi.org/10.3389/fmars.2021.592784.

Gutiérrez, M.H., Pantoja, S., Quiñones, R.A. and González, R.R. (2010). First record of filamentous fungi in the coastal upwelling ecosystem off central Chile. Gayana, 74(1): 66–73. https://doi.org/10.4067/S0717-65382010000100010.

Gutiérrez, M.H., Vera, J., Srain, B., Quiñones, R.A., Wörmer, L., Hinrichs, K.U. and Pantoja-Gutiérrez, S. (2020). Biochemical fingerprints of marine fungi: Implications for trophic and biogeochemical studies. Aquat. Microb. Ecol., 84: 75–90. https://doi.org/10.3354/ame01927.

Helly, J.J. and Levin, L.A. (2004). Global distribution of naturally occurring marine hypoxia on continental margins. Deep-Sea Res. I: Oceanogr. Res. Pap., 51(9): 1159–1168. https://doi.org/10.1016/j.dsr.2004.03.009.

Jebaraj, C., Raghukumar, C., Behnke, A. and Stoeck, T. (2010). Fungal diversity in oxygen-depleted regions of the Arabian Sea revealed by targeted environmental sequencing combined with cultivation. FEMS Microbiol. Ecol., 71(3): 399–412. https://doi.org/10.1111/j.1574-6941.2009.00820.x.

Jebaraj, C.S., Forster, D., Kauff, F. and Stoeck, T. (2012). Molecular diversity of fungi from marine oxygen-deficient environments (ODEs). Biol. Mar. Fungi, 53: 189–208. https://doi.org/10.1007/978-3-642-23342-5_10.

Levin, L.A. (2018). Manifestation, drivers, and emergence of open ocean deoxygenation. Annu. Rev. Mar. Sci., 10: 229–260. https://doi.org/10.1146/annurev-marine-121916-063359.

Li, J., Xie, N., Liu, X., Bai, M., Hunt, D.E. and Wang, G. (2024). Oxygen levels differentially attenuate the structure and diversity of microbial communities in the oceanic oxygen minimal zones. Sci. Tot. Environ., 948: 174934. https://doi.org/10.1016/j.scitotenv.2024.174934.

Li, W., Wang, M.M., Wang, X.G., Cheng, X.L., Guo, J.J. et al. (2016). Fungal communities in sediments of subtropical Chinese seas as estimated by DNA metabarcoding. Sci. Rep., 6: 26528. https://doi.org/10.1038/srep26528.

Long, A.M., Jurgensen, S.K., Petchel, A.R., Savoie, E.R. and Brum, J.R. (2021). Microbial ecology of oxygen minimum zones amidst ocean deoxygenation. Front. Microbiol., 12: 748961. https://doi.org/10.3389/fmicb.2021.748961.

Maeda, K., Spor, A., Edel-Hermann, V., Heraud, C., Breuil, M.C. et al. (2015). N_2O production, a widespread trait in fungi. Sci. Rep., 5: 9697. https://doi.org/10.1038/srep09697.

Manohar, C.S. and Raghukumar, C. (2013). Fungal diversity from various marine habitats deduced through culture-independent studies. FEMS Microbiol. Lett., 341(1): 69–78. https://doi.org/10.1111/1574-6968.12071.

Manohar, C.S., Boekhout, T., Müller, W.H. and Stoeck, T. (2014). *Tritirachium candoliense* sp. nov., a novel basidiomycetous fungus isolated from the anoxic zone of the Arabian Sea. Fungal Biol., 118(2): 139–149. https://doi.org/10.1016/j.funbio.2013.10.010.

Manohar, C.S., Menezes, L.D., Ramasamy, K.P. and Meena, R.M. (2015). Phylogenetic analyses and nitrate-reducing activity of fungal cultures isolated from the permanent, oceanic oxygen minimum zone of the Arabian Sea. Can. J. Microbiol., 61(3): 217–226. https://doi.org/10.1139/cjm-2014-0462.

Moffitt, S.E., Moffitt, R.A., Sauthoff, W., Davis, C.V., Hewett, K. and Hill, T.M. (2015). Paleoceanographic insights on recent oxygen minimum zone expansion: Lessons for modern oceanography. PLoS One, 10(1). https://doi.org/10.1371/journal.pone.0115246.

Møller, T.E., Le Moine Bauer, S., Hannisdal, B., Zhao, R., Baumberger, T. et al. (2022). Mapping microbial abundance and prevalence to changing oxygen concentration in deep-sea sediments using machine learning and differential abundance. Front. Microbiol., 13: 804575. https://doi.org/10.3389/fmicb.2022.804575.

Morgan-Smith, D., Clouse, M., Herndl, G.J. and Bochdansky, A. (2013). Diversity and distribution of microbial eukaryotes in the deep tropical and subtropical North Atlantic Ocean. Deep-Sea Research Part I: Oceanographic Research Papers, 78: 58–69. https://doi.org/10.1016/j.dsr.2013.04.010.

Nagahama, T., Takahashi, E., Nagano, Y., Abdel-Wahab, M.A. and Miyazaki, M. (2011). Molecular evidence that deep-branching fungi are major fungal components in deep-sea methane cold-seep sediments. Environ. Microbiol., 13(8): 2359–2370.

Nagano, Y. and Nagahama, T. (2012). Fungal diversity in deep-sea extreme environments. Fungal Ecol., 5(4): 463–471. https://doi.org/10.1016/j.funeco.2012.01.004.

Nagano, Y., Nagahama, T., Hatada, Y., Nunoura, T., Takami, H. et al. (2010). Fungal diversity in deep-sea sediments—The presence of novel fungal groups. Fungal Ecol., 3(4): 316–325. https://doi.org/10.1016/j.funeco.2010.01.002.

Naqvi, S.W.A., Naik, H., Pratihary, A., D'Souza, W., Narvekar, P.V. et al. (2006). Coastal versus open-ocean denitrification in the Arabian Sea. Biogeosciences, 3(4): 621–633. https://doi.org/10.5194/bg-3-621-2006.

Papiol, V., Hendrickx, M.E. and Serrano, D. (2017). Effects of latitudinal changes in the oxygen minimum zone of the northeast Pacific on the distribution of bathyal benthic decapod crustaceans. Deep Sea Res. Pt. II Top Stud. Oceanogr., 137: 1–15. https://doi.org/10.1016/j.dsr2.2016.04.023.

Paulmier, A. and Ruiz-Pino, D. (2009). Oxygen minimum zones (OMZs) in the modern ocean. Prog. Oceanogr., 80(3-4): 113–128. https://doi.org/10.1016/j.pocean.2008.08.001.

Peng, X. and Valentine, D.L. (2021). Diversity and N_2O production potential of fungi in an oceanic oxygen minimum zone. J. Fungi, 7(3): 212. https://doi.org/10.3390/jof7030212.

Peng, X. and Valentine, D.L. (2025). Expanded genetic and functional diversity of oceanic fungi. Microbiome, 13: 179. https://doi.org/10.1186/s40168-025-02162-2.

Posadas, J., Velez, P., Pajares, S., Gasca-Pineda, J. and Espinosa-Asuar, L. (2024). Fungal diversity in sediments of the eastern tropical Pacific oxygen minimum zone revealed by metabarcoding. PLOS One, 19(5). https://doi.org/10.1371/journal.pone.0301605.

Rabalais, N.N., Díaz, R.J., Levin, L.A., Turner, R.E., Gilbert, D. and Zhang, J. (2010). Dynamics and distribution of natural and human-caused coastal hypoxia. Biogeosciences, 7(2): 585–619. https://doi.org/10.5194/bg-7-585-2010.

Rédou, V., Navarri, M., Meslet-Cladière, L., Barbier, G. and Burgaud, G. (2015). Species richness and adaptation of marine fungi from deep-subseafloor sediments. Appl. Environ. Microbiol., 81(10): 3571–3583. https://doi.org/10.1128/AEM.04064-14.

Shoun, H., Kim, D.H., Uchiyama, H. and Sugiyama, J. (1992). Denitrification by fungi. FEMS Microbiol. Lett., 94(3): 277–281. https://doi.org/10.1111/j.1574-6968.1992.tb05331.x.

Stief, P., Fuchs-Ocklenburg, S., Kamp, A., Manohar, C.S., Houbraken, J. et al. (2014). Dissimilatory nitrate reduction by *Aspergillus terreus* isolated from the seasonal oxygen minimum zone in the Arabian Sea. BMC Microbiol., 14(1): 29. https://doi.org/10.1186/s12866-014-0029-9.

Takishita, K., Yubuki, N., Kakizoe, N., Inagaki, Y. and Maruyama, T. (2007). Diversity of microbial eukaryotes in sediment at a deep-sea methane cold seep: Surveys of ribosomal DNA libraries from raw sediment samples and two enrichment cultures. Extremophiles, 11(4): 563–576. https://doi.org/10.1007/s00792-007-0068-z.

Thaler, A.D., Van Dover, C.L. and Vilgalys, R. (2012). Ascomycete phylotypes recovered from a Gulf of Mexico methane seep are identical to an uncultured deep-sea fungal clade from the Pacific. Fungal Ecol., 5(2): 270–273. https://doi.org/10.1016/j.funeco.2011.07.002.

Thompson, M.A., Valentine, D.L. and Peng, X. (2024). Size fractionation informs microbial community composition and interactions in the eastern tropical North Pacific Ocean. FEMS Microbes, 5: xtae028. https://doi.org/10.1093/femsmc/xtae028.

Ulloa, O., Canfield, D.E., DeLong, E.F., Letelier, R.M. and Stewart, F.J. (2012). Microbial oceanography of anoxic oxygen minimum zones. Proc. Natl. Acad. Sci., USA. 109(40): 15996–16003. https://doi.org/10.1073/pnas.1205009109.

Velez, P., Salcedo, D.L., Espinosa-Asuar, L., Gasca-Pineda, J., Hernández-Monroy, A. and Soto, L.A. (2022). Fungal diversity in sediments from deep-sea extreme ecosystems: Insights into low- and high-temperature hydrothermal vents, and an oxygen minimum zone in the southern Gulf of California, Mexico. Front. Mar. Sci., 9: 802634. https://doi.org/10.3389/fmars.2022.802634.

Vera, J., Gutiérrez, M.H., Palfner, G. and Pantoja, S. (2017). Diversity of culturable filamentous Ascomycetes in the eastern South Pacific Ocean off Chile. World J. Microbiol. Biotechnol., 33: 157. https://doi.org/10.1007/s11274-017-2321-7.

3

Fungal Diversity in Marine Sandy Beaches

An Integrated Ecological Perspective

Patricia Velez[1,*] and *Aurora Rivera-Zizumbo*[1,2]

Sandy beaches function as ecotones, situated within a complex web of environmental gradients influenced by numerous abiotic factors such as temperature, wave dynamics, sand grain size, and salinity. These dynamic and often harsh conditions restrict microbial colonization to a limited group of specialized species. Among these specialized microbes, obligate arenicolous marine fungi are a unique ecological assemblage of heterotrophic organisms specifically adapted to life in sandy substrates. These fungi have developed diverse strategies to withstand fluctuating intertidal environments and are distributed globally across tropical, subtropical, and temperate coastal zones. Likewise, facultative marine fungi can thrive in sandy beaches, adapting to the dynamic conditions of both terrestrial and marine environments. Overall, these microeukaryotes inhabit a vulnerable ecotone, facing increasing risks from both local and non-local pressures. Herein, research on fungal communities in sandy beaches is discussed, including morphological and genetic adaptations of fungi to beach ecosystems, and variations in species composition over time and space. This review highlights the need for an interdisciplinary approach to gain an integrated insight of beach ecology to the conservation and management of coastal ecosystems under increasing anthropogenic and climate pressures.

[1] Instituto de Biología, Universidad Nacional Autónoma de México, Mexico City, Mexico.
[2] Posgrado en Ciencias Biológicas, Universidad Nacional Autónoma de México, Mexico City, Mexico.
* Corresponding author: pvelez@ib.unam.mx

DOI: 10.1201/9781003768708-4

1. Introduction

Approximately half of the world's ice-free ocean coastline is covered by sandy beaches (Defeo et al., 2021). These dynamic ecosystems dominate the coastal landscape in temperate and tropical zones, extending over about 75% of the world's coasts (McLachlan, 2001; Brown and McLachlan, 2002). Their dynamism is defined by the interaction of three factors that constantly modify the physical environment and determine the classification of the beach morphodynamic type: waves, the tidal regime and the nature of the sand (Short, 1996; Defeo and McLachlan, 2005a, b; Defeo et al., 2009). Overall, sandy beaches may be classified into: (1) reflective beaches, those that are characterized by long swell period, small waves with low turbulence; coarse sand, and steep intertidal slopes with narrow or null surfzone, and (2) dissipative beaches, those that present a gentle slope and extensive surf zone that dissipate the waves energy, finer sediments and larger tide ranges (McLachlan and Turner, 1994; Harris et al., 2014) (Fig. 1).

In these ecosystems there is a convergence of materials and organisms that move from adjacent coastal dunes and from the adjacent seas or oceans. Therefore, a diverse community of bacteria, archaea, protists, microalgae, fungi and meiofauna inhabit the sand grains contributing to the flow of matter and energy of the ecosystem (Defeo et al., 2009; Okamoto et al., 2022). Photosynthetic protists like diatoms and dinoflagellates, cyanobacteria and chlorophytes act as primary producers and represent an important resource to the micro- and macrofauna at the beach. Microorganisms like fungi, bacteria, and some protists participate in the degradation of organic matter, supplied by the land and the sea, and nutrient cycling (Radziejewska et al., 2017; Okamoto et al., 2022).

The economies of coastal nations are dependent on beach recreation-tourism activities (McLachlan et al., 2013). Despite their apparently resilient nature, sandy marine beaches face significant ecological impacts and escalating risks from anthropogenic activities. Human population growth and coastal development

Fig. 1. Example of the two main beach morphodynamics: A-B) dissipative coast profiles in Celestún and Coatzacoalcos (Gulf of Mexico), and C-D) reflective beaches in Ensenada and Acapulco (Pacific Ocean), Mexico.

directly contribute to habitat loss and degradation, posing severe threats to the unique biodiversity inhabiting these dynamic ecotones (McLachlan and Brown, 2006). Furthermore, pollution from several sources like marine litter, sewage discharge, and industrial activities, combined with the intensifying effects of climate change, jeopardize the ecological integrity of these ecosystems (Schlacher et al., 2008; Defeo et al., 2009). Presently, human footprint extends to every beach, denoting the intricate relationship between their ecology and human activities cannot be overlooked. Hence, to ensure the continued ecological functioning of the world's sandy shores it is crucial to adopt a broader social-ecological system viewpoint (Defeo et al., 2021).

2. Microbial Assemblages in Coastal Beaches

Coastal sandy beaches create complex and fluctuating environments that host a diverse array of microbial life. These microbial assemblages, comprising bacteria, archaea, fungi, protists, and microalgae (Fig. 2), play critical roles in ecosystem functioning, including nutrient cycling, organic matter decomposition, and sediment stabilization (McLachlan and Defeo, 2018). The distribution and composition of these communities are governed by environmental gradients such as salinity, moisture, temperature, and grain size, which vary spatially across beach profiles (e.g., supralittoral, intertidal, subtidal zones) and temporally with tidal cycles and seasonal changes (Schneider et al., 2011). Despite their ecological importance, microbial communities in sandy beaches remain understudied compared to those in other coastal systems like estuaries or mangroves.

Particularly, obligate arenicolous marine fungi are microscopic organisms adapted to live in the intertidal zones of sandy beaches, where they play crucial roles in degrading cellulose-rich materials and supporting nutrient cycling in

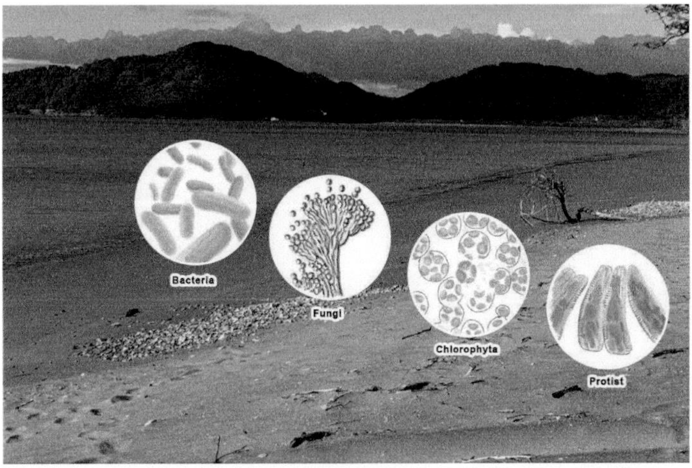

Fig. 2. Schematic representation of microbial communities in a sandy beach ecosystem, highlighting major microbial groups, including bacteria, fungi, microalgae, and protists.

coastal ecosystems (Kohlmeyer and Kohlmeyer, 1979). These fungi co-occur with terrestrial, freshwater, and further marine fungal species that are introduced to sandy beaches via several mechanisms, and survive for short or long periods. So, the fungal community in sandy beaches contains both obligate marine species and facultative or terrestrial-derived fungi, forming a diverse and ecologically heterogeneous assemblage (revised by Velez et al., 2022).

Recent studies have highlighted the adaptability of microbial taxa to the extreme and transient conditions characteristic of sandy shores. For instance, obligate arenicolous fungi have evolved morphological (e.g., spore appendages, early discussed by Kohlmeyer, 1966; thick, hard, carbonaceous ascomata walls (Kohlmeyer and Kohlmeyer, 1979) and genetic genes related with osmotic homeostasis (Velez et al., 2015c) traits that enable them to persist in shifting and extreme salinity and temperature conditions, and high UV radiation levels (Jones et al., 2019). Fungi in sandy beaches play a key role maintaining the ecological balance and resilience of sandy beach ecosystems, by stabilizing organic matter and interacting with other microbial and benthic communities, even establishing mutualistic relationships with the vegetation associated with coastal dunes (Beena et al., 2000; Gladfelter et al., 2019). So, understanding these microbial assemblages is essential not only for characterizing beach biodiversity, but also for assessing the impacts of anthropogenic pressures such as pollution, climate change, and coastal development, which may disrupt these finely balanced microbial ecosystems.

In terms of their taxonomic diversity, obligate arenicolous include limited genera such as: *Arenariomyces*, *Carbosphaerella*, *Corollospora*, *Kohlmeyeriella*, *Lindra*, and *Lulwoidea*. All are members of Ascomycota, suggesting that a restricted set of phylogenetic lineages of the Ascomycota have successfully colonized sandy beaches. Remarkably, *Corollospora* is the largest genus within this group, accommodating 27 species. Among these, most species are known to have only teleomorphs, but three species are anamorphic species having only asexual states, and four species were described having both teleomorph and anamorph (Nakagiri et al., 2025). Facultative arenicolous marine fungi represent a taxonomically more diverse group containing members of the Microascales, Pleosporales, and Torpedosporales (Ascomycota); as well as Agaricales (Basidiomycota). This assemblage of fungi is primarily known to inhabit submerged substrata (e.g., wood, seaweeds, seagrasses, etc. (revised in Velez et al., 2022).

3. Protecting Marine Fungi in Coastal Beaches

Marine fungal diversity in sandy beaches is affected simultaneously by both: natural and anthropogenic impacts, endangering their persistence and vital ecological roles. Factors such as coastal development, pollution (e.g., oil spills, wastewater discharges, and chemical contaminants), and habitat modification through beach nourishment or recreational use profoundly disturb the delicate balance of microbial communities, frequently resulting in a decline in fungal diversity (Kohlmeyer and Kohlmeyer, 1979; Pang et al., 2016).

Beach use affects the diversity and composition of the overall arenicolous marine fungal community. Several investigations have shown that fungal diversity in beaches responds to human stressors, often leading to an altered community structure, and the directional replacement of species. This phenomenon is frequently coupled to the proliferation of opportunistically pathogenic taxa, examples include reports in tourist beaches in the Mediterranean coast (Salvo and Fabiano, 2007), Greece (Papadakis et al., 1997), Mexico (Velez et al., 2021), Hawaii (Kishimoto and Baker, 1969), and South Carolina (Stevens et al., 2012) in the USA, among others. This constitutes a potential health issue, restricting beach recreational activities and posing a risk to users.

In addition, the effect of anthropogenic (presence of main and highly polluted river mouths, nearshore marine environments, and core industrial and port developments) (Velez et al., 2013, 2015a), versus natural stressors (storms and hurricanes) (Velez et al., 2015a, b, 2021) has been also documented. Research efforts conducted at the small spatial scale in marine arenicolous fungal communities on an impacted Pacific tourist sandy beach have revealed diversity shifts in response to storm season. The findings suggest that while some arenicolous fungi persist as inoculum, human activities can favor opportunistic terrestrial pathogens, highlighting the need for better understanding and management of these neglected but ecologically vital beach inhabitants (Velez et al., 2021). Also, the vulnerability of arenicolous key saprobic species to anthropogenic disturbances like beach grooming has been discussed, warning against this common practice in tourist beaches since it leads to the directional reduction of substrate-specific obligate marine arenicolous fungal groups. In this regard, coastal zoning, through the delimitation of areas for conservation, recreation, and other uses, could establish undisturbed buffer zones serving as marine fungal inocula banks.

Climate change significantly impacts sandy marine beaches through accelerated erosion due to sea-level rise and increased storm intensity, threatening their ecological functions and physical integrity (Vousdoukas et al., 2020). These alterations directly affect beach biota, leading to habitat loss and shifts in species distribution, with long-term consequences for coastal ecosystems (Defeo et al., 2009). For sand fungi, climate change intensifies these threats by altering temperature regimes, and salinity patterns, and sea level rise, which can shift their spatial distribution and abundance (Gladfelter et al., 2019). Particularly, the resilience of obligate marine arenicolous fungi is challenged by their dependence on specific substrates and environmental conditions (Jones et al., 2019), which may become scarce or altered under changing environmental conditions, and should be addressed to mitigate their impacts.

4. Integrated Insight of Beach Ecology

Sandy beaches provide essential ecosystem services, including recreation, food, coastal protection, pollution assimilation, biodiversity maintenance, water storage, and nutrient cycling. However, these vital services are increasingly threatened by diverse anthropogenic perturbations (Defeo et al., 2021). Complex coastal environments, characterized by diverse nearshore habitats such as mangroves, shoals, lagoons, and river mouths, foster high marine fungal diversity, consistent with the observed relationship between fungal and plant diversity (Tsui et al., 1998). These neighboring areas likely support a rich mycobiota by providing varied substrates and heterogeneous conditions conducive to ascomycete development on sandy beaches. So, an integrated ecological perspective is urgent to recognize beaches as interconnected systems linked to adjacent terrestrial and marine ecosystems through nutrient fluxes, organic matter deposition, and species migrations (Defeo et al., 2009). Specifically, the integration of marine arenicolous fungi to beach monitoring and management policies would promote sustainable practices and will contribute to feed predictive models of beach ecosystem dynamics under anthropogenic pressures and climate change (Gladfelter et al., 2019). In this sense, major threats such as: coastal squeeze (constrained beach coverage due to recreational, urban, and industrial development on land, coupled to the migration of the seaward boundary in response to erosion), sand-mining practices, algal blooms, pollution (oil spills, plastics, sewage, etc.), and grooming should be carefully addressed.

Presently, general beach monitoring involves regular collection and laboratory analysis of water samples for fecal indicator bacteria to assess recreational water quality and issue public advisories or closures (e.g., U.S. Environmental Protection Agency, 2014). Nonetheless, ecological monitoring of beaches, particularly utilizing native microbial species, is crucial for accurately assessing ecosystem health and detecting subtle environmental changes often missed by traditional methods. At the large temporal scale (over a decade), a positive correlation between beach disturbance level and obligate marine arenicolous fungal species frequencies of occurrence has been demonstrated (González and Hanlin, 2010). This poses species such as *Corollospora maritima* as a bioindicator for sandy beach degradation, providing a useful reference framework for beach monitoring and management.

5. Conclusion

Marine arenicolous fungi constitute a diverse ecological group of microfungi, characterized by their restricted distribution to sandy beach ecosystems. These frequently overlooked heterotrophs are adapted to cope with intertidal extreme and fluctuating conditions, providing key ecological functions in addition to further potential applications as bioindicators for beach health, offering valuable insights into the impacts of anthropogenic pressures on sandy beaches. However, their diversity is subtly declining in response to anthropogenic stressors. Even though fungi have unique conservation demands (Arnolds, 2001), the specific dangers to marine mycobiota have not been comprehensively identified. So, a holistic social-ecological system approach is crucial for sandy beaches, integrating a multidimensional framework that extensively considers the native microbial community, including marine arenicolous fungi. Key impacts such as coastal squeeze, pollution, beach grooming, and climate stressors should be addressed, to ensure the long-term resilience and functionality of these vital coastal ecosystems. In addition, the generation of baseline data (monitoring programs) at multiple spatial and temporal scales is crucial, since disturbances act across several dimensions. These efforts could provide early indicators of ecosystem stress and degradation.

Acknowledgements

The authors thank the Programa de Apoyo a Proyectos de Investigación e Innovación Tecnológica DGAPA-PAPIIT IN203524, and the Secretaría de Ciencia, Humanidad, Tecnología e Innovación (SECIHTI) for funding. Aurora Rivera-Zizumbo is a student from the Posgrado en Ciencias Biológicas, Universidad Nacional Autónoma de México (UNAM) and received a fellowship (1326072) from CONAHCYT.

References

Arnolds, E. (2001). The future of fungi in Europe: Threats, conservation, and management. pp. 64–80. *In*: Moore, D., Nauta, M.M., Evans, S.E. and Rotheroe, M. (eds.). Fungal Conservation: Issues and Solutions. Cambridge University Press, Cambridge.

Beena, K.R., Raviraja, N.S., Arun, A.B. and Sridhar, K.R. (2000) Diversity of arbuscular mycorrhizal fungi on the coastal sand dunes of the west coast of India. Curr. Sci., 79(10): 1459–1466.

Brown, A.C. and McLachlan, A. (2002). Sandy shore ecosystems and the threats facing them: Some predictions for the year 2025. Environ. Conserv., 29(1): 62–77.

Defeo, O. and McLachlan, A. (2005a). Patterns, processes, and regulatory mechanisms in sandy beach macrofauna: A multi-scale analysis. Mar. Ecol. Prog. Ser., 295: 1–20.

Defeo, O. and McLachlan, A. (2005b). Coastal Ecology: An Introduction to the Ecology of Coastal Habitats. Oxford University Press.

Defeo, O., McLachlan, A., Schoeman, D.S., Schlacher, T.A., Dugan, J., Jones, A. et al. (2009). Threats to sandy beach ecosystems: A review. Estuar. Coast. Shelf. Sci., 81(1): 112. https://doi.org/10.1016/j.ecss.2008.09.022.

Defeo, O., McLachlan, A., Armitage, D., Elliott, M. and Pittman, J. (2021). Sandy beach social-ecological systems at risk: Regime shifts, collapses, and governance challenges. Front. Ecol. Environ., 19(10): 564–573. https://doi.org/10.1002/fee.2406.

Gladfelter, A.S., James, T.Y. and Stajich, J.E. (2019). Marine fungi. Curr. Biol., 29(6): R191–R195. https://doi.org/10.1016/j.cub.2019.01.048.

González, M.C. and Hanlin, R.T. (2010). Potential use of marine arenicolous ascomycetes as bioindicators of ecosystem disturbance on sandy Cancun beaches: *Corollospora maritima* as a candidate species. Bot. Mar., 53(6): 577–580. https://doi.org/10.1515/bot.2010.073.

Harris, L., Campbell, E.E., Nel, R. and Schoeman, D. (2014). Rich diversity, strong endemism, but poor protection: Addressing the neglect of sandy beach ecosystems in coastal conservation planning. Divers. Distrib., 20: 1120–1135. https://doi.org/10.1111/ddi.12226.

Jones, E.B.G., Alias, S.A., Pang, K.L. and Boekhout, T. (2019). Marine Fungi: And Fungal-like Organisms. De Gruyter. https://doi.org/10.1515/9783110648255.

Kishimoto, R.A. and Baker, G.E. (1969). Pathogenic and potentially pathogenic fungi isolated from beach sands and selected soils of Oahu, Hawaii. Mycologia, 61(3): 537–548.

Kohlmeyer, J. (1966). Ecological observations on arenicolous marine fungi. Z. Allg. Mikrobiol., 6(2): 95–106.

Kohlmeyer, J. and Kohlmeyer, E. (1979). Marine Mycology: The Higher Fungi. Academic Press.

McLachlan, A. (2001). Coastal beach ecosystems. pp. 741–751. *In*: Lewin, R. (ed.). Encyclopedia of Biodiversity. New York, Academic Press.

McLachlan, A. and Brown, A.C. (2006). The Ecology of Sandy Shores. Academic Press.

McLachlan, A. and Defeo, O. (2018). The Ecology of Sandy Shores (3rd Edn.). Academic Press.

McLachlan, A. and Turner, I. (1994). The Interstitial Environment of Sandy Beaches. Mar. Ecol., 15: 177–211.

McLachlan, A., Defeo, O., Jaramillo, E. and Short, A.D. (2013). Sandy beach conservation and recreation: guidelines for optimising management strategies for multi-purpose use. Ocean Coast Manage., 71: 256–268. https://doi.org/10.1016/j.ocecoaman.2012.10.005.

Nakagiri, A., Miyazaki, K. and Shino, R. (2025). Anamorph diversity and species differentiation with shifting of life cycle types in *Corollospora*, a genus of marine ascomycetes; with an emendation of teleomorph of *Varicosporina prolifera* and a description of a new anamorphic species. Mycoscience, 66(4): 1–16. https://doi.org/10.47371/mycosci.2025.04.004.

Okamoto, N., Keeling, P.J., Leander, B.S. and Tai, V. (2022). Microbial communities in sandy beaches from the three domains of life differ by microhabitat and intertidal location. Mol. Ecol., 31: 3210–3227. https://doi.org/10.1111/mec.16453.

Pang, K.L., Overy, D.P., Jones, E.G., da Luz Calado, M., Burgaud, G. et al. (2016). 'Marine fungi' and 'marine-derived fungi' in natural product chemistry research: Toward a new consensual definition. Fungal Biol. Rev., 30(4): 163–175. https://doi.org/10.1016/j.fbr.2016.12.001.

Papadakis, J.A., Mavridou, A., Richardson, S.C., Lampiri, M. and Marcelou, U. (1997). Bather-related microbial and yeast populations in sand and seawater. Water Res., 31(4): 799–804.

Radziejewska, T., Kotta, J. and Kotwick, L. (2017). Sandy coasts. pp. 457–482. *In*: Snoeijs-Leijonmalm, P., Schubert, H. and Radziejewska, T. (eds.). Biological Oceanography of the Baltic Sea. Netherlands, Springer. https://doi.org/10.1007/978-94-007-0668-2_12.

Salvo, V.S. and Fabiano, M. (2007). Mycological assessment of sediments in Ligurian beaches in the Northwestern Mediterranean: Pathogens and opportunistic pathogens. J. Environ. Manage., 83(3): 365–369. https://doi.org/10.1016/j.jenvman.2006.04.001.

Schlacher, T.A., Schoeman, D.S., Dugan, J.E., Lastra, M., Jones, A., Scapini, F. and McLachlan, A. (2008). Sandy beach ecosystems: Key features, management challenges, climate change impacts, and sampling issues. Mar. Ecol., 29(Suppl. 1): 70–90. https://doi.org/10.1111/j.1439-0485.2007.00204.x.

Schneider, D., Engelhaupt, M., Allen, M., Huber, J.A. and Kelley, S.T. (2011). Biogeography and diversity patterns of intertidal microbial communities along a latitudinal gradient. FEMS Microbiology Ecology, 77(2): 190–203. https://doi.org/10.1111/j.1574-6941.2011.01109.x.

Short, A.D. (1996). The role of wave height, period, slope, tide range, and embaymentisation in beach classification: A review. Rev. Chil. Hist. Nat., 69: 589–604.

Stevens, J.L., Evans, G.E. and Aguirre, K.M. (2012). Human beach use affects abundance and identity of fungi present in sand. J. Coast. Res., 28(4): 787–792. https://doi.org/10.2307/23259019.

Tsui, K.M., Fryar, S.C., Hodgkiss, I.J., Hyde, K.D., Poonyth, A.D. and Taylor, J.E. (1998). The effect of human disturbance on fungal diversity in the tropics. Fungal Divers., 1: 19–26.

U.S. Environmental Protection Agency (2014). National Beach Guidance and Required Performance Criteria for Grants, 2014 Edition. EPA-823-B-14-001 (Retrieved from https://www.regulations.gov/document/EPA-HQ-OW-2015-0641-0017).

Velez, P., González, M.C., Rosique-Gil, E., Cifuentes, J., del Rocío Reyes-Montes, M. et al. (2013). Community structure and diversity of marine ascomycetes from coastal beaches of the southern Gulf of Mexico. Fungal Ecology, 6(6): 513–521. https://doi.org/10.1016/j.funeco.2013.10.002.

Velez, P., Gonzalez, M.C., Capello-Garcia, S., Rosique-Gil, E. and Hanlin, R.T. (2015a). Diversity of marine ascomycetes from the disturbed sandy beaches of Tabasco, Mexico. J. Mar. Biol. Ass. UK., 95(5): 897–903. https://doi.org/10.1017/S0025315414002112.

Velez, P., Gonzalez, M.C., Cifuentes, J., Rosique-Gil, E. and Hanlin, R.T. (2015b). Diversity of sand inhabiting marine ascomycetes in some tourist beaches on Cozumel Island, Mexico. Mycoscience, 56(2): 136–140. https://doi.org/10.1016/j.myc.2014.04.007.

Velez, P., Alejandri-Ramírez, N.D., González, M.C., Estrada, K.J., Sanchez-Flores, A. and Dinkova, T.D. (2015c). Comparative transcriptome analysis of the cosmopolitan marine fungus *Corollospora maritima* under two physiological conditions. G3 Genes|Genomes|Genetics, 5(9): 1805–1814. https://doi.org/10.1534/g3.115.019620.

Velez, P., Walker, A.K., Gasca-Pineda, J., Barrios, A., Divanli, D. et al. (2021). Fine-scale temporal variation of intertidal marine fungal community structure: Insights from an impacted Baja California sandy beach in Mexico. Mar. Biodiver., 51(1): 6. https://doi.org/10.1007/s12526-020-01148-7.

Velez, P., Walker, A.K., González, M.C., Narayanan, S.S.S. and Nakagiri, A. (2022). In depth review of the ecology of arenicolous marine fungi. Fungal Ecology, 60: 101164. https://doi.org/10.1016/j.funeco.2022.101164.

Vousdoukas, M.I., Ranasinghe, R., Mentaschi, L., Plomaritis, T.A., Athanasiou, P. et al. (2020). Sandy coastlines under threat of erosion. Nat. Clim. Chang., 10: 260–263. https://doi.org/10.1038/s41558-020-0697-0.

Genomics and Phylogeny

4

Marine Fungal Genomics and Biosynthetic Gene Clusters

Micael F.M. Gonçalves, Nuno Reisinho* and *Tiago M. Gonçalves*

Marine fungi represent an underexplored source of genetic and chemical diversity, with vast potential for natural product discovery. Advances in sequencing and genome mining technologies have revealed that marine fungi harbor a remarkable abundance of biosynthetic gene clusters (BGCs), many of which are distinct or different from those found in terrestrial species. These clusters encode the enzymatic machinery for the biosynthesis of secondary metabolites, such as polyketides, non-ribosomal peptides, terpenes, and hybrid molecules. By interfering with microbial interactions, chemical defense, and adaptation to extreme conditions, these metabolites play important ecological roles in marine environments.

The integration of bioinformatics pipelines with multi-omics approaches and innovative biosynthetic cluster activation strategies has accelerated the identification and functional characterization of marine fungal BGCs. These methods have started to reveal new compounds with antibacterial, antifungal, and anticancer properties by unlocking cryptic biosynthetic pathways. Examples that demonstrate both evolutionary conservation and diversification of biosynthetic potential in marine fungi include highly conserved BGCs for metabolites like clavaric acid and choline derivatives, as well as more divergent clusters that may encode analogs, such as Squalestatin S1 and Metachelin C. The identification of these compounds underscores the pharmaceutical and biotechnological potential of these organisms.

Centre for Environmental and Marine Studies, Department of Biology, University of Aveiro, 3810-193 Aveiro, Portugal.
* Corresponding author: mfmg@ua.pt

DOI: 10.1201/9781003768708-6

This chapter emphasizes the key role that BGC pathways play in fostering natural product innovation, highlighting current knowledge, methodological advancements such as bioinformatics analysis, and future perspectives in marine fungal genomics.

1. Introduction

Marine fungi represent an underexplored reservoir of biological and chemical diversity, increasingly recognized for their ecological roles and biosynthetic capabilities. These organisms inhabit a wide range of marine environments, from coastal mangroves and coral reefs to deep-sea sediments and hydrothermal vents, and contribute significantly to nutrient cycling, symbiosis, and organic matter degradation (Pang et al., 2016; Gonçalves et al., 2022). Advances in next-generation sequencing (NGS) technologies have opened new avenues for investigating their genomes, unveiling the presence of numerous biosynthetic gene clusters (BGCs) responsible for producing diverse secondary metabolites with pharmaceutical and biotechnological potential (Keller, 2019; Ji et al., 2025).

Traditionally, the discovery of fungal natural products has relied heavily on culture-dependent methods. However, many marine fungi remain uncultivable or produce metabolites only under specific environmental cues, leaving a substantial portion of their biosynthetic potential silent or cryptic under standard laboratory conditions (Rutledge and Challis, 2015; Zhang et al., 2024). Genomics, combined with transcriptomics and metabolomics, now provides a powerful framework to systematically identify and activate these silent BGCs, even from non-model organisms. The application of genome mining tools such as antiSMASH, PRISM, and SMURF has revealed that marine-derived fungal genomes often contain a significantly higher number and diversity of BGCs compared to their terrestrial relatives (Khaldi et al., 2010; Navarro-Muñoz et al., 2020; Blin et al., 2021; Gonçalves et al., 2021; Wei et al., 2025).

This chapter provides an overview of current insights into marine fungal genomics, emphasizing the identification, classification, and functional analysis of BGCs. We discuss key computational and experimental approaches to unlock the biosynthetic potential of marine fungi and explore the ecological relevance and biomedical applications of their secondary metabolites. By integrating genomic information with functional assays, researchers can better understand the adaptive evolution of marine fungal metabolism and harness it for natural product discovery.

2. Genomic Diversity of Marine Fungi

Marine fungi exhibit remarkable genomic diversity, reflecting their adaptation to a wide range of ecological niches and stressors in marine environments. Unlike terrestrial fungi, many marine species are exposed to high salinity, variable pressure and temperature, low nutrient availability, and complex biotic interactions, all of which have shaped their genomic architectures in unique ways (Zhong et al., 2024; Ji et al., 2025).

Marine fungi can be broadly classified into two groups: obligate marine fungi, which require a marine environment to complete their life cycle, and facultative marine fungi, which originate from terrestrial habitats but have adapted to marine conditions (Pang et al., 2016). While the genomes of many facultative marine fungi closely resemble their terrestrial relatives, obligate marine fungi often possess distinct genomic signatures, including expanded gene families related to osmotic stress response, halotolerance, and secondary metabolite biosynthesis (Jones and Pang, 2012; Amend et al., 2019).

Recent comparative genomics studies have revealed that marine fungal genomes tend to harbor a high number of BGCs, many of which are unique or poorly conserved among terrestrial species. This diversity suggests that marine fungi have evolved novel biosynthetic pathways, possibly as a result of intense ecological competition and niche specialization in marine ecosystems (Kempken, 2023). Moreover, gene clusters involved in polyketide and non-ribosomal peptide biosynthesis are often found in larger numbers and with greater structural diversity in marine-derived strains, particularly those isolated from extreme environments like deep-sea sediments or hypersaline habitats.

With the advancement of long-read sequencing technologies and hybrid genome assembly strategies, high-quality genomes of marine fungi are now becoming increasingly available. These advances facilitate the accurate prediction and annotation of BGCs and allow researchers to explore genomic islands, gene synteny, and horizontal gene transfer events, phenomena that appear to play important roles in shaping the secondary metabolism of marine fungi.

3. Biosynthetic Gene Clusters

3.1 What are Biosynthetic Gene Clusters?

BGCs are physically co-localized groups of genes within a genome that collectively encode the enzymatic machinery required for the biosynthesis of specialized or secondary metabolites (Fig. 1). Unlike primary metabolites, which are essential for cell survival (e.g., amino acids, nucleotides), secondary metabolites often serve ecological or defensive functions. BGCs typically include genes for core biosynthetic enzymes, tailoring enzymes (e.g., oxidases, glycosyltransferases), transporters, and regulatory proteins (Medema et al., 2015; Keller, 2019; Wei et al., 2025).

BGCs are highly modular and often encode complex enzyme systems such as:

- **Polyketide synthases (PKSs):** These enzymes assemble polyketide compounds through sequential condensation of acyl-CoA precursors, resulting in structurally diverse molecules with antibiotic, anticancer, or immunosuppressive activities (Cox, 2007).
- **Non-ribosomal peptide synthetases (NRPSs):** NRPSs synthesize peptide-based compounds independently of the ribosome, often incorporating unusual amino acids or modified residues, as seen in siderophores, toxins, and antifungals (Süssmuth and Mainz, 2017).

- **Terpene synthases (TSs):** Terpenes are derived from isoprene units and include volatile and aromatic compounds with ecological signaling roles, such as pheromones or antifoulants (Quin et al., 2014).
- **Hybrid clusters:** These BGCs combine PKS and NRPS modules or integrate additional biosynthetic pathways, leading to hybrid compounds with highly complex structures and multifunctional properties (Wei et al., 2025).

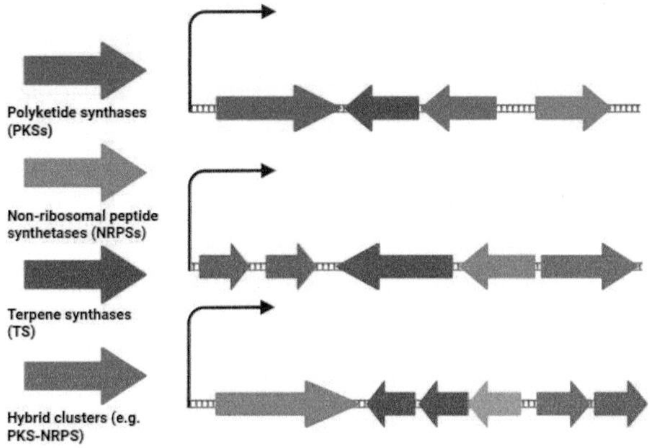

Fig. 1. Schematic representation of different types of biosynthetic gene clusters (BGCs).

Each row illustrates a distinct BGC class, including PKSs, NRPSs, TSs, and hybrid clusters (e.g., PKS-NRPS). Arrows in different shades represent genes encoding biosynthetic core enzymes, tailoring enzymes, regulatory proteins, and transporters, organized contiguously within the genome.

In marine-derived fungi, genome mining has revealed a substantial abundance and diversity of BGCs, many of which are either novel or significantly divergent from their terrestrial counterparts. This suggests that marine fungi represent a vast and underexplored reservoir of chemodiversity with unique ecological adaptations.

Figure 2 illustrates the distribution of BGC classes across sequenced and annotated marine fungal genomes. The most prevalent BGC types identified include Type I (T1PKS), NRPS, and terpene synthases, underscoring the dominance of these pathways in marine fungal secondary metabolism. The relative abundance of T1PKS and NRPS clusters, in particular, aligns with their central roles in producing bioactive compounds such as antibiotics, siderophores, and toxins.

To further investigate the conservation of biosynthetic pathways, Fig. 3 presents a heatmap of the 30 most representative BGCs across marine fungal species, showing both their prevalence and degree of sequence similarity. Among these, clusters associated with clavaric acid, and choline derivatives stand out due to their broad distribution and high sequence conservation, suggesting a widespread capacity among diverse marine fungi to produce these metabolites. In contrast, other clusters, such as those for mellein, 1,3,5,8-tetrahydroxynaphthalene, asperlactone, and aspergillic acid, appear to be species-specific or genus-specific. The chemical structures of the natural products mentioned are shown in Fig. 4.

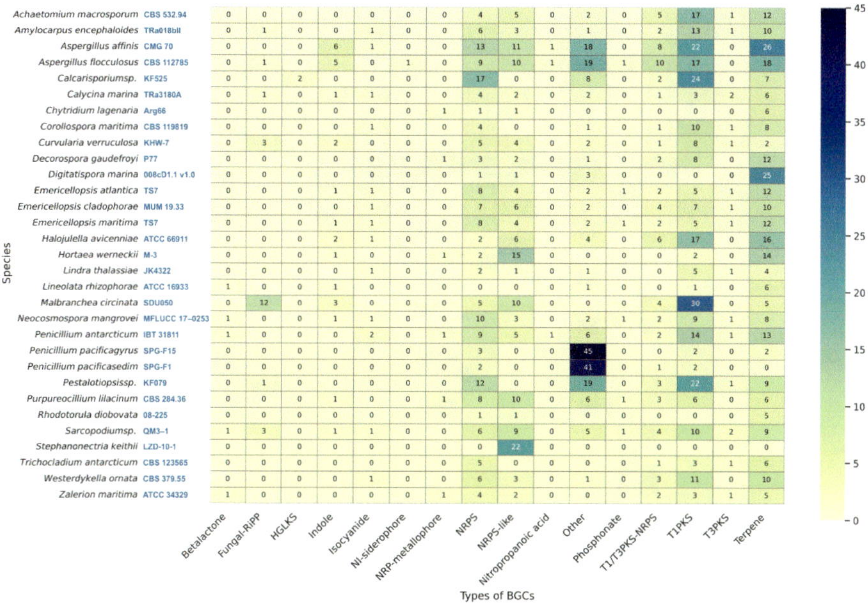

Fig. 2. Heatmap showing the distribution of BGCs across marine fungal species with available genomes and annotations. Each row represents a fungal isolate, identified by species name and strain designation. Columns correspond to different BGC types, ordered alphabetically. The numbers within each cell indicate the count of BGCs of that type found in the respective species. "Other" includes hybrid or rare BGCs found in a limited number of species. Color intensity reflects the relative abundance of BGCs, using a YlGnBu color scale.

Clavaric acid is a triterpenoid compound with promising anticancer potential, primarily through its action as a farnesyltransferase inhibitor, blocking a critical step in the maturation of oncogenic Ras proteins (Lingham et al., 1998). The conserved presence of its BGC across marine fungi may indicate important ecological or signaling roles, possibly related to microbial competition or defense.

Choline-based metabolites contribute to osmotic balance and membrane fluidity (Eghtesadi et al., 2024), which are essential under the high-salinity and pressure conditions found in many marine environments. The conserved nature of these BGCs across fungal genomes reinforces their physiological importance in adapting to marine life.

Mellein is a polyketide lactone produced by various fungi, including marine-derived strains, and exhibits antimicrobial and antifungal activity (Reveglia et al., 2020). Its BGC has been detected in all sequenced genomes of *Emericellopsis* species, suggesting a core ecological role in microbial competition or defense, helping these species to inhibit rival microorganisms under challenging ecological conditions.

1,3,5,8-Tetrahydroxynaphthalene (THN) is a polyketide intermediate in fungal melanin biosynthesis, contributing to protection against UV radiation, oxidative stress, and other environmental challenges (Lino and Manini, 2022). THN-related

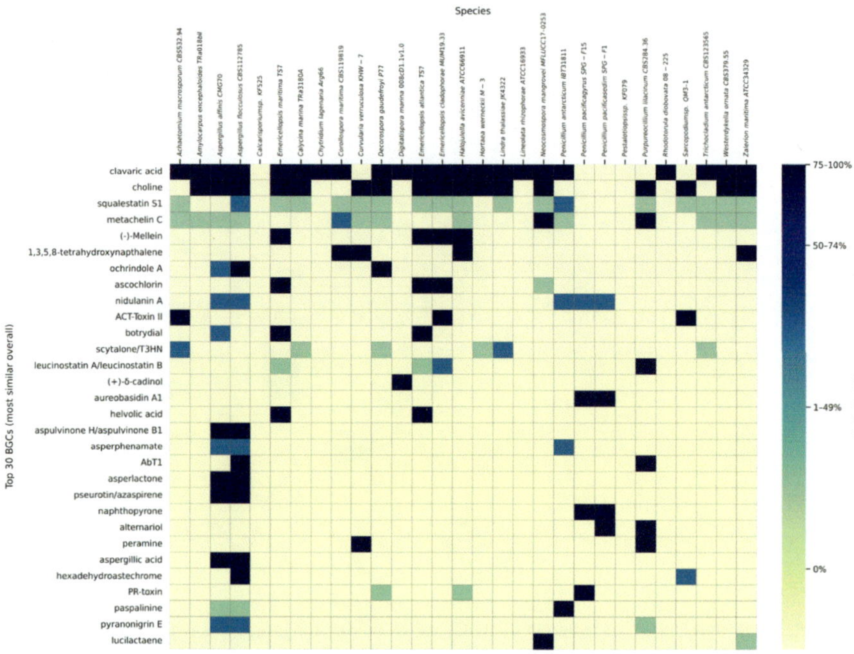

Fig. 3. Heatmap showing the similarity levels of the 30 most representative BGCs across marine fungal species with available genomes and annotations. Each cell indicates the degree of similarity based on sequence comparison. Species names are displayed at the top, and BGCs are listed on the left. Color intensity increases with higher similarity using the YlGnBu color palette.

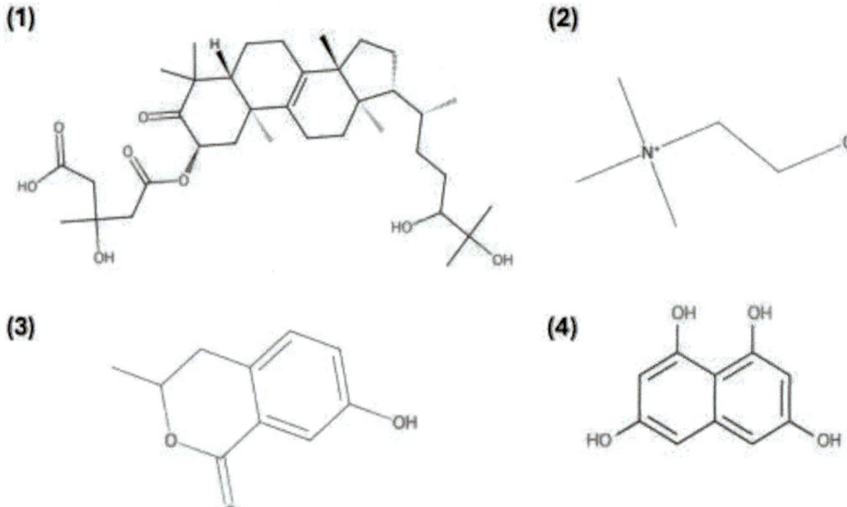

Fig. 4. Chemical structures of selected natural products: (1) Clavaric acid, (2) Choline, (3) Mellein, (4) 1,3,5,8-Tetrahydroxynaphthalene.

BGCs have been detected in a subset of marine fungal species, often correlating with darkly pigmented cultures (e.g., *Zalerion maritima*, *Corollospora maritima*), reflecting melanin accumulation. The presence of this BGC appears to be species-specific, suggesting a specialized ecological role rather than a broadly conserved function across marine fungi. In marine environments, THN-derived melanin may enhance survival under high light exposure, salinity fluctuations, and other stress conditions, providing a selective advantage in these challenging habitats.

In contrast, gene clusters corresponding to Squalestatin S1 and Metachelin C are also detected in nearly all analyzed genomes but exhibit low sequence similarity, suggesting they may encode structural analogues rather than exact biosynthetic matches.

Altogether, these patterns highlight a dual trend in marine fungal BGCs: the conservation of key biosynthetic capacities and the evolutionary divergence of cluster sequences, potentially giving rise to chemical novelty. This emphasizes the importance of coupling genomics with metabolomics and functional validation to fully unravel the natural product potential of marine fungi (Fig. 4).

3.2 *Ecological Function of Secondary Metabolites in Marine Environments*

Secondary metabolites produced by marine fungi play pivotal roles in mediating interactions within the marine ecosystem. Unlike terrestrial environments, marine habitats are characterized by high biodiversity, nutrient competition, and extreme or variable physicochemical conditions. In this context, fungal metabolites may function as:

- **Chemical defenses:** Many marine fungal metabolites exhibit potent antimicrobial, antifungal, or cytotoxic activity. These compounds help fungi deter competing microorganisms or predators (Hang and Jiang, 2024).
- **Allelopathic agents:** Some secondary metabolites modulate the growth of neighboring organisms, such as algae or bacteria, thereby influencing community structure and niche occupation (Imhoff, 2016).
- **Symbiosis Facilitators:** Fungi associated with marine invertebrates (e.g., algae, sponges, and corals) may produce metabolites that protect the host from pathogens or modulate host immune responses, contributing to mutualistic interactions (Gonçalves et al., 2021).
- **Stress Response Regulators:** Secondary metabolites can mitigate oxidative stress, salinity, or UV exposure, allowing fungi to survive in harsh or fluctuating marine conditions (Conrado et al., 2022).

The ecological roles of these compounds may also be reflected in the regulation and activation of BGCs, many of which are only expressed under specific environmental conditions or in the presence of other organisms. This phenomenon highlights the adaptive value of having multiple, often silent, BGCs that can be selectively turned on in response to ecological cues.

4. Bioinformatics Tools in BGC Discovery

The integration of advanced bioinformatics and multi-omics approaches has revolutionized the study of BGCs in marine fungi. Given the complexity, and often silent expression of BGCs, computational biology tools are essential for their detection, classification, and functional annotation. Furthermore, transcriptomic and proteomic datasets provide the means to verify BGC activity under specific conditions, linking genotype to phenotype.

The identification and annotation of BGCs in fungal genomes largely rely on automated tools designed to recognize conserved motifs, domain architectures, and gene cluster organization. Some of the most widely used software platforms include:

- **antiSMASH (Antibiotics & Secondary Metabolite Analysis Shell):** Perhaps the most comprehensive tool, antiSMASH predicts known classes of BGCs such as PKS, NRPS, terpenes, and RiPPs. It provides domain-level annotations, cluster boundaries, and comparative analyses with characterized BGCs in the MIBiG database (Blin et al., 2021).
- **fungiSMASH:** A fungal-specific version of antiSMASH that is optimized for fungal genomes, accounting for their unique cluster structures and tailoring enzymes.
- **SMURF:** One of the earliest tools developed for filamentous fungi, SMURF uses probabilistic models to identify core biosynthetic genes and associated tailoring enzymes (Khaldi et al., 2010).
- **PRISM (Prediction Informatics for Secondary Metabolomes):** Provides chemical structure prediction of BGC products based on domain architecture. It can be particularly useful for identifying novel or hybrid clusters (Skinnider et al., 2017).
- **BiG-SCAPE and CORASON:** These tools facilitate the comparative analysis of BGCs across multiple genomes. BiG-SCAPE clusters BGCs into gene cluster families (GCFs), while CORASON reconstructs evolutionary relationships among BGCs (Navarro-Muñoz et al., 2020).
- **Genome Mining Platforms:** Tools like Clinker, MultiGeneBlast, and EFI-GNT help visualize synteny and gene neighborhood conservation, supporting the annotation of uncharacterized clusters.

The use of long-read sequencing technologies (e.g., PacBio, Oxford Nanopore) has further improved genome assemblies, enabling the recovery of complete BGCs that were previously fragmented in short-read datasets (Sereika et al., 2025).

5. Expression and Activation of Cryptic BGCs

Despite the large number of BGCs revealed through genome mining in marine fungi, only a small fraction is expressed under standard laboratory conditions.

These "cryptic" or "silent" BGCs represent a major untapped reservoir of chemical diversity. Understanding and manipulating their expression is therefore a key frontier in fungal genomics and natural product discovery.

5.1 Challenges in Expressing Inactive BGCs Under Laboratory Conditions

Fungal BGCs are often tightly regulated and context dependent. In marine fungi, many BGCs remain transcriptionally silent due to the absence of natural ecological stimuli, such as competition, symbiosis, or environmental stress, in axenic cultures (Keller, 2019). Regulatory bottlenecks include:

- Promoter repression under laboratory growth conditions.
- Lack of specific transcriptional activators or environmental cues.
- Post-transcriptional and translational limitations, where mRNA is produced but not translated into active enzymes.
- Epigenetic silencing, including heterochromatin formation over BGC loci, particularly in secondary metabolite-rich fungi (Cichewicz, 2010).

These challenges necessitate both analytical (e.g., omics profiling) and experimental (e.g., induction strategies) approaches to unlock BGC expression.

5.2 Activation Techniques for Cryptic BGCs

Several methods have been developed to stimulate the expression of cryptic BGCs in fungi, often mimicking ecological or physiological triggers:

- **Co-cultivation (microbial interactions):** Cultivating marine fungi alongside bacteria or other fungi can simulate natural competition or symbiosis, triggering interspecific signaling that activates silent BGCs. For example, co-culture with marine actinomycetes led to the production of new polyketides and alkaloids in fungal strains (Bertrand et al., 2014).
- **Chemical Elicitation:** The addition of small molecules such as sodium butyrate, suberoylanilide hydroxamic acid (SAHA), or 5-azacytidine can alter chromatin structure or signaling pathways, resulting in the activation of silent genes (Williams et al., 2008). These compounds act as histone deacetylase (HDAC) inhibitors or DNA methylation inhibitors, suspending epigenetic repression.
- **Epigenetic Modulation:** Targeted manipulation of chromatin-modifying genes, such as deletion of *LaeA* or overexpression of global regulators like *VeA*, has been used to modulate BGC activity. Genetic engineering of epigenetic switches has shown promise in terrestrial and marine-derived fungi (Pfannenstiel and Keller, 2019).
- **Media Engineering and Stress Induction:** Altering salinity, pH, carbon/nitrogen sources, or applying oxidative or UV stress can mimic marine

environmental pressures, leading to conditional BGC expression. This is often combined with transcriptomic monitoring to identify responsive clusters.

5.3 Heterologous Expression of Marine Fungal BGCs

When native expression systems fail, heterologous expression offers a powerful alternative. BGCs can be cloned and transferred into genetically tractable model hosts such as:

- *Aspergillus nidulans* or *A. oryzae* (well-developed tools for fungal promoters, gene editing, and chromatin remodeling).
- *Saccharomyces cerevisiae* (used for expressing small or modular clusters with synthetic promoters).
- *Penicillium chrysogenum* or *Fusarium* species, depending on compatibility and target product.

Heterologous expression involves:

- Reconstructing entire BGCs with appropriate promoters and terminators.
- Codon optimization, especially when expressing in distantly related hosts.
- Fusion of regulatory elements, like strong constitutive or inducible promoters, to bypass native repression.
- Integration of multi-omics data (e.g., transcriptomics to guide construct design, proteomics to confirm expression).

These approaches allow not only the production of novel compounds under controlled conditions but also biosynthetic pathway elucidation, enabling future synthetic biology applications (Alberti et al., 2017).

6. Natural Products Derived from Marine Fungi

Marine fungi have emerged as prolific sources of structurally diverse and biologically potent secondary metabolites. These natural products (NPs) play key ecological roles in marine ecosystems but also hold immense potential in biotechnology and drug discovery.

Several classes of bioactive secondary metabolites have been isolated from marine-derived fungal strains, including species from *Aspergillus*, *Penicillium*, *Fusarium*, and *Trichoderma*. Notable examples include antibiotics (e.g., stachybotrin D), antitumor agents (e.g., plinabulin, ophiobolins, and psychrophilins), antifungals (e.g., griseofulvin, and penicillides), anti-inflammatory (e.g., flavoglaucin, spiromaterpenes, and amphichopyrones), insecticidal (e.g., asperalins), and antivirals (e.g., talaromycins) (Zorofchian Moghadamtousi et al., 2015; Imhoff, 2016; Qin et al., 2009; Nies et al., 2020; Shreadah et al., 2020; Yuan et al., 2022; Zeng et al., 2024).

6.1 Relationship between Chemical Structure and BGC Type

Marine fungal BGCs typically encode enzymes that determine the backbone and modification patterns of the natural product:

- PKS clusters produce polyketides, characterized by their polyacyl carbon backbones and often macrocyclic lactone rings (e.g., *Aspergillus* sp. F00785 from marine sediments produces asperentin (a polyketide with antifungal activity against *Colletotrichum gleosporioides* and *Botrytis cinerea*) (Tang et al., 2014), and lovastatin analogues have also been reported from marine-derived *Aspergillus terreus* (Na et al., 2024)).
- NRPS clusters give rise to peptidic scaffolds, often cyclized and with heteroatoms (e.g., *Penicillium* sp. from deep ocean sediment yields roquefortine C (a diketopiperazine alkaloid with neurotoxic activity) (Du et al., 2009), while *Aspergillus* sp. from sponges produces gliotoxin (Wong Chin et al., 2025)).
- Hybrid PKS–NRPS clusters synthesize complex mixed scaffolds (e.g., Fusarin derivatives and steckfusarins from marine algicolous fungus *Penicillium steckii* SCSIO41040 combine a polyketide backbone with peptide-derived moieties (Song et al., 2023)).
- Terpene synthase clusters are responsible for cyclic sesquiterpenes and diterpenes (e.g., Eremophilane-type sesquiterpenes isolated from the marine sediment fungus *Emericellopsis maritima* BC17 (Virués-Segovia et al., 2023)).
- Indole alkaloids derive from indole prenyltransferase-containing BGCs (e.g., Tryptoquivalines and brevianamides from marine *Aspergillus fumigatus* exhibit cytotoxic and antiviral activities (Afiyatullov et al., 2005)).
- The presence of tailoring enzymes, like halogenases, methyltransferases, and epoxidases, adds structural diversity and bioactivity. (e.g., Halogenated azaphilones from marine *Chaetomium globosum* (chlorinated polyketides) (Gao et al., 2023)).

6.2 Potential for Pharmaceutical and Biotechnology Applications

Marine fungal NPs represent a relatively underexploited but highly promising reservoir for:

- New antibiotics to counteract multidrug resistance (MDR), particularly against methicillin-resistant *Staphylococcus aureus* (MRSA), *Enterococcus*, and *Acinetobacter* species (e.g., Aspergillomarasamine A from *Aspergillus versicolor* acts as a metallo-β-lactamase inhibitor restoring carbapenem activity (Rotondo and Wright, 2024)).
- Anticancer agents with novel mechanisms of action (e.g., Plinabulin, a diketopiperazine from *Aspergillus* sp., disrupts microtubules and is in clinical trials (Mohanlal et al., 2018)).
- Antioxidants, anti-inflammatories, and UV-protective agents for cosmeceutical and nutraceutical development (e.g., Chaetopyranin (a polyketide from *C. globosum*) exhibits strong antioxidant activity (Wang et al., 2006)).

- Biopesticides and algicides for marine aquaculture and agriculture (e.g., Spirocyclic drimanes from marine *Stachybotrys chartarum* display algicidal activity (Wu et al., 2014)).

The ability to activate silent BGCs, combined with synthetic biology tools, is enabling the development of engineered strains for overproduction, structural analog synthesis, and biosynthetic optimization.

Fungi from extreme marine habitats such as deep-sea sediments, hydrothermal vents, cold seeps, and high-pressure zones have attracted growing attention due to their unique genetic and metabolic features. Metagenomic and single-cell genomics approaches are further expanding our view of unculturable fungal diversity, revealing thousands of unexplored BGCs in under-characterized clades.

7. Challenges and Future Perspectives

Despite the promising biosynthetic potential of marine fungi, several technical limitations continue to hinder full genomic and metabolic exploitation:

- **Genome Assembly and Annotation:** High levels of genome fragmentation in marine-derived strains, often due to complex secondary metabolite gene clusters and repetitive sequences, complicate accurate BGC prediction (Kautsar et al., 2020).
- **Species Identification and Cultivation:** Many marine fungi remain uncultured or are difficult to grow under laboratory conditions. Taxonomic ambiguity due to limited reference genomes also hinders comparative analysis (Jones and Pang, 2012).
- **Low Expression of BGCs:** Many BGCs remain transcriptionally silent or weakly expressed under standard culture conditions, making metabolite detection and characterization challenging (Zhang et al., 2024).
- **Limited Functional Validation:** Even when BGCs are identified, linking them to specific products often requires complex heterologous expression systems or in vivo gene knockouts, which remain technically demanding for non-model marine species.

8. Conclusion

Marine fungal genomics is entering a transformative era, driven by innovations in sequencing technologies, bioinformatics, and systems biology. As we continue to mine fungal genomes for BGCs, a vast and largely untapped reservoir of chemically novel and biologically active natural products is being revealed.

BGCs are at the heart of this discovery process, acting as molecular blueprints for secondary metabolite biosynthesis. The identification, characterization, and engineering of these clusters have already led to promising lead compounds with antimicrobial, anticancer, and anti-inflammatory activities.

However, unlocking the full biosynthetic potential of marine fungi will require overcoming persistent barriers. Interdisciplinary collaboration, open-access databases, and responsible bioprospecting frameworks are essential to ensure that future research is both impactful and sustainable.

Ultimately, the integration of genomics, transcriptomics, synthetic biology, and marine ecology holds the promise of transforming marine fungi from obscure microorganisms into cornerstones of next-generation drug discovery and biotechnology.

Acknowledgements

Micael F.M. Gonçalves thanks the FCT – Fundação para a Ciência e a Tecnologia I.P., under the project/grant UID/50006 + LA/P/0094/2020 (doi.org/10.54499/LA/P/0094/2020) and his contract 2022.00758.CEECIND/CP1720/CT0051 (doi.org/10.54499/2022.00758.CEECIND/CP1720/CT0051).

The authors recognize the use of artificial intelligence tools to assist in language editing and improve manuscript readability.

References

Afiyatullov, S.S., Kalinovskii, A.I., Pivkin, M.V., Dmitrenok, P.S. and Kuznetsova, T.A. (2005). Alkaloids from the marine isolate of the fungus *Aspergillus fumigatus*. Chem. Nat. Compd., 41: 236–238. https://doi.org/10.1007/s10600-005-0122-y.

Alberti, F., Foster, G.D. and Bailey, A.M. (2017). Natural products from filamentous fungi and production by heterologous expression. Appl. Microbiol. Biotechnol., 101: 493–500. https://doi.org/10.1007/s00253-016-8034-2.

Amend, A.S., Barshis, D.J. and Oliver, T.A. (2019). Coral-associated marine fungi form novel lineages and heterogeneous assemblages. Fungal Ecol., 39: 200–208. https://doi.org/10.1038/ismej.2011.193.

Bertrand, S., Bohni, N., Schnee, S., Schumpp, O., Gindro, K. and Wolfender, J.L. (2014). Metabolite induction via microorganism co-culture: A potential way to enhance chemical diversity for drug discovery. Biotechnol. Adv., 32(6): 1180–1204. https://doi.org/10.1016/j.biotechadv.2014.03.001.

Blin, K., Shaw, S., Steinke, K., Villebro, R., Ziemert, N., Lee, S.Y., Medema, M.H. and Weber, T. (2021). antiSMASH 6.0: Improving cluster detection and comparison capabilities. Nucleic Acids Res., 49(W1): W29–W35. https://doi.org/10.1093/nar/gkab335.

Cichewicz, R.H. (2010). Epigenome manipulation as a pathway to new natural product scaffolds and their congeners. Nat. Prod. Rep., 27(1): 11–22. https://doi.org/10.1039/b920860g.

Conrado, R., Gomes, T.C., Roque, G.S.C. and De Souza, A.O. (2022). Overview of bioactive fungal secondary metabolites: Cytotoxic and antimicrobial compounds. Antibiot., 11(11): 1604. https://doi.org/10.3390/antibiotics11111604.

Cox, R.J. (2007). Polyketides, proteins, and genes in fungi: Programmed nano-machines begin to reveal their secrets. Org. Biomol. Chem., 5(13): 2010–2026. https://doi.org/10.1039/b704420h.

Du, L., Li, D., Zhu, T., Cai, S., Wang, F., Xiao, X. and Gu, Q. (2009). New alkaloids and diterpenes from a deep ocean sediment derived fungus *Penicillium* sp. Tetrahedron, 65: 1033–1039. https://doi.org/10.1016/j.tet.2008.11.078.

Eghtesadi, N., Olaifa, K., Pham, T.T., Capriati, V., Ajunwa, O.M. and Marsili, E. (2024). Osmoregulation by choline-based deep eutectic solvent induces electroactivity in *Bacillus subtilis* biofilms. Enz. Microbial. Technol., 180: 110485. https://doi.org/10.1016/j.enzmictec.2024.110485.

Gao, Z.J., Cao, L.L., Ren, H.P., Yu, H. and Wang, Y. (2023). Novel chlorinated and nitrogenated azaphilones with cytotoxic activities from the marine algal-derived fungus *Chaetomium globosum* 2020HZ23. Front Microbiol., 14: 1252563. https://doi.org/10.3389/fmicb.2023.1252563.

Gonçalves, M.F.M., Esteves, A.C. and Alves, A. (2022). Marine fungi: Opportunities and challenges. Encyclopedia, 2(1): 559–577. https://doi.org/10.3390/encyclopedia2010037.

Gonçalves, M.F.M., Hilário, S., Tacão, M., Van de Peer, Y., Alves, A. and Esteves, A.C. (2021). Genome and metabolome MS-based mining of a marine strain of *Aspergillus affinis*. J. Fungi, 7(12): 1091. https://doi.org/10.3390/jof7121091.

Gonçalves, M.F.M., Hilário, S., Van de Peer, Y., Esteves, A.C. and Alves, A. (2021). Genomic and metabolomic analyses of the marine fungus *Emericellopsis cladophorae*: Insights into saltwater adaptability mechanisms and its biosynthetic potential. J. Fungi, 8(1): 31. https://doi.org/10.3390/jof8010031.

Hang, S., Lu, H. and Jiang, Y. (2024). Marine-derived metabolites act as promising antifungal agents. Mar. Drugs, 22(4): 180. https://doi.org/10.3390/md22040180.

Imhoff, J.F. (2016). Natural products from marine fungi: Still an underrepresented resource. Mar. Drugs, 14(1): 19. https://doi.org/10.3390/md14010019.

Ji, R., Zha, X. and Zhou, S. (2025) Marine fungi: A prosperous source of novel bioactive natural products. Curr. Med. Chem., 32(5): 992–1006. https://doi.org/10.2174/0109298673266304231015070956.

Jones, E.B.G. and Pang, K.L. (2012). Marine Fungi: And Fungal-like Organisms. Walter de Gruyter GmBH & Co KG, Göttingen, Germany, 543 pp.

Kautsar, S.A., Blin, K., Shaw, S., Navarro-Muñoz, J.C., Terlouw, B.R. et al. (2020). MIBiG 2.0: A repository for biosynthetic gene clusters of known function. Nuc. Acids Res., 48(D1): D454–D458. https://doi.org/10.1093/nar/gkz882.

Keller, N.P. (2019). Fungal secondary metabolism: Regulation, function, and drug discovery. Nat. Rev. Microbiol., 17(3): 167–180. https://doi.org/10.1038/s41579-018-0121-1.

Kempken, F. (2023). Marine fungi: A treasure trove of novel natural products and for biological discovery. PLoS Pathog., 19(9): e1011624. https://doi.org/10.1371/journal.ppat.1011624.

Khaldi, N., Seifuddin, F.T., Turner, G., Haft, D., Nierman, W.C. et al. (2010). SMURF: Genomic mapping of fungal secondary metabolite clusters. Fungal Gen. Biol., 47(9): 736–741. https://doi.org/10.1016/j.fgb.2010.06.003.

Lingham, R.B., Silverman, K.C., Jayasuriya, H., Kim, B.M., Amo, S.E. et al. (1998). Clavaric acid and steroidal analogues as Ras-and FPP-directed inhibitors of human farnesyl-protein transferase. J. Med. Chem., 41(23): 4492–4501. https://doi.org/10.1021/jm980356+.

Lino, V. and Manini, P. (2022). Dihydroxynaphthalene-based allomelanins: A source of inspiration for innovative technological materials. ACS Omega, 7(18): 15308–15314. https://doi.org/10.1021/acsomega.2c00641.

Mohanlal, R.W., Lloyd, K. and Huang, L. (2018). Plinabulin, a novel small molecule clinical stage IO agent with anti-cancer activity, to prevent chemo-induced neutropenia and immune related AEs. J. Clin. Oncol., 36: 126. https://doi.org/10.1200/JCO.2018.36.5_suppl.126.

Na, H., Zheng, Y.Y., Jia, Y., Feng, J., Huang, J. et al. (2024). Screening and genetic engineering of marine-derived *Aspergillus terreus* for high-efficient production of lovastatin. Microb. Cell. Fact., 23(1): 134. https://doi.org/10.1186/s12934-024-02396-z.

Navarro-Muñoz, J.C., Selem-Mojica, N., Mullowney, M.W., Kautsar, S.A., Tryon, J.H. et al. (2020). A computational framework to explore large-scale biosynthetic diversity. Nat. Chem. Biol., 16(1): 60–68. https://doi.org/10.1038/s41589-019-0400-9.

Nies, J., Ran, H., Wohlgemuth, V., Yin, W.B. and Li, S.M. (2020). Biosynthesis of the prenylated salicylaldehyde flavoglaucin requires temporary reduction to salicyl alcohol for decoration before reoxidation to the final product. Org. Lett., 22: 2256–2260. https://doi.org/10.1021/acs.orglett.0c00440.

Pang, K.L., Overy, D.P., Jones, E.B.G., Calado, M.L., Burgaud, G. et al. (2016). "Marine fungi" and "marine-derived fungi" in natural product chemistry research: Toward a new consensual definition. Fungal Biol. Rev., 30(4): 163–175. https://doi.org/10.1016/j.fbr.2016.08.001.

Pfannenstiel, B.T. and Keller, N.P. (2019). On top of biosynthetic gene clusters: How epigenetic machinery influences secondary metabolism in fungi. Biotech. Adv., 37(6): 107345. https://doi.org/10.1016/j.biotechadv.2019.02.001.

Quin, M.B., Flynn, C.M. and Schmidt-Dannert, C. (2014). Traversing the fungal terpenome. Nat. Prod. Rep., 31(10): 1449–1473. https://doi.org/10.1039/c4np00075g.

Reveglia, P., Masi, M. and Evidente, A. (2020). Melleins: Intriguing natural compounds. Biomolecules, 10(5): 772. https://doi.org/10.3390/biom10050772.

Rotondo, C.M. and Wright, G.D. (2024). Efficacy of aspergillomarasmine A/meropenem combinations with and without avibactam against bacterial strains producing multiple β-lactamases. AAC, 68(9): e00272-24. https://doi.org/10.1128/aac.00272-24.

Rutledge, P.J. and Challis, G.L. (2015). Discovery of microbial natural products by activation of silent biosynthetic gene clusters. Nat. Rev. Microbiol., 13(8): 509–523. https://doi.org/10.1038/nrmicro3496.

Sereika, M., Mussig, A.J., Jiang, C., Knudsen, K.S., Jensen, T.B.N. et al. (2025). Genome-resolved long-read sequencing expands known microbial diversity across terrestrial habitats. Nat. Microbiol., 10: 2018–2030. https://doi.org/10.1038/s41564-025-02062-z.

Shreadah, M.A., El Moneam, N.M., El-Assar, S.A. and Nabil-Adam, A. (2020). Metabolomics and pharmacological screening of *Aspergillus versicolor* isolated from *Hyrtios erectus* Red Sea sponge; Egypt. Curr. Bioactive Comp., 16(7): 1083–1102. https://doi.org/10.2174/157340721566619111122711.

Skinnider, M.A., Merwin, N.J., Johnston, C.W. and Magarvey, N.A. (2017). PRISM 3: Expanded prediction of natural product chemical structures from microbial genomes. Nuc. Acids Res., 45(W1): W49–W54. https://doi.org/10.1093/nar/gkx320.

Song, Y., She, J., Chen, W., Wang, J., Tan, Y. et al. (2023). New fusarin derivatives from the marine algicolous fungus *Penicillium steckii* SCSIO41040. Mar. Drugs, 21(10): 532. https://doi.org/10.3390/md21100532.

Süssmuth, R.D. and Mainz, A. (2017). Nonribosomal peptide synthesis: Principles and prospects. Angew. Chemie., 56(14): 3770–3821. https://doi.org/10.1002/anie.201609079.

Tang, Q., Guo, K., Li, X.Y., Zheng, X.Y., Kong, X.J. et al. (2014). Three new asperentin derivatives from the algicolous fungus *Aspergillus* sp. F00785. Mar. Drugs, 12(12): 5993–6002. https://doi.org/10.3390/md12125993.

Virués-Segovia, J.R., Millán, C., Pinedo, C., González-Rodríguez, V.E., Papaspyrou, S. et al. (2023). New eremophilane-type sesquiterpenes from the marine sediment-derived fungus *Emericellopsis maritima* BC17 and their cytotoxic and antimicrobial activities. Mar. Drugs, 21(12): 634. https://doi.org/10.3390/md21120634.

Wang, S., Li, X.M., Teuscher, F., Li, D.L., Diesel, A. et al. (2006). Chaetopyranin, a benzaldehyde derivative, and other related metabolites from *Chaetomium globosum*, an endophytic fungus derived from the marine red alga *Polysiphonia urceolata*. J. Nat. Prod., 69(11): 1622–1625. https://doi.org/10.1021/np060248n.

Williams, R.B., Henrikson, J.C., Hoover, A.R., Lee, A.E. and Cichewicz, R.H. (2008). Epigenetic remodeling of the fungal secondary metabolome. Org. Biomol. Chem., 6(11): 1895–1897. https://doi.org/10.1039/b804701d.

Wei, B., Ying, T.T., Lv, H.W., Zhou, Z.Y., Cai, H., Hu, G.A. et al. (2025). Global analysis of fungal biosynthetic gene clusters reveals the diversification of diketopiperazine biosynthesis. Bioresour. Technol., 422: 132218. https://doi.org/10.1016/j.biortech.2025.132218.

Wong Chin, J.M., Jeewon, R., Fahad Alrefaei, A., Puchooa, D., Bahorun, T. and Neergheen, V.S. (2025). Marine-derived fungi from the genus *Aspergillus* (*Ascomycota*) and their anticancer properties. Mycology, 16(2): 545–592. https://doi.org/10.1080/21501203.2024.2402309.

Wu, B., Oesker, V., Wiese, J., Malien, S., Schmaljohann, R. and Imhoff, J.F. (2014). Spirocyclic drimanes from the marine fungus *Stachybotrys* sp. strain MF347. Mar. Drugs, 12(4): 1924–1938. https://doi.org/10.3390/md12041924.

Yuan, S.W., Chen, L.T., Wu, Q.L., Jiang, M.H., Guo, H. et al. (2022). Genome mining of alpha-pyrone natural products from ascidian-derived fungus Amphichordafelina SYSU-MS7908. Mar. Drugs, 20: 294. https://doi.org/10.3390/md20050294.

Zeng, Y.J., Lu, T.T., Ren, S.Y., Hu, Z.B., Fang, J. et al. (2024). Biosynthesis of ester-bond containing quinolone alkaloids with (3R,4S) stereoconfiguration. Org. Lett., 26: 6692–6697. https://doi.org/10.1021/acs.orglett.4c02372.

Zhang, Z.F., Liu, F., Liu, L.R., Li, M., Cai, L. et al. (2024). Culturing the uncultured marine fungi in the omics age: Opportunities and challenges. Fungal Biol. Rev., 48: 100353.

Zhong, M., Li, Y., Deng, L., Fang, J. and Yu, X. (2024). Insight into the adaptation mechanisms of high hydrostatic pressure in physiology and metabolism of hadal fungi from the deepest ocean sediment. Msystems., 9(1): e01085-23. https://doi.org/10.1128/msystems.01085-23.

Zorofchian Moghadamtousi, S., Nikzad, S., Abdul Kadir, H., Abubakar, S. and Zandi, K. (2015). Potential antiviral agents from marine fungi: An overview. Mar. Drugs, 13(7): 4520–4538. https://doi.org/10.3390/md13074520.

5

Biodiversity and Systematics of Marine Fungi in Thailand

Mark S. Calabon,[1,2,*] *Jan Felnesh Exe Bagacay,*[1,2] *Lilcah Angelique D. Opiña,*[1,2] *Fred Ric C. Sara*[1] and *Majoerly Augustine B. Siccio*[1]

Marine fungi represent a vital, yet underexplored, component of tropical aquatic biodiversity, with Thailand serving as one of the key centers for their study in Asia. This chapter provides an overview of the biodiversity and systematics of marine fungi in Thailand, emphasizing their taxonomic diversity, molecular phylogenetics, geographical distribution, and substrates. Marine fungi in Thailand exhibit remarkable diversity and geographical variation across the country's coastal provinces. Surveys conducted along both the Andaman Sea and the Gulf of Thailand coasts have documented distinct regional patterns in species richness and substrate preference. Ranong Province supports the highest diversity, with 83 recorded species across 65 genera, followed by Trat (74 species, 56 genera). The majority of marine fungi are associated with mangrove ecosystems, colonizing senescent and decaying wood, driftwood, submerged seedlings, petioles, fronds, and roots of dominant mangrove species such as *Rhizophora apiculata*, *Avicennia marina*, and *Nypa fruticans.* Species of *Halosarpheia*, *Lignincola*, *Aniptodera*, *Saagaromyces*, *Savoryella*, and *Aigialus* represent the most speciose genera. To date, 334

[1] Division of Biological Sciences, College of Arts and Sciences, University of the Philippines Visayas, Miagao, Iloilo, Philippines.

[2] Graduate School, University of the Philippines Visayas, Miagao, Iloilo, Philippines.

* Corresponding author: mscalabon@up.edu.ph

DOI: 10.1201/9781003768708-7

marine fungal taxa have been recorded from Thailand, predominantly within the Ascomycota (317 species), with Sordariomycetes (178 species) and Dothideomycetes (98 species) as the dominant classes. Advances in molecular phylogenetics have further clarified the evolutionary relationships among key lineages of approximately 126 marine fungal species in Thailand and previously unrecognized diversity within tropical aquatic environments. Overall, this chapter underscores the significance of Thailand's coastal and mangrove ecosystems as reservoirs of marine fungal biodiversity and provides a framework for future taxonomic and molecular studies in tropical marine mycology.

1. Introduction

Marine fungi are a globally distributed ecological group, but their diversity and abundance are strongly shaped by geography and substrate specificity. They thrive across a wide range of habitats, from intertidal zones, mangrove forests, salt marshes, and open ocean to even deepsea environments (Raghukumar, 2017; Gonçalves et al., 2022). Marine fungi have been recorded on substrates such as submerged wood, decaying plant matter, algae, sea foam, and sediments. They also serve as important saprobes, pathogens, endosymbionts, and mutualists (Hyde et al., 1998; Shearer et al., 2007; Jones et al., 2019). Among the groups of marine fungi, lignicolous species have received the greatest attention. Lignicolous fungi are those that colonize submerged decaying wood but can also occur on herbaceous material like algae and sea grasses (Pang and Mitchell, 2005; Abreu et al., 2025). Large amounts of lignocellulose constantly enter marine environments through mangrove areas and coastal vegetation. Lignicolous fungi are the main organisms that undertake the decomposition of this recalcitrant material, especially in areas where marine borers cannot survive (Hyde et al., 1998). They are particularly well documented in temperate regions (Hughes, 1974; Jones, 1985; Koch and Jones, 1983; Koch and Petersen, 1996; Kohlmeyer and Kohlmeyer, 1979; Schaumann, 1975; Petersen and Koch, 1997) and in tropical mangrove ecosystems (Kohlmeyer, 1984; Jones and Alias, 1997; Sarma and Hyde, 2001; Schmidt and Shearer, 2003, 2004). Tropical habitats were found to support a higher diversity of marine fungi, but this also reflects the intensity of studies focused in this region (Jones et al., 2019). Overall, these studies demonstrate the ecological specialization of marine fungi and their crucial role in wood decomposition and nutrient cycling within coastal habitats.

Thailand has made significant contributions to the study of marine fungi. Its various habitats, particularly mangrove forests, intertidal zones, and coral reefs (Jones et al., 2006), provide ideal environments for marine fungi to flourish. This diversity prompted a continuous series of investigations that has spanned several decades. Early work by Kohlmeyer (1984) was followed by important surveys by Koch (1986), Hyde (1988, 1989a, b, 1992, 1995), Hyde et al. (1990), Chalermpongse (1991), Hyde and Jones (1992), Sundari et al. (1996), and Ito et al. (2001), which collectively established a baseline understanding of Thai marine fungal diversity.

Building on this foundation, Jones et al. (2006) introduced two novel mangrove-associated species and assembled an account of the marine fungal diversity at the time. More novel species have since been identified in the succeeding years (Dayarathne et al., 2016; Zhang et al., 2018; Zhang et al., 2019a; Zhang et al., 2019; Asghari et al., 2023; Dayarathne et al., 2020b; Wijesinghe et al., 2023; Preedanon et al., 2024; Mukhopadhyay et al., 2025; Apurillo et al., 2025), reinforcing Thailand's status as a hotspot for marine fungal biodiversity. These efforts have laid the groundwork for systematic and molecular studies, which continue to reveal the hidden diversity and evolutionary significance of marine fungi in the region.

2. Biodiversity

With its expansive geography and predominantly tropical and subtropical climates, Asia offers an extensive range of ecological niches, making it a global hotspot for fungal diversity. In 2017, the majority of new fungal species worldwide were recorded from Asia, accounting for about 35% of all new introductions (Hyde et al., 2020).

Regions in China, particularly Yunnan Province and the Greater Mekong Subregion, are also recognized as significant biodiversity hotspots for fungi (Ren et al., 2022), while areas like Zhanjiang Bay have shown that Ascomycota and Basidiomycota are the dominant phyla in marine sediments (Gao et al., 2024). Moreover, the utilization of ribosomal RNA internal transcribed spacer-2 (ITS2) metabarcoding revealed the fungal diversity across the coastal and mangrove areas of South China, reporting a high number of Ascomycota Operational Taxonomic Units (OTUs) (Li et al., 2016).

Meanwhile, fungal diversity studies in India were initiated in the mid-20th century, laying the taxonomic foundation by documenting species from decaying wood and mangroves in coastal regions (Borse, 2002; Kohlmeyer and Vittal, 1986). Since the late 1990s, India has expanded its marine fungal knowledge beyond traditional saprobes to include endophytic microfungi from mangroves, and through continued use of molecular methods, India leads in the number of documented mangrove-associated fungal taxa globally (Jones et al., 2006).

On the other hand, Japanese mycologists made notable contributions beginning in the mid-20th century, when lignicolous marine fungi from driftwood and submerged wood panels along the Japanese coast were cultured and catalogued (Tubaki, 1963, 1968; Nakagiri, 1989, 1993). Their work led to discovering numerous species from genera *Ceriosporopsis*, *Corollospora*, *Clavatospora*, and *Zalerion* (Tubaki, 1968), contributing significantly to the global understanding of marine fungal diversity, particularly in the Pacific.

Other Southeast Asian countries, such as the Philippines, Malaysia, and Vietnam, have also made notable contributions, often detailing the taxonomy and distribution of fungi on decaying wood and within mangrove plant tissues (Calabon et al., 2018; Su et al., 2014; Alias et al., 2010; Pang et al., 2010).

In contrast, there are historically fewer studies on the marine fungi of Thailand (Hyde et al., 1990). Early studies of fungal diversity in Thailand were documented by

Kohlmeyer (1984), wherein 15 species from Chonburi Province were documented. Koch (1986) reported 16 species from Phuket Island, two of which were novel species. Following these studies, Hyde (1989) observed 38 species of mangrove fungi in Phang Nga Bay, 26 of which were new records in the country. Apart from the documentation of marine fungi, several studies also reported fungal species from soil. Ito et al. (2001) documented 42 mangrove species collected from mangrove forest soils in the Phang-Nga and Ranong Research Center.

Comprehensive studies were made to further discover the marine fungal diversity in the country in the following years. In a survey of marine hyphomycetes across different sites in Southern and Eastern Thailand, a total of 44 taxa were collected, wherein one taxon was described as new to science (Chatmala et al., 2004). This survey suggests that previous studies may have underestimated the true number of fungal taxa in the region. Jones et al. (2006) were the first to catalog frequently occurring species and collect data from different provinces. This pioneering study reported 154 species of marine fungi, which comprised 116 species of Ascomycota, 3 Basidiomycota, 28 anamorphic fungi, and 7 Stramenophiles.

In the provinces of Chanthaburi, Rayong, and Chon Buri, Eastern Thailand, 31 species from 152 marine fungal isolates were reported by Dethoup and Manoch (2009), spanning 26 genera. In Southern Thailand, a total of 112 higher marine fungi, representing 79 ascomycetes, 3 basidiomycetes and 30 anamorphic fungi, were recorded from the Hat Khanom-Mu Ko Thale Tai National Park in Nakhon Si Thammarat Province (Sakayaroj et al., 2011). A study by Suetrong et al. (2017) also added to the growing list by investigating species collected from intertidal and mangrove forests in Eastern (Trat Province) and Southern Thailand (Prachuap Khiri Khan and Phetchaburi Provinces). A total of 99 fungal species were recorded, including 84 Ascomycota, 2 Basidiomycota, and 13 asexual fungi (Suetrong et al., 2017).

Thailand ranks among the most extensively studied regions for marine fungi in Asia, with numbers comparable to neighboring Asian countries. India currently has 207 marine species recorded, comprising 14 species of Labyrinthulomycota, 4 Chytridiomycota, 4 Oomycota, 139 Ascomycota, 3 Basidiomycota, and 43 mitosporic fungi (Borse et al., 2013). In Malaysia, 139 species have been reported from mangroves, including 115 species of Ascomycota, 2 species of Basidiomycota, and 22 anamorphic taxa (Alias et al., 2010). Similarly, China has reported 213 marine fungal species, with 117 species of Ascomycota, 1 species of Basidiomycota, 5 species of Oomycetes, 42 yeasts, and 48 hyphomycetous species (Hu et al., 2013). This high number recorded for China also includes at least 128 species from Hong Kong (Jones and Vrijmoed, 2003; Hu et al., 2013). In Japan, 112 marine fungi have been reported, with 65 from Ascomycota, 1 from Basidiomycota, and the remaining 18 species were anamorphic fungi (Abdel-Wahab, 2011). Among these records from various countries, submerged woody substrates support the majority of species (Alias et al., 2010; Abdel-Wahab, 2011; Hu et al., 2013), but a few species were recorded from substrates like deep sea sediments (Borse et al., 2013), sand, and seafoam (Abdel-Wahab, 2011). This collective body of work reflects the significant progress being made to document the marine fungal diversity in Asia.

The last decade has seen a substantial surge in the documentation of Thailand's marine fungal biodiversity, driven by extensive surveys across multiple provinces and diverse habitats. Mangrove and estuarine ecosystems have proven to be significant hotspots, yielding numerous novel taxa in the Diatrypaceae family, including the establishment of two new genera, *Halodiatrype* and *Halocryptovalsa*, and five new species from provinces like Phang-nga, Krabi, and Ranong (Dayarathne et al., 2020b; Mukhopadhyay, 2025). Concurrently, studies on palmicolous fungi advanced the documentation of marine fungal diversity. Surveys across provinces, including Ranong, Chanthaburi, and Krabi, led to the establishment of the novel family Striatiguttulaceae, which comprises two new genera, *Striatiguttula* and *Longicorpus* (Zhang, 2019b). This work also included the reclassification of a *Trichocladium*-like species into *Savoryella nypae* (Zhang et al., 2019a). Beyond introductions of novel families, other provincial surveys yielded unique taxa such as *Myrmecridium hydei* from Prachuap Khiri Khan and *Helicascus satunensis* from Satun (Asghari et al., 2023; Preedanon et al., 2024).

Globally, an updated count of 2,253 species of marine fungi is recorded, which spans across 36 classes, 117 orders, 301 families, and 881 genera (Jones et al., 2019). Booth and Kenkel (1986) suggested that sea temperature affects the geographic distribution of marine fungi. Hughes (1974) then published distribution maps of marine fungi based on average temperature range, which prompted several studies showing world distribution maps of marine fungi.

Marine fungal diversity varies significantly across global regions, with temperate and tropical environments showing substantially more documented communities compared to underexplored polar regions. Studies from the temperate North Carolina coast identified 2,305 OTUs, and the tropical Mauritian coast revealed 807 OTUs, with both regions being dominated by the phylum Ascomycota (Picard, 2017; Wong Chin et al., 2022). This tropical/temperate pattern, where Ascomycota often includes terrestrial immigrants like *Aspergillus* and *Penicillium* (Mouton et al., 2012), aligns with the global assessment that aquatic fungal biodiversity is greatest in temperate and tropical areas (Shearer et al., 2007). In contrast, a study in the High Arctic reported only 113 OTUs from marine sediments, with 62 belonging to Ascomycota, confirming the view that biodiversity is least in boreal regions and highlighting the need for further exploration in polar areas (Zhang et al., 2015; Hagestad et al., 2019).

3. Systematics

3.1 *Taxonomic Classification*

Marine fungi in Thailand are predominantly represented by Ascomycota (317 species), with few collections from Basidiomycota (15 species) and Mucoromycota (2 species: *Cunninghamella bertholletiae*, *C. blakesleeana* (Loilong et al., 2012)). Within Ascomycota, the most frequently reported classes are Sordariomycetes (178 species), Dothideomycetes (98 species), and Eurotiomycetes (26 species) (Fig. 1). Other classes are represented by fewer taxa, including Pichiomycetes (five species: *Candida metapsilosis*, *C. parapsilosis*, *C, spencermartinsiae*, *C, tropicalis*, *C. zeylanoides* (Kaewkrajay et al., 2020)), Saccharomycetes (*Kodamaea ohmeri*, *Meyerozyma caribbica*, *M. guilliermondii*, *Wickerhamomyces anomalus* (Kaewkrajay et al., 2020)), and Leotiomycetes (*Halenospora varia*, *Vibrissea nypicola* (Dethoup and Manoch, 2009; Jones et al., 2006b; Loilong et al., 2012).

The Sordariomycetes form the largest group of marine fungi in Thailand. Among these, the order Microascales is the most species-rich (61 species), followed by Xylariales (34 species), Hypocreales (20 species), and Lulworthiales (13 species) (Fig. 1). The other 15 orders each include fewer than 10 species. At the family level, Halosphaeriaceae dominate with 61 species, followed by Diatrypaceae (15 species), Xylariaceae (14 species), and Lulworthiaceae (12 species). The remaining 34 families are represented by fewer than 10 species each. At the genus level, *Corollospora* is the most speciose with 8 species, followed by *Aniptodera* (7 species), *Savoryella* (6 species), *Linocarpon* (5 species), and *Nemania* (5 species), while the other genera contain fewer than 5 species (Table 1). The taxon *Mangrovispora pemphii* has been placed within the Sordariomycetes, but its precise familial and ordinal affinities remain unresolved, and it is currently treated as Sordariomycetes genera *incertae sedis*.

Within the Dothideomycetes, taxa are distributed across 15 orders, 44 families, and 67 genera. The order Pleosporales is the most speciose with 65 species, followed by Dyfrolomycetales (8 species) and Tubeufiales (5 species), while the remaining 12 orders contain fewer than 5 species each (Fig. 1). The most diverse families include Pleurotremataceae (8 species), Morosphaeriaceae (6 species), and both Leptosphaeriaceae and Tubeufiaceae (5 species each), with the remaining 39 families represented by fewer than five species. Four taxa remain classified within Dothideomycetes genera *insertae sedis*, namely *Caryosporella rhizophorae* (Hyde et al., 1990; Hyde, 1994; Suetrong et al., 2017), *Clavariopsis bulbosa* (Dethoup and Manoch, 2009), *Monodictys pelagica* (Jones et al., 2006b), and *Passeriniella savoryellopsis* (Hyde et al., 1990; Hyde, 1994; Suetrong et al., 2017) (Table 1).

The order Eurotiales is the dominant lineage in Eurotiomycetes, accounting for 22 species recorded from Thailand. The family Aspergillaceae is particularly diverse, comprising 20 species. This includes representatives of *Aspergillus* (15 species), *Hamigera* (two species: *H. avellanea* and *H. terricola*), and *Penicillium* (two species: *P. citrinum* and *P. oxalicum*) (Dethoup and Manoch, 2009; Chalearmsrimuang et al., 2019; Wingfield et al., 2024). The family Trichocomaceae is represented by *Talaromyces tratensis* and *T. stipitatus* (Chalearmsrimuang et al., 2019). Other orders of Eurotiomycetes are represented more sparsely: Pyrenulales by a single species, *Pyrenographa xylographoides* (Pyrenulaceae) (Jones et al., 2006b; Suetrong et al., 2017), and Sclerococcales by *Sclerococcum haliotrephum* and *S. mangrovei* (Dactylosporaceae) (Kohlmeyer, 1984; Koch, 1986; Dethoup and Manoch, 2009; Hyde et al., 1990; Hyde, 1994; Sakayaroj et al., 2011; Loilong et al., 2012; Suetrong et al., 2017).

Only a limited number of Basidiomycota have so far been identified from marine environments in Thailand. Most belong to the class Agaricomycetes (2 species), with

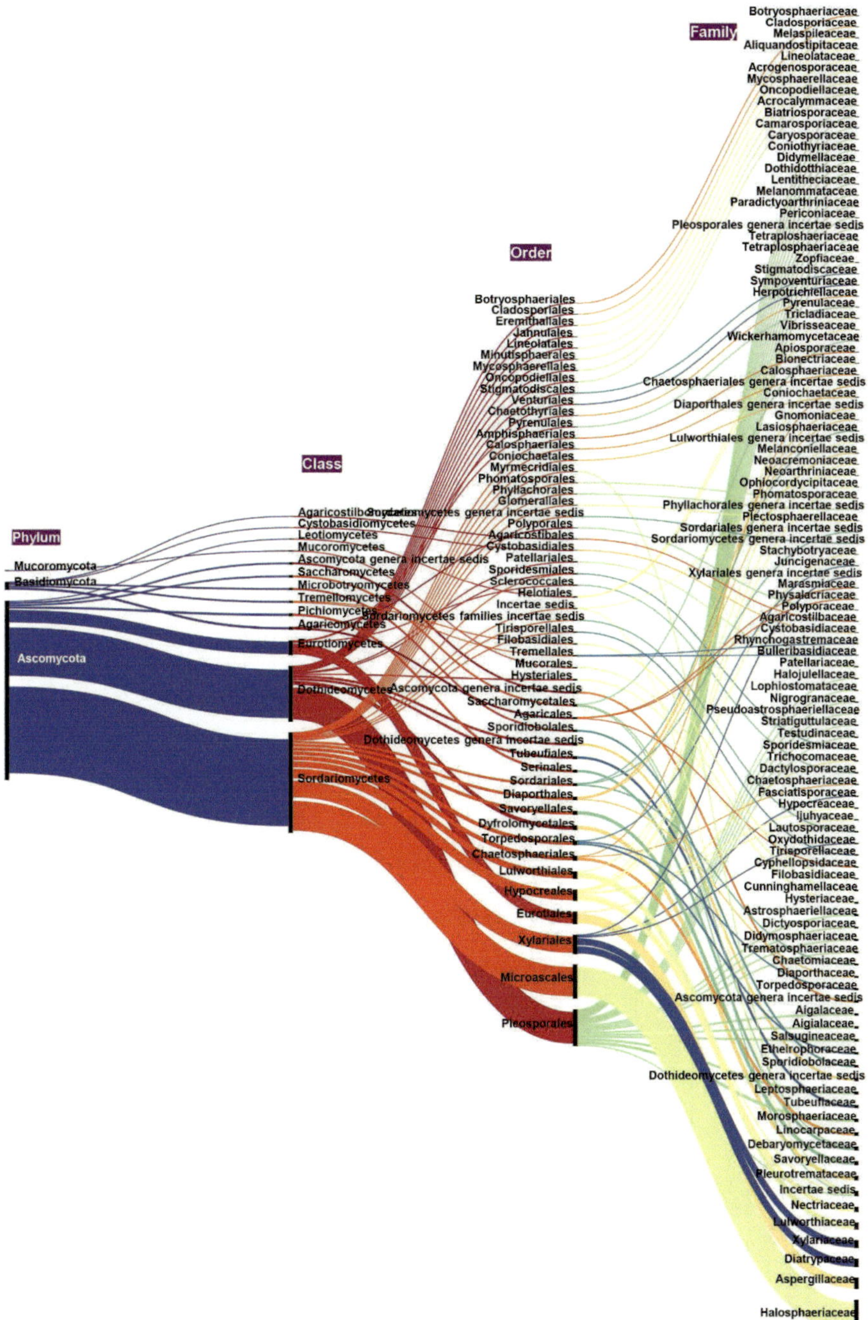

Fig. 1. Taxonomic classification of marine fungi recorded in Thailand.

Table 1. Marine fungi recorded from Thailand with details on habitat, life mode, substrate and/or host, and distribution.

Taxon	Habitat	Life Mode	Substrate/Host	Province	References
ASCOMYCOTA Caval.-Sm. **Ascomycota genera *incertae sedis***					
Dryosphaera navigans Jørg, Koch & E.B.G. Jones	Intertidal	Saprobic	Driftwood	Phuket	Koch (1986)
Dryosphaera tropicalis Kohlm. & Volkm.-Kohlm.	Intertidal	Saprobic	Wood	Phuket	Kohlmeyer and Volkmann-Kohlmeyer (1993)
Helicorhoidion nypicola K.D. Hyde & Goh		Saprobic	Decaying intertidal petioles and fronds of *Nypa fruticans*; wood	Trang, Satun, Surat Thani, Samut Prakarn, Chachoengsao, Samut Sakorn, Samut Songkhram	Pilantanapak et al. (2005); Jones et al. (2006b); Loilong et al. (2012)
Nypaella frondicola K.D. Hyde & B. Sutton	Mangrove/ Intertidal	Saprobic	*Nypa fruticans*	Samut Songkhram	Jones et al. (2006b)
DOTHIDEOMYCETES O.E. Erikss. & Winka **Botryosphaeriales** C.L. Schoch, Crous & Shoemaker **Botryosphaeriaceae** Theiss. & Syd.					
Lasiodiplodia krabiensis Dayar.* (holotype)	Submerged	Saprobic	Decaying wood of *Bruguiera* sp.	Krabi	Dayarathne et al. (2020a)
Cladosporiales Abdollahz. & Crous **Cladosporiaceae** Chalm. & R.G. Archibald					
Cladosporium sphaerospermum Penz.*	Seagrass	Endophytic	*Enhalus acoroides*	Nakhon Si Thammarat	Sakayaroj et al. (2010)
Dothideomycetes genera *incertae sedis*					
Bactrodesmium linderi (J.L. Crane & Shearer) M.E. Palm & E.L. Stewart	Intertidal	Saprobic	Decaying wood		Suetrong et al. (2017)
Caryosporella rhizophorae Kohlm.	Intertidal	Saprobic	Senescent and decaying wood	Trat. Phang-Nga. Ranong	Hyde et al. (1990); Hyde (1994); Suetrong et al. (2017)
Clavariopsis bulbosa Anastasiou.	Intertidal	Saprobic	Twigs	Chanthaburi, Cho Buri	Dethoup and Manoch (2009)
Monodictys pelagica (T. Johnson) E.B.G. Jones	Intertidal	Saprobic	Wood	Phuket, Trang	Jones et al. (2006b)

Passeriniella savoryellopsis K.D. Hyde & Mouzouras (Holotype)	Intertidal; mangrove	Saprobic	Senescent and decaying wood	Ranong, Trat	Hyde et al.(1990); Hyde (1994); Suetrong et al. (2017)
Dyfrolomycetales K.L. Pang, K.D. Hyde & E.B.G. Jones **Pleurotremataceae** Walt, Watson					
Dyfrolomyces marinosporus (K.D. Hyde) K.D. Hyde, K.L. Pang, Alias, Suetrong & E.B.G. Jones (as *Saccardoella marinospora*)	Intertidal		Decaying wood		Suetrong et al. (2017)
Dyfrolomyces neothailandicus Dayar., E.B.G. Jones & K.D. Hyde* (holotype)	Submerged	Saprobic	Decaying wood of *Rhizophora* sp.	Krabi	Dayarathne et al. (2020a)
Melomastia rhizophorae (K.D. Hyde) W.L. Li, Maharachch. & Jian K. Liu (as *Dyfrolomyces rhizophorae*)	Intertidal		Decaying wood		Suetrong et al. (2017)
Dyfrolomyces thailandicus (Dayar., E.B.G. Jones & K.D. Hyde) Dayar., E.B.G. Jones, Maharachch., Devadatha, V.V. Sharma, Khongph., Chomnunti & K.D. Hyde (as *Pleurotrema thailandicum*)* (holotype)	Submerged	Saprobic	Decaying wood of *Avicennia marina* and *Rhizophora* sp.	Ranong	Dayarathne et al. (2020a)
Melomastia mangrovei (K.D. Hyde) W.L. Li, Maharachch. & Jian K. Liu (as *Saccardoella mangrovei*) holotype	Mangrove	Saprobic	Roots of *Rhizophora apiculata*; decaying wood	Nakhon Si Thammarat; Ranong	Hyde (1992); Hyde (1994); Sakayaroj et al. (2011)
Melomastia marinospora (K.D. Hyde) W.L. Li, Maharachch. & Jian K. Liu (as *Saccardoella marinospora*) holotype	Intertidal	Saprobic	Senescent and decaying wood; *Nypa fruticans*	Trat, Prachuap Khiri Khan, Samut Songkhram	Suetrong et al. (2017); Hyde (1992); Pilantanapak et al. (2005)
Melomastia phetchaburiensis (Dayar., E.B.G. Jones & K.D. Hyde) W.L. Li, Maharachch. & Jian K. Liu (as *Dyfrolomyces phetchaburiensis*)* holotype	Mangrove	Saprobic	submerged wood of *Rhizophora apiculata*	Phetchaburi	Hyde et al. (2017)

Table 1 contd. ...

...Table 1 contd.

Taxon	Habitat	Life Mode	Substrate/Host	Province	References
Melomastia rhizophorae (K.D. Hyde) W.L. Li, Maharachch. & Jian K. Liu (as *Saccardoella rhizophorae*) (Holotype)	Intertidal	Saprobic	Senescent and decaying wood; *Nypa fruticans*	Trat; Prachuap Khiri Khan; Nakhon Si Thammarat; Ranong, Phuket	Hyde (1992); Sakayaroj et al. (2011); Suetrong et al. (2017)
Eremithallales Lücking & Lumbsch **Melaspileaceae** Walt, Watson					
Melaspilea mangrovei Vrijmoed, K.D. Hyde & E.B.G. Jones	Mangrove/ Intertidal	Saprobic	Decaying wood	Phuket	Jones et al. (2006b)
Hysteriales Lindau **Hysteriaceae** Chevall					
Hysterium rhizophorae Dayar. & K.D. Hyde*	Mangrove	Saprobic	on bark attached to the living plant of *Avicennia* sp.	Phetchaburi	Dayarathne et al. (2020a)
Ericboehmia thailandica (Jayasiri & K.D. Hyde) Gardiennet, Lechat & J. Fourn (*Hysterium thailandicum*)*	Intertidal	Saprobic	Decaying wood of unidentified mangrove	Ranong	Dayarathne et al. (2020a)
Rhytidhysteron Bruguierae Dayar.* (Holotype)	Intertidal	Saprobic	Decaying wood of *Bruguiera* sp.	Phetchaburi	Dayarathne et al. (2020a)
Jahnulales K.L. Pang, Abdel-Wahab, El-Shar., E.B.G. Jones & Sivichai **Aliquandostipitaceae** Inderb.					
Xylomyces rhizophorae Kohlm. & Volkm.-Kohlm	Intertidal	Saprobic	mangrove wood	Phang-Nga	Jones et al. (2006b)
Lineolatales Crous, Spatafora, Haridas & I.V. Grig. **Lineolataceae** Crous, Spatafora, Haridas & I.V. Grig.					
Lineolata rhizophorae (Kohlm. & E. Kohlm.) Kohlm. & Volkm.-Kohlm. (as *Didymosphaeria rhizophorae*)	Intertidal	Saprobic	Senescent and decaying wood; Roots of *Rhizophora apiculata*	Prachuap Khiri Khan, Phetchaburi, Phuket, Ranong, Trang, Narathiwat, Trat; Nakhon Si Thammarat	Hyde et al. (1990); Hyde (1994); Sakayaroj et al. (2011); Suetrong et al. (2017)

Minutisphaerales Raja, Oberlies, Shearer & A.N. Mill. **Acrogenosporaceae** Jayasiri & K.D. Hyde					
Acrogenospora sphaerocephala (Berk. & Broome) M.B. Ellis		Saprobic	Decaying intertidal petioles and fronds of *N. fruticans*	Trang	Chatmala et al. (2004); Jones et al. (2006b); Loilong et al. (2012)
Mycosphaerellales P.F. Cannon **Mycosphaerellaceae** Lindau					
Piricauda pelagica T. Johnson	Intertidal	Saprobic	Senescent and decaying wood	Prachuap Khiri Khan	Suetrong et al. (2017)
Oncopodiellales Y.R. Sun, Yong Wang bis & K.D. Hyde **Oncopodiellaceae** Y.R. Sun, Yong Wang bis & K.D. Hyde					
Oncopodiella lignicola S. Mukhop., Phukhams, & E.B.G. Jones* (holotype)	Mangrove	Saprobic	decaying submerged stems of Hibiscus tiliaceus	Prachuap Khiri Khan	Mukhopadhyay et al. (2025)
Patellariales D. Hawksw. & O.E. Erikss. **Patellariaceae** Corda					
Patellaria apiculatae Dayar. & K.D. Hyde	Intertidal	Saprobic	Decaying wood of *Rhizophora apiculata*	Phetchaburi	Dayarathne et al. (2020a)
Patellaria atrata (Hedw.) Fr.	Intertidal	Saprobic	Senescent and decaying wood	Phetchaburi	Suetrong et al. (2017)
Pleosporales Luttr. ex M.E. Barr **Acrocalymmaceae** Crous & Trakun.					
Acrocalymma estuarina M.S. Calabon. E.B.G. Jones & K.D. Hyde* (holotype)	Mangrove	Saprobic	submerged decaying wood	Samut Songkhram	Manawasinghe et al. (2024)
Aigialaceae Suetrong, Sakay., E.B.G. Jones, Kohlm., Volkm.-Kohlm. & C.L. Schoch					
Aigialus grandis Kohlm. & S. Schatz*	Mangrove, intertidal	Saprobic	Mangrove wood; Roots of *Rhizophora apiculata*	Phang-Nga, Ranong, Phuket, Ranong, Krabi; Trat	Hyde (1989); Hyde et al. (1990); Hyde (1994); Suetrong et al. (2017); Dayarathne et al. (2020a)

Table 1 contd. ...

...Table 1 contd.

Taxon	Habitat	Life Mode	Substrate/Host	Province	References
Aigialus mangrovis Borse*	Mangrove	Saprobic	decaying wood of *Rhizophora* sp.; Senescent and decaying wood	Phetchaburi; Trat	Suetrong et al. (2017); Dayarathne et al. (2020a)
Aigialus parvus S. Schatz & Kohlm.	mangrove	saprobic	Twigs; root of *Rhizophora apiculata*; decaying wood	Chon Buri, Ranong, Phetchaburi, Trat	Dethoup and Manoch (2009); Hyde et al. (1990); Suetrong et al. (2017); Dayarathne et al. (2020a)
Aigialus striatisporus K.D. Hyde (as *Aigialus striatispora*)	Mangrove	Saprobic	wood; Roots of *Rhizophora apiculata*	Ranong	Hyde et al. (1990); Hyde (1994)
Ascocratera manglicola Kohlm.*	Mangrove	Saprobic	submerged wood of *Rhizophora* sp., decaying wood of *Rhizophora* sp.; Roots of *Rhizophora apiculata*	Phang-Nga, Ranong, Trat, Prachuap Khiri Khan, Phetchaburi, Krabi	Hyde (1989); Hyde et al. (1990); Hyde (1994); Suetrong et al. (2017)
Rimora mangrovei (Kohlm. & Vittal) Kohlm.. Volkm.-Kohlm., Suetrong, Sakay. & E.B.G. Jones (as *Lophiostoma mangrovei*)	Mangrove	Saprobic	Roots of *Rhizophora apiculata*; *Nypa fruticans*	Ranong, Phuket, Narathiwat	Hyde et al. (1990)
Neoastrosphaeriella phoenicis S.N. Zhang, E.B.G. Jones & Jian K. Liu* (holotype)	Mangrove	Saprobic	petiole of *Phoenix paludosa*	Ranong	Zhang et al. (2020)
Rimora mangrovei (Kohlm. & Vittal) Kohlm., Volkm.-Kohlm., Suetrong, Sakay. & E.B.G. Jones (as *Lophiostoma mangrovei*)*	Mangrove; Intertidal	Saprobic	decomposing wood of Ceriops tagal; Senescent and decaying wood; Decaying intertidal petioles and fronds of *N. fruticans*	Trat, Prachuap Khiri Khan, Phetchaburi; Samut Prakarn	Loilong et al. (2012); Suetrong et al. (2017); Apurillo et al. (2025)
Astrosphaeriellaceae Phookamsak & K.D. Hyde					
Astrosphaeriella asiana (K.D. Hyde) Aptroot & K.D. Hyde (*Lophiostoma asiana*) (Holotype)	Mangrove	Saprobic	Mangrove wood *Rhizophora apiculata*; roots of *Aegiceras corniculatum* and *Sonneratia alba*	Ranong	Hyde et al. (1990)

Astrosphaeriella mangrovei (Kohlm. & Vittal) Aptroot & K.D. Hyde (as *Lophiostoma mangrovei*)	Mangrove/ Intertidal	Saprobic	Mangrove wood *Rhizophora apiculata*	Ranong, Samut Songkhram, Phuket	Hyde et al. (1990); Hyde (1994)
Astrosphaeriella nypae K.D. Hyde	Mangrove	Saprobic	Decaying intertidal petioles and fronds of *N. fruticans*	Trang, Satun, Samut Prakarn, Chachoengsao; Samut Songkhram	Jones et al. (2006b); Loilong et al. (2012)
Biatriosporaceae K.D. Hyde					
Biatriospora marina K.D. Hyde & Borse	Mangrove	Saprobic	Mangrove wood *Rhizophora apiculata*	Ranong	Hyde et al. (1990)
Camarosporiaceae Wanas., Wijayaw., K.D. Hyde & Crous					
Camarosporium roumeguerei Sacc.	Mangrove/ Intertidal	Saprobic	Mangrove wood	Trang	Jones et al. (2006b)
Caryosporaceae Huang Zhang, K.D. Hyde & H.A. Ariy.					
Acrocordiopsis patilii Borse & K.D. Hyde*	mangrove	saprobic	Decaying submerged wood of *Xylocarpus* sp., *Rhizophora apiculata*	Ranong	Dayarathne et al. (2020a)
Coniothyriaceae W.B. Cooke					
Coniothyrioides thailandica Wijes., M.S. Calabon, E.B.G. Jones & K.D. Hyde* (holotype, new genus)	Saltmarsh	Saprobic	submerged decaying wood	Pranburi	Wijesinghe et al. (2023)
Dictyosporiaceae Boonmee & K.D. Hyde					
Cattanea heptaspora Garov	Intertidal	Saprobic	*Nypa fruticans*	Samut Songkhram	Pilantanapak et al. (2005)
Dictyosporium elegans Corda	Intertidal	Saprobic	*Nypa fruticans*	Trang, Samut Songkhram	Pilantanapak et al. (2005); Jones et al. (2006b)
Dictyosporium heptasporum (Garov.) Damon (as *Cattanea heptaspora*)	Intertidal	Saprobic	*Nypa fruticans*	Samut Songkhram	Jones et al. (2006b)
Didymellaceae Gruyter, Aveskamp & Verkley					
Didymella Avicenniae S.D. Patil & Borse	Mangrove	Saprobic	Mangrove wood *Rhizophora apiculata*	Ranong	Hyde et al. (1990)

Table 1 contd. ...

...Table 1 contd.

Taxon	Habitat	Life Mode	Substrate/Host	Province	References
Didymosphaeriaceae Munk					
Didymosphaeria maritima (P, Crouan & H. Crouan) Sacc. (as *Sphaeria maritima*)		Saprobic	Twigs	–	Dethoup and Manoch (2009)
Paraconiothyrium brasiliense Verkley	Sea cucumber		*Holothuria scabra*	Phang Nga	Wingfield et al. (2024)
Pseudopithomyces maydicus (Sacc.) Jun F. Li, H.A. Ariy. & K.D. Hyde (as *Clasterosporium maydicum*)	Sea cucumber		*Holothuria scabra*	Phang Nga	Wingfield et al. (2024)
Dothidotthiaceae Crous & A.J.L. Phillips					
Belizeana tuberculata Kohlm. & Volkm.-Kohlm.	Intertidal/ Mangrove	Saprobic	Senescent and decaying wood; Roots of *Rhizophora apiculata*	Prachuap Khiri Khan; Ranong	Hyde et al. (1990); Hyde (1994); Suetrong et al. (2017)
Halojulellaceae Suetrong, K.D. Hyde & E.B.G. Jones					
Halojulella Avicenniae (Borse) Suetrong, K.D. Hyde & E.B.G. Jones (as *Pleospora Avicenniae*)	Intertidal	Saprobic	Senescent and decaying wood	Phetchaburi	Suetrong et al. (2017)
Halojulella Avicenniae (Borse) Suetrong, K.D. Hyde & E.B.G. Jones (as *Pleospora Avicenniae*)	Mangrove/ Intertidal	Saprobic	Decaying mangrove	Nakhon Si Thammarat; Phuket	Jones et al. (2006b); Sakayaroj et al. (2011)
Pleosporales *incertae sedis*					
Acuminatispora palmarum S.N. Zhang, K.D. Hyde & Jian K, Liu* (holotype) (new genus)	Mangrove	Saprobic	submerged on decayed rachis of *Phoenix paludosa;* on decayed petiole of *Nypa fruticans*	Trat; Chanthaburi, Ranong	Zhang et al. (2018)
Pseudochaetosphaeronema Bruguierae Apurillo, Phukhams., E.B.G. Jones & K.D. Hyde* (holotype)	Mangrove	Saprobic	aerial dead branch of *Bruguiera* cylindrica	Prachuap Khiri Khan	Apurillo et al. (2025)
Pseudochaetosphaeronema pandanicola Tibpromma & K.D. Hyde* (holotype)	Sea cucumber		*Holothuria scabra*	Phang Nga	Wingfield et al. (2024)

Lentitheciaceae Y, Zhang ter, C.L. Schoch, J. Fourn., Crous & K.D. Hyde					
Poaceascoma halophilum Dayar. & K.D. Hyde* (holotype)	Intertidal	Saprobic	submerged wood of *Rhizophora apiculata*	Prachuap Khiri Khan	Hyde et al. (2017)
Leptosphaeriaceae M.E. Barr					
Leptosphaeria australiensis (Cribb & J.W. Cribb) S. Hughes (as *Metasphaeria australiensis*)	Mangrove/ Intertidal	Saprobic	Senescent and decaying wood; Decaying intertidal petioles and fronds of *N. fruticans*; Roots and seedlings of *Rhizophora apiculata*	Trat, Prachuap Khiri Khan, Phetchaburi; Ranong; Samut Prakarn	Hyde (1989); Hyde et al. (1990); Hyde (1994); Loilong et al. (2012); Suetrong et al. (2017)
Leptosphaeria Avicenniae Kohlm. & E. Kohlm.	Mangrove	Saprobic	Mangrove wood	Phang-Nga, Ranong	Hyde (1989)
Leptosphaeria nypicola K.D. Hyde & Alias		Saprobic	Decaying intertidal petioles and fronds of *N. fruticans*	Trang, Satun, Surat Thani, Samut Prakarn, Chachoengsao, Samut Sakorn	Loilong et al. (2012)
Leptosphaeria peruviana Speg.	Intertidal/ Mangrove	Saprobic	*Nypa fruticans*; Senescent and decaying wood	Samut Songkhram; Prachuap Khiri Khan, Phetchaburi	Pilantanapak et al. (2005); Jones et al. (2006b); Suetrong et al. (2017)
Lignincola conchicola Jian K. Liu, E.B.G. Jones & K.D. Hyde	Mangrove	Saprobic	adhesive disc of invertebrate, fronds of *Phoenix paludosa* submerged in m	Ranong	Liu et al. (2011)
Lophiostomataceae Luerss.					
Quintaria lignatilis (Kohlm.) Kohlm. & Volkm.-Kohlm. (as *Trematosphaeria lignatilis*) (Holotype)	Intertidal; Mangrove	Saprobic	Senescent and decaying wood; Roots of *Rhizophora apiculata*	Trat, Prachuap Khiri Khan, Phetchaburi; Nakhon Si Thammarat Ranong; Phang-Nga, Phuket, Trang	Hyde (1994); Sakayaroj et al. (2011); Suetrong et al. (2017)
Vaginatispora acrostichi S. Mukhop., Phukhams. & E.B.G. Jones* (holotype)	Mangrove	Saprobic	decaying submerged stems of *Acrostichum aureum*	Prachuap Khiri Khan	Mukhopadhyay et al. (2024)

Table 1 contd. ...

...Table 1 contd.

Taxon	Habitat	Life Mode	Substrate/Host	Province	References
Melanommataceae G. Winter					
Bicrouania maritima (P. Crouan & H. Crouan) Kohlm. & Volkm.-Kohlm. (as *Sphaeria maritima*)	Intertidal	Saprobic	Twigs	Chon Buri	Dethoup and Manoch (2009)
Morosphaeriaceae Suetrong, Sakay., E.B.G. Jones & C.L. Schoch					
Helicascus kanaloanus Kohlm.	Mangrove	Saprobic	Mangrove wood	Phang-Nga, Ranong, Trat	Hyde (1989); Hyde et al.(1990)
Helicascus mangrovei Preedanon, Suetrong & Sakay.	Intertidal	Saprobic	Senescent and decaying wood	Trat	Suetrong et al. (2017)
Helicascus nypae K.D. Hyde	Intertidal	Saprobic	Decaying intertidal petioles and fronds of *N. fruticans*; wood	Samut Songkhram; Trang, Surat Thani; Phuket	Pilantanapak et al. (2005); Jones et al. (2006b); Loilong et al. (2012)
Helicascus satunensis Preedanon. Suetrong & Sakay.* (holotype)	Mangrove	Saprobic	on dead fruits of *Nypa fruticans*	Satun Province	Preedanon et al. (2024)
Morosphaeria ramunculicola (K.D. Hyde) Suetrong, Sakay., E.B.G. Jones & C.L. Schoch (as *Massarina ramunculicola*)	Mangrove (Nypa)	Saprobic	*Nypa fruticans*; Seedlings of *Rhizophora apiculata*	Phuket, Ranong	Hyde et al. (1990); Hyde (1994)
Morosphaeria velatispora (K.D. Hyde & Borse) Suetrong, Sakay., E.B.G. Jones & C.L. Schoch (as *Massarina velatispora*) (Holotype)	Mangrove / Intertidal	Saprobic	Mangrove wood; *Nypa fruticans*; Roots of *Rhizophora apiculata*	Phang-Nga, Ranong; Nakhon Si Thammarat; Trat; Ranong	Hyde (1989); Hyde et al. (1990); Hyde (1994); Sakayaroj et al. (2011) Suetrong et al. (2017)
Nigrogranaceae Jaklitsch & Voglmayr					
Nigrograna rhizophorae Dayar., E.B.G. Jones & K.D. Hyde	Submerged	Saprobic	Decaying wood of *Rhizophora* sp.	Phetchaburi	Dayarathne et al. (2020a)
Nigrograna mackinnonii (Borelli) Gruyter, Verkley & Crous (*Pyrenochaeta mackinnonii*)	Sea cucumber		*Holothuria scabra*	Phang Nga	Wingfield et al. (2024)

Paradictyoarthriniaceae Doilom, Jian K. Liu & K.D. Hyde					
Paradictyoarthrinium salsipaludicola Prematunga & Htet* (holotype)	Mangrove	Saprobic	decaying wood	Prachuap Khiri Khan	Htet et al. (2023)
Periconiaceae Nann.					
Periconia prolifica Anastasiou	Intertidal	Saprobic	Senescent and decaying wood; Roots of *Rhizophora apiculata*; driftwood	Phetchaburi, Chon Buri, Pattani, Phang-Nga, Phuket, Ranong, Rayong, Narathiwat, Prachuap Khiri Khan, Trat, Chanthaburi, Trang	Kohlmeyer (1984); Koch (1986); Hyde et al. (1990); Hyde (1994); Chatmala et al. (2004); Suetrong et al. (2017)
Pleosporales genera *incertae sedis*					
Plectophomella nypae K.D. Hyde & B. Sutton	Intertidal	Saprobic	*Nypa fruticans*	Samut Songkhram	Pilantanapak et al. (2005); Jones et al. (2006b)
Pseudoastrosphaeriellaceae Phookamsak & K.D. Hyde					
Carinispora nypae K.D. Hyde (holotype)	Intertidal	Saprobic	Decaying intertidal petioles and fronds of *N. fruticans*	Trang, Satun	Loilong et al. (2012)
Carinispora velatispora K.D. Hyde (holotype)	Intertidal	Saprobic	Decaying intertidal petioles and fronds of *N. fruticans*	Samut Sakorn	Loilong et al. (2012)
Acrocordiopsis patilii Borse & K.D. Hyde (holotype)	Intertidal/ Mangrove	Saprobic	Senescent and decaying wood; Roots of *Rhizophora apiculata*	Prachuap Khiri Khan; Ranong, Phuket	Hyde et al. (1990); Hyde (1994); Suetrong et al. (2017)
Salsugineaceae K.D. Hyde & Tibpromma					
Salsuginea phoenicis S.N. Zhang, E.B.G. Jones, K.D. Hyde & Jian K. Liu* (holotype)	Intertidal	Saprobic	decaying petiole of *Phoenix paludosa*	Chanthaburi	Jones et al. (2020)
Salsuginea ramicola K.D. Hyde (holotype)	Intertidal/ Mangrove	Saprobic	*Nypa fruticans*; Senescent and decaying wood	Samut Songkhram; Trat, Prachuap Khiri Khan, Phetchaburi	Pilantanapak et al. (2005); Jones et al. (2006b); Suetrong et al. (2017)

Table 1 contd. ...

...Table 1 contd.

Taxon	Habitat	Life Mode	Substrate/Host	Province	References
Salsuginea rhizophorae Dayar., E.B.G. Jones & K.D. Hyde* (holotype)	Intertidal	Saprobic	Decaying wood of *Rhizophora* apiculata	Ranong	Dayarathne et al. (2020a)
Striatiguttulaceae S.N. Zhang, K.D. Hyde & Jian K. Liu					
Longicorpus striatisporus (K.D. Hyde) S.N. Zhang, K.D. Hyde & Jian K. Liu (as *Trematosphaeria striatispora*)* (new genus)	Mangrove	Saprobic	on rachis of *Phoenix paludosa;* on decayed rachis of *Nypa fruticans*; Senescent and decaying wood; Decaying intertidal petioles and fronds of *N. fruticans*	Chachoengsao, Chanthaburi, Krabi, Ranong, Samut Prakan, Samut Songkhram, Satun, Surat Thani, Trat, Trang	Hyde et al. (1990); Pilantanapak et al. (2005); Loilong et al. (2012); Suetrong et al. (2017); Zhang et al. (2019b)
Striatiguttula nypae S.N. Zhang, K.D. Hyde & Jian K. Liu* (holotype) (new genus)	Mangrove	Saprobic	on submerged decaying rachis of *Nypa fruticans*	Krabi; Ranong	Zhang et al. (2019b)
Testudinaceae Arx					
Halotestudina muriformis Dayar, & K.D. Hyde* (holotype)	Mangrove	Saprobic	Decaying wood of *Rhizophora* sp. and *Bruguiera* sp.	Krabi	Dayarathne et al. (2020a)
Verruculina enalia (Kohlm.) Kohlm. & Volkm.-Kohlm. (as *Didymosphaeria enalia*)*	Intertidal/ Mangrove	Saprobic	Senescent and decaying wood; Decaying intertidal petioles and fronds of *N. fruticans*; Roots of *Rhizophora apiculata*; stump of Ceriops sp., dead prop roots of *Rhizophora* sp.; on bark attached to the living *Rhizophora* sp.	Chanthaburi, Chon Buri, Krabi, Nakhon Si Thammarat, Phetchaburi, Prachuap Khiri Khan, Ranong, Rayong, Samut Songkhram, Satun, Trat	Hyde (1989); Dethoup and Manoch (2009); Hyde et al. (1990); Hyde (1994); Pilantanapak et al. (2005); Sakayaroj et al. (2011); Loilong et al. (2012); Suetrong et al. (2017); Dayarathne et al. (2020a)
Tetraplosphaeriaceae Kaz, Tanaka & K. Hiray,					
Tetraploa maritima R. Asghari, Phukhams, & K.D. Hyde* (holotype)	Intertidal	Saprobic	decaying *Areca* wood	Prachuap Khiri Khan	Asghari et al. (2025)
Tetraploa aristata Berk. & Broome		Saprobic		Narathiwat	Chatmala et al. (2004)

Trematosphaeriaceae K.D. Hyde, Y, Zhang ter, Suetrong & E.B.G. Jones					
Halomassarina thalassiae (Kohlm. & Volkm.-Kohlm.) Suetrong, Sakay., E.B.G. Jones, Kohlm., Volkm.-Kohlm. & C.L. Schoch (as *Massarina thalassiae*)	Intertidal/ Mangrove	Saprobic	Senescent and decaying wood; Decaying intertidal petioles and fronds of *N. fruticans*; Roots of *Rhizophora apiculata*	Trat, Prachuap Khiri Khan; Ranong	Hyde (1994); Loilong et al. (2012); Suetrong et al. (2017)
Trematosphaeria lineolatispora K.D. Hyde (holotype)	Intertidal	Saprobic	*Nypa fruticans*	Samut Songkhram	Pilantanapak et al. (2005); Jones et al. (2006b)
Trematosphaeria mangrovis Kohlm.	Intertidal	Saprobic	Decaying intertidal petioles and fronds of *N. fruticans*	Phang-Nga, Samut Songkhram, Phuket, Trat, Prachuap Khiri Khan	Pilantanapak et al. (2005); Jones et al. (2006b); Loilong et al. (2012); Suetrong et al.(2017)
Zopfiaceae Arnaud					
Coronopapilla mangrovei (K.D. Hyde) Kohlm. & Volkm.-Kohlm. (as *Caryospora mangrovei*)	Mangrove	Saprobic	buttress root of *Xylocarpus granatum*	Phang-Nga	Hyde (1989)
Stigmatodiscales Voglmayr & Jaklitsch **Stigmatodiscaceae** Voglmayr & Jaklitsch					
Stigmatodiscus mangrovei (Dayar., E.B.G. Jones & K.D. Hyde) M.S. Calabon, E.B.G. Jones & K.L. Pang (as *Asterodiscus mangrovei*)* holotype		Saprobic	Unidentified mangrove trees or mangrove wood	Krabi	Dayarathne et al. (2020a)
Tubeufiales Boonmee, Rossman & K.D. Hyde **Tubeufiaceae** M.E. Barr					
Helicoma hyalonemum G.Z. Zhao, Xing Z, Liu & W.P. Wu (holotype)		Saprobic	Decaying intertidal petioles and fronds of *N. fruticans*	Surat Thani	Loilong et al. (2012)
Helicoma pannosum (Berk. & M.A. Curtis) Y.Z. Lu & K.D. Hyde (as *Drepanospora pannosa*)		Saprobic	Decaying intertidal petioles and fronds of *N. fruticans*	Surat Thani	Loilong et al. (2012)
Helicosporium hongkongense C.K.M. Tsui, Goh, K.D. Hyde & Hodgkiss		Saprobic	Decaying intertidal petioles and fronds of *N. fruticans*	Surat Thani	Loilong et al. (2012)

Table 1 contd. ...

...Table 1 contd.

Taxon	**Habitat**	**Life Mode**	**Substrate/Host**	**Province**	**References**
*Kamalomyces mangrovei** Dayar, & K.D. Hyde (holotype)	Mangrove	Saprobic	submerged wood of mangrove; on decaying wood of *Xylocarpus* sp.	Ranong	Dayarathne et al. (2020a); Hyde et al. (2019)
*Parahelicomyces dictyosporus** M.S. Calabon, E.B.G. Jones & K.D. Hyde (holotype)	Saltmarsh	Saprobic	submerged decaying culms of *Spartina* sp.	Prachuap Khiri Khan	Liu et al. (2024)
Venturiales Y, Zhang ter, C.L. Schoch & K.D. Hyde **Sympoventuriaceae** Y, Zhang ter, C.L. Schoch & K.D. Hyde					
UPDATED: *Ochroconis musae* (G.Y. Sun & Lu Hao) Samerp, & de Hoog (as *Scolecobasidium musae*)*	Sea cucumber		*Holothuria scabra*	Phang Nga	Wingfield et al. (2024)
EUROTIOMYCETES O.E. Erikss, & Winka **Chaetothyriales** M.E. Barr **Herpotrichiellaceae** Munk.					
Cladophialophora bantiana (Sacc,) de Hoog, Kwon-Chung & McGinnis (as *Torula bantiana*)*	Sea cucumber		*Holothuria scabra*	Phang Nga	Wingfield et al. (2024)
Eurotiales G.W. Martin ex Benny & Kimbr, **Aspergillaceae** Link					
Aspergillus flavus Link	Sea cucumber		*Holothuria scabra*	Phang Nga	Wingfield et al. (2024)
Aspergillus fumigatus Fresen.*	Sea cucumber		*Holothuria scabra*	Phang Nga	Wingfield et al. (2024)
Aspergillus nidulans (Eidam) G. Winter (as *Sterigmatocystis nidulans*)*	Intertidal	Saprobic	Twigs; Mycale armata	Chon Buri	Dethoup and Manoch (2009); Chalearmsrimuang et al. (2019)
Aspergillus nomius Kurtzman, B.W. Horn & Hesselt* (holotype)	Sea cucumber		*Holothuria scabra*	Phang Nga	Wingfield et al. 2024)
Aspergillus oryzae (Ahlb,) Cohn (as *Eurotium oryzae*)*	Sea cucumber		*Holothuria scabra*	Phang Nga	Wingfield et al. (2024)

Aspergillus penicillioides Speg.*	Seagrass	Endophytic	*Enhalus acoroides*	Nakhon Si Thammarat; Surat Thani	Sakayaroj et al. (2010)
Aspergillus stellatus Curzi	Intertidal	Saprobic	Twigs; *Xestospongia testudinaria*	Chon Buri	Dethoup and Manoch (2009); Chalearmsrimuang et al.(2019)
Aspergillus terreus Thom*	Sea cucumber		*Holothuria scabra*	Phang Nga	Wingfield et al. (2024)
Aspergillus unguis (Émile-Weill & L, Gaudin) Thom & Raper (as *Sterigmatocystis unguis*)*	Sea cucumber		*Holothuria scabra*	Phang Nga	Wingfield et al. (2024)
Aspergillus foveolatus Y. Horie*	Submerged		*Xestospongia testudinaria*	Chonburi	Chalearmsrimuang et al. (2019)
Aspergillus rugulosus Thom & Raper (as *Emericella rugulosa*)	Submerged		*Acanthella* sp.	Chonburi	Chalearmsrimuang et al. (2019)
Aspergillus chevalieri (L, Mangin) Thom & Church*(as *Eurotium chevalieri*)	Submerged		*Rhabdermia* sp.	Phang Nga	Chalearmsrimuang et al. (2019)
Hamigera avellanea Stolk & Samson*	Submerged		*Acanthella* sp.	Chonburi	Chalearmsrimuang et al. (2019)
Hamigera terricola S.W. Peterson, Jurjević, Bills, Stchigel, Guarro & F.E. Vega*	Submerged		*Xestospongia testudinaria*	Chonburi	Chalearmsrimuang et al. (2019)
Aspergillus fischeri Wehmer*(as Neosartorya fischeri; Neosartorya pseudofischeri)	Submerged		*Rhabdermia* sp.	Phang Nga	Chalearmsrimuang et al. (2019)
Aspergillus quadricinctus E. Yuill* (as *Neosartorya quadricincta*)	Submerged		*Xestospongia testudinaria*	Chonburi	Chalearmsrimuang et al. (2019)
Aspergillus tsunodae (Yaguchi, Abliz & Y. Horie) Samson, Visagie & Houbraken*(as *Neosartorya tsunodae*)	Submerged		*Aka coralliphaga*	Phang Nga	Chalearmsrimuang et al. (2019)
Talaromyces aculeatus (Raper & Fennell) Samson, N. Yilmaz, Frisvad & Seifert* (as *Penicillium aculeatum*)	Submerged		*Xestospongia testudinaria*	Chonburi	Chalearmsrimuang et al. (2019)

Table 1 contd. ...

...Table 1 contd.

Taxon	Habitat	Life Mode	Substrate/Host	Province	References
Penicillium citrinum Thom*	Sea cucumber		*Holothuria scabra*	Phang Nga	Wingfield et al. (2024)
Penicillium oxalicum Currie & Thom*	Sea cucumber		*Holothuria scabra*	Phang Nga	Wingfield et al. (2024)
Trichocomaceae E. Fisch,					
Talaromyces stipitatus (Thom ex C.W. Emmons) C.R. Benj., (as *Penicillium stipitatum*)	Submerged		*Stylissa flabelliformis*	Chonburi	Chalearmsrimuang et al. (2019)
Talaromyces tratensis Manoch, Dethoup & N. Yilmaz	Submerged		Mycale sp.	Chonburi	Chalearmsrimuang et al. (2019)
Pyrenulales Fink ex D, Hawksw, & O.E. Erikss, **Pyrenulaceae** Rabenh					
Pyrenographa xylographoides Aptroot (holotype)	Intertidal	Saprobic	Senescent and decaying wood	Phuket, Trat	Jones et al. (2006b); Suetrong et al. (2017)
Sclerococcales Réblová, Unter. & W. Gams **Dactylosporaceae** Bellem, & Hafellner					
Sclerococcum haliotrephum (Kohlm. & E, Kohlm.) Ertz & Diederich (as *Buellia haliotrepha*)	Mangrove	Saprobic	Roots of *Rhizophora apiculata*; Driftwood; mangrove wood/twigs; Decaying intertidal petioles and fronds of *N. fruticans*	Chanthaburi, Chon Buri, Krabi, Nakhon Si Thammarat, Phetchaburi, Phang-Nga, Phuket, Prachuap Khiri Khan, Ranong, Rayong, Satun, Trat, Trang	Kohlmeyer (1984); Koch (1986); Dethoup and Manoch (2009); Hyde et al. (1990); Hyde (1994); Sakayaroj et al. (2011); Loilong et al. (2012); Suetrong et al. (2017)
Sclerococcum mangrovei (E.B.G. Jones, Alias, Abdel-Wahab & S.Y. Hsieh) Ertz & Diederich (as *Dactylospora mangrovei*)	Intertidal		Decaying wood	Trat, Phetchaburi; Phuket	Jones et al. (1999); Suetrong et al. (2017)
LEOTIOMYCETES O.E. Erikss, & Winka **Helotiales** Nannf, **Tricladiaceae** P.R. Johnst, & Baschien					
Halenospora varia (Anastasiou) E.B.G. Jones (as *Zalerion varia*)	Intertidal	Saprobic	Twigs; Decaying intertidal petioles and fronds of *N. fruticans*	Chon Buri; Satun, Samut Prakarn, Chachoengsao	Dethoup and Manoch (2009); Loilong et al. (2012)

Helotiales Nannf. **Vibrisseaceae** Locq,					
Vibrissea nypicola K.D. Hyde & Alias	Mangrove	Saprobic	Decaying intertidal petioles and fronds of *N. fruticans*	Trang, Satun, Samut Prakarn, Chachoengsao; Samut Songkhram, Narathiwat	Jones et al. (2006b); Loilong et al. (2012)
PICHIOMYCETES M, Groenew., Hittinger, Opulente & A. Rokas **Serinales** M. Groenew., Hittinger, Opulente & A. Rokas **Debaryomycetaceae** Kurtzman & M. Suzuki					
Candida metapsilosis Tavanti, A, Davidson, Gow, M, Maiden & Odds*	Zoanthid		Unknown zoanthid	Chonburi	Kaewkrajay et al. (2020)
Candida parapsilosis (Ashford) Langeron & Talice (as *Monilia parapsilosis*)*	Corals		*Sinularia* sp.	Chonburi	Kaewkrajay et al. (2020)
Candida spencermartinsiae Statzell, Scorzetti & Fell*	Corals		*Klyxum* sp.	Chonburi	Kaewkrajay et al. (2020)
Candida tropicalis (Castell,) Berkhout (as *Oidium tropicale*)*	Zoanthid		Unknown zoanthid	Chonburi	Kaewkrajay et al. (2020)
Candida zeylanoides (Castell,) Langeron & Guerra (as *Monilia zeylanoides*)*	Corals		Unknown coral	Chonburi	Kaewkrajay et al. (2020)
Meyerozyma caribbica (Vaughan-Mart., Kurtzman, S.A. Mey. & E.B. O'Neill) Kurtzman & M. Suzuki (as *Pichia caribbica*)	Zoanthid		Unknown zoanthid; *Sarcophyton* sp.	Chonburi	Kaewkrajay et al. (2020)
Meyerozyma guilliermondii (Wick,) Kurtzman & M. Suzuki (as *Pichia guilliermondii*)*	Zoanthid		Unknown zoanthid	Chonburi	Kaewkrajay et al. (2020)
SACCHAROMYCETES G, Winter **Saccharomycetales** Luerss, Saccharomycetales *incertae sedis*					
Kodamaea ohmeri (Etchells & T.A. Bell) Y, Yamada, Tom, Suzuki, M. Matsuda & Mikata (*Endomycopsis ohmeri*)*	Zoanthid		Unknown zoanthid	Chonburi	Kaewkrajay et al. (2020)

Table 1 contd. ...

...Table 1 contd.

Taxon	Habitat	Life Mode	Substrate/Host	Province	References
Wickerhamomycetaceae Kurtzman, Robnett & Bas,					
Wickerhamomyces anomalus (E.C. Hansen) Kurtzman, Robnett & Bas.-Powers (as *Saccharomyces anomalus*)*	Zoanthid		Unknown zoanthid	Chonburi	Kaewkrajay et al. (2020)
SORDARIOMYCETES O.E. Erikss. & Winka **Amphisphaeriales** D. Hawksw. & O.E. Erikss. **Apiosporaceae** K.D. Hyde, J. Fröhl., Joanne E. Taylor & M.E. Barr					
Apiospora xenocordella (Crous) Pintos & P. Alvarado (as *Arthrinium xenocordella*)	Submerged		Unidentified marine sponge	Chonburi	Chalearmsrimuang et al. (2019)
Calosphaeriales M.E. Barr **Calosphaeriaceae** Munk					
Jattaea mucronata Dayar. & K.D. Hyde*		Saprobic	On the bark of *Rhizophora* mucronata	Krabi	Abdel-Wahab et al. (2017)
Sordariales Chadef, ex D, Hawksw. & O.E. Erikss. **Chaetomiaceae** G, Winter					
Humicola alopallonella Meyers & R.T. Moore (as *Trichocladium alopallonellum*)	Mangrove/ Intertidal	Saprobic	Roots of *Rhizophora apiculata*	Chon Buri, Phang-Nga, Ranong, Samut Songkhram, Phuket, Trang	Kohlmeyer (1984); Hyde et al. (1990); Hyde (1994)
Trichocladium lignicola I, Schmidt	Intertidal	Saprobic	Senescent and decaying wood	Trat	Suetrong et al. (2017)
Trichocladium melhae E.B.G. Jones, Abdel-Wahab & Vrijmoed	Intertidal, Mangrove	Saprobic	Driftwood, mangrove wood	Narathiwat, Rayong, Samut Prakan, Songkhla, Trang, Trat	Chatmala et al. (2004); Jones et al. (2006b)
Chaetosphaeriales Huhndorf, A.N. Mill. & F.A. Fernández **Chaetosphaeriaceae** Réblová, M.E. Barr & Samuels					
Pseudophialocephala humicola (S.C. Jong & E.E. Davis) M.S. Calabon, E.B.G. Jones & K.D. Hyde (as *Chloridium humicola*)*	Saltmarsh	Saprobic	decaying leaf blades of *Spartina* sp.	Prachuap Khiri Khan	Manawasinghe et al. (2022)
Thozetella nivea (Berk.) Kuntze (as *Thozetia nivea*)		Saprobic	Decaying intertidal petioles and fronds of *N. fruticans*	Surat Thani	Loilong et al. (2012)

Linocarpaceae S, Konta & K.D. Hyde 2017					
Linocarpon angustatum K.D. Hyde & Alias	Intertidal/ Mangrove	Saprobic	Decaying intertidal petioles and fronds of *N. fruticans*; Senescent and decaying wood	Samut Sakorn, Samut Songkhram, Satun, Surat Thani, Trang, Trat	Hyde (1988); Pilantanapak et al. (2005); Loilong et al. (2012); Suetrong et al. (2017)
Linocarpon appendiculatum K.D. Hyde	Mangrove	Saprobic	Decaying intertidal petioles and fronds of *N. fruticans*	Chachoengsao, Samut Prakan, Samut Sakhon, Samut Songkhram, Satun, Surat Thani, Trang	Pilantanapak et al. (2005); Loilong et al. (2012)
Linocarpon bipolare K.D. Hyde	Intertidal/ Mangrove	Saprobic	Senescent and decaying wood; Decaying intertidal petioles and fronds of *N. fruticans*	Samut Songkhram, Surat Thani, Trang, Trat	Jones et al. (2006b); Loilong et al. (2012); Suetrong et al. (2017)
Linocarpon longisporum K.D. Hyde	Intertidal	Saprobic	Decaying intertidal petioles and fronds of *N. fruticans*	Satun	Loilong et al. (2012)
Linocarpon nipae (Henn,) K.D. Hyde (as *Ophiobolus nipae*)	Intertidal/ Mangrove	Saprobic	*Nypa fruticans*	Chachoengsao, Phang Nga, Samut Prakan, Samut Sakhon, Samut Songkhram, Surat Thani	Pilantanapak et al. (2005); Jones et al. (2006b); Loilong et al. (2012)
Neolinocarpon globosicarpum K.D. Hyde	Mangrove	Saprobic	Mangrove wood	Samut Songkhram	Jones et al. (2006b)
Chaetosphaeriales genera *incertae sedis*					
Camarops ustulinoides (Henn,) Nannf, (as *Nummularia ustulinoides*)	Mangrove	Saprobic	Decaying wood of *Rhizophora* sp.	Prachuap Kiri Khan	Dayarathne et al. (2020a)
Coniochaetales Huhndorf, A.N. Mill. & F.A. Fernández **Coniochaetaceae** Malloch & Cain					
Coniochaeta krabiensis Dayar, & E.B.G. Jones* (holotype)	Submerged	Saprobic	on decaying submerged wood	Phang-nga; Krabi	Dayarathne et al. (2020a)
Diaporthales Nannf, **Diaporthaceae** Höhn. ex Wehm,					
Diaporthe marina Dayar,* (holotype)	mangrove	Saprobic	on decaying submerged wood of *Rhizophora* sp.	Krabi	Dayarathne et al. (2020a)

Table 1 contd. ...

...Table 1 contd.

Taxon	Habitat	Life Mode	Substrate/Host	Province	References
Diaporthe salinicola Dayar* (holotype)	mangrove	Saprobic	on decaying wood of *Xylocarpus* sp.; on decaying submerged wood	Ranong; Krabi	Dayarathne et al. (2020a)
Phomopsis mangrovei K.D. Hyde	Intertidal/ Mangrove	Saprobic	Roots of *Rhizophora apiculata*	Ranong	Hyde et al. (1990); Hyde (1994)
Diaporthales genera *incertae sedis*					
Hypophloeda rhizospora K.D. Hyde & E.B.G. Jones (holotype)	Mangrove	Saprobic	Roots of *Rhizophora apiculata*	Phuket, Ranong	Hyde et al. (1990); Hyde (1994)
Diaporthales Nannf. **Gnomoniaceae** G. Winter					
Neptunella longirostris (Cribb & J.W. Cribb) K.L. Pang & E.B.G. Jones (as *Gnomonia longirostris*)	Mangrove	Saprobic	Roots of *Rhizophora apiculata*	Ranong	Hyde (1994)
Microascales Luttr, ex Benny & R.K. Benj, **Halosphaeriaceae** E. Müll. & Arx ex Kohlm.					
Aniptodera chesapeakensis Shearer & M.A. Mill.	Intertidal/ Mangrove	Saprobic	Senescent and decaying wood; Decaying intertidal petioles and fronds of *N. fruticans*; Roots of *Rhizophora apiculata*; driftwood	Chachoengsao, Chon Buri, Narathiwat, Pattani, Phang-Nga, Phuket, Prachuap Khiri Khan, Ranong, Samut Prakan, Samut Sakhon, Samut Songkhram, Satun, Surat Thani, Trat, Trang	Koch (1986); Hyde et al. (1990); Hyde (1994); Pilantanapak et al. (2005); Loilong et al. (2012); Suetrong et al. (2017)
Aniptodera intermedia K.D. Hyde & Alias		Saprobic	Decaying intertidal petioles and fronds of *N. fruticans*	Trang, Satun, Samut Prakarn	Loilong et al. (2012)
Aniptosporopsis lignatilis (K.D. Hyde) K.L. Pang, C.L. Lu, W.T. Ju & E.B.G. Jones (as *Aniptodera lignatilis*)	Intertidal	Saprobic	Senescent and decaying wood	Prachuap Khiri Khan	Suetrong et al. (2017)
Paraaniptodera longispora (K.D. Hyde) K.L. Pang, C.L. Lu, W.T. Ju & E.B.G. Jones (*Aniptodera longispora*)	Intertidal	Saprobic	Senescent and decaying wood; Roots of *Rhizophora apiculata*	Phang-Nga, Phuket, Prachuap Khiri Khan, Ranong, Trat, Trang	Hyde (1989); Hyde et al. (1990); Hyde (1994); Suetrong et al. (2017)

Aniptodera mangrovei K.D. Hyde	Intertidal/ Mangrove	Saprobic	Senescent and decaying wood; Roots of *Rhizophora apiculata*	Trat, Prachuap Khiri Khan; Ranong	Hyde et al. (1990); Hyde (1994); Suetrong et al. (2017)
Aniptodera nypae K.D. Hyde	Intertidal	Saprobic	Decaying intertidal petioles and fronds of *N. fruticans*	Chachoengsao, Samut Prakan, Samut Sakon, Samut Songkhram, Samut Songkrham, Satun, Surat Thani, Trang	Pilantanapak et al. (2005); Jones et al. (2006b); Loilong et al. (2012)
Aniptodera salsuginosa Nakagiri & Tad, Ito	Intertidal	Saprobic	Senescent and decaying wood	Trat	Suetrong et al. (2017)
Antennospora quadricornuta (Cribb & J.W. Cribb) T.W. Johnson (as *Halosphaeria quadricornuta*)	Intertidal	Saprobic	Senescent and decaying wood; Roots of *Rhizophora apiculata*	Chanthaburi, Chon Buri, Narathiwat, Pattani, Phetchaburi, Phang-Nga, Phuket, Ranong, Rayong, Trat	Dethoup and Manoch (2009); Hyde (1994); Suetrong et al. (2017)
UPDATED: *Haiyanga salina* (Meyers) K.L. Pang & E.B.G. Jones (as *Antennospora salina)*	Intertidal	Saprobic	Driftwood	Phuket, Chonburi	Kohlmeyer (1984); Koch (1986)
Arenariomyces parvulus Jørg, Koch (Holotype)	Intertidal	Saprobic	Driftwood, sand	Phuket	Koch (1986)
Arenariomyces trifurcatus Höhnk	Intertidal	Saprobic	Twigs; Driftwood; spores in foam and washed-up detritus	Chon Buri, Rayong; Phuket	Kohlmeyer (1984); Koch (1986); Dethoup and Manoch (2009)
Bathyascus grandisporus K.D. Hyde	Mangrove	Saprobic	Roots of *Rhizophora apiculata*	Ranong	Hyde et al.(1990); Hyde (1994)
Carbosphaerella leptosphaerioides I. Schmidt	Intertidal	Saprobic	–	Phuket, Trat	Jones et al. (2006b)
Ceriosporopsis halima Linder	Intertidal	Saprobic	–	Phuket, Narathiwat	Jones et al. (2006b)
Cirrenalia macrocephala (Kohlm.) Meyers & R.T. Moore (as *Helicoma macrocephalum*)		Saprobic	Decaying intertidal petioles and fronds of *N. fruticans*	Satun	Loilong et al. (2012)

Table 1 contd. ...

...Table 1 contd.

Taxon	**Habitat**	**Life Mode**	**Substrate/Host**	**Province**	**References**
Cirrenalia pseudomacrocephala Kohlm.	Intertidal/ Mangrove	Saprobic	Roots of *Rhizophora apiculata*	Phang-Nga, Ranong	Hyde (1989); Hyde et al. (1990, 1993); Hyde (1994)_ Hyde et al. (1990); Hyde (1989)
Clavatospora bulbosa (Anastasiou) Nakagiri & Tubaki (as *Clavariopsis bulbosa*)	Mangrove/ Intertidal	Saprobic	Wood	Chon Buri, Narathiwat, Pattani, Phang-Nga, Phangnga, Phuket, Prachuap Khiri Khan, Ranong, Songkhla, Trat	Koch (1986); Chatmala et al. (2004)
Corollospora besarispora Sundari	Intertidal	Saprobic	sand grain	Phuket	Sundari et al. (1996)
Corollospora cinnamomea Jørg. Koch (Holotype)	Intertidal	Saprobic	Driftwood, sand	Trang, Phuket	Koch (1986)
UPDATED: *Tokurathelia colossus* (Nakagiri & Tokura) M.F. Caeiro, E. Azevedo & P. Correia (as *Corollospora colossus* Nakagiri & Tokura [as '*colossa*']	Intertidal	Saprobic	Wood	Trang, Phuket, Trat	Jones et al. (2006b)
UPDATED: *Nakagariella filiformis* (Nakagiri) P. Correia, E. Azevedo & M.F. Caeiro (as *Corollospora filiformis*)	Intertidal	Saprobic	Wood	Chon Buri, Phuket, Narathiwat, Phang Nga, Trat	Jones et al. (2006b)
UPDATED: *Shirahamella gracilis* (Nakagiri & Tokura) E. Azevedo, P. Correia & M.F. Caeiro (as *Corollospora gracilis*)	Intertidal	Saprobic	–	Phuket, Trang	Jones et al. (2006b)
Corollospora maritima Werderm	Intertidal	Saprobic	Driftwood; spores in foam and washed-up detritus	Phuket; Chon Buri	Kohlmeyer (1984); Koch (1986)
UPDATED: *Ajigaurospora pseudopulchella* (Nakagiri & Tokura) E, Azevedo, P. Correia & M.F. Caeiro (as *Corollospora pseudopulchella*)	Intertidal	Saprobic	–	Trang	Jones et al. (2006b)

UPDATED: *Keraliethelia pulchella* (Kohlm., I. Schmidt & Nair) P. Correia, E. Azevedo & M.F. Caeiro (as *Corollospora pulchella*)	Intertidal	Saprobic	Driftwood	Phuket, Chonburi	Kohlmeyer (1984); Koch (1986)
Corollosporella ramulosa (Meyers & Kohlm.) E, Azevedo, P. Correia & M.F. Caeiro (as *Varicosporina ramulosa* Meyers & Kohlm.)	Intertidal	Saprobic	Twigs	Cho Buri, Rayong	Dethoup and Manoch (2009)
Cucullosporella mangrovei (K.D. Hyde & E.B.G. Jones) K.D. Hyde & E.B.G. Jones (as *Cucullospora mangrovei*)	Mangrove/ Intertidal	Saprobic	Senescent and decaying wood; Roots of *Rhizophora apiculata*	Prachuap Khiri Khan, Phang-Nga, Ranong, Phuket, Nakhon Si Thammarat	Hyde (1989); Hyde et al. (1990); Hyde (1994); Sakayaroj et al. (2011); Suetrong et al. (2017)
Cucurbitinus constrictus (I, Schmidt) L.L. Liu & Z.Y. Liu (as *Trichocladium constrictum*)	Mangrove	Saprobic	Mangrove wood	Trat	Jones et al. (2006b)
Halosarpheia fibrosa Kohlm. & E. Kohlm	Intertidal	Saprobic	Senescent and decaying wood	Phetchaburi	Suetrong et al. (2017)
UPDATED: *Praelongicaulis kandeliae* (Abdel-Wahab & E.B.G. Jones) E.B.G. Jones, Abdel-Wahab & K.L. Pang (as *Halosarpheia kandeliae*)	mangrove	Saprobic	Decaying mangrove	Nakhon Si Thammarat, Ranong	Jones et al. (2006b); Sakayaroj et al. (2011)
Halosarpheia marina (Cribb & J.W. Cribb) Kohlm. (as *Gnomonia marina*)	Intertidal/ Mangrove	Saprobic	Senescent and decaying wood; Decaying intertidal petioles and fronds of *N. fruticans*; Roots of *Rhizophora apiculata*	Phang-Nga, Phetchaburi, Ranong, Satun	Hyde (1989); Hyde et al. (1990); Hyde (1994); Loilong et al. (2012); Suetrong et al. (2017)
Halosarpheia minuta W.F. Leong	Intertidal	Saprobic	Senescent and decaying wood	Trat, Nakhon Si Thammarat, Phang-Nga, Ranong, Trang, Chon Buri	Hyde (1989); Hyde et al. (1990); Sakayaroj et al. (2011); Suetrong et al. (2017)

Table 1 contd. ...

...Table 1 contd.

Taxon	Habitat	Life Mode	Substrate/Host	Province	References
UPDATED: *Antennospora quadricornuta* (Cribb & J.W. Cribb) T.W. Johnson (as *Halosphaeria quadricornuta*)	Mangrove		Twigs	Chon Buri	Kohlmeyer (1984); Dethoup and Manoch (2009)
Halosphaeria quadri-remis (Höhnk) Kohlm. (as *Palomyces quadri-remis* Höhnk [as '*quadriremis*']	Intertidal	Saprobic	Twigs; *Nypa fruticans*	Chon Buri; Samut Songkhram	Dethoup and Manoch (2009); Pilantanapak et al. (2005)
Halosphaeriopsis mediosetigera (Cribb & J.W. Cribb) T.W., Johnson (as *Trichocladium achrasporum*)		Saprobic	Decaying intertidal petioles and fronds of *N. fruticans*; wood	Satun, Trat	Chatmala et al. (2004); Jones et al. (2006b); Loilong et al. (2012)
Keraliethelia hydei Bagacay, Opiña, & M.S. Calabon* (holotype)	Intertidal	Saprobic	Driftwood	Phuket	Hongsanan et al. (2025)
Kochiella crispa (Kohlm.) Sakay., K.L. Pang & E.B.G. Jones (as *Remispora crispa*)	Intertidal	Saprobic	Mangrove wood	Phuket	Jones et al. (2006b)
Lignincola laevis Höhnk	Intertidal	Saprobic	Decaying intertidal petioles and fronds of *N. fruticans*; wood; Roots of *Rhizophora apiculata*; driftwood; dead branch and prop roots of *Rhizophora* sp.	Bangkok, Chachoengsao, Chon Buri, Nakhon Si Thammarat, Pattani, Phetchaburi, Phang-Nga, Phuket, Prachuap Khiri Khan, Ranong, Samut Prakarn, Samut Songkhram, Satun, Surat Thani, Trang, Trat	Kohlmeyer (1984); Hyde (1994); Pilantanapak et al. (2005); Sakayaroj et al. (2011); Loilong et al. (2012); Suetrong et al. (2017)
Lignincola nypae K.D. Hyde & Alias (holotype)		Saprobic	Decaying intertidal petioles and fronds of *N. fruticans*	Trang, Satun, Surat Thani, Samut Sakorn	Loilong et al. (2012)
Lignincola tropica Kohlm.	Mangrove	Saprobic	Mangrove wood	Ranong	Hyde et al. (1990); Hyde et al. (1990)
Nais inornata Kohlm.	Intertidal	Saprobic	Twigs	Chon Buri; Pattani, Ranong	Dethoup and Manoch (2009); Jones et al. (2006b)

Natantispora retorquens (Shearer & J.L. Crane) J, Campb., J.L. Anderson & Shearer (as *Halosarpheia retorquens*)	Mangrove/ Intertidal	Saprobic	Senescent and decaying wood; Decaying intertidal petioles and fronds of *N. fruticans*	Trat, Trang, Surat Thani	Jones et al. (2006b); Loilong et al. (2012); Suetrong et al. (2017)
Nectria marina Dayar, & K.D. Hyde* (holotype)	mangrove	Saprobic	on submerged unknown mangrove wood	Krabi	Dayarathne et al. (2020a)
Neptunella longirostris (Cribb & J.W. Cribb) K.L. Pang & E.B.G. Jones (as *Gnomonia longirostris*)	Intertidal	Saprobic	Decaying intertidal petioles and fronds of *N. fruticans*; Senescent and decaying wood	Narathiwat, Phetchaburi, Phuket, Prachuap Khiri Khan, Ranong, Samut Prakan, Samut Sakhon, Samut Songkhram, Satun, Trang, Trat	Hyde et al.(1990); Hyde (1994); Pilantanapak et al. (2005); Loilong et al. (2012); Suetrong et al. (2017)
Oceanitis cincinnatula (Shearer & J.L. Crane) J. Dupont & E.B.G. Jones (as *Halosarpheia cincinnatulum*)	Intertidal/ Mangrove	Saprobic	Senescent and decaying wood; Decaying intertidal petioles and fronds of *N. fruticans*	Trat, Trang, Satun, Ranong	Jones et al. (2006b); Loilong et al. (2012); Suetrong et al. (2017)
Okeanomyces cucullatus (Kohlm.) K.L. Pang & E.B.G. Jones (as *Remispora cucullata*)	Mangrove	Saprobic	Mangrove wood	Ranong	Hyde et al. (1990)
Okeanomyces marinus Calabon, E.B.G. Jones, Boonmee, K.D. Hyde* (holotype)		Saprobic	Decaying wood	Phuket	Hyde et al. (2020a)
Ophiodeira monosemeia Kohlm. & Kohlm. Volkm.-Kohlm.	Mangrove	Saprobic	Mangrove wood	Ranong	Hyde et al. (1990); Hyde (1994)
Panorbis viscosus (I, Schmidt) J. Campb., J.L. Anderson & Shearer (as *Halosarpheia viscosa*)		Saprobic	Decaying intertidal petioles and fronds of *N. fruticans*; Roots of *Rhizophora apiculata*	Satun, Samut Prakarn, Pattani, Phang-Nga, Ranong	Hyde (1989); Hyde et al. (1990); Hyde (1994); Loilong et al. (2012)
Paraaniptodera longispora (K.D. Hyde) K.L. Pang, C.L. Lu, W.T. Ju & E.B.G. Jones (as *Aniptodera longispora*)		Saprobic	Decaying intertidal petioles and fronds of *N. fruticans*	Trang	Loilong et al. (2012)
Pseudolignincola siamensis Chatmala & E.B.G. Jones (holotype)	Mangrove	Saprobic	*Nypa fruticans*	Trang	Jones et al. (2006a)

Table 1 contd. ...

...Table 1 contd.

Taxon	Habitat	Life Mode	Substrate/Host	Province	References
Sordariomycetes *incertae sedis*					
Rimaconus multiguttulatus Dayar, & K.D. Hyde* (holotype)	Submerged	Saprobic	on submerged wood of *Bruguiera* sp.	Ranong	Dayarathne et al. (2020a)
Myrmecridiales Crous **Myrmecridiales *incertae sedis***					
Myrmecridium hydei Asghari, Phukhams. & E.B.G. Jones*(holotype)	Mangrove	Saprobic	*decaying branches of Areca*	Prachuap Khiri Khan	Asghari et al. (2023)
Sordariales Chadef, ex D, Hawksw. & O.E. Erikss **Lasiosphaeriaceae** Nannf					
Zopfiella latipes (N, Lundq.) Malloch & Cain (as *Tripterospora latipes*)	Intertidal	Saprobic	*Nypa fruticans*	Samut Songkhram	Pilantanapak et al. (2005); Jones et al. (2006b)
Sordariomycetes families *incertae sedis* **Lautosporaceae** Kohlm., Volkm.-Kohlm. & O.E. Erikss,					
Lautospora gigantea K.D. Hyde & E.B.G. Jones (Holotype)	Mangrove	Saprobic	Mangrove wood	Phuket, Ranong	Hyde et al. (1990)
Lautospora simillima Kohlm., Volkm.-Kohlm. & O.E. Erikss.	Intertidal	Saprobic	Senescent and decaying wood	Trat, Prachuap Khiri Khan	Suetrong et al. (2017)
Lulworthiales Kohlm., Spatafora & Volkm.-Kohlm. **Lulworthiaceae** Kohlm., Spatafora & Volkm.-Kohlm.					
Cumulospora marina I, Schmidt	Mangrove	Saprobic	decayed seeds of *Heritiera littoralis*, wood	Phuket, Ranong	Chatmala et al. (2004); Jones et al. (2006b)
Moromyces varius (Chatmala & Somrith,) Abdel-Wahab, K.L. Pang, Nagah., Abdel-Aziz & E.B.G. Jones (as *Cumulospora varia*) (Holotype)	Intertidal	Saprobic	Driftwood	Trang, Trat, Pattani, Rayong	Chatmala et al. (2004)

Hydea pygmea (Kohlm.) K.L. Pang & E.B.G. Jones (as *Cirrenalia pygmea*)	Intertidal/ Mangrove	Saprobic	Driftwood; tube of a dead prop root (*Rhizophora* sp.)	Phuket Chon Buri, Krabi, Phang-Nga, Ranong, Samut Songkhram, Phuket, Trang, Trat, Narathiwat	Kohlmeyer (1984); Koch (1986); Hyde 1989; Hyde et al. (1990); Chatmala et al. (2004); Sakayaroj et al. |(2011)
Lindra thalassiae Orpurt, Meyers, Boral & Simms		Saprobic	Decaying mangrove	Nakhon Si Thammarat Province	Sakayaroj et al. (2011)
Lulworthia kniepii (Ade & Bauch) Petr. (as *Ophiobolus kniepii*)	Mangrove/ Intertidal	Saprobic	Mangrove wood/?	Trang, Phuket	Jones et al. (2006b)
Matsusporium tropicale (Kohlm.) E.B.G. Jones & K.L. Pang (as *Cirrenalia tropicalis*)	Intertidal/ Mangrove	Saprobic	Senescent and decaying wood; Roots of *Rhizophora apiculata*	Chanthaburi, Narathiwat, Pattani, Phang-Nga, Phuket, Ranong, Trat	Hyde (1989); Hyde et al. (1990); Hyde (1994); Chatmala et al. (2004); Suetrong et al. (2017)
Moleospora maritima Abdel-Wahab, Abdel-Aziz & Nagah.	Intertidal	Saprobic	Senescent and decaying wood	Trat	Suetrong et al. (2017)
Moromyces varius (Chatmala & Somrith,) Abdel-Wahab, K.L. Pang, Nagah., Abdel-Aziz & E.B.G. Jones (as *Cumulospora varia*)	Intertidal	Saprobic	Senescent and decaying wood	Trat	Suetrong et al. (2017)
Paralulworthia halima (Anastasiou) M, Gonç., A, Abreu & A. Alves (as *Papulaspora halima*)	Intertidal	Saprobic	Wood	Ranong	Chatmala et al. (2004); Jones et al. (2006b)
Sammeyersia grandispora (Meyers) Sheng Y., Guc, E.B.G. Jones & K.L. Pang (as *Lulworthia grandispora*)*	Intertidal/ mangrove	Saprobic	Decaying intertidal petioles and fronds of *N. fruticans*; Senescent and decaying wood; Twigs; Seedlings of *Rhizophora apiculata*	Chanthaburi, Chon Buri, Nakhon Si Thammarat, Phetchaburi, Phang-Nga, Phuket, Prachuap Khiri Khan, Ranong, Rayong, Samut Prakan, Samut Sakhon, Samut Songkhram, Satun, Trat, Trang	Hyde (1989); Dethoup and Manoch (2009); Hyde et al.(1990); Hyde (1994); Pilantanapak et al. (2005); Sakayaroj et al. (2011); Loilong et al. (2012); Suetrong et al. (2017; Chethana et al. (2021)

Table 1 contd. ...

...Table 1 contd.

Taxon	Habitat	Life Mode	Substrate/Host	Province	References
Zalerion maritima (Linder) Anastasiou (as *Helicoma maritimum*)	Intertidal	Saprobic	Driftwood	Phuket	Jones et al. (2006b)
Halenospora varia (Anastasiou) E.B.G. Jones (as *Zalerion varia*)	Mangrove/ Intertidal	Saprobic	Wood	Chanthaburi, Chon Buri, Narathiwat, Pattani, Prachuap Khiri Khan, Phuket, Ranong, Rayong, Samut Prakan, Samut Songkhram, Songkhla, Trat, Trang	Chatmala et al. (2004); Jones et al. (2006b)
Diaporthales Nannf. **Melanconiellaceae** Senan., Maharachch, & K.D. Hyde					
Pseudomelanconiella mangrovei Apurillo, Phukhams., E.B.G. Jones & K.D. Hyde* (holotype, new genus)	Mangrove	Saprobic	decomposing branch of *Avicennia marina*	Prachuap Khiri Khan	Apurillo et al. (2025)
Hypocreales Lindau **Nectriaceae** Tul. & C. Tul.					
Chaetopsina aurantisalinicola Dayar., E.B.G. Jones & K.D. Hyde* (holotype)	mangrove	Saprobic	on decaying submerged wood of *Rhizophora* sp.	Ranong	Dayarathne et al. (2020a)
Fusarium citri M.M. Wang, Qian Chen & L. Cai*	Sea cucumber		*Holothuria scabra*	Phang Nga	Loilong et al. (2012)
Fusarium equiseti (Corda) Sacc,* (as *Selenosporium equiseti*)	Sea cucumber		*Holothuria scabra*	Phang Nga	Loilong et al. (2012)
Fusarium pernambucanum A.C.S. Santos, C.S. Lima, P.V., Tiago & N.T. Oliveira*	Sea cucumber		*Holothuria scabra*	Phang Nga	Loilong et al. (2012)
Fusarium sulawesiense Maryani, Sand.-Den., L, Lombard, Kema & Crous*	Sea cucumber		*Holothuria scabra*	Phang Nga	Loilong et al. (2012)
Fusicolla gigantispora Dayar, & K.D. Hyde* (holotype)	mangrove	Saprobic	on dead bark of *Avicennia marina*; on dead bark of *Bruguiera* sp.	Krabi	Dayarathne et al. (2020a)

Neocosmospora rhizophorae Dayar,* (holotype)	Mangrove	Saprobic	Bark of *Rhizophora* mangle	Krabi	Dayarathne et al. (2020a)
Fusarium neocosmosporiellum O'Donnell & Geiser (as *Neocosmospora vasinfecta*)*,	Submerged		Mycale sp.	Chonburi	Chalearmsrimuang et al. (2019)
Xenoacremonium brunneosporum Dayar., E.B.G. Jones & K.D. Hyde* (holotype)	mangrove	Saprobic	on aerial bark of *Rhizophora* mucronata	Krabi	Dayarathne et al. (2020a)
Neoacremoniaceae L.W. Hou, L, Cai & Crous					
Neoacremonium taiwanense (K.L. Pang, Alias & E.B.G. Jones) L.W. Hou, L. Cai & Crous (as *Sedecimiella taiwanensis*)	Intertidal	Saprobic	Senescent and decaying wood	Prachuap Khiri Khan, Phetchaburi	Suetrong et al. (2017)
Sordariomycetes family *incertae sedis* **Neoarthriniaceae** S, Mukhop., Phukhams. & E.B.G. Jones					
Neoarthrinium aquaticum S, Mukhop., Phukhams, & E.B.G. Jones* (holotype)	Mangrove	Saprobic	decaying submerged stems of *Acrostichum aureum*	Prachuap Khiri Khan	Mukhopadhyay et al. (2025)
Stachybotryaceae L, Lombard & Crous					
Albifimbria verrucaria (Alb, & Schwein,) L, Lombard & Crous (as *Peziza verrucaria*)	Sea cucumber		*Holothuria scabra*	Phang Nga	Loilong et al. (2012)
Ophiocordycipitaceae G.H. Sung, J.M. Sung, Hywel-Jones & Spatafora					
Purpureocillium lilacinum (Thom) Luangsa-ard, Houbraken, Hywel-Jones & Samson* (as *Penicillium lilacinum*)	Mangrove	Saprobic	wood of *Bruguiera* sp.	Krabi	Dayarathne et al. (2020a)
Xylariales Nannf. **Diatrypaceae** Nitschke					
Cryptosphaeria bathurstensis (K.D. Hyde & Rappaz) Dayar, & K.D. Hyde (as *Eutypa bathurstensis)*	Intertidal	Saprobic	Senescent and decaying wood	Trat, Prachuap Khiri Khan, Phetchaburi; Phuket, Trang	Jones et al. (2006b); Suetrong et al. (2017)
Cryptovalsa mangrovei Abdel-Wahab & Inderb.	Intertidal	Saprobic	Senescent and decaying wood	Phetchaburi; Phuket	Jones et al. (2006b); Suetrong et al. (2017)

Table 1 contd. ...

...Table 1 contd.

Taxon	**Habitat**	**Life Mode**	**Substrate/Host**	**Province**	**References**
Diatrype mangrovei Dayarathne & K.D.Hyde	Submerged	Saprobic	Decaying wood of *Bruguiera* cylindrica	Krabi	Dayarathne et al. (2020a)
Eutypella naqsii K.D. Hyde	Intertidal	Saprobic	Senescent and decaying wood	Phetchaburi; Phuket	Jones et al. (2006b); Suetrong et al. (2017)
Halocryptovalsa salicorniae Dayar. & K.D. Hyde* (holotype)	Intertidal	Saprobic	Decayed wood of *Avicennia* sp.	Phang-nga	Dayarathne et al. (2020b)
*Halodiatrype Avicenniae** Dayar. & K.D. Hyde (holotype)	Submerged	Intertidal	Decaying wood of *Rhizophora* sp.	Phetchaburi	Dayarathne et al. (2016); Dayarathne et al. (2020a)
Halodiatrype salinicola Dayar, & K.D. Hyde	Intertidal	Saprobic	Wood	Phang-nga	Dayarathne et al. (2016)
Halodiatrype mangrovei (K.D. Hyde) Dayar. & K.D. Hyde (as *Cryptosphaeria mangrovei*)	Mangrove	Saprobic	Seedlings of *Rhizophora apiculata*	Ranong	Hyde (1994)
Pedumispora rhizophorae K.D. Hyde & E.B.G. Jones (Holotype)	Mangrove	Saprobic	Mangrove wood; Seedlings of *Rhizophora apiculata*	Ranong	Hyde and Jones (1992); Hyde (1994)
Peroneutypa hibisci Apurillo, Phukhams., E.B.G. Jones & K.D. Hyde* (holotype)	Mangrove	Saprobic	decomposing branches of Hibiscus tiliaceus	Prachuap Khiri Khan	Apurillo et al. (2025)
Rhizophila marina K.D. Hyde & E.B.G. Jones	Mangrove/ intertidal	Saprobic	Seedlings of *Rhizophora apiculata*; Senescent and decaying wood	Ranong, Chanthaburi, Phang-Nga, Phuket, Ranong, Trang, Prachuap Khiri Khan, Trat	Hyde (1989); Hyde et al. (1990); Hyde (1994); Suetrong et al. (2017)
Fasciatisporaceae S.N. Zhang, Jian K., Liu & K.D. Hyde					
Fasciatispora lignicola Alias, E.B.G. Jones & Kuthub,	Intertidal	Saprobic	*Nypa fruticans*; Mangrove wood *Rhizophora apiculata*	Samut Songkhram	Hyde et al. (1990); Hyde (1994); Pilantanapak et al. (2005)
Fasciatispora nypae K.D. Hyde		Saprobic	Decaying intertidal petioles and fronds of *N. fruticans*	Satun, Samut Prakarn, Chachoengsao	Loilong et al. (2012)
Oxydothidaceae S, Konta & K.D. Hyde					
Ophiodeira monosemeia Kohlm. & Volkm.-Kohlm.	Mangrove	Saprobic	Roots of *Rhizophora apiculata*	Ranong	Hyde (1994)

Oxydothis nypae K.D. Hyde & Nakagiri	Intertidal/ Mangrove	Saprobic	Decaying intertidal petioles and fronds of *N. fruticans*; wood	Chachoengsao, Samut Prakan, Samut Songkhram, Satun, Trang	Hyde and Nakagiri (1989); Pilantanapak et al. (2005); Loilong et al. (2012)
Phomatosporales Senan., Maharachch, & K.D. Hyde **Phomatosporaceae** Senan, & K.D. Hyde					
Phomatospora nypicola K.D. Hyde & Alias		Saprobic	Decaying intertidal petioles and fronds of *N. fruticans*	Satun, Surat Thani, Samut Prakarn, Samut Sakorn	Loilong et al. (2012)
Phyllachorales M.E. Barr **Phyllachorales genera *incertae sedis***					
Marinosphaera mangrovei K.D. Hyde	Intertidal/ mangrove	Saprobic	Decaying intertidal petioles and fronds of *N. fruticans*; Senescent and decaying wood; twig; Roots of *Rhizophora apiculata*	Chanthaburi, Chon Buri, Nakhon Si Thammarat, Narathiwat, Phang-Nga, Phetchaburi, Phuket, Prachuap Khiri Khan, Ranong, Rayong, Samut Prakan, Satun, Trang, Trat	Hyde (1989); Dethoup and Manoch (2009); Hyde et al. (1990); Hyde (1994); Sakayaroj et al. (2011); Loilong et al. (2012); Suetrong et al. (2017)
Glomerallales **Plectosphaerellaceae** W, Gams, Summerb. & Zare					
Musicillium theobromae (Turconi) Zare & W, Gams* (as *Stachylidium theobromae*)	Intertidal	Saprobic	decaying *Areca* wood	Prachuap Khiri Khan	Asghari et al. (2025)
Savoryellales Boonyuen, Suetrong, Sivichai, K.L. Pang & E.B.G. Jones **Savoryellaceae** Jaklitsch & Réblová					
Bactrodesmium linderi (J.L. Crane & Shearer) M.E. Palm & E.L. Stewart (as *Trichocladium linderi*)	Intertidal	Saprobic	Senescent and decaying wood	Pattani, Phang-Nga, Phangnga, Phuket, Ranong, Trang, Trat	Chatmala et al. (2004; Suetrong et al. (2017)
Savoryella appendiculata K.D. Hyde & E.B.G. Jone	Intertidal/ Mangrove	Saprobic	Senescent and decaying wood	Narathiwat, Trat	Jones et al. (2006b); Suetrong et al. (2017)

Table 1 contd. ...

...Table 1 contd.

Taxon	Habitat	Life Mode	Substrate/Host	Province	References
Savoryella lignicola E.B.G. Jones & R.A. Eaton	Intertidal/ Mangrove	Saprobic	Senescent and decaying wood; Decaying intertidal petioles and fronds of *N. fruticans*; Driftwood	Chachoengsao, Nakhon Si Thammarat, Phetchaburi, Phang-Nga, Prachuap Khiri Khan, Phuket, Ranong, Samut Prakarn, Samut Songkhram, Satun, Surat Thani, Trang, Trat	Koch (1986); Hyde et al. (1990); Hyde (1994); Sakayaroj et al. (2011); Loilong et al. (2012); Suetrong et al. (2017)
Savoryella longispora E.B.G. Jones & K.D. Hyde	Mangrove/ Intertidal	Saprobic	Roots of *Rhizophora apiculata*; Senescent and decaying wood	Phetchaburi, Phuket, Prachuap Khiri Khan, Ranong, Trat	Hyde (1994); Suetrong et al. (2017)_
Savoryella melanospora Abdel-Wahab & E.B.G. Jones	Intertidal	Saprobic	Senescent and decaying wood	Trat	Suetrong et al. (2017)
Savoryella nypae (K.D. Hyde & Goh) S.N. Zhang, K.D. Hyde & Jian K., Liu (as *Trichocladium nypae*)*	Mangrove/ Intertidal	Saprobic	submerged petiole of *Nypa fruticans*; submerged seeds of *Avicennia marina*; Senescent and decaying wood; Decaying intertidal petioles and fronds of *N. fruticans*; submerged seeds of *Avicennia marina*	Chachoengsao, Chanthaburi, Narathiwat, Phangnga, Phang-Nga, Ranong, Samut Prakan, Samut Sakon, Samut Songkhram, Satun, Songkhla, Surat Thani, Trat, Trang	Zhang et al. (2019a)_ Pilantanapak et al. (2005)_ Manawasinghe et al. (2022); Suetrong et al. (2017); Pilantanapak et al. (2005); Loilong et al. (2012); Jones et al. (2006b) Chatmala et al. (2004)
Savoryella paucispora (Cribb & J.W. Cribb) J, Koch (as *Leptosphaeria paucispora*)	Mangrove / Intertidal	Saprobic	Senescent and decaying wood; Decaying intertidal petioles and fronds of *N. fruticans*; twigs; Driftwood	Chachoengsao, Chanthaburi, Chon Buri, Nakhon Si Thammarat, Phuket, Samut Prakan, Samut Sakhon, Samut Songkhram, Satun, Trat	Koch (1986); Dethoup and Manoch (2009); Sakayaroj et al. (2011); Loilong et al. (2012); Suetrong et al. (2017)
Sordariales Chadef, ex D, Hawksw, & O.E. Erikss. **Sordariales genera *incertae sedis***					
Biconiosporella corniculata Schaumann	Intertidal	Saprobic	–	Phuket, Trat	Jones et al. (2006b)

Sordariomycetes genera *incertae sedis*					
Mangrovispora pemphii K.D. Hyde & Nakagiri	Mangrove/ Intertidal	Saprobic	Senescent and decaying wood	Nakhon Si Thammarat, Prachuap Khiri Khan, Phuket, Trat	Jones et al. (2006b); Sakayaroj et al. (2011); Suetrong et al. (2017)
Sporidesmiales Crous **Sporidesmiaceae** Fr.					
Sporidesmium eupatoriicola M.B. Ellis			Twigs		Dethoup and Manoch (2009)
Tirisporellales Suetrong, E.B.G. Jones & K.L. Pang **Tirisporellaceae** Suetrong, E.B.G. Jones & K.L. Pang					
Tirispora unicaudata E.B.G. Jones & Vrijmoed		Saprobic	Decaying intertidal petioles and fronds of *N. fruticans*	Satun	Loilong et al. (2012)
Tirisporella beccariana (Ces,) E.B.G. Jones, K.D. Hyde & Alias* (as *Sphaeria beccariana*)	Mangrove/ Intertidal	Saprobic	Senescent and decaying wood; intertidal petiole of *Nypa fruticans* Decaying intertidal petioles and fronds of *N. fruticans*	Chachoengsao, Narathiwat, Phetchaburi, Phang-Nga, Ranong, Samut Prakarn, Samut Sakorn, Satun, Surat Thani, Trang, Trat	Loilong et al. (2012); Suetrong et al. (2017); Hyde et al. (2020b)
Torpedosporales E.B.G. Jones, Abdel-Wahab & K.L. Pang **Etheirophoraceae** Rungjind., Somrith. & Suetrong					
Etheirophora bijubata Kohlm. & Volkm.-Kohlm.	Intertidal	Saprobic	Senescent and decaying wood	Trat	Suetrong et al. (2017)
Etheirophora blepharospora (Kohlm. & E, Kohlm.) Kohlm. & Volkm.-Kohlm. (as *Keissleriella blepharospora*)	Intertidal	Saprobic	Senescent and decaying wood	Trat	Suetrong et al. (2017)
Swampomyces aegyptiacus Abdel-Wahab, El-Shar. & E.B.G. Jones	Intertidal		Decaying wood		Suetrong et al. (2017)
Swampomyces triseptatus K.D. Hyde & Nakagiri	Mangrove/ Intertidal	Saprobic	Roots of *Rhizophora apiculata*	Ranong	Hyde (1994)

Table 1 contd. ...

...Table 1 contd.

Taxon	Habitat	Life Mode	Substrate/Host	Province	References
Juncigenaceae E.B.G. Jones, Abdel-Wahab & K.L. Pang					
Fulvocentrum aegyptiacum (Abdel-Wahab, El-Shar. & E.B.G. Jones) E.B.G. Jones & Abdel-Wahab (as *Swampomyces aegyptiacus*)	Intertidal	Saprobic	Senescent and decaying wood	Prachuap Khiri Khan	Suetrong et al. (2017)
Glomerulispora mangrovei Abdel-Wahab & Nagah.	Intertidal		Decaying wood		Suetrong et al. (2017)
Torpedosporaceae E.B.G. Jones & K.L. Pang					
Torpedospora mangrovei (Abdel-Wahab & Nagah.) E.B.G. Jones & Abdel-Wahab (as *Glomerulispora mangrovei*)	Intertidal	Saprobic	Senescent and decaying wood	Trat	Suetrong et al. (2017)
Torpedospora radiata Meyers	Intertidal	Saprobic	Driftwood, twigs	Chanthaburi, Chon Buri, Narathiwat, Nakhon Si Thammarat, Pattani, Phuket, Rayong, Trat, Trang	Koch (1986); Dethoup and Manoch (2009); Sakayaroj et al. (2011)
Halodiatrype salinicola Dayar, & K.D. Hyde* *(holotype)*	Mangrove	Saprobic	on submerged decayed wood of *Rhizophora* sp.	Krabi	Dayarathne et al. (2020a) Dayarathne et al. (2016)
Halorosellinia krabiensis Dayar, & K.D. Hyde* (holotype)	Intertidal	Saprobic	Decaying wood of unidentified mangrove	Krabi	Dayarathne et al. (2020a)
Halorosellinia oceanica (S, Schatz) Whalley, E.B.G. Jones, K.D. Hyde & Læssøe (as *Hypoxylon oceanicum*)	Mangrove/ Intertidal	Saprobic	*Nypa fruticans*; Roots of *Rhizophora apiculata*; Senescent and decaying wood	Chanthaburi, Krabi, Nakhon Si Thammarat, Phetchaburi, Phang-Nga, Phuket, Prachuap Khiri Khan, Ranong, Samut Songkhram, Trat, Trang	Hyde (1989); Hyde (1994); Pilantanapak et al. (2005); Sakayaroj et al. (2011; Suetrong et al. (2017)
Halorosellinia rhizophorae Dayar., E.B.G. Jones & K.D. Hyde* (holotype)	Mangrove	Saprobic	Decaying wood of *Xylocarpus* sp.	Krabi, Phang Nga	Dayarathne et al. (2020a)
Halorosellinia xylocarpi Dayar, & K.D. Hyde* *(holotype)*	Mangrove	Saprobic	On decaying submerged wood of *Xylocarpus*	Ranong	Dayarathne et al. (2020a)

Hypoxylon aurantium Dayar., E.B.G. Jones & K.D. Hyde* *(holotype)*	Mangrove	Saprobic	on submerged decaying wood of *Rhizophora apiculata*	Krabi	Dayarathne et al. (2020a)
Hypoxylon hypomiltum Mont,	Mangrove	Saprobic	Roots of *Rhizophora apiculata*	Ranong	Hyde (1994)
Hypoxylon mangrovei Dayar., E.B.G. Jones & K.D. Hyde* (holotype)	Mangrove	Saprobic	On decaying wood of *Rhizophora*	Prachaub Kiri Khan	Dayarathne et al. (2020a)
Nemania hydei Apurillo, Phukhams. & E.B.G. Jones* (holotype)	Mangrove	Saprobic	dead branch of *Avicennia marina*	Prachuap Khiri Khan	Apurillo et al. (2024)
Nemania maritima Y.M. Ju & J.D. Rogers* (holotype)	Intertidal	Saprobic	Senescent and decaying wood; on submerged root of *Rhizophora* sp.	Krabi, Phetchaburi, Prachuap Khiri Khan, Trat	Jones et al. (2006b); Suetrong et al. (2017); Dayarathne et al. (2020a)
Nemania phetchaburiensis Dayar., E.B.G. Jones & K.D. Hyde* *(holotype)*	Submerged	Saprobic	On decaying wood of *Rhizophora*	Phetchaburi	Dayarathne et al. (2020a)
Nemania viridi Dayar, & K.D. Hyde* *(holotype)*	Submerged	Saprobic	On submerged decaying wood of *Rhizophora*	Phetchaburi	Dayarathne et al. (2020a)
Rosellinia maritima R, Asghari, Phukhams. & K.D. Hyde* (holotype)	Intertidal	Saprobic	decaying *Areca* wood	Prachuap Khiri Khan	Asghari et al. (2025)
Xylariales genera *incertae sedis*					
Anthostomella nypae K.D. Hyde, B.S. Lu & Alias	Mangrove	Saprobic	Decaying intertidal petioles and fronds of *N. fruticans*; wood	Chachoengsao, Phang-Nga, Samut Prakarn, Satun, Trang	Jones et al. (2006b); Loilong et al. (2012)
BASIDIOMYCOTA R.T. Moore **AGARICOMYCETES** Doweld **Agaricales** Underw. **Marasmiaceae** Roze ex Kühner					
Calathella mangrovei E.B.G. Jones & Agerer	Mangrove	Saprobic	*Nypa fruticans*; wood	Ranong; Nakhon Si Thammarat	Jones et al. (2006b); Sakayaroj et al. (2011)

Table 1 contd. ...

...Table 1 contd.

Taxon	Habitat	Life Mode	Substrate/Host	Province	References
Cyphellopsidaceae Jülich					
Halocyphina villosa Kohlm. & E, Kohlm.	Mangrove/ Intertidal	Saprobic	Decaying intertidal petioles and fronds of *N. fruticans*; Senescent and decaying wood; Roots of *Rhizophora apiculata*; Driftwood; stump of Ceriops sp., dead prop roots of *Rhizophora* sp.	Chanthaburi, Chon Buri, Nakhon Si Thammarat, Narathiwat, Phang-Nga, Phuket, Ranong, Surat Thani, Trang, Trat	Suetrong et al. (2017); Loilong et al. (2012); Hyde (1994); Hyde et al. 1990); Hyde (1994); Koch (1986); Kohlmeyer (1984); (Sakayaroj et al. (2011); Hyde (1989)
Nia vibrissa R.T. Moore & Meyers	Mangrove	Saprobic	Mangrove wood	Phuket, Trang; Nakhon Si Thammarat	Jones et al. (2006b); Sakayaroj et al. (2011)
Physalacriaceae Corner					
Physalacria maipoensis Inderb, & Desjardin	Intertidal	Saprobic	Senescent and decaying wood	Trat	Suetrong et al. (2017)
Polyporales Gäum, **Polyporaceae** Fr. ex Corda					
Cyanoporus fuligo (Berk. & Broome) Y.C. Dai, W., Li Mao & Yuan Yuan (as *Polyporus fuligo*)		Saprobic	Decaying intertidal petioles and fronds of *N. fruticans*	Trang, Samut Prakarn	Loilong et al. (2012)
AGARICOSTILBOMYCETES R, Bauer, Begerow, J.P., Samp., M, Weiss & Oberw, **Agaricostilbales** Oberw, & R, Bauer **Agaricostilbaceae** Locq,					
Sterigmatomyces halophilus Fell	Corals		*Sarcophyton* sp.	Chonburi	Kaewkrajay et al. (2020)
CYSTOBASIDIOMYCETES R, Bauer, Begerow, J.P., Samp., M, Weiss & Oberw. **Cystobasidiales** R, Bauer, Begerow, J.P., Samp., M, Weiss & Oberw. **Cystobasidiaceae** Gäum,					

Cystobasidium calyptogenae (Nagah., Hamam., Nakase & Horikoshi) Yurkov, Kachalkin. H.M., Daniel, M. Groenew., Libkind, V. de García, Zalar, Gouliam., Boekhout & Begerow (as *Rhodotorula calyptogenae*)*	Corals		*Klyxum* sp.; *Cladiella* sp.	Chonburi	Kaewkrajay et al. (2020)
MICROBOTRYOMYCETES R, Bauer, Begerow, J.P., Samp., M, Weiss & Oberw, miicrobotryomycetes **Sporidiobolales** Doweld **Sporidiobolaceae** R.T. Moore					
Rhodosporidiobolus fluvialis (Fell, Kurtzman, Tallman & J.D., Buck) Q.M., Wang, F.Y., Bai, M. Groenew, & Boekhout (as *Rhodosporidium fluviale*)*	Corals; Zoanthid		*Cladiella* sp.; Unknown zoanthid	Chonburi	Kaewkrajay et al. (2020)
Rhodotorula diobovata (S.Y., Newell & I.L., Hunter) Q.M., Wang, F.Y., Bai, M, Groenew. & Boekhout (as *Rhodosporidium diobovatum*)*	Corals/ Zoanthid		*Sinularia* sp.; *Cladiella* sp.; *Klyxum* sp. Unknown zoanthid	Chonburi	Kaewkrajay et al. (2020)
Rhodotorula mucilaginosa (A, Jörg,) F.C., Harrison (as *Torula mucilaginosa*)*	Corals/ Zoanthid		*Sarcophyton* sp.; *Cladiella* sp.; Unknown zoanthid;*Pocillopora damicornis*;Unknown coral	Chonburi	Kaewkrajay et al. (2020)
Rhodotorula toruloides (I. Banno) Q.M. Wang. F.Y. Bai, M. Groenew. & Boekhout (as *Rhodosporidium toruloides*)*	Zoanthid		Unknown zoanthid	Chonburi	Kaewkrajay et al. (2020)
TREMELLOMYCETES Doweld **Filobasidiales** Jülich **Filobasidiaceae** L.S. Olive					
Filobasidium uniguttulatum Kwon-Chung*	Corals		*Pocillopora damicornis*	Chonburi	Kaewkrajay et al. (2020)

Table 1 contd. ...

...Table 1 contd.

Taxon	Habitat	Life Mode	Substrate/Host	Province	References
Naganishia liquefaciens (Saito & M, Ota) Xin Zhan Liu, F.Y. Bai, M. Groenew. & Boekhout (as *Torulopsis liquefaciens*)*	Corals		*Sinularia* sp.; *Pocillopora damicornis*	Chonburi	Kaewkrajay et al. (2020)
Tremellales Fr. **Bulleribasidiaceae** Xin Zhan Liu, F.Y. Bai, M. Groenew. & Boekhout					
Vishniacozyma victoriae (M.J., Montes, Belloch, Galiana, M.D. García, C, Andrés, S. Ferrer. Torr.-Rodr. & J. Guinea) Xin Zhan Liu, F.Y. Bai, M. Groenew, & Boekhout	Corals		Unknown coral	Chonburi	Kaewkrajay et al. (2020)
Rhynchogastremaceae Oberw, & B, Metzler					
Papiliotrema laurentii (Kuff,) Xin Zhan Liu, F.Y. Bai, M. Groenew, & Boekhout (as *Torula laurentii*)*	Corals/ Zoanthid		*Sinularia* sp.; Pavona sp.; Pavona decussata;*Cladiella* sp.; Unknown zoanthid	Chonburi	Kaewkrajay et al. (2020)
MUCOROMYCOTA **MUCOROMYCETES** Doweld **Mucorales** Dumort, **Cunninghamellaceae** Naumov ex R.K. Benj.					
Cunninghamella elegans Lendn, (as *Cunninghamella bertholletiae* Stadel)	Sea cucumber		*Holothuria scabra*	Phang Nga	Loilong et al. (2012)
Cunninghamella blakesleeana Lendn.	Sea cucumber		*Holothuria scabra*	Phang Nga	Loilong et al. (2012)

additional representatives in Agaricostilbomycetes (1 species), Cystobasidiomycetes (1 species), Microbotryomycetes (1 species), and Tremellomycetes (2 species). These taxa are distributed across seven orders: Agaricales and Sporidiobolales (4 species combined), Filobasidiales and Tremellales (2 species each), and Polyporales, Agaricostilbales, and Cystobasidiales (1 species each). To date, 13 genera of marine basidiomycetes have been reported from Thailand: *Calathella, Halocyphina, Nia, Physalacria, Cyanoporus, Sterigmatomyces, Cystobasidium, Rhodosporidiobolus, Rhodotorula, Filobasidium, Naganishia, Papiliotrema*, and *Vishniacozyma* (Kohlmeyer, 1984; Koch, 1986; Hyde, 1989; Hyde et al., 1990; Hyde, 1994; Jones et al., 2006b; Sakayaroj et al., 2011; Loilong et al., 2012; Suetrong et al., 2017; Kaewkrajay et al., 2020).

3.2 *Molecular Phylogenetics*

To date, approximately 126 marine fungal species have been sequenced in Thailand, representing a wide taxonomic breadth, particularly within the Ascomycota (Figs. 2, 3). The majority of sequenced taxa belong to the classes Dothideomycetes and Sordariomycetes, with fewer representatives of Eurotiomycetes, Saccharomycetes, and Pichiomycetes.

Ascomycota: Dothideomycetes is the most diverse group, with about 40 species spanning 34 genera, 25 families, and 11 orders. The order Pleosporales is especially well represented (26 species), followed by Dyfrolomycetales and Hysteriales (3 species each). Other lineages include Tubeufiales (*Parahelicomyces dictyosporus, Kamalomyces mangrovei*) and single representatives from Botryosphaeriales, Cladosporiales, Oncopodiellales, Patellariales, Stigmatodiscales (*Asterodiscus mangrovei*), and Venturiales (*Scolecobasidium musae*).

Sordariomycetes is similarly diverse, comprising around 45 species distributed across 33 genera, 17 families, and 12 orders. The richest lineages include Hypocreales (14 species) and Xylariales (13 species). Other recorded orders are Diaporthales (4), Microascales (3), Chaetosphaeriales (2), Calosphaeriales (1), Coniochaetales (1), Glomerallales (1), Lulworthiales (1), Myrmecridiales (1), Savoryellales (1), Tirisporellales (1), and two taxa *Neoarthrinium aquaticum* and *Rimaconus multiguttulatus*, under *incertae sedis*.

Eurotiomycetes accounts for approximately 21 species (8 genera, 3 families, 2 orders). Most belong to Eurotiales, with Aspergillaceae being dominant (16 species). Within this family, *Aspergillus* (13 species) and *Penicillium* (3 species) are particularly common, along with *Hamigera* (2 species). Additional taxa include *Talaromyces* (Trichocomaceae) and a single representative of Chaetothyriales, *Cladophialophora bantiana*.

Yeast lineages are represented by both Pichiomycetes and Saccharomycetes. The Pichiomycetes include *Candida tropicalis, C. metapsilosis, C. zeylanoides, C. parapsilosis, C. spencermartinsiae, Meyerozyma guilliermondii*, and *M. caribbica* (Serinales, Debaryomycetaceae). Saccharomycetes is represented by *Wickerhamomyces anomalus* and *Kodamaea ohmeri* (Wickerhamomycetaceae, Saccharomycetales).

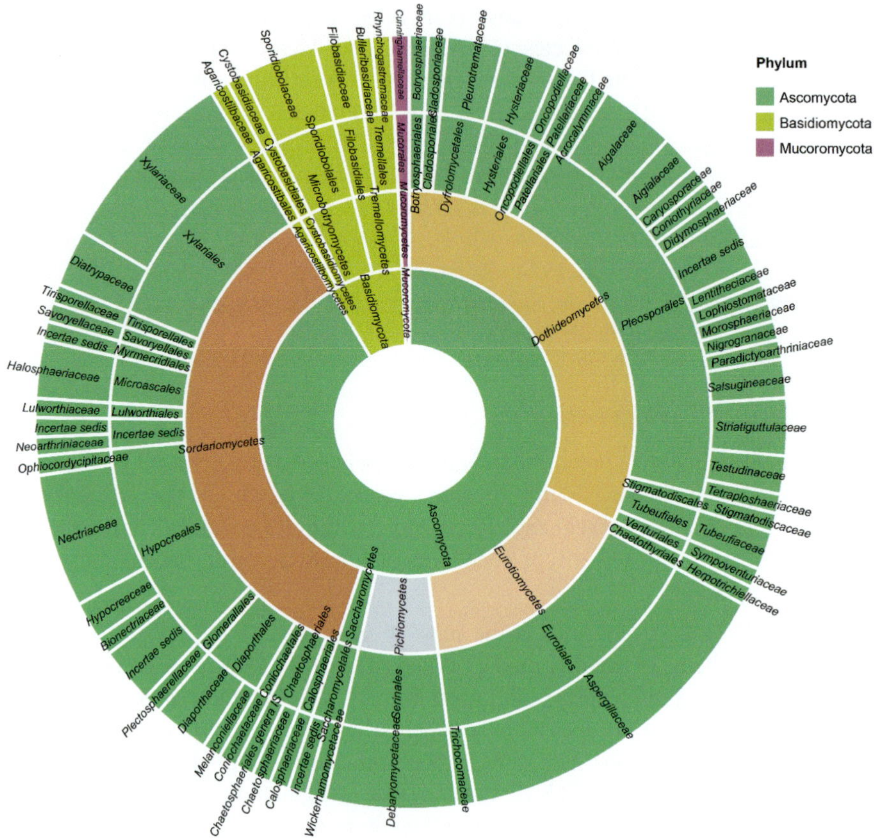

Fig. 2. Geographic distribution of marine fungi in Thailand based on species with available molecular data.

Basidiomycota and Mucoromycota: For Basidiomycota, 10 marine fungal species have been sequenced in Thailand, spanning four major classes: Agaricostilbomycetes, Cystobasidiomycetes, Microbotryomycetes, and Tremellomycetes (Kaewkrajay et al., 2020). *Agaricostilbomycetes* is represented by a single species, *Sterigmatomyces halophilus* (Agaricostilbaceae, Agaricostilbales), while *Cystobasidiomycetes* include *Cystobasidium calyptogenae* (Cystobasidiales, Cystobasidiaceae). *Microbotryomycetes* show greater diversity, with four species of Sporidiobolaceae (Sporidiobolales): *Rhodotorula mucilaginosa*, *R. diobovata*, *R. toruloides*, and *Rhodosporidiobolus fluvialis*. The most diverse class is Tremellomycetes, comprising four species across five genera, three families, and two orders: *Filobasidium uniguttulatum* and *Naganishia liquefaciens* (Filobasidiaceae, Filobasidiales), *Vishniacozyma victoriae* (Bulleribasidiaceae, Tremellales), and *Papiliotrema laurentii* (Rhynchogastremaceae, Tremellales).

Cunninghamella blakesleeana (Cunninghamellaceae, Mucorales, Mucoromycetes), isolated from *Holothuria scabra*, is currently the only species within Mucoromycota that has been reported with available molecular sequence data from marine environments in Thailand (Wingfield et al., 2024).

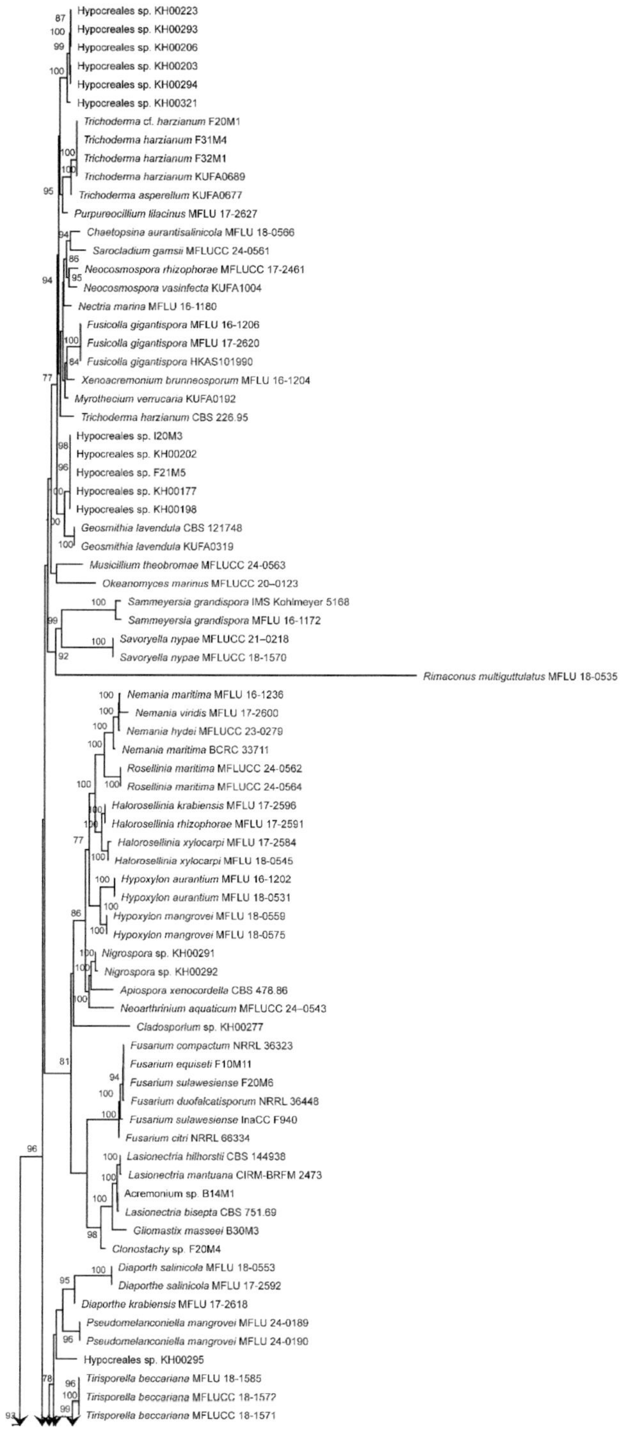
Hypocreales sp. KH00223
Hypocreales sp. KH00293
Hypocreales sp. KH00206
Hypocreales sp. KH00203
Hypocreales sp. KH00294
Hypocreales sp. KH00321
Trichoderma cf. harzianum F20M1
Trichoderma harzianum F31M4
Trichoderma harzianum F32M1
Trichoderma harzianum KUFA0689
Trichoderma asperellum KUFA0677
Purpureocillium lilacinus MFLU 17-2627
Chaetopsina aurantisalinicola MFLU 18-0566
Sarocladium gamsii MFLUCC 24-0561
Neocosmospora rhizophorae MFLUCC 17-2461
Neocosmospora vasinfecta KUFA1004
Nectria marina MFLU 16-1180
Fusicolla gigantispora MFLU 16-1206
Fusicolla gigantispora MFLU 17-2620
Fusicolla gigantispora HKAS101990
Xenoacremonium brunneosporum MFLU 16-1204
Myrothecium verrucaria KUFA0192
Trichoderma harzianum CBS 226.95
Hypocreales sp. I20M3
Hypocreales sp. KH00202
Hypocreales sp. F21M5
Hypocreales sp. KH00177
Hypocreales sp. KH00198
Geosmithia lavendula CBS 121748
Geosmithia lavendula KUFA0319
Musicillium theobromae MFLUCC 24-0563
Okeanomyces marinus MFLUCC 20–0123
Sammeyersia grandispora IMS Kohlmeyer 5168
Sammeyersia grandispora MFLU 16-1172
Savoryella nypae MFLUCC 21–0218
Savoryella nypae MFLUCC 18-1570
Rimaconus multiguttulatus MFLU 18-0535
Nemania maritima MFLU 16-1236
Nemania viridis MFLU 17-2600
Nemania hydei MFLUCC 23-0279
Nemania maritima BCRC 33711
Rosellinia maritima MFLUCC 24-0562
Rosellinia maritima MFLUCC 24-0564
Halorosellinia krabiensis MFLU 17-2596
Halorosellinia rhizophorae MFLU 17-2591
Halorosellinia xylocarpi MFLU 17-2584
Halorosellinia xylocarpi MFLU 18-0545
Hypoxylon aurantium MFLU 16-1202
Hypoxylon aurantium MFLU 18-0531
Hypoxylon mangrovei MFLU 18-0559
Hypoxylon mangrovei MFLU 18-0575
Nigrospora sp. KH00291
Nigrospora sp. KH00292
Apiospora xenocordella CBS 478.86
Neoarthrinium aquaticum MFLUCC 24–0543
Cladosporium sp. KH00277
Fusarium compactum NRRL 36323
Fusarium equiseti F10M11
Fusarium sulawesiense F20M6
Fusarium duofalcatisporum NRRL 36448
Fusarium sulawesiense InaCC F940
Fusarium citri NRRL 66334
Lasionectria hilhorstii CBS 144938
Lasionectria mantuana CIRM-BRFM 2473
Acremonium sp. B14M1
Lasionectria bisepta CBS 751.69
Gliomastix masseei B30M3
Clonostachy sp. F20M4
Diaporth salinicola MFLU 18-0553
Diaporthe salinicola MFLU 17-2592
Diaporthe krabiensis MFLU 17-2618
Pseudomelanconiella mangrovei MFLU 24-0189
Pseudomelanconiella mangrovei MFLU 24-0190
Hypocreales sp. KH00295
Tirisporella beccariana MFLU 18-1585
Tirisporella beccariana MFLUCC 18-157?
Tirisporella beccariana MFLUCC 18-1571

Fig. 3 contd. ...

...Fig. 3 contd.

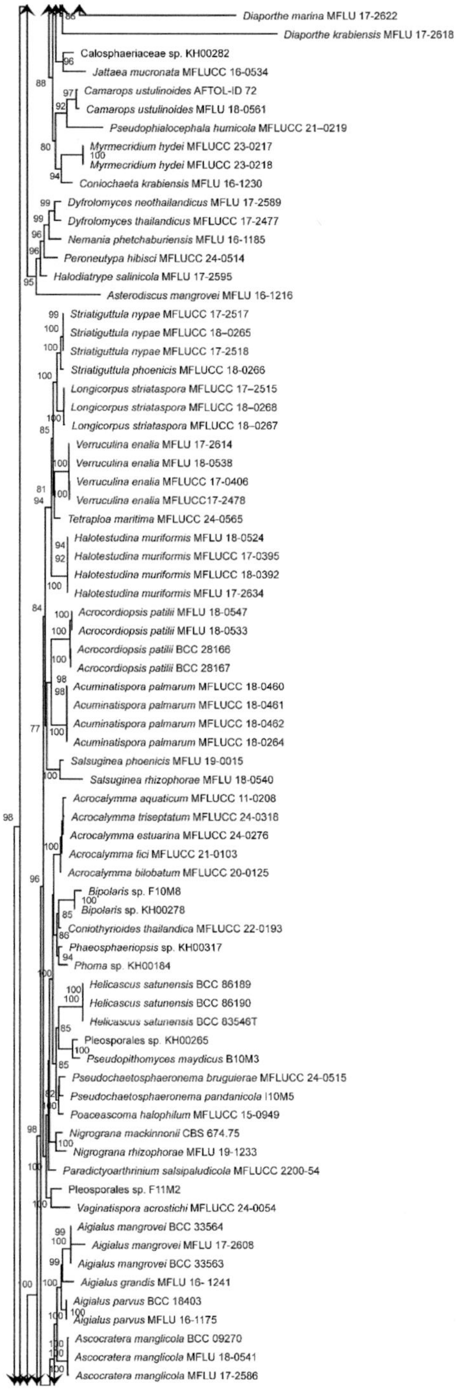

Fig. 3 contd. ...

...Fig. 3 contd.

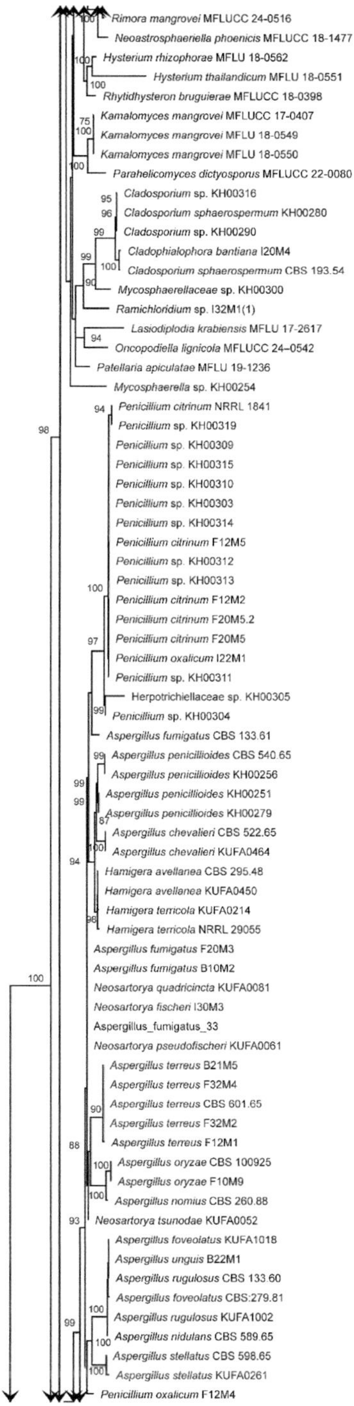

Fig. 3 contd. ...

...Fig. 3 contd.

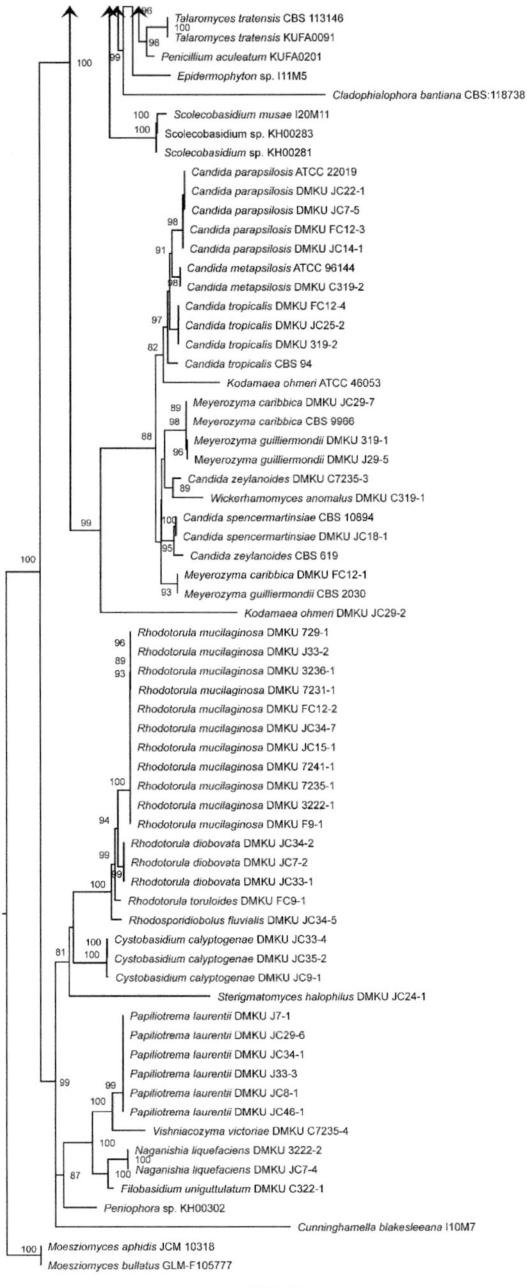

Fig. 3. Phylogenetic tree generated from maximum likelihood (ML) analysis based on ITS, LSU, SSU, *tef1-α*, *rpb2,* and *tub2* sequence data for the marine fungal taxa observed in Thailand. Bootstrap support value for ML higher than 75% is indicated above the nodes in this order. The tree is rooted to *Moesziomyces aphidis* (JCM 10318) and *M. bullatus* (GLM-F105777). Bar = 0.3 estimated number of nucleotide substitutions per site per branch.

Unidentified Marine Fungi: Several studies have reported marine fungal isolates that could not be identified to the species level, but were classified up to higher taxonomic ranks.

Sakayaroj et al. (2010) isolated diverse marine fungi associated with the seaweed *Enhalus acoroides*, which were identified only to the genus or higher taxonomic levels. These included representatives of the genera *Bipolaris*, *Cladosporium*, *Mycosphaerella*, *Nigrospora*, *Penicillium*, *Peniophora*, *Phaeosphaeriopsis*, *Phoma*, and *Scolecobasidium*. In addition, isolates belonging to the families Calosphaeriaceae, Mycosphaerellaceae, and Herpotrichiellaceae, as well as to the orders Hypocreales and Pleosporales were also recorded.

Similarly, Wingfield et al. (2024) reported several unidentified marine fungal isolates obtained from the sea cucumber *Holothuria scabra*. These included *Absidia* sp., *Acremonium* sp., *Bipolaris* sp., *Clonostachys* sp., *Epidermophyton* sp., *Hypocreales* sp., *Nectria* sp., *Paraphaeosphaeria* sp., *Pleosporales* sp., and *Ramichloridium* sp.

4. Geographical Distribution and Substrate

Marine fungi have been documented across multiple coastal provinces of Thailand, revealing distinct regional variations in species richness (Fig. 4). Among these, Ranong Province hosts the highest diversity, with 83 recorded species, followed by Trat (69 species) and Phang Nga (68 species).

Ranong Province, located along the Andaman Sea coast in southern Thailand, represents one of the country's richest areas for marine fungi, with approximately 83 species (65 genera) reported. Extensive surveys in its marine habitats have revealed remarkable diversity across a range of substrates, including senescent and decaying wood, driftwood, seedlings, submerged petioles, fronds, and roots of dominant mangrove species such as *Rhizophora apiculata*, *Avicennia marina*, *Nypa fruticans*, *Phoenix paludosa*, *Xylocarpus* spp., and *Sonneratia alba*. The most speciose genera, each represented by three species, include *Halosarpheia*, *Lignincola*, *Aniptodera*, *Saagaromyces*, *Savoryella*, and *Aigialus*. Genera with two species each include *Melomastia*, *Leptosphaeria*, *Kallichroma*, *Halorosellinia*, and *Morosphaeria*, while the remaining 54 genera are represented by a single species each.

Trat Province, along the eastern Gulf of Thailand, also supports high marine fungal diversity. Surveys have recorded 74 species (56 genera) in coastal and mangrove habitats. The genus *Savoryella* is the most species-rich with four species, followed by *Aniptodera*, *Saagaromyces*, and *Aigialus*, each with three species. Several genera, including *Halosarpheia*, *Kallichroma*, *Sclerococcum*, *Etheirophora*, *Linocarpon*, and *Torpedospora*, are represented by two species each, while the remaining genera are mostly monotypic, highlighting specialized ecological niches. The majority of fungi were collected from senescent and decaying mangrove wood, with common hosts including *Rhizophora apiculata*, *Nypa fruticans*, *Ceriops tagal*, and *Avicennia marina*. Fungi were also frequently observed on decaying intertidal petioles and fronds, roots and seedlings of *R. apiculata*, driftwood,

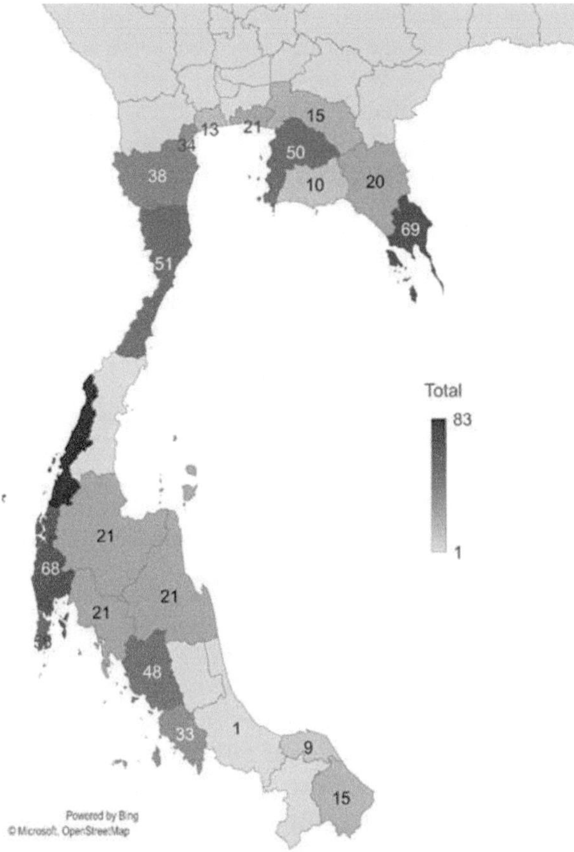

Fig. 4. Species richness of marine fungi across 21 coastal provinces in Thailand.

twigs, and the submerged rachis of *P. paludosa*. A few isolates were obtained from submerged seeds of *A. marina* and dead prop roots.

Phang Nga Province and other moderately diverse areas, including Phuket (56 species), Prachuap Khiri Khan (51 species), and Chon Buri (50 species). Provinces such as Trang (48 species), Phetchaburi (38 species), and Samut Songkhram (34 species) also exhibit considerable fungal diversity. Lower diversity is reported from Satun (33 species), Surat Thani, Samut Prakan, Nakhon Si Thammarat, and Krabi (21 species each), while Chanthaburi (20 species), Narathiwat and Chachoengsao (15 each), and Samut Sakhon (13 species) are moderately studied. The least documented provinces include Rayong (10 species), Pattani (9 species), and Songkhla (1 species), which likely reflects limited sampling rather than true species scarcity.

Collectively, these records demonstrate that Thai marine fungi predominantly colonize mangrove-associated substrates, with decaying wood, roots, petioles, and fronds serving as the principal habitats. Occasional occurrences on submerged seeds, buttress roots, driftwood, and even sponge and sea cucumber hosts suggest ecological plasticity and adaptation to variable marine conditions.

5. Conclusion and Future Perspectives

Marine fungi in Thailand exhibit extensive taxonomic diversity, thereby contributing significantly to the understanding of tropical marine ecosystems. In the last decade, substantial progress has been made in molecular-based taxonomy and phylogeny. With the application of modern phylogenetic techniques, taxonomic ambiguities were resolved, leading to the establishment of numerous new fungal families and genera to correctly accommodate both newly discovered and reevaluated existing species (Zhang, 2019b; Jones et al., 2020; Wijesinghe, 2023; Mukhopadhyay et al., 2025). The current trajectory suggests that research into Thai marine fungi will continue to expand, with numerous unknown marine fungal taxa expected to be discovered.

Extensive studies have documented fungal diversity from lignicolous substrates in the country, particularly from mangrove areas and intertidal zones (Jones et al., 2006). In contrast, other substrates such as sediments, chitinous exoskeletons, algae, and other herbaceous materials remain underrepresented. Other avenues for exploration include endolithic environments such as the pores in calcareous shells, rocks, and sand grains (Gleason et al., 2007), and marine-derived fungi associated with animals like sponges, nematodes, diatoms and corals (Jones et al., 2019). While marine fungi are widely recognized as key decomposers in marine ecosystems (Raghukumar, 2017; Gonçalves et al., 2022), their roles as pathogens, endobionts, and mutualists also warrant further examination. Other marine environments, including open waters and the deep sea, should also be considered, as these are large reservoirs for microbes and are likely to harbor more novel taxa (Jones et al., 2019). Expanding research into these underexplored habitats will present methodological challenges, as the majority of marine fungi are unculturable using traditional methods (Jones et al., 2019; Gonçalves et al., 2022). Isolation and culture methods must be more strategic, accounting for factors such as substrate specificity, interdependence with other organisms, dormancy and low relative abundance (Zhang et al., 2024), to more effectively capture the true diversity of marine fungi. Overcoming these challenges will require the integration of genomic, transcriptomic, and metagenomic approaches such as the use of environmental DNA (eDNA) metabarcoding, which enables the detection of unculturable fungal species.

With advances in isolation, cultivation and identification comes the opportunity to explore marine fungi as sources of novel bioactive compounds. In the ongoing search for new natural products, approximately 38% of 22,000 reported bioactive compounds are fungal in origin (Shin, 2020). Marine-derived fungi are known to produce metabolites that exhibit a range of antibiotic, antiviral, and anticancer activities (Pang et al., 2016; Jones et al., 2019). This shows the high potential for discovery within the largely unexplored marine fungi and the need for continued research.

Beyond diversity studies and bioprospecting, defining the ecological roles of marine fungi is crucial for understanding the resilience and overall health of Thailand's marine ecosystems. This work necessitates a detailed investigation into their local symbiotic relationships (e.g., seagrasses), encompassing both host-pathogen and mutualistic interactions. Such ecological knowledge is pertinent given the ongoing challenges of climate change, specifically ocean warming and acidification. On a global scale, the significance of marine fungi extends to global biogeochemical cycles, particularly the Marine Carbon Cycle (MCC). They are recognized as significant heterotrophic biomass contributors, with standing stocks comparable to bacteria and archaea (Amend et al., 2019), suggesting a substantial role in the biological carbon pump that sequesters carbon to the deep ocean. To accurately model their contribution to nutrient fluxes, especially under changing climatic conditions, it is essential to understand how their abundance correlates with factors like increased particulate organic carbon (POC) availability and salinity fluctuations.

The ecological resilience of marine fungi is rooted in their remarkable physiological adaptability, a trait that holds immense potential for applied science. Their ability to adapt and thrive under high salinity, fluctuations in temperature, and extreme pH conditions enables the production of extracellular enzymes and secondary metabolites with high functional stability, making marine fungi a valuable resource for biotechnology. This provides fundamental insights that researchers can use to potentially engineer transgenic plants capable of surviving similar harsh environments and adapting to new climate conditions.

In summary, marine fungi of Thailand remain understudied, yet they represent a diverse and ecologically vital group of organisms with potential biotechnological applications. Continued integration of molecular systematics, functional studies in ecology, and other ecological investigations will not only expand the understanding of these groups of fungi, but will also uncover their contributions to ecosystem stability and resilience under fluctuating environmental conditions.

References

Abdel-Wahab, M. (2011). Lignicolous marine fungi from Yokosuka, Japan. Bot. Mar., 54: 209–221. https://doi.org/10.1515/bot.2011.013.

Abdel-Wahab, M.A., Dayarathne, M.C., Suetrong, S., Guo, S.Y., Alias, S.A. et al. (2017). New saprobic marine fungi and a new combination. Bot. Mar., 60(4): 469–488. https://doi.org/10.1515/bot-2016-0118.

Abreu, A.C., Gonçalves, M.F.M., Esteves, A.C. and Alves, A. (2025). Diversity and temporal succession of early-colonizing fungi in wood baits from an estuarine environment, with description of *Penicillium alavariense* sp. Nov. Fungal Biol., 129(7): 101645. https://doi.org/10.1016/j.funbio.2025.101645.

Alias, S., Zainuddin, N. and Jones, E. (2010). Biodiversity of marine fungi in Malaysian mangroves. Bot. Mar., 53(6): 545–554.

Apurillo, C.C.S., Phukhamsakda, C., Hyde, K.D., Thiyagaraja, V. and Jones, E.B.G. (2025). New fungal genus, three novel species, and one new record from mangroves, with reclassification of *Melanconiella* (Melanconiellaceae) species. MycoKeys, 116: 25–52. https://doi.oog/10.3897/mycokeys.116.137351.

Apurillo, C.C.S., Phukhamsakda, C., Mukhopadhyay, S., Gunarathne, A. and Jones, E.B.G. (2024). *Nemania hydei* sp. nov. (Xylariaceae) from *Avicennia marina* in Central Thailand. N.Z.J. Bot., 62: 103–118. https://doi.org/10.1080/0028825X.2023.2289420.

Asghari, R., Phukhamsakda, C., Jones, E.B.G., Bahkali, A., Apurillo, C.C.S. et al. (2025). Morphology and phylogeny reveal two new species and host records of hyphomycetous fungi on *Areca* species from marine habitats in Thailand. MycoKeys, 118: 179–206. https://doi.org/10.3897/mycokeys.118.147229.

Asghari, R., Phukhamsakda, C., Karimi, O., Apurillo, C.C.S. and Jones, E.B.G. (2023). *Myrmecridium hydei*, a novel marine species from Thailand. Phytotaxa, 625: 265–279. https://doi.org/10.11646/phytotaxa.625.3.3.

Booth, T. and Kenkel, N. (1986). Ecological studies of lignicolous marine fungi: A distribution model based on ordination and classification. pp. 97–310. *In*: Raghukumar, C. (ed.). *The Biology of Marine Fungi*. Springer.

Borse, B. (2002). Marine fungi from India—XI: A check list. J. Indian Bot. Soc., 81: 83–92.

Borse, B.D., Borse, K.N., Pawar, N.S. and Tuwar, A.R. (2013). Marine Fungi from India—XII. A revised check list. Indian J. Geo-Mar. Sci., 42(1): 110–119.

Calabon, M., Sadaba, R.B. and Campos, W. (2018). Fungal diversity of mangrove-associated sponges from New Washington, Aklan, Philippines. Mycology, 10: 6–21. https://doi.org/ https://doi.org/10.1080/21501203.2018.1518934.

Chalearmsrimuang, T., Ismail, S.I., Mazlan, N., Suasaard, S. and Dethoup, T. (2019). Marine-derived fungi: A promising source of halo tolerant biological control agents against plant pathogenic fungi. J. Pure Appl. Microbiol., 13: 209–223. https://doi.org/10.22207/JPAM.13.1.22.

Chatmala, I., Sakayaroj, J., Somrithipol, S. and Phongpaichit, S. (2004). Marine hyphomycetes of Thailand and *Cumulospora varia* sp. nov. Fungal Diver., 17: 1–9.

Chethana, K.W.T., Niranjan, M., Dong, W., Samarakoon, M.C., Bao, D.F. et al. (2021). AJOM new records and collections of fungi: 101–150. Asian J. Mycol., 4: 113–260. https://doi.org/10.5943/ajom/4/1/8.

Dayarathne, M.C., Jones, E.B.G., Maharachchikumbura, S.S.N., Devadatha, B., Sarma, V.V. et al. (2020a). Morpho-molecular characterization of microfungi associated with marine based habitats. Mycosphere, 11: 1–188. https://doi.org/10.5943/mycosphere/11/1/1.

Dayarathne, M.C., Phookamsak, R., Hyde, K.D., Manawasinghe, I.S., To-Anun, C. et al. (2016). Halodiatrype, a novel diatrypaceous genus from mangroves with *H. salinicola* and *H. avicenniae* spp. nov. Mycosphere, 7. https://doi.org/10.5943/mycosphere/7/5/7.

Dayarathne, M.C., Wanasinghe, D.N., Devadatha, B., Abeywickrama, P., Gareth Jones, E.B. et al. (2020b). Modern taxonomic approaches to identifying diastrypaceous fungi from marine habitats, with a novel genus *Halocryptovalsa* Dayarathne & K.D. Hyde, gen. nov. Cryptogam Mycol., 41: 21–67. https://doi.org/10.5252/cryptogamie-mycologie2020v41a3.

Dethoup, T. and Manoch, L. (2009). Diversity of marine fungi in Eastern Thailand. Kasetsart J. (Nat. Sci.), 43: 100–106.

Gao, M., Liu, B., Li, J., Deng, Y., Zhang, Y. et al. (2024). Diversity and distribution of fungi in the marine sediments of Zhanjiang Bay, China. J. Fungi, 10: 867. https://doi.org/10.3390/jof10120867.

Gleason, F.H., Gadd, G.M., Pitt, J.I. and Larkum, A.W.D. (2017). The roles of endolithic fungi in bioerosion and disease in marine ecosystems. I. General concepts. Mycology, 8(3): 205–215. https://doi.org/10.1080/21501203.2017.1352049.

Gonçalves, M.F.M., Esteves, A.C. and Alves, A. (2022). Marine fungi: Opportunities and challenges. Encyclopedia, 2(1): 559–577. https://doi.org/10.3390/encyclopedia2010037.

Hagestad, O.C., Andersen, J.H., Altermark, B., Hansen, E. and Rämä, T. (2019). Cultivable marine fungi from the Arctic Archipelago of Svalbard and their antibacterial activity. Mycology, 11(3): 230–242. https://doi.org/10.1080/21501203.2019.1708492.

Hongsanan, S., Khuna, S., Manawasinghe, I., Tibpromma, S., Chethana, K. et al. (2025). Mycosphere Notes 521–571: A special edition of fungal biodiversity to celebrate Kevin D. Hyde's 70th birthday and his exceptional contributions to Mycology. Mycosphere, 16: 1–178. https://doi.org/10.5943/mycosphere/16/2/1.

Htet, Z.H., Prematunga, C., Mapook, A., Jones, E.B.G. and Chethana, K.W.T. (2023). Taxonomy and phylogeny of *Paradictyoarthrinium salsipaludicola* sp. nov. (Paradictyoarthriniaceae, Pleosporales) from mangroves. Phytotaxa, 620: 283–292. https://doi.org/10.11646/phytotaxa.620.4.4.

Hu, D.M., Liu, F. and Cai, L. (2013). Biodiversity of aquatic fungi in China. Mycology, 4: 125–168. https://doi.org/10.1080/21501203.2013.835752.

Hughes, G.C. (1974). Geographical distribution of the higher marine fungi. Veröffentlichender Institute für Meeresforschung. Bremerhaven Supplement, 5: 419–441.

Hyde, K.D. (1988). The genus Linocarpon from the mangrove palm *Nypa fruticans*. Mycol. Soc. Japan, 29: 339–350.

Hyde, K.D. (1989). Caryospora mangrovei sp. nov. and notes on marine fungi from Thailand. Trans. Mycol. Soc. Japan, 30: 333–341.

Hyde, K.D. (1992). The genus *Saccardoella* from intertidal mangrove wood. Mycologia, 84: 803–810. https://doi.org/10.2307/3760393.

Hyde, K.D. (1994). The distribution of intertidal fungi on Rhizophora apiculata. pp. 643–652. *In*: Morton, B. (ed.), The Marine Biology of the South China Sea I & II (2 vols.). Hong Kong University Press, Hong Kong.

Hyde, K.D. and Jones, E.B.G. (1992). Intertidal mangrove fungi: Pedumispora gen. nov. (Diaporthales). Mycol. Res., 96. https://doi.org/10.1016/S0953-7562(09)81001-0.

Hyde, K.D. and Nakagiri, A. (1989). A new species of *Oxydothis* from the mangrove palm *Nypa fruticans*. Trans. Mycol. Soc. Japan, 30: 69–75.

Hyde, K.D., Chalermpongse, A. and Boonthavikoon, T. (1990). Ecology of intertidal fungi at Ranong mangrove, Thailand. Trans. Mycol. Soc. Japan, 31: 17–27.

Hyde, K.D., Jeewon, R., Chen, Y.J., Bhunjun, C.S., Calabon, M.S. et al. (2020a). The numbers of fungi: Is the descriptive curve flattening? Fungal Diver., 103: 219–271. https://doi.org/10.1007/s13225-020-00458-2.

Hyde, K.D., Jones, E.B.G., Leaño, E., Pointing, S.B., Poonyth, A.D. et al. (1998). Role of fungi in marine ecosystems. Biodivers. Conserv., 7: 1147–1161. https://doi.org/10.1023/A:1008823515157.

Hyde, K.D., Norphanphoun, C., Abreu, V.P., Bazzicalupo, A., Thilini Chethana, K.W. et al. (2017). Fungal diversity notes 603–708: Taxonomic and phylogenetic notes on genera and species. Fungal Diver., 87: 1–235. https://doi.org/10.1007/s13225-017-0391-3.

Hyde, K.D., Norphanphoun, C., Maharachchikumbura, S.S.N., Bhat, D.J., Jones, E.B.G. et al. (2020b). Refined families of Sordariomycetes. Mycosphere, 11: 305–1059. https://doi.org/10.5943/mycosphere/11/1/7.

Hyde, K.D., Tennakoon, D.S., Jeewon, R., Bhat, D.J., Maharachchikumbura, S.S.N. et al. (2019). Fungal diversity notes 1036–1150: Taxonomic and phylogenetic contributions on genera and species of fungal taxa. Fungal Diver., 96: 1–242. https://doi.org/10.1007/s13225-019-00429-2.

Ito, T., Nakagiri, A., Tantichaoren, M. and Manoch, L. (2001). Mycobiota of mangrove forests in Thailand. Research Communications, Institute for Fermentation, Osaka, 20: 50–60.

Jones, E.B.G., Abdel-Wahab, M.A., Alias, S.A. and Hsieh, S.Y. (1999). *Dactylospora mangrovei* sp. nov. (Discomycetes, Ascomycota) from mangrove wood. Mycoscience, 40: 317–320. https://doi.org/10.1007/BF02463875.

Jones, E.B.G., Chatmala, I. and Pang, K.L. (2006a). Two new genera isolated from marine habitats in Thailand: *Pseudolignincola* and *Thalespora* (Halosphaeriales, Ascomycota). Nova Hedwigia, 83: 219–232. https://doi.org/10.1127/0029-5035/2006/0083-0219.

Jones, E.B.G., Devadatha, B., Abdel-Wahab, M.A., Dayarathne, M.C., Zhang, S.N. et al. (2020). Phylogeny of new marine Dothideomycetes and Sordariomycetes from mangroves and deep-sea sediments. Bot. Mar., 63: 155–181. https://doi.org/10.1515/bot-2019-0014.

Jones, E.B.G., Pang, K.L., Abdel-Wahab, M.A., Scholz, B., Hyde, K.D. et al. (2019). An online resource for marine fungi. Fungal Diver., 96: 347–433. https://doi.org/10.1007/s13225-019-00426-5.

Jones, E.B.G., Pilantanapak, A., Chatmala, I., Sakayaroj, J., Phongpaichit, S. et al. (2006b). Thai marine fungal diversity. Songklanakarin J. Sci. Technol. (SJST), 28: 687–708.

Kaewkrajay, C., Chanmethakul, T. and Limtong, S. (2020). Assessment of diversity of culturable marine yeasts associated with corals and zoanthids in the gulf of Thailand, South China Sea. Microorganisms, 8: 474. https://doi.org/10.3390/microorganisms8040474.

Koch, J. (1986). Some lignicolous marine fungi from Thailand, including two new species. Nord. J. Bot., 6: 497–499. https://doi.org/10.1111/j.1756-1051.1986.tb00905.x.

Kohlmeyer, J. (1984). Tropical marine fungi. Mar. Ecol., 5: 329–378. https://doi.org/10.1111/j.1439-0485.1984.tb00130.x.

Kohlmeyer, J. and Vittal, B. (1986). *Lophiostoma mangrovis*, a new marine ascomycete from the tropics. Mycologia, 78: 485–489.

Kohlmeyer, J. and Volkmann-Kohlmeyer, B. (1993). A new marine *Dryosphaera* (Ascomycotina) from the tropics. Can. J. Bot., 71: 992–995. https://doi.org/10.1139/b93-113.

Liu, J.K., Jones, G.E.B., Chukeatirote, E., Bahkali, A.H. and Hyde, K.D. (2011). *Lignincola conchicola* from palms with a key to the species of *Lignincola*. Mycotaxon, 117: 343–349. https://doi.org/10.5248/117.343.

Liu, S.L., Wang, X.W., Li, G.J., Deng, C.Y., Rossi, W. et al. (2024). Fungal diversity notes 1717–1817: Taxonomic and phylogenetic contributions on genera and species of fungal taxa. Fungal Diver., 124: 1–216. https://doi.org/10.1007/s13225-023-00529-0.

Loilong, A., Sakayaroj, J., Rungjindamai, N., Choeyklin, R. and Jones, E.B.G. (2012). Biodiversity of fungi on the palm *Nypa fruticans*. pp. 273–290. *In*: Jones, E.B.G. and Pang, K.-L. (eds.). Marine Fungi and Fungal-like Organisms. De Gruyter, Berlin, Boston. https://doi.org/10.1515/9783110264067.273.

Manawasinghe, I.S., Calabon, M.S., Jones, E.B.G., Zhang, Y.X., Liao, C.F. et al. (2022). Mycosphere notes 345–386. Mycosphere, 13: 454–557. https://doi.org/10.5943/mycosphere/13/1/3.

Manawasinghe, I.S., Hyde, K.D., Wanasinghe, D.N., Karunarathna, S.C., Maharachchikumbura, S.S.N. et al. (2024). Fungal diversity notes 1818–1918: Taxonomic and phylogenetic contributions on genera and species of fungi. Fungal Diver., 130: 1–261. https://doi.org/10.1007/s13225-024-00541-y.

Mouton, M., Postma, F., Wilsenach, J. and Botha, A. (2012). Diversity and characterization of culturable fungi from marine sediment collected from St. Helena Bay, South Africa. Microb. Ecol., 64(2): 311–319. https://doi.org/10.1007/s00248-012-0035-9.

Mukhopadhyay, S., Bhunjun, C.S., Phukhamsakda, C., Apurillo, C.C.S., Al-Otibi, F. et al. (2024). Introducing a new species, *Vaginatispora acrostichi* (Lophiostomataceae), based on morphology and multigene phylogeny. Bot. Mar., 67: 401–410. https://doi.org/10.1515/bot-2024-0008.

Mukhopadhyay, S., Phukhamsakda, C., Bhunjun, C.S., Bahkali, A.H., Jones, E.B.G. and Hyde, K.D. (2025). Exploring fungi on mangrove associates: Novel family Neoarthriniaceae and two novel species in *Neoarthrinium* and *Oncopodiella*. Mycol. Prog., 24: 18. https://doi.org/10.1007/s11557-025-02035-w.

Nakagiri, A. (1989). Marine fungi in sea form from Japanese coast. Res. Commun., 14: 52–79.

Nakagiri, A. (1993). Intertidal mangrove fungi from Iriomote Island. Res. Commun., 16: 24–62.

Pang, K., Sharuddin, S., Alias, S., Nor, N. and Awaluddin, H. (2010). Diversity and abundance of lignicolous marine fungi from the east and west coasts of Peninsular Malaysia and Sabah (Borneo Island). Bot. Mar., 53: 539–550.

Pang, K.L. and Mitchell, J. (2005). Molecular approaches for assessing fungal diversity in marine substrata. Bot. Mar., 48: 332–347. https://doi.org/10.1515/BOT.2005.046.

Pang, K.L., Overy, D.P., Jones, E.B.G., da Luz Calado, M., Burgaud, G. et al. (2016). "Marine fungi" and "marine-derived fungi" in natural product chemistry research: Toward a new consensual definition. Fungal Biol. Rev., 30: 163–175. https://doi.org/10.1016/j.fbr.2016.08.001.

Petersen, K.R.L. and Koch, J. (1997). Substrate preference and vertical zonation of lignicolous marine fungi on mooring posts of oak (*Querens* sp.) and larch (*Larix* sp.) in Svanemøllen Harbour, Denmark. Bot. Mar., 40: 451–464. https://doi.org/10.1515/botm.1997.40.1-6.451.

Picard, K.T. (2017). Coastal marine habitats harbor novel early-diverging fungal diversity. Fungal Ecol., 25: 1–13. https://doi.org/10.1016/j.funeco.2016.10.006.

Pilantanapak, A., Jones, E.B.G. and Eaton, R.A. (2005). Marine fungi on Nypa fruticans in Thailand. Bot. Mar., 48: 365–373. https://doi.org/10.1515/bot.2005.049.

Preedanon, S., Klaysuban, A., Suetrong, S., Pracharoen, O., Promchoo, W. et al. (2024). Morphological, molecular, and 3D synchrotron X-ray tomographic characterizations of *Helicascus satunensis* sp. nov., a novel mangrove fungus. PeerJ., 12: e18341. https://doi.org/10.7717/peerj.18341.

Raghukumar, S. (2017). Fungi in Coastal and Oceanic Marine Ecosystems. Springer, New York, 378 pp. https://doi.org/10.1007/978-3-319-54304-8.

Ren, G., Wanasinghe, D.N., Roberto, A., Hyde, K., Yasanthika, E. et al. (2022). Taxonomic novelties of woody litter fungi (Didymosphaeriaceae, Pleosporales) from the Greater Mekong Subregion. Biology (Basel), 11: 1660. https://doi.org/10.3390/biology11111660.

Sakayaroj, J., Preedanon, S., Supaphon, O., Jones, E.B.G. and Phongpaichit, S. (2010). Phylogenetic diversity of endophyte assemblages associated with the tropical seagrass *Enhalus acoroides* in Thailand. Fungal Diver., 42: 27–45. https://doi.org/10.1007/s13225-009-0013-9.

Sakayaroj, J., Supaphon, O., Jones, E.B.G. and Phongpaichit, S. (2011). Diversity of higher marine fungi at Hat Khanom-Mu Ko Thale Tai National Park, Southern Thailand. Songklanakarin J. Sci. Technol., (SJST), 33: 15–22.

Sarma, V.V. and Hyde, K.D. (2001). A review on frequently occurring fungi in mangroves. Fungal Divers., 8: 1–34.

Schaumann, K. (1975). Ökologische Untersuchungen über höhere Pilze im Meer- und Brackwasser der Deutschen Bucht unter besonderer Berücksichtigung der holz-besiedelnden Arten. Veröff. Inst. Meeresforsch. Bremerh., 15: 79–182.

Schmit, J.P. and Shearer, C.A. (2003). A checklist of mangrove-associated fungi, their geographical distribution and knownhost plants. Mycotaxon, 80: 423–477.

Schmit, J.P. and Shearer, C.A. (2004). Geographic and host distribution of lignicolous mangrove microfungi. Bot. Mar., 47: 496–500.

Shearer, C.A., Descals, E., Kohlmeyer, B., Kohlmeyer, J., Marvanová, L. et al. (2007). Fungal diversity in aquatic habitats. Biodivers. Conserv., 16 (1): 49–67. https://doi.org/10.1007/s10531-006-9120-z.

Shin, H.J. (2020). Natural products from marine fungi. Mar. Drugs, 18: 230. https://doi.org/10.3390/md18050230.

Su, G., Dueñas, K., Roderno, K., Sison, M., Su, M.L.S. et al. (2014). Distribution and diversity of marine fungi in Manila Bay, Philippines. Res. Rev. Biol., 4: 4166–4173.

Suetrong, S., Preedanon, S., Klaysuban, A., Gundool, W., Unagul, P. et al. (2017). Distribution and occurrence of manglicolous marine fungi from eastern and southern Thailand. Bot. Mar., 60: 503–514. https://doi.org/10.1515/bot-2016-0107.

Sundari, R., Vikineswary, S., Yusoff, M. and Jones, E.B.G. (1996). *Corollospora besarispora*, a new arenicolous marine fungus from Malaysia. Mycol. Res., 100: 1259–1262. https://doi.org/10.1016/S0953-7562(96)80190-0.

Tubaki, K. (1968). Studies on the Japanese marine fungi lignicolous group II. Publ. Seto. Mar. Biol. Lab., 15: 357–372.

Wijesinghe, S.N., Calabon, M.S., Xiao, Y., Jones, E.B.G. and Hyde, K.D. (2023). A novel coniothyrium-like genus in Coniothyriaceae (Pleosporales) from salt marsh ecosystems in Thailand. Stud. Fungi. Studies in Fungi, 8: 1–10. https://doi.org/10.48130/SIF-2023-0006.

Wingfield, L.K., Atcharawiriyakul, J. and Jitprasitporn, N. (2024). Diversity and characterization of culturable fungi associated with the marine sea cucumber Holothuria scabra. PLoS One, 19: e0296499. https://doi.org/10.1371/journal.pone.0296499.

Wong Chin, J.M., Puchooa, D., Bahorun, T., Neergheen, V.S., Aullybux, A.A. et al. (2022). Metabarcoding assessment of fungal diversity in brown algae and sponges of Mauritius. Front. Microbiol., 13: 1003790. https://doi.org/10.3389/fmicb.2022.1003790.

Zhang, S.N., Abdel-Wahab, M.A., Jones, E.B.G., Hyde, K.D. and Liu, J.K.J. (2019a). Additions to the genus *Savoryella* (Savoryellaceae), with the asexual morphs *Savoryella nypae* comb. nov. and *S. sarushimana* sp. nov. Phytotaxa, 408: 195–207. https://doi.org/10.11646/phytotaxa.408.3.4.

Zhang, S.N., Hyde, K.D., Jones, G.E.B., Cheewangkoon, R. and Liu, J.K. (2018). *Acuminatispora palmarum* gen. et sp. nov. from mangrove habitats. Mycol. Prog., 17: 1173–1188. https://doi.org/10.1007/s11557-018-1433-2.

Zhang, S.N., Hyde, K.D., Jones, E.B.G., Cheewangkoon, R. and Liu, J.K. (2020). Additions to *Fissuroma* and *Neoastrosphaeriella* (Aigialaceae, Pleosporales) from palms. Mycosphere, 11: 269–284. https://doi.org/10.5943/mycosphere/11/1/5.

Zhang, S.N., Hyde, K.D., Jones, E.B.G., Jeewon, R., Cheewangkoon, R. et al. (2019b). Striatiguttulaceae, a new pleosporalean family to accommodate *Longicorpus* and *Striatiguttula* gen. Nov. From palms. MycoKeys, 49: 99–129. https://doi.org/10.3897/mycokeys.49.30886.

Zhang, T., Fei Wang, N., Qin Zhang, Y., Liu, H.Y. and Lu, L.Y. (2015). Diversity and distribution of fungal communities in the marine sediments of Kongsfjorden, Svalbard (High Arctic). Sci. Rep., 5: 14524. https://doi.org/10.1038/srep14524.

Zhang, Z.F., Liu, F., Liu, L.R., Li, M., Cai, L. et al. (2024) Culturing the uncultured marine fungi in the omics age: Opportunities and challenges. Fungal Biol. Rev., 48: 100353. https://doi.org/10.1016/j.fbr.2023.100353.

Human and Ecosystem Health

6

The Bioactive Potential of Marine Fungi

Priyanka Shenoy, N.,[1,*] *Sneha Bhaskar,*[2] *Shailasree Sekhar*[3] and *K. Ramachandra Kini*[2]

Marine fungi are the potent producers of diversified metabolites with different functional attributes. These metabolites can be classified based on their bio molecular structures which includes bioactive pigments, polysaccharides, peptides, novel enzymes, biocatalysts, secondary metabolites. These bioactive compounds derived from marine fungi are studied for their biological activities like anti microbial, anti-protozoal, anti-malarial, anti-inflammatory and so on. These metabolites find applications in various fields including medicine, bioremediation, agriculture, industries like pharmaceutical, food, cosmetics, and many more. Secondary metabolic pathways also play a pivotal role in the synthesis of varieties of metabolites through biosynthetic pathways which includes alkaloid biosynthetic pathway, terpenoid pathways, flavonoid metabolic pathways, etc. Gene(s) coding for key enzymes from these pathways also play a significant role in the expression of specific bioactive metabolites. Evolved metabolic pathways in fungi have led to the adaptations in marine extreme environments. Structural elucidation of metabolites derived from marine fungi aids in novel drug discovery. Fungal endophytes derived

[1] SBRR Mahajana's first grade college, Mysore, India.
[2] Department of Studies in Biotechnology, University of Mysore, Manasagangotri, Mysore – 570 006, India.
[3] Division of Biochemistry, School of Life Science, JSS Academy of Higher Education and Research, Mysore, India.
* Corresponding author: priyankapravanvivek83741@gmail.com

DOI: 10.1201/9781003768708-9

from marine organisms also contribute to the production of novel bioactive compounds. The combination of marine fungal metabolites and bioinformatic tools associated with potent biosynthetic gene cluster identification are also overviewed.

1. Introduction

Earth's biosphere contains 70% of marine environment. Oceans contain diversified varieties of fungi and sponges and are recognised to be a treasure store for valuable natural bioactive compounds. In addition to marine fungi and sponges, oceans comprise molluscs, tunicates, cyanobacteria, bacteria, and a wide variety of algal species. Marine fungi are evenly distributed in marine habitats like hydrothermal vents, deep-sea sediments, salty marshes, sandy beaches, and arctic ice. The marine environment exhibits unique chemical and physical characteristics like temperature, salinity, and nutrients. Fungi derived from aquatic and marine ecosystems are resistant to high salinity, ultraviolet light, and hydrostatic pressure; require lesser substrates for growth; and have astonishing functional roles in synthesising novel metabolites in these extreme conditions. According to the recent discovery marine-derived fungi are potent producers of biologically active molecules when compared to terrestrial fungal habitats due to their unique metabolism and secondary metabolite pathways. Microorganisms derived from halobiont environments have proved to be significant synthesizers of novel drug metabolites, agrochemicals, and fragrances with different properties. Important anti-tumour drug Yondelis and painkiller Prialt, isolated from marine fungi are currently market-based valued products (Proksch et al., 2010). Biotransformations play a significant role in the multi folded increase in production of marine-derived unique products. Compounds like alkaloids, terpenoids, flavonoids, melanins, derived enzymes, peptides, lipids, isoprenoids, steroids, dyes, fusidic acid, penicillins, and cephalosporins show unique properties like anti-cytotoxicity, anti-inflammatory, anti-allergic, anti-malarial, and many other properties are found in marine fungi (Fig. 1). Cephalosporin C was reported as the first bioactive compound derived from marine fungus *Cephalosporium* sp. from the Sardinian coast. Marine-derived metabolites are released in higher proportion from marine fungi, which have sensible properties like oral bioavailability and unique physicochemical features, making them pharmacologically advantageous (Jin et al., 2016).

Marine-derived fungi are classified into two types: (1) obligate marine fungi, which require saline conditions for their growth and sporulation; (2) facultative marine fungi, which are isolated from terrestrial and freshwater environments and have the ability to grow and sporulate in marine ecosystems too. Outlined are the three ways of the current definition of marine fungi: (1) Ability to sporulate/grow in marine conditions; (2) Symbiotic mutualistic interaction with other marine organisms; (3) Adaptation and evolution at the genome level or biological activity in marine habitats are the three ways of the current definition of marine fungi (Le Strat et al., 2022).

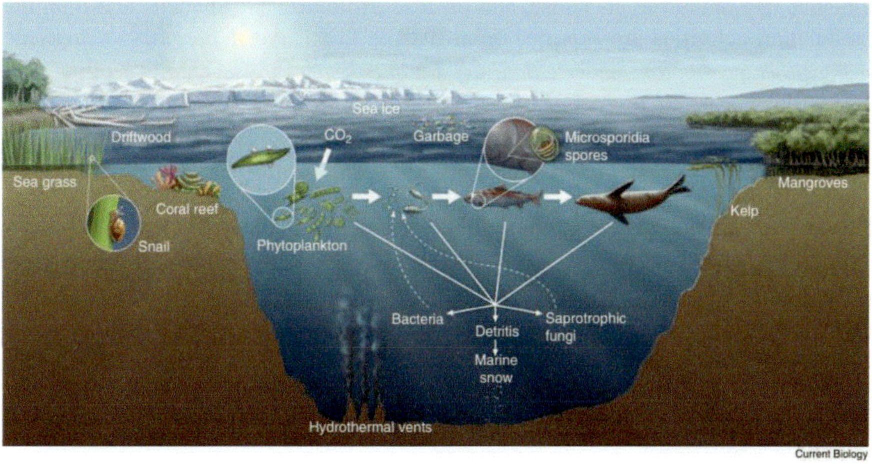

Fig. 1. Diversified marine fungal habitat (Gladfelter et al., 2019).

Fungi derived from marine environments also play an important role in marine food webs by decomposing dead and decaying matter and thereby recycling the nutrients, making them available for other marine organisms. In addition, these fungi also live in mutualistic association with other marine-derived habitats by providing beneficial nutrient uptake and defense mechanisms. Updated detailed information about the marine fungal classification, which includes Basidiomycota, Ascomycota, Mucoromycota, Blastocladiomycota, Chytridiomycota, and fungus-like organisms like Thraustochytriales, and presently 1,257 marine fungal species listed with 5 phyla, 20 classes, 168 families, 74 orders, and 539 genera, can be retrieved through the online resource (www.marinefungi.org) (Jones et al., 2019).

This chapter focuses on bioactive metabolites and their potential significance in various areas like pharmaceuticals and drugs, bioremediation, enzyme production, and textile dyes. Marine fungal genomics are also highlighted for natural product discovery and enabling tools for comparisons of Biosynthetic Gene Clusters (BGCs) which provide insights about antimicrobial resistance and novel metabolites.

2. Marine Fungal Polysaccharides

Polysaccharides are storage and structural biomolecules having unique structures and diversities that can be extracted from different marine species like microalgae, marine endophytic fungi, seaweeds, corals, diatoms, and many other marine calcified organisms. Carbohydrates share the highest proportion in marine biomass. Different types of polysaccharides like alginates, gum Arabic, konjaç flour, pullulan, dextrans, and pectin are highly applicable in food industries and are isolated from marine-derived fungi and associated endophytes. Recent evidence shows that only 1% of the marine microbes have been identified or cultured for their potential applications, and the rest are underexplored. There is more scope for identification of novel polysaccharides synthesised from marine-derived fungi. Some of the basic sugar moieties like mannose, galactose, and glucose exhibiting anti-diabetic and immune modulation activities are reported from marine fungi like *Aspergillus ochraceus*, *Aspergillus versicolor*, and *Penicillium janthinellum.* Polysaccharides often play a significant role in protecting the marine fungi from the external environment (Jeewon et al., 2023). Capsular polysaccharide production was reported in the marine filamentous fungus *Leptosphaeria albopunctata.* Xylose and maltose aid in significant production of diversified polysaccharides in Helotium sp. and Claviceps fusiformis marine fungi (Szaniszlo et al., 1968). *Fusarium equiseti* ANP2, a filamentous fungus isolated from Krishna estuarine mangrove regions, exhibited the highest viscosity and was reported to have exo-polysaccharides with antioxidant effects (Prathyusha et al., 2018) (Table 1).

Table 1. List of polysaccharides derived from marine fungi and their bioactivities.

Taxon	**Bioactivity**	**References**
Alternaria sp. SP-32	cytotoxic activity and antioxidant	Chen et al. (2016)
Epicoccum nigrum JJY-40	*in vitro* antioxidant	Sun et al. (2011)
Arthrospira platensis	antioxidant activity	Lin et al. (2022)
Aureobasidium melanogenum	immunomodulatory and antioxidant	
Fusarium oxysporum	antioxidant	
Hansfordia sinuosae	antitumour	
Phoma herbarum	immunomodulatory	
Keissleriella sp. YS4108	antioxidant activity	Sun et al. (2005)
Phoma herbarum YS 4108	antioxidant activity	Yang et al. (2005)
Aspergillus pseudoglaucus	immunomodulatory	Peng et al. (2023)
A. terreus	antioxidant	
A. versicolor	immunomodulatory	
Rhodotorula sp. RY1801	immunomodulatory	

3. Peptides from Marine Fungi

Peptides are the class of bioactive molecules having pharmacological significance. Peptides are made up of an array of amino acids with a remarkable molecular weight of a range between 500–5000 Da (Hafez Ghoran et al., 2023). Peptides and amino acids contain a nitrogen source, which plays a vital role in physiological functions. Structural elucidation and functional characteristics of different derived peptides can be analysed through various analytical techniques. The structural and functional significance of the peptides is totally dependent upon the composition and the sequence of amino acids.

Marine-derived fungal peptides are the active biomolecules that play a significant role in various fields like drug industry, cosmetics, agriculture, fermentation technology, and food technology.

Fungal polyketides, peptides, and amino acids derived from marine environments exhibit unique structural and physico-chemical properties. Organisms like marine algae, sponges, coral, and mangrove plants exhibit symbiotic association with diversified fungal classes like *Aspergillus*, *Acremonium*, *Beauveria*, *Eurotium*, *Penicillium*, *Simplicillium*, and *Trichoderma*, without causing deleterious effects but producing different classes of peptides like depsipeptides and cyclic and linear peptides. Several of these potential bioactive peptides exhibited various curative medicinal properties like anti-tumour, antiviral, anticoagulant, antioxidant, and immunoinflammatory effects. Marine organisms and their associated fungi produce notable peptides with the aid of certain large domain enzymes like non-ribosomal peptide synthetases (NRPS) or polyketide synthase hybrids (PKS) (Hafez Ghoran et al., 2023) (Fig. 2).

Youssef et al. (2019) have reported 131unique peptides from 17 diversified marine fungal genera *Acremonium*, *Ascotricha*, *Aspergillus*, *Asteromyces*, *Ceratodictyon*, *Clonostachys*, *Emericella*, *Exserohilum*, *Metarrhizium*, *Microsporum*, *Penicillium*, *Scytalidium*, *Simplicillium*, *Stachylidium*, *Talaromyces*, *Trichoderma*, and *Zygosporium* and reviewed different biological activities like cytotoxic, antimicrobial, antiviral, antidiabetic, anti-inflammatory, and lipid-lowering (Fig. 3). Polypeptides derived from marine fungi exhibited lesser harmful effects when compared with processed drugs. The marine-derived fungus K063 strain associated with the red algae *Ceratodictyon spongiosum* was reported for the presence of novel peptides Dictyonamides A and B, and seragakinone A (Komatsu et al., 2001). Marine sponge-derived *Acremonium* sp. was found to be a significant producer of a unique class of efrapeptin G peptides and exhibited anti-tumour properties against H125 cells (Boot et al., 2007). Marine fungi *Aspergillus allahabadii* and *A. ochraceopetaliformis*, showed the presence of cyclopentapeptides and

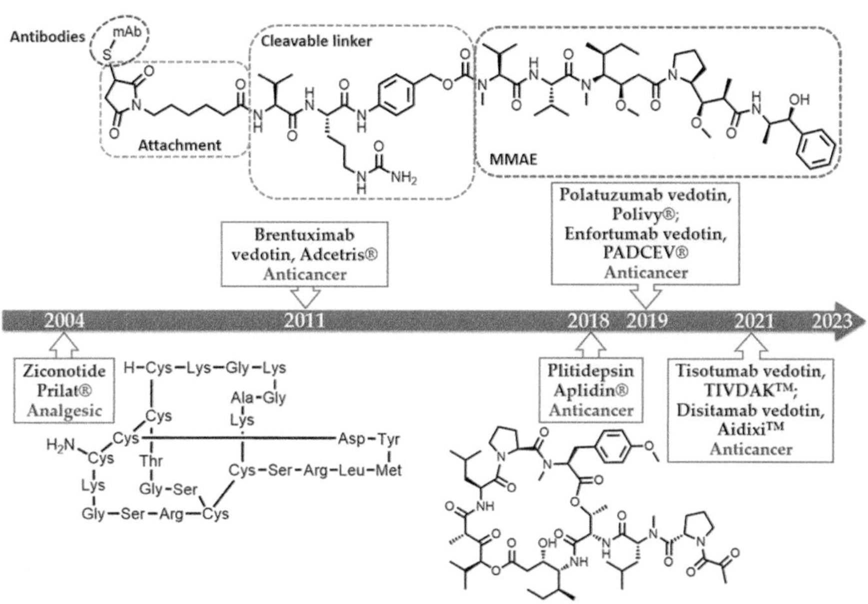

Fig. 2. Use of marine peptides as a drug approved by the Food and Drug Administration with the year of approval (Hafez Ghoran et al., 2023).

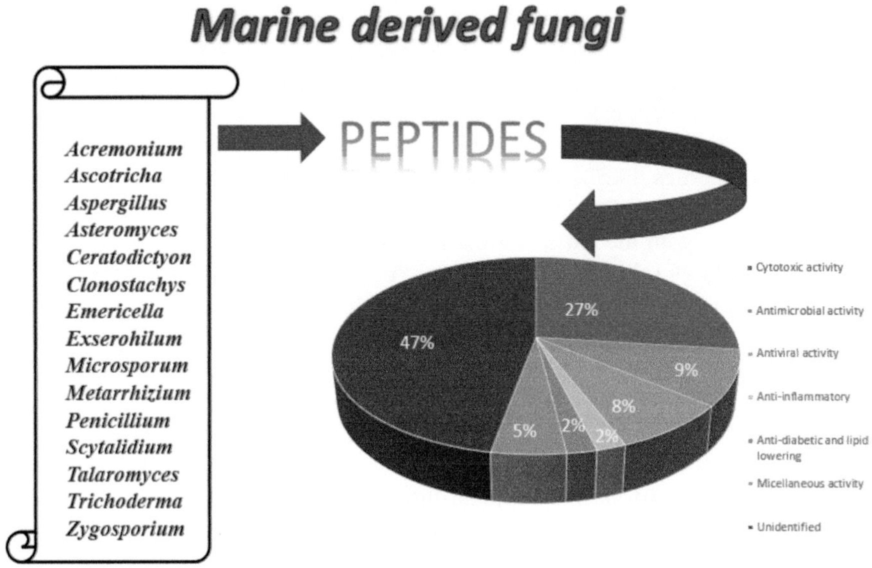

Fig. 3. Schematic representation of natural peptides derived from marine fungi showcasing various bioactivities (Youssef et al., 2019).

cyclodepsihexapeptide, respectively. The isolated peptides showed inhibitory effects on different classes of enzymes like isocitrate lyase and sortase A (Hwang et al., 2019).

A cyclic pentapeptide, Asperpeptide A, from *Aspergillus* species exhibited antimicrobial activity against bacterial species, *Bacillus cereus* and *Staphylococcus epidermidis*. Similarly, *Aureobasidium pullulans* derived from the marine environment synthesised the bioactive peptide Aureobasidin A, a class of cyclic octadepsipeptides reported for its anti microbial action against *C. albicans* and *C. neoformans*. Anti-malarial activity was reported from Cordyheptapeptide A and Cyclic heptapeptide compounds, which are produced by *Cordyceps* sp. Anti-micro algal activity is observed from the bioactive peptides Exumolides A and B isolated from *Scytalidium* sp. Glomosporin and Petriellin A demonstrated an antagonist effect against the pathogenic variant of *Aspergillus*. The compounds were isolated from *Glomospora* sp. and *Petriellasordida* sp., respectively (Ribeiro et al., 2022). *Penicillium chyrysogenum* DXY-1 from marine environments was meant for the synthesis of cyclic dipeptide having anti-quorum sensing activity against *Chromobacterium violaceum* CV026 and *Pseudomonas aeruginosa* PA01 (Yu et al., 2021) (Table 2).

4. Marine Fungal Pigments

Biological pigments are the source of natural-coloured compounds, which are also known as biochromes and are insoluble in water. Natural pigments are harmless and are obtained from plants, animals, algae, fungi, bacteria, and many other living

Table 2. Some of the marine fungi and bioactive peptides with various bioactivities.

Taxon	Peptide Name	Biological Activity	References
Asteromyces cruciatus	Lajollamide A	Antibacterial	Youssef et al. (2019)
Ceratodictyon spongiosum	Dictyonamides A and B	Cyclin-dependent kinase 4 inhibitor	
Clonostachys rogersoniana	Clonostachysins A and B	Dinoflagellate *Prorocentrum micans* inhibitor	
Emericella unguis	Unguisins A and B	Antimicrobial activity against MRSA	
Exserohilum rostratum	Rostratins A–D	Cytotoxic activity	
Microsporum cf. *gypseum*	Microsporins A and B	Inhibition of histone deacetylase	
Metarhizium sp.	Destruxin A and B	Mycotoxin	
Penicillium purpurogenum	Penicimutide	Inhibited growth of HeLa cells	
Scytalidium sp.	Halovirs A–E	*In vitro* antiviral activity against herpes simplex viruses	
Simplicillium obclavatum	Simplicilliumtides A–M	Antimicrobial, antifouling, cytotoxic, and acetylcholinesterase inhibitory activity	
Stachylidium sp.	Endolides A–D	Toxic to silkworms and mice	
Talaromyces sp.	Talaropeptides A–D	Antibacterial activity	
Trichoderma	Peptaibols	Antimicrobial activity	
Zygosporium masonii	Zygosporamide	Cytotoxic activity	
Aspergillus fumigatus, Nectria inventa	Diketopiperazines	Antiprotozoal	Kang et al. (2015)
Trichoderma sp.	Trichoderin A, A1, B	Antituberculosis	
Aspergillus terreus	Asperterrestide A	Antiviral	

creatures. Basically, the pigments are classified according to their sources into two types: (1) Organic pigments; (2) Inorganic pigments. Organic pigments are also known as natural pigments which are derived from natural habitats, whereas, inorganic pigments are known as synthetic or artificial pigments. Pigments and colourants are applicable in various industries like cosmetics, pharmaceuticals, dye industries, food industries, textiles, and many other industrial sectors (Fig. 4).

Some of the predominant examples for pigments are carotenoids, melanins, riboflavins, azaphilones, and quinones, Bioactive pigments and colourants produced from the microbes and plants are in high demand due to their high stability, availability, downstream processing, and cost = effectiveness and are considered to be the best replacement for synthetic or artificial colourants. Carmine, which is the most stable natural red-coloured pigment, is found to be resistant to light and heat (Lebeau et al., 2017).

Marine fungal-derived natural pigments are generally secondary metabolites produced during biotic and abiotic stress conditions like pathogen infection,

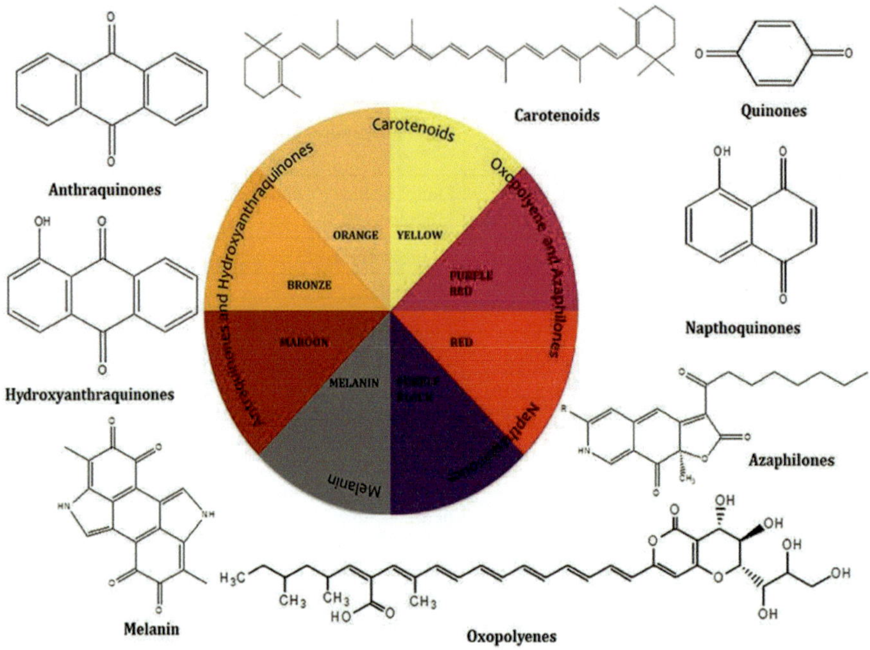

Fig. 4. Different classes of bioactive pigments and colorants associated with fungi (Kalra et al., 2020).

temperature, climatic conditions, salt tolerance, and many other factors. Some of the marine fungi, like *Alternaria*, *Aspergillus*, *Eurotium*, *Emericella*, *Fusarium*, *Halorosellinia*, *Microsphaeropsis*, *Monodictys*, *Paecilomyces*, and *Penicillium* have been reported for the bioproduction of pigments like carotenoids and polyketides.

Some marine fungi associated with higher plants, algae, phytoplankton, corals, seaweeds are known for the synthesis of unique and diversified pigments commonly shared in both organisms. Some of the examples include chaetoglobin, chaetoglobosin, and chaetomugilin pigments isolated from *Chaetomium globosum*, which are associated with coral rubbles. Likewise, anthraquinone, naphthoquinone, and hydroxyanthraquinone compounds are synthesised by *Fusarium* species, which are ecological symbionts with mangrove marshes (Meruvu et al., 2021). *Talaromyces albobiverticillius* strains isolated from the marine ecosystem of Reunion Island, Indian Ocean, showed the presence of red- and orange-coloured pigments.

A list of natural colourants and pigments like mitorubrin, monascorubrin, monascorubramine, monascin, and xanthomonasin are identified in *Penicillium* sp. (Venkatachalam et al., 2018).

Aspergillus sydowii, *Rhodotorula evergladensis*, and *Talaromyces coalescens* fungi isolated from marine ecological niches were found to be the source for synthesis of pigments with antibacterial and antioxidant properties (Lingappa et al., 2022).

Some of the epiphytic fungi like *Aspergillus niger*, *A. flavus*, *A. terreus*, *A. fumigatus*, *Cladosporium oxysporum*, *Curvularia chlamydospora*, *Alternaria alternata*, *Penicillium chrysogenum*, and *Rhizopus stolonifera*, isolated from marine macroalgae were reported to have melanin pigment with antimicrobial efficacy (Mansour et al., 2025).

Marine sponge-associated fungi *Trichoderma parareesei* also synthesised natural pigments with antimicrobial activity against multi drug resistant pathogens like *Salmonella typhi* and *Escherichia coli* (Sibero et al., 2016).

Marine-derived filamentous fungi are the best source for the production of vital pigments known as anthroquinones. Anthraquinones are naturally occurring phenolic derivatives and are also known as anthracenedione (Table 3). These are aromatic derivatives synthesised by the acetate-malonate pathway that are found in most plants, fungi, bacteria, and algae have different biological activities like anti-inflammatory, anti microbial, antioxidant, and anti-osteoporosis activities.

5. Marine Fungal Biocatalysts and Novel Enzymes

Biocatalysts are proteins, specifically enzymes, that gear up the rate of biochemical reactions and react with the specific type of substrate molecule. They play a significant role in life processes like cellular metabolism, biochemical reactions, and performing transformations chemically on organic compounds.

Marine fungi are involved in the synthesis of different groups of enzymes of industrial significance. Specific substrates like pectin, starch, malt extract, casein, wheat bran, xylan, olive oil, and sugarcane bagasse are supplemented for the enzyme production derived from marine fungi. Carbon and nitrogen supplements also play a key role for the growth of fungi.

Marine fungi are halotolerant and can resist sodium levels by captivating the sodium ions from their vacuoles and through the efflux mechanisms by ATPase, which indirectly leads to the promotion of novel characteristics to the cells by producing novel enzymes. Marine fungi grow and survive in saline conditions corresponding to increase in temperature and the phenomena is known as the *Phoma* pattern which was first reported in mariner-derived *Phoma* sp. Such characteristics may be exhibited by the presence of diversified thermos- and saline-tolerance enzymes in the species, which can be of high market value. Solid state fermentation and submerged fermentation can be used to isolate highly relevant enzymes and bioactive compounds from marine-derived fungus. Generally, because of less water consumption, solid-state fermentation is preferred.

Generally different classes of enzymes have been studied in marine fungal species depending upon their cost effectiveness, specificity, and speed. In order to evaluate and study multiple groups of enzymes and their corresponding fungal strains, two different methods are: (1) High throughput screening (HTS); (2) Medium-throughput screening (MTS).

The genus *Aspergillus* was found to be a model organism and the largest producer of beneficial enzymes among fungal varieties. Halotolerant lignocellulolytic enzymes from the fungus *Pestalotiopsis* sp., isolated from a mangrove host, have also been reported.

Table 3. List of anthraquinone-producing marine-derived fungi (Fouillaud et al., 2016).

Marine Fungi	Host Organism	Pigmented Compounds
Alternaria eichorniae	Marine plant pathogen	4-deox-bostrycin, bostrycin
Alternaria SK11	Mangrove plant	(+) α S-alterporriol C, 6-methylquinizarin, austrocortinin, and alterporriol S
Aspergillus glaucus	Mangrove sediment	Catenarin, emodin, cynodontin, erythroglaucin, helminthosporin, questin, physcion, rubrocristin, and variecolorquinone A
Aspergillus sp. 05F16	Algal endophyte	1-deoxytetrahydrobostrycin and tetrahydrobostrycin
Curvularia lunata	Marine organism endophyte	Cytoskyrin A and lunatin
Eurotium cristatum	Marine organism endophyte	Catenarin, emodin, erythroglaucyn, physcion, physcion anthrone, questin, and rubrocristin
Eurotium rubrum	Marine plant endophyte	6,3-O-(α-D-ribofuranosyl)-questin, and questin
Fusarium sp. No. B77	Mangrove Plant endophyte	5-acetyl-2-methoxy-1,4,6-trihydroxy-anthraquinone
Fusarium sp. No. ZH-210	Mangrove sediment	Fusaquinon A, B, and C
Halorosellinia sp. (No. 1403)	Marine plant-derived	1,4,5,6,7,9-hexahydroxy-2-methoxy-7-methyl-5β,9β, 8aβ,6α,10aα—hexahydroanthracene10(10aH)-one, austrocortirubin, and demethoxyaustrocortirubin, Hydroxy-9,10-anthraquinone SZ-685C
Nigrospora spp.	Marine plant/ Organism endophyte	1-deoxytetrahydrobostrycin, 4-deoxybostrycin, anthraquinone, SZ-685C Bostrycin, and 4a-epi-9α-methoxydihydrodeoxybostrycin, and 10-deoxybostrycin
Nigrospora sp. MA75	Marine	4-deoxybostrycin anthraquinone SZ-685C and Bostrycin
Paecilomyces sp.	Mangrove	Chrysophanol and Emodin
Penicillium citrinum PSU-F51	Marine organism endophyte	Chrysophanol, Citreorosein, Emodin, and Penicillanthranins A and B
P. chrysogenum	Salt Lake	Skyrin
P. oxalicum 2-HL-M-6	Mangrove sediment	Aloe emodin, chrysophanol, citreorosein, citreorosein-3-o-sulfate, emodin, emodin-3-o-sulfate, and isorhodoptilometrin
Phomopsis sp. PSU-MA214	Mangrove plant endophyte	Phomopsanthraquinone, 1-hydroxy-3-methoxy-6-methylanthraquinone, Ampelanol, and Macrosporin
Trichoderma aureoviride PSU-F95	Marine organism endophyte	Coniothranthraquinone 1 and trichodermaquinone
Xylaria sp. 2508	Marine	Altersolanol A, bostrycin, deoxybostrycin, and xylanthraquinone

Marine ecosystems can be affected by the release of aromatic hydrocarbons like polycyclic aromatic hydrocarbons, leading to environmental pollution and disturbing the marine food chain and ecosystem. Bioremediation plays a significant role wherein some of the novel enzymes isolated from marine fungi have a capacity to degrade these toxic pollutants.

Bonugli-Santos et al. (2015) have reported that PAH degradation involves two different pathways, like the ligninolytic pathway and the monooxygenase pathway of cytochrome P-450 in filamentous fungi. Monooxygenases are intracellular enzymes, whereas ligninolytic enzymes are extracellular, which include manganese-dependent peroxidase, lignin peroxidase, and laccase.

Marine-derived fungal genera *Penicillium* sp. FM16 and FM02 and *Trichoderma* sp. FM14 have shown significant tolerance and degradative capability of pyrene compounds (Nogueira et al., 2024). Marine-derived fungal strains *Aspergillus sclerotiorum* CBMAI 849 and *Mucor racemosus* CBMAI 847 have also been reported as potent degraders of the pyrene synthetic compounds pyrenylsulfate and benzopyrene to benzopyrenyl sulfate (Passarini et al., 2011). Marine-derived *Aspergillus* sp. BAP14 has significantly removed benzo[a]pyrene by 30–60% (Bonugli-Santos et al., 2015) (Fig. 5).

Synthetic dyes are the harmful recalcitrant moieties that are deposited in the marine system and which affect marine inhabitants leading to imbalance in the aquatic ecosystem by interfering in the photosynthesis and diffusion of the gases, which affects human health.

Some of the newer cost-effective technologies include biological agents and genetically modified organisms, which strongly influence the degradation of azo dyes.

Congo red, brilliant green, and RBBR synthetic dyes are decolourised and degraded by halotolerant, acidophilic, and alkaliphilic marine-derived fungus. Some examples include *Pestalotiopsis* sp. J63 and *Penicillium janthinellum* P1, which are reported to decolourise azure B dye.

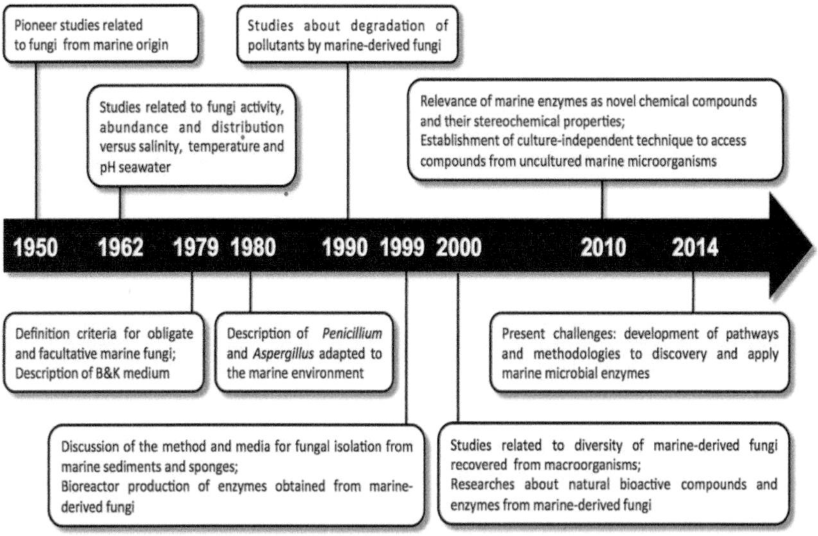

Fig. 5. Scientific events of marine-derived fungi and their potent enzymes (Bonugli-Santos et al., 2015).

Tinctoporellus filamentous fungi associated with coral and zoanthids were found to degrade RBBR synthetic dye. Similarly, *Penicillium citrinum* CBMAI 853 was also reported as a decolourizer of RBBR.

Twenty-five different varieties of chitinases and their types were produced by marine fungal endophytes like *Aspergillus, Beauveria Bassiana, Fusarium, Lecanicillium lecanii, Metarhizium anisopliae,* and *Trichoderma harzianum* derived from seaweeds. These fungi are also commercially significant with reported bioactivities like anti-tumour activity, anti microbial activity, and wound healing, and applications in drug delivery and wastewater treatment (Venkatachalam et al., 2015). Proteases are the most widely used enzymes in lather production and the detergent and pharmaceutical industries. *Aureobasidium pullulans*, a marine-derived yeast producing a potent alkaline enzyme, is used in the detergent industry due to its resistance to various extreme conditions. Proteases-producing marine fungi such as *Armillaria mellea*, *Aspergillus ochraceus*, *Cochliobolus lunatus*, *Cordyceps militaris*, *Cordyceps sinensis*, *Fusarium oxysporum*, *Fusarium* sp., *Flammulina*, *Ganoderma lucidum*, *Penicillium chrysogenum*, *Pleurotus ostreatus*, *Rhizopus chinensis*, *Schizophyllum*, *Tricholoma saponaceum*, and *Velutipes* are reported to have fibrinolytic activity (Jeevithan et al., 2024).

Fats and oils are degraded by a group of enzymes known as lipases, which are isolated and synthesised commercially by a group of different marine-derived fungi. They are highly useful in the detergent, paper, cosmetics, medicine, and food industries. *Penicillium oxalicum* and *Aspergillus flavus* are reported for the synthesis of lipases. Seawater isolate *Aspergillus awamori* BTMFW032 is known to degrade 92% of oil and fat content in the effluent by lipase production (Basheer et al., 2011).

The chitinase gene was isolated from the marine endophyte *Aspergillus fumigatus* df347 and was cloned and expressed in *E. coli* for increased multiple-fold production of the chitinase enzyme (Wu et al., 2022).

Xylanases are the group of novel enzymes having greater applications in various fields like paper and pulp, food and feed, bioethanol production, bleaching industries, and many more (Table 4).

Inducible xylanase synthesis plays a pivotal role in breaking down a complex polysaccharide, xylan. As xylan polysaccharides cannot easily pass through cell walls, xylanases play an important role in the hydrolysis of complex xylan into simpler molecules like xylose and xylo-oligomers, which can easily penetrate and promote the synthesis of xylanases. Some important xylanase-producing marine fungi like *Aspergillus cf. tubingensis* LAMAI 31, *Bartalinia robillardoides* LF550, *Calcarisporium* sp. KF525, *Penicillium pinophilum* LF458, *Pestalotiopsis* sp. KF079, *Scopulariopsis brevicaulis* LF580, and *Tritirachium* sp. LF562 are also reported (Qeshmi et al., 2019).

Marine-derived fungi are considered as a potential producer of novel diversified enzymes with multiple biotechnological applications due to their stability and tolerance under extreme physiological conditions. This adaptability makes them a promising source of enzymes for future research development. Bioprospecting of newer enzymes, enzyme and metabolic engineering, and codon optimisation processes can lead to enhanced synthesis of enzymes.

Table 4. List of industrially important enzymes derived from different classes of marine fungi.

Enzyme	Marine Fungi	Substrate	Industrial Application	References
Cellulase	*Trichoderma viride* ATCC	Wheat bran	Biofuel production, textiles, food processing, and animal feed	Shah et al. (2021)
Esterase	*Aspergillus niger* I-1472	Sugar beet pulp	Pharmaceuticals and food industry	
Invertase	*Aspergillus niger*	Polyurethane foam	Food and beverage production	
Lipase	*A. niger* NCIM1207	Olive oil and wheat bran	Food and detergent industries	
Phytase	*A. niger*	Soybean milk and wheat bran	Animal feed and the food industry	
Polygalacturonase	*A. niger*	Coffee pulp and wheat bran	Food industry and plant biology	
Tannase	*A. niger* Aa 20	Polyurethane foam	Food, beverage, and pharmaceutical industry	
Laccase	*Flavodon flavus*, *Cerrena unicolor*, *Mucor*, *racemosus*, and *Trematosphaeria mangrovei*	Sugarcane, bagasse, glycine, fructose, wheat bran + glucose, sucrose, peptone, and yeast extract	The food, cosmetics, and pharmaceutical industries	Arifeen and Liu (2018)
Lignin	*Flavodon flavus*	Sugarcane bagasse	Biofuels, bioplastics, construction and building materials, and heavy metal removal	
Peroxidase	*Mucor racemosus*	Malt extract	Medical diagnostics, cosmetics, biosensors, paper, and detergents	
Protease	*Rhodotorula mucilaginosa* sp. and *Acremonium* sp.	Skimmed milk and Sabouraud Cactus and pear	Detergent, food, leather, and pharmaceuticals	

6. Novelty of Marine Endophytic Fungi

Endophytic fungi are those that are symbiotically associated with marine plants, algae, seagrasses, mangroves, corals, and sponges and are also distributed in the internal lining of tissues of invertebrates without causing any harmful effects (Fig. 6). Intercellular and intracellular distribution of these endophytes depends on the age and type of host system and on biotic and abiotic environmental conditions. Marine coasts may harbour a variety of endophytes, which are the repositories for natural bioactives.

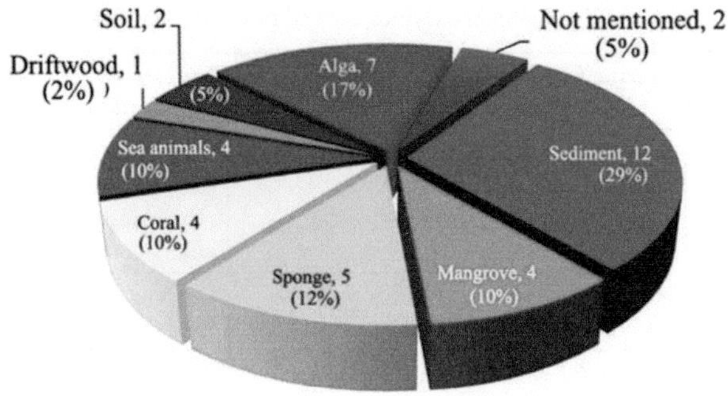

Fig. 6. Isolation rate of marine fungal endophytes from different marine habitats (El-Bondkly et al., 2021).

Due to mutual interaction between the host and the endophyte, these fungi influence the host life cycle and in turn are reported to release secondary metabolites in response to the chemical signal produced by the host. These secondary metabolites are produced through the secondary metabolism involving different pivotal metabolic pathways, which are quite common in both hosts and their endophytes. Some of the pathways include mevalonate and non-mevalonate pathway producing terpenes, similarly other pathways include alkaloid, flavonoid, polyketide, non-ribosomal peptide, shikimate, phenazine biosynthetic pathways. Key enzymes involved in these biosynthetic pathways play a crucial role in the catalysis of specific reactions and thereby lead to the improved production of diversified metabolites. Along with the synthesis of secondary metabolites, different volatile and smaller molecules, including esters, alcohols, hydrocarbons, acids, ketones, and enzymes like amylases, pectinases, xylanases, polygalacturonases, ligninase, cellulases, and so on are produced by endophytes.

These compounds, including enzymes, have a wide array of applications, including drugs, the food industry, cosmetics, dyes, bioremediation, textiles, paper and pulp, wastewater management, beverages, and diagnostics. The endophytic fungal bioactive compounds also have bioactivities like anti-tumour, anti-diabetic, anti-malarial, anti-inflammatory, and antimicrobial activities. Some compounds are known to show effective immunemodulatory effects, and several molecules are patented for their efficacy and novelty. Till date, 14 bioactive marine drugs have been approved, and 30 different marine bioactives are under clinical trials (Gonçalves et al., 2022).

6.1 Bioactive Potential of Fungal Endophytes

Marine-derived red algae *Laurencia obtuse*-associated endophytic fungi *Penicillium aculeatum* was shown to produce sulfonyl metabolites pensulfonoxy and pensulfonamide, which are reported to have antibacterial and antifungal inhibitory activity against *Escherichia coli* and *Candida albicans*, respectively. Meanwhile, pensulfonamide was also reported as an effective drug for its cytotoxic effects.

Fusarium equiseti showed a high-level inhibition of HCV protease, among two others endophytic fungi, *Scopulariopsis fusca* and *Geotrichum candidum*, isolated from the marine brown alga *Padina pavonica*.

The endophytic fungi *Eurotium chevalieri*, *Alternaria alternata*, and *Penicillium crustosum* were isolated from the coral *Litophyton arboreum*. The metabolites isolated from *Alternaria alternata*, alternariol-9-methyl ether3-O-sulphate, alternariol-9-methyl ether, alternariol, maculosin, and maculosin-5 were further studied and explored for their antiviral activities against hepatitis C virus (HCV) by specifically inhibiting HCV NS3 NS4A protease. Similarly, anticancer and antimicrobial activities were also exhibited by microbial metabolites obtained from marine-derived fungi (Elkhateeb and Daba, 2022).

Antidiabetic biological activity was reported from the metabolite compounds produced from marine-derived endophytes by inhibitory action towards various diabetes-related enzymes including α-glucosidase, dipeptidyl peptidase-4 (DPP-4), protein tyrosine phosphatase 1B (PTP1B), fatty acid-binding protein 4 (FABP4), and glycogen synthase kinase-3β (GSK-3β). Predominant endophytes derived from marine ecosystems, like *Aspergillus* sp., *Alternaria* sp., *Daldinia* sp., *Eurotium* sp., *Meira* sp., *Mycosphaerella* sp., *Penicillium* sp., *Talaromyces* sp., and *Trichoderma* sp., showed downregulatory efficacy against α-glucosidase enzymes, and similarly, many other endophytic fungal species were also reported for their anti-diabetic effects by enzyme inhibition (Wang et al., 2025).

Marine-derived sponges harbour various endophytic fungi, including *Aspergillus* and *Penicillium* sp., which were isolated using suitable growth media and further analysed for their potential abilities, like anti-inflammatory, and cytotoxic effects and free radical scavenging activities (Sajeevan et al., 2020).

The ethyl acetate extracts of five endophytic fungi, *Candida railenensis*, *Cystobasidium minutum*, *Grammothele fuligo*, *Pichia kudriavzevii*, *Rigidoporus vinctus* isolated from marine algae, were screened for antimicrobial activity against *P. aeruginosa* and high cytotoxic activity against the MG63 cell line. Among which *Cystobasidium minutum* showed promising antimicrobial activity, and *Pichia kudriavzevii* showed a potent cytotoxic effect (Harikrishnan et al., 2021) (Table 5).

Table 5. List of marine endophytic fungi and their characteristics (Behera and Das, 2023).

Marine Endophytic Fungi	Marine Source	Characteristics
Aspergillus, Cladosporium, Exserohilum, Penicillium, Trichoderma	Marine sponges and corals	Organic carbon degrader
Fusarium, Phialemoniopsis, Trichoderma	Surface sediment	Stable iron-oxide nanoparticles production
Pseudogymnoascus, Penicillium	Marine soil, sediments, and plants	Active production of lipase and protease enzymes
Aspergillus	Mangrove Sediment	Synthesis of extracellular laccase and manganese peroxidase
Alternaria	Mangrove water	Biosorbent for metal chelation
Cladosporium, Coprinellus, Eurotium, Penicillium, Tolypocladium	Marine sponge Grantia compressa	Secondary metabolite production
Aspergillus, Cladosporium, Parengyodontium	Sea water	Protease, amylase, phytase, and cellulase production

Table 6. List of novel metabolites from marine endophytic fungi (El-Bondkly et al., 2021).

Marine Endophytic Fungi	Host	Activity	Metabolite
Alternaria sp.	*Sarcophyton*	Antiviral and cytotoxic	Tetrahydroaltersolanols and dihydroaltersolanol A
Aspergillus wentii	*Sargassum* sp.	Cytotoxicity	Wentilactone B
Aspergillus sp.	*X. testudinaria*	Antibacterial	Bisabolane
Daldinia eschscholtzii	*Xestospongia* sp.	Antimicrobial	Karimanone
Paecilomyces variotii	marine algae	Cytotoxic	Indole derivatives
Penicillium citreonigrum	*Pseudoceratina purpurea*	Protein kinase Inhibition	Citreonigrin B
P. pinophilum	*gorgonian corals*	Antiviral and cytotoxic	Pinophilins D-F
Penicillium sp.	*Kandelia candel*	Acetylcholine sterase inhibition	Arigsugacin I
Pestalotia sp.	Brown algae	Antibiotic	Chlorinated bezophenone
Xylaria sp.	mangrove	Acetyl choline esterase inhibition	Xyloketals

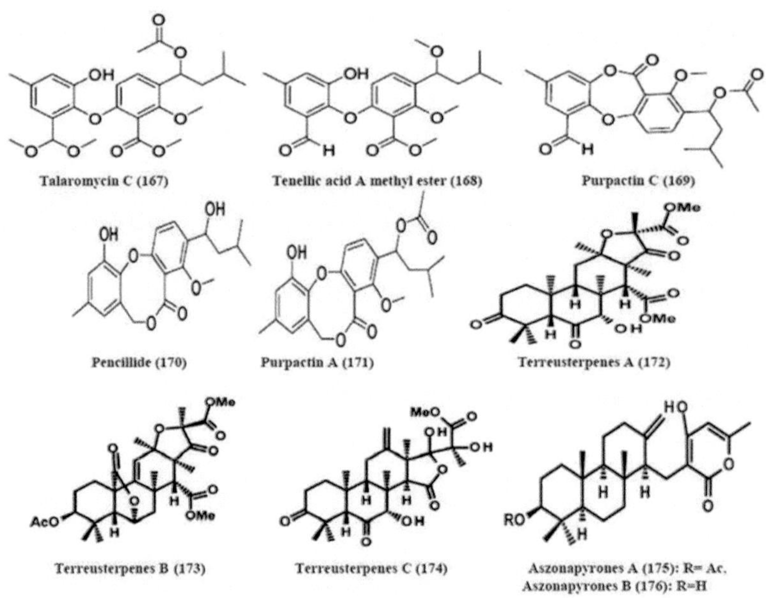

Fig. 7. Some of the secondary metabolites from marine-derived endophytic fungi (El-Bondkly et al., 2021).

Alternaria alternata, *Aspergillus niger*, and *Penicillium rubens* endophytes isolated from mangrove plants (*Avicennia marina*) have revealed the presence of several unique metabolites with antiproliferative activity against tumor cells (Al-Rajhi et al., 2022) (Table 6; Fig. 7).

7. Metabolomic Analysis of Marine-derived Fungi

Traditional gene analysis methods involving bioactive gene identification, isolation, characterisation, and expression are laborious and time-consuming. To overcome this drawback, the current approach is to use high-throughput whole genome sequencing (WGS), omics studies like genomic, proteomic, transcriptomic, metabolomic, and epigenomic analysis. These techniques provide detailed knowledge about the organisms, which helps in identifying novel unravelled metabolites and their BGCs. These BGCs have a vital role in a variety of biological interactions like communication, defence, and competition. Some of the BGCs include PKS gene clusters, Non-Ribosomal Peptide Synthetase (NRPS) gene clusters, and Terpenoid Biosynthetic Clusters. Most of the metabolites obtained from these clusters are mainly used as drugs. These clusters can be further investigated in understanding the microbial resistance mechanisms.

Bioinformatic tools for BGC identification include Antibiotic and Secondary Metabolite Analysis Shell (antiSMASH), Significantly Mutated Region Finder (SMuRF), AUGUSTUS, Minimum information about a Biosynthetic Gene Cluster (MIBiG), Big SLiCE, Biosynthetic Gene Cluster Family's Database, (BiG-FAM), ClusterMine 360, DoBISCUIT, Integrated Microbial Genomes: Atlas of BGC (IMG-ABC), and several others (Fig. 8).

Whole genome sequencing data of the marine-derived *Aspergillus niger* L14 strain analysed by antiSMASH recorded the presence of 69 secondary metabolite BGCs, which include 18 NRPSs, 17 PKSs, 9 terpenes, 21 NRPS-likes, 2 indoles, 1 betalactone, and 1 siderophore (Wang et al., 2022). Similarly, terrestrial endophyte *Talaromyces* sp., isolated from a medicinal plant and analysed for BGCs through funannotate and antiSMASH analysis, showed the presence of 76 BGCs, including 10 NRPSs, 28 T1 polyketide synthase (T1 PKS), 1 indole, 9 terpenes, 12 NRPs-like, 4 T1 PKS and NRPs-like, 2 T1 PKS Indole, 3 NRPs T1 PKS, 1 NRPs-like terpene, 1 indole NRPs, 2 betalactones, 1 fungal-RiPP T1 PKS, 1 phosphonate, and 1 other type (Shenoy et al., 2023). Genome mining analysis of a lichen symbiont, *Xanthoria elegans*, predicted several novel polyketides and ambiguous metabolites in the clusters. Assembly of the genome and *de novo* sequencing of the species predicted the genome size to be 44.63 Mb with 40.69% GC and 207 scaffolds. A total of 9,581 genes were found involved in the synthesis of terpenes and polyketide groups of compounds. Some of the predicted clusters, like NR PKS, HR-PKSs, and hybrid PKS-PKSs, were analysed (Yuan et al., 2023).

WGS and antiSMASH analysis of an endophyte *Fusarium* sp., derived from a coastal plant, *Rumex madaio*, found 37 SM BGCs, including 17 NRPSs, 3 terpene synthases (Ts), 13 PKSs, 1 hybrid indole + NRPS, and 3 hybrid NRPS + PKS. These investigations further aid in the improved target identification with specificity (Liu et al., 2022). A unique fungal *Penicillium* sp., derived from a sponge subjected to WGS analysis was predicted to have 34 BGCs. Novel compounds pyridinones and penicidihydro pyridone–B were extracted, and structural characterization was done by NMR and MS. Molecular docking based on structure found that neo-pyridinones effectively block PD-1 protein (programmed cell death) (Chen et al., 2022).

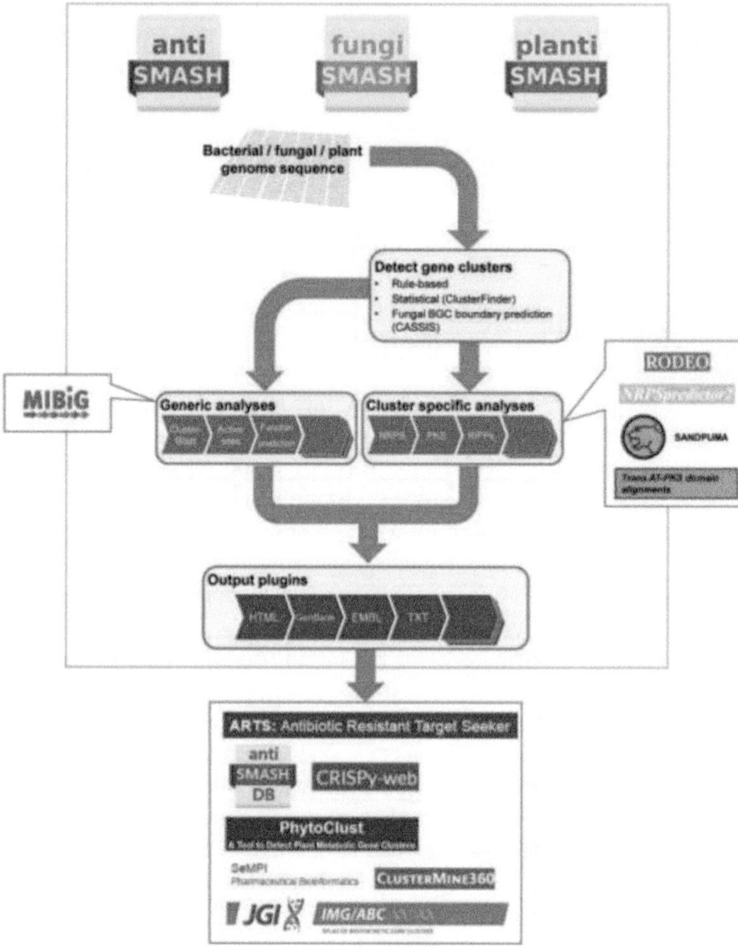

Fig. 8. General workflow of antiSMASH (Blin et al., 2019).

Marine fungal endophytes produce secondary metabolites and bioactive compounds at a very low concentration. Metabolic engineering of the pathways involved in the synthesis of these compounds may result in the improved production of the desired products. With the use of genetic modifications, it is now possible to produce alkaloids, flavanoids, terpenoids, and other important bioactives at a large scale. To enhance the production of metabolites, two ways can be followed. The first method is to modify the expression of one or more genes. Secondly, attempts can be made to modify the expression of a key gene that controls various gene clusters. Endophytes can be used to achieve constant expression of the biosynthetic genes required for sustained production of desired bioactive secondary metabolites. Expression of certain genes encoding biosynthetic enzymes or regulatory proteins can alter various secondary metabolite pathways.

Automated genome mining pipeline strategies like antiSMASH aid in the identification of specific core enzymes by sequence similarity analysis of enzymes/genes involved in metabolic pathways, which cannot be done through manual methods. Comprehensive biosynthetic data analysis of marine fungal endophytes can be carried out by genome mining depending upon natural product biosynthesis, which is highly novel, diversified, and specific in each organism.

In the future marine-derived endophytic fungi and other marine-derived organisms algae, and bacteria, can be further investigated for the presence of unknown BGCs, which are highly potent and can be identified by highly emerging *in silico* databases as described above and validated through *in vitro* studies for the presence of metabolites and their clusters.

8. Conclusion and Outlook

The present book chapter provides insights on some of the marine-derived fungal metabolites and compounds having bioactive potential. Marine fungi are known to produce novel compounds which are rare in terrestrial habitats. These metabolites are also applicable in life threatening diseases based on their mode of action. Screening of marine fungi and their derived metabolites for bioactivities are basically limited to the lab-scale exploration. Clinical trials and industrial scale up production of these bioactives are essential for commercial availability to combat various diseases. Isolation and culturing of marine fungi still remain a big challenge to the researchers.

Advanced *in silico* and omics tools approach helps in the identification of unknown functional and silent gene clusters in existing marine fungi. Genome mining and annotation aids in the target gene identification, which acts as a new gateway for further molecular studies like gene cloning, expression and characterization. Structural elucidation of these marine-derived fungal metabolites can be done by NMR and X-Ray diffraction studies. Bioactive metabolite libraries can be developed by analysing and combining the metabolites coded by biosynthetic gene clusters present in marine fungi. Some of the novel metabolites from marine fungi can also be used against antimicrobial resistance (AMR) organisms.

Due to the inaccessible zones in marine environments, its fungal diversity still remains unexplored and though there are many advanced *in silico* tools for the marine fungal systematics, complete detailed knowledge about metabolic networks is yet to be studied.

References

Al-Rajhi, A.M., Mashraqi, A., Al Abboud, M.A., Shater, A.R.M., Al Jaouni, S.K. et al. (2022). Screening of bioactive compounds from endophytic marine-derived fungi in Saudi Arabia: Antimicrobial and anticancer potential. Life, 12(8): 1182. https://doi.org/10.3390/life12081182.

Arifeen, M.Z.U. and Liu, C.H. (2018). Novel enzymes isolated from marine-derived fungi and its potential applications. United J. Biochem. Biotechnol., 1: 1–11. https://doi.org/10.22259/2637-5362.0204001.

Basheer, S.M., Chellappan, S., Beena, P.S., Sukumaran, R.K., Elyas, K.K. and Chandrasekaran, M. (2011). Lipase from marine *Aspergillus awamori* BTMFW032: Production, partial purification, and application in oil effluent treatment. New Biotechnol., 28(6): 627–638. https://doi.org/10.1016/j.nbt.2011.04.007.

Behera, A.D. and Das, S. (2023). Ecological insights and potential application of marine filamentous fungi in environmental restoration. Rev. Environ. Sci. Bio. Technol., 22(2): 281–318. https://doi.org/10.1007/s11157-023-09655-2.

Blin, K., Kim, H.U., Medema, M.H. and Weber, T. (2019). Recent development of antiSMASH and other computational approaches to mine secondary metabolite biosynthetic gene clusters. Briefings Bioinform., 20(4): 1103–1113. https://doi.org/10.1093/bib/bbx146.

Bonugli-Santos, R.C., dos Santos Vasconcelos, M.R., Passarini, M.R., Vieira, G.A., Lopes, V.C. et al. (2015). Marine-derived fungi: diversity of enzymes and biotechnological applications. Front. Microbiol., 6: 269. https://doi.org/10.3389/fmicb.2015.00269.

Boot, C.M., Amagata, T., Tenney, K., Compton, J.E., Pietraszkiewicz, H. et al. (2007). Four classes of structurally unusual peptides from two marine-derived fungi: Structures and bioactivities. Tetrahedron, 63(39): 9903–9914. https://doi.org/10.1016/j.tet.2007.06.034.

Chen, C., Qi, J., He, Y., Lu, Y. and Wang, Y. (2022). Genomic and chemical profiling of B9, a unique Penicillium fungus derived from sponge. J. Fungi, 8(7): 686. https://doi.org/10.3390/jof8070686.

Chen, Y., Mao, W.J., Yan, M.X., Liu, X., Wang, S.Y. et al. (2016). Purification, chemical characterization, and bioactivity of an extracellular polysaccharide produced by the marine sponge endogenous fungus *Alternaria* sp. SP-32. Mar. Biotechnol., 18: 301–313. https://doi.org/10.1007/s10126-016-9696-6.

El-Bondkly, E.A.M., El-Bondkly, A.A.M. and El-Bondkly, A.A.M. (2021). Marine endophytic fungal metabolites. A whole new world of pharmaceutical therapy exploration. Heliyon, 7(3). https://doi.org/10.1016/j.heliyon.2021.e06362.

Elkhateeb, W. and Daba, G.M. (2022). Marine endophytes, a natural novel source for a treasure of bioactive compounds. J. Adv. Microbiol. Res., 5(018): 2. http://dx.doi.org/10.24966/AMR-694X/100018.

Fouillaud, M., Venkatachalam, M., Girard-Valenciennes, E., Caro, Y. and Dufossé, L. (2016). Anthraquinones and derivatives from marine-derived fungi: Structural diversity and selected biological activities. Mar. Drugs, 14(4): 64. https://doi.org/10.3390/md14040064.

Gladfelter, A.S., James, T.Y. and Amend, A.S. (2019). Marine fungi. Curr. Biol., 29(6): R191–R195. https://doi.org/10.1016/j.cub.2019.02.009.

Gonçalves, M.F., Esteves, A.C. and Alves, A. (2022). Marine fungi: Opportunities and challenges. Encycl., 2(1): 559–577. https://doi.org/10.3390/encyclopedia2010037.

Hafez Ghoran, S., Taktaz, F., Sousa, E., Fernandes, C. and Kijjoa, A. (2023). Peptides from marine-derived fungi: Chemistry and biological activities. Mar. Drugs, 21(10): 510. https://doi.org/10.3390/md21100510.

Harikrishnan, M., Prakash, P., Jayabaskaran, C. and Bhat, S.G. (2021). Multi-functional bioactive secondary metabolites derived from endophytic fungi of marine algal origin. Curr. Res. Microb. Sci., 2: 100037. https://doi.org/10.1016/j.crmicr.2021.100037.

Hwang. J.Y., Lee, J.H., Park, S.C., Lee, J., Oh, D.C. et al. (2019). New peptides from the marine-derived fungi *Aspergillus allahabadii* and *Aspergillus ochraceopetaliformis*. Mar. Drugs, 17(9): 488. https://doi.org/10.3390/md17090488.

Jeevithan, L., Diao, X., Hu, J., Elango, J., Wu, W. et al. (2024). Recent advancement of novel marine fungi derived secondary metabolite fibrinolytic compound FGFC in biomedical applications: A review. Front. Cell Infect Microbiol., 14: 1422648. https://doi.org/10.3389/fcimb.2024.1422648.

Jeewon, R., Aullybux, A.A., Puchooa, D., Nazurally, N., Alrefaei, A.F. and Zhang, Y. (2023). Marine microbial polysaccharides: An untapped resource for biotechnological applications. Mar. Drugs, 21(7): 420. https://doi.org/10.3390/md21070420.

Jin, L., Quan, C., Hou, X. and Fan, S. (2016). Potential pharmacological resources: Natural bioactive compounds from marine-derived fungi. Mar. Drugs, 14(4): 76. https://doi.org/10.3390/md14040076.

Jones, E.G., Pang, K.L., Abdel-Wahab, M.A., Scholz, B., Hyde, K.D. et al. (2019). An online resource for marine fungi. Fungal Divers., 96: 347–433. https://doi.org/10.1007/s13225-019-00426-5.

Kalra, R., Conlan, X.A. and Goel, M. (2020). Fungi as a potential source of pigments: Harnessing filamentous fungi. Front. Chem., 8: 369. https://doi.org/10.3389/fchem.2020.00369.

Kang, H.K., Seo, C.H. and Park, Y. (2015). Marine peptides and their anti-infective activities. Mar. Drugs, 13(1): 618–654. https://doi.org/10.3390/md13010618.

Komatsu, K., Shigemori, H. and Kobayashi, J.I. (2001). Dictyonamides A and B, new peptides from marine-derived fungus. J. Org. Chem., 66(18): 6189–6192.

Le Strat, Y., Ruiz, N., Fleurence, J., Pouchus, Y.F., Déléris, P. and Dumay, J. (2022). Marine fungal abilities to enzymatically degrade algal polysaccharides, proteins, and lipids: A review. J. Appl. Phycol., 34(3): 1131–1162. https://doi.org/10.1007/s10811-022-02713-2.

Lebeau, J., Venkatachalam, M., Fouillaud, M., Petit, T., Vinale, F., Dufossé, L. and Caro, Y. (2017). Production and new extraction method of polyketide red pigments produced by ascomycetous fungi from terrestrial and marine habitats. J. Fungi, 3(3): 34. https://doi.org/10.3390/jof3030034.

Lin, Y., Yang, J., Luo, L., Zhang, X., Deng, S. et al. (2022). Ferroptosis related immunomodulatory effect of a novel extracellular polysaccharides from marine fungus *Aureobasidium melanogenum*. Mar. Drugs, 20(5): 332. https://doi.org/10.3390/md20050332.

Lingappa, S., Ganapathi, S., Awasthi, S. and Somasundaram, S.T. (2022). Isolation and characterization of marine fungi for their bioactive pigments. Adv. Biores., 13(4): 62–75. DOI: 10.15515/abr.0976-4585.13.4.6275.

Liu, Y., Xu, M., Tang, Y., Shao, Y., Wang, H. and Zhang, H. (2022). Genome features and AntiSMASH analysis of an endophytic strain *Fusarium* sp. R1. Metabolites, 12(6): 521. https://doi.org/10.3390/metabo12060521.

Mansour, R.I., Elshoubaky, G.A. and Attia, E.A. (2025). Exploring antioxidant and cytotoxicity prospective of melanin pigment produced by fungi associated with marine macroalgae. Microb. Biosyst., J., 10(1). https://dx.doi.org/10.21608/mb.2025.352914.1229.

Meruvu, H. and Dos Santos, J.C. (2021). Colors of life: A review on fungal pigments. Crit. Rev. Biotechnol., 41(8): 1153–1177. https://doi.org/10.1080/07388551.2021.1901647.

Nogueira, O.M.N., Bernal, S.P.F., Peres, C.K., Boroski, M. and Passarini, M.R.Z. (2024). Isolation of marine-derived filamentous fungi and their potential application for bioremediation process. Braz. J. Microbiol., 55(4): 3403–3412. https://doi.org/10.1007/s42770-024-01536-2.

Passarini, M.R., Sette, L.D. and Rodrigues, M.V. (2011). Improved extraction method to evaluate the degradation of selected PAHs by marine fungi grown in fermentative medium. J. Braz. Chem. Soc., 22: 564–570. https://doi.org/10.1590/S0103-50532011000300022.

Peng, B., Liu, Y., Lin, Y., Kraithong, S., Mo, L. et al. (2023). A new exopolysaccharide of marine coral-associated *Aspergillus pseudoglaucus* SCAU265: Structural characterization and immunomodulatory activity. J. Fungi, 9(11): 1057. https://doi.org/10.3390/jof9111057.

Prathyusha, A.M.V.N., Sheela, G.M. and Bramhachari, P.V. (2018). Chemical characterization and antioxidant properties of exopolysaccharides from mangrove filamentous fungi *Fusarium equiseti* ANP2. Biotechnol. Rep., 19: e00277. https://doi.org/10.1016/j.btre.2018.e00277.

Proksch, P., Putz, A., Ortlepp, S. Kjer, J. and Bayer, M. (2010). Bioactive natural products from marine sponges and fungal endophytes. Phytochem. Rev., 9: 475–489. https://doi.org/10.1007/s11101-010-9178-9.

Qeshmi, F.I., Homaei, A., Fernandes, P., Hemmati, R., Dijkstra, B.W. and Khajeh, K. (2020). Xylanases from marine microorganisms: A brief overview on scope, sources, features, and potential applications. Biochim. Biophys. Acta, Proteins Proteomics, 1868(2): 140312. https://doi.org/10.1016/j.bbapap.2019.140312.

Ribeiro, R., Pinto, E., Fernandes, C. and Sousa, E. (2022). Marine cyclic peptides: Antimicrobial activity and synthetic strategies. Mar. Drugs, 20(6): 397. https://doi.org/10.3390/md20060397.

Sajeevan, T.P., Umar, M.D., Dhaneesha, M., Joseph, R., Ravinesh, R. and Sajeevan, T.P. (2020). Endophytic fungi isolated from the marine sponges as a source of potential bioactive compounds. J. Aquat. Biol. and Fisheries, 8: 58–66.

Shah, F., Ranawat, B. and Mishra, S. (2021). Marine fungal biocatalysts: The bionomical standpoint. In: Fungi Bio-Prospects in Sustainable Agriculture, Environment, and Nano-technology. Academic Press, pp. 45–67. https://doi.org/10.1016/B978-0-12-821925-6.00003-4.

Shenoy, P.N., Bhaskar, S., Manu, M., Likitha, M.P., Geetha, N. et al. (2023). Genome mining and AntiSMASH analysis of an Endophytic *Talaromyces* sp. reveal biosynthetic pathway gene clusters for novel bioactive compounds. J. Appl. Biol. Biotechnol., 12(3): 229–235. https://doi.org/10.7324/JABB.2024.154874.

Inflammation-related disorders are among the most prevalent health challenges worldwide (Tsoupras et al., 2018). While inflammation represents a critical defense mechanism, its chronic dysregulation contributes to pathologies such as cardiovascular diseases (CVDs), autoimmune disorders, and cancer. Lipids are central to both the initiation and resolution of inflammation, participating in processes such as apoptosis, phagocytosis, and immune cell-signaling. Metabolic reprogramming of lipid pathways has been shown to strongly influence the outcome of inflammatory immune responses (Tsoupras et al., 2024, 2018).

Several lipid bioactives derived from marine fungi demonstrate pronounced activity against thrombo-inflammatory processes. These include lipid-soluble vitamins (e.g., A, D, carotenoids), unsaturated fatty acids (UFAs, particularly ω-3 PUFAs), amphiphilic polar lipids (e.g., phospholipids and glycolipids), and diverse secondary metabolites including alkaloids, terpenes, and terpenoids (Deshmukh et al., 2024). Because of their unique structural and functional profiles, marine fungal lipids hold considerable promise for biotechnology. Current research is centered on their extraction, characterization, and usage in pharmaceuticals, nutraceuticals, and cosmeceuticals. However, the lipid diversity of many fungal species remains unexplored, underscoring the need for more comprehensive investigations into their biochemical composition and bioactivity (Astudillo et al., 2023).

Marine fungal lipids, ranging from terpenes and polyketides to sterols, constitute a wide class of hydrophobic molecules synthesized from acetyl-CoA through diverse biosynthetic pathways. For instance, the total lipid content in *Clonostachys rosea* has been estimated at 7.9–31.1% of dry weight. The following sections provide a detailed overview of their composition, including fatty acids, polar lipids, neutral lipids, and lipid-derived bioactives (Shi and Tu, 2015).

2.1 *Fatty Acid Composition*

Fatty acids, consisting of a carboxyl group attached to a hydrocarbon chain, are essential as both energy reserves and structural components of biological membranes. They regulate cellular metabolism, hormone responsiveness, and intercellular communication. Based on the degree of unsaturation, they are broadly categorized into saturated fatty acids (SFAs), monounsaturated fatty acids (MUFAs), and PUFAs (Calder, 2015).

Marine fungi are typically rich in SFAs. High levels of palmitic (C16:0, 12.0–78.12%) and stearic acid (C18:0, 1.34–35.77%) have been reported, with particularly elevated SFA contents observed in *P. puberulum* and Anamorphic fungi (89.34–90.39%). In *Verticillium tenerum*, SFAs accounted for the entire fatty acid profile. Furthermore, MUFA levels are comparatively lower (8.38–19.57%), with oleic acid (C18:1) as the dominant representative (7.87–56.73%) (Papikinou et al., 2024). PUFAs constitute a high proportion (24.19–55.30%) in certain species, although key cardioprotective fatty acids like eicosapentaenoic acid (EPA) and docosahexaenoic acid (DHA) were absent in some studies. Interestingly, other detected PUFAs include α-linolenic acid (C18:3 n-6, 0–9.42%) and linoleic acid (C18:2 n-6, 17.19–57.90%) (Papikinou et al., 2024). Overall, a comparative overview is presented in Table 1.

Table 1. Composition of fatty acids (% of total fatty acids) of different marine fungi species (Papikinou et al., 2024) ([1]ND stands for Not Defined; [2]TL stands for Total Lipids; [3]DW stands for Dry Weight).

Marine Fungi Species	Composition of Fatty Acids (% of Total FAs)								
	C:16	C:18	C: 18:1	C: 18:2	C: 18:3	SFAs	MUFAs	PUFAs	TL[2] % (DW[3])
A. caelatus	15.12–15.44	6.98–9.12	15.62–25.38	49.47–54.11	2.79–5.29	–	–	–	–
E. cladophorae	15.24–16.70	7.96–19.06	17.76–18.94	40.59–50.17	3.53–4.91	26.06–36.26	18.41–19.57	44.40–55.30	–
A. flavus	13.60–16.34	8.97–12.67	31.04–40.60	29.32–39.20	0.13–1.55	–	–	–	–
A. fumigatus	17.78–19.18	5.07–7.67	19.93–28.35	45.40–54.00	ND[1]	–	–	–	–
P. islandicum	21.70–40.13	1.34–2.84	12.34–56.73	57.90		23.04–42.97	–	–	–
Z. maritima	24.92–30.36	17.09–35.77	7.87–14.27	17.19–31.83	6.48–9.42	43.62–67.12	8.38–15.18	24.19–40.91	–
A. nomius	18.23–19.87	5.40–6.78	28.34–34.84	35.86–42.84	0–0.90	–	–	–	–
A. ochraceus	19.28–21.12	6.79–8.43	32.06–34.56	36.34–38.32	0–1.25	–	–	–	–
A. parasiticus	15.64–16.44	9.90–15.30	15.14–27.90	40.26–51.52	0.71–4.33	–	–	–	–
P. puberulum	69.18	20.16	10.65	–		89.34	10.65	–	–
C. rosea	12.0–25.0	4.0–12.0	39.0–77.0	–		–	–	–	7.9–31.1
V. tenerum	78.18	21.81	–	–		99.99	–	–	–
Anamorphic fungi (Agonomycetes)	33.54–79.12	9.94–11.27	–	–		33.54–90.39	–	–	–

2.2 Polar Lipid and Neutral Lipid Composition

Polar lipids (PLs) are amphiphilic molecules composed of hydrophilic headgroups and hydrophobic fatty acid tails. They include phospholipids and glycolipids, both of which are fundamental components of cellular membranes. Phospholipids may be classified as glycerophospholipids, with a glycerol backbone, or sphingolipids, with a sphingosine backbone. Glycolipids, on the other hand, are carbohydrate-linked polar lipids that include galactolipids and sulfolipids (Deshmukh et al., 2024; Tsoupras et al., 2022).

Marine-derived PLs have been increasingly associated with anti-inflammatory and antithrombotic properties, particularly in the context of CVDs (Deshmukh et al., 2024; Tsoupras et al., 2022). In marine fungi, their concentrations range from 4.0 to 32.0 mg/g DW, with glycolipids and phospholipids constituting 2.0–6.0% and 6.0–18.0% of TLs, respectively. Within the phospholipid fraction, phosphatidylcholine (PCs) is predominant (78.0%), followed by phosphatidylethanolamine (PE, 6.0%) and phosphatidylinositol (PI $\leq$ 1.0%). Betaine glycerophospholipids have also been reported at levels comparable to PE.

Neutral lipids serve primarily as energy storage molecules and include triacylglycerols (TAGs), diacylglycerols (DAGs), and wax esters. Among these, TAGs composed of glycerol esterified with three fatty acids represent the dominant class in marine fungi, accounting for 77.0–91.0% of total neutral lipids. Fatty acid methyl esters (FAMEs), produced via methanol transesterification, have been quantified at 0.1–3.0 mg/g DW (Deshmukh et al., 2024; Tsoupras et al., 2022). In Table 2, the representative values of marine-derived PLs and neutral lipids are presented.

Table 2. Composition of polar and neutral lipid bioactives from marine fungus strains (Papikinou et al., 2024).

Polar lipids (mg/g DW)	4.0–32.0
Phospholipids (% of TLs)	6.0–18.0
Glycolipids (% of TLs)	2.0–6.0
PC (% of PLs)	78.0
Betaine-DGTS (% of PLs)	6.0
PE (% of PLs)	6.0
PI (% of PLs)	0–1.0
TAGS (% of TLs)	77.0–91.0
FAMEs (mg/g DW)	0.1–3
Sterols μg/ g DW	10–600
Ergosterol (% of sterols)	85.0
Dehydrostellasterol (% of sterols)	–

2.3 *Other Lipid and Lipid-Derived Bioactives*

Marine fungi produce a wide variety of lipid-derived secondary metabolites, most notably terpenes and terpenoids. Terpenes are hydrocarbons composed of repeating isoprene units (C_5H_8), while terpenoids are their oxygenated derivatives. Their biosynthesis originates from acetyl-CoA via the mevalonate pathway, followed by cyclization, oxidation, and functional group modifications. These processes generate structurally diverse metabolites, many of which serve as precursors for essential compounds such as sterols, steroids, retinoids, and carotenoids (Ninkuu et al., 2021).

In addition to terpenes and terpenoids, marine fungi also produce sterols, quinones, carotenoids, and phenolic compounds. Sterols, which are characterized by a tetracyclic cyclopentanoperhydrophenanthrene nucleus, are critical for maintaining membrane integrity and fluidity. Ergosterol is the most abundant sterol in marine fungi, often accompanied by dehydrostellasterol, together comprising up to 85% of the total sterol fraction. Quinones, cyclic diones with redox activity, play central roles in electron transport and cellular respiration, while phenolic compounds, aromatic metabolites bearing hydroxyl groups, contribute to pigmentation and antioxidant defense (Ninkuu et al., 2021).

Marine fungal yeasts, particularly species of the genus *Rhodotorula*, are promising producers of bioactive carotenoids such as β-carotene, lycopene, astaxanthin, torulene, and torularhodin. For example, *Rhodotorula mucilaginosa* cultivated with an aqueous extract of the brown seaweed *Macrocystis pyrifera* produced lycopene ($\approx 38.4 \pm 9.4\%$), β-carotene ($\approx 21.8 \pm 1.5\%$), and astaxanthin ($\approx 1.8 \pm 0.3\%$) of the total pigment content under optimized culture conditions, reaching a total carotenoid yield of $1.84\,mg\,L^{-1}$ (Leyton et al., 2019). Additionally, other marine *Rhodotorula* strains have been similarly optimized for carotenoid production by adjusting important optimization parameters, including temperature, pH, and carbon/nitrogen sources (Zhao et al., 2019).

These carotenoids are bioactives with multiple health benefits. Their strong antioxidant capacity stems from their ability to quench reactive oxygen species (ROS) and singlet oxygen, thereby reducing oxidative damage, while they also exert anti-inflammatory, antimicrobial, and cardioprotective effects. Notably, carotenoids from marine fungi may possess enhanced stability or unique structural adaptations to extreme marine environments (e.g., varying salinity, pressure, and light exposure), which could influence their stability and bioactivity. From a biotechnological standpoint, fungal carotenoid production offers advantages such as rapid growth rates, the ability to use low-cost or waste substrates, and tighter control of culture conditions.(Mapelli-Brahm et al., 2023).

Carotenoids such as β-carotene, astaxanthin, and torularhodin demonstrate vast cardioprotective potential through multiple mechanisms. Their antioxidant activity reduces lipid peroxidation, a critical driver of atherosclerosis, while also improving endothelial function by maintaining nitric oxide (NO) bioavailability. Astaxanthin has been shown to lower blood pressure, decrease oxidative stress markers, and modulate inflammatory pathways associated with cardiovascular disease. β-Carotene serves as a vitamin A precursor, supporting immune regulation and vascular health, whereas torularhodin exhibits potent free radical scavenging and anti-inflammatory effects. Collectively, these findings point to marine fungal carotenoids as promising nutraceutical candidates for CVD prevention and management (Mapelli-Brahm et al., 2023).

Phenolic compounds from marine fungi are also emerging as important bioactives. They encompass diverse structural classes, including simple phenols, flavonoids, glycosides, and anthraquinones, and are well-recognized for their antioxidant and anti-inflammatory action. Extracts from marine fungi such as *Aspergillus versicolor*, *Engyodontium album*, *Gymnascella dankaliensis*, and *Chaetomium globosum* have demonstrated high total phenolic content (TPC) and strong radical scavenging activity in assays such as 2,2-diphenyl-1-picrylhydrazyl (DPPH) and thiobarbituric acid reactive substances (TBARS), in many cases outperforming standard antioxidants (Vitale et al., 2020). In particular, several phenolic metabolites from *A. versicolor*, including newly identified molecules, displayed scavenging potency comparable to or greater than ascorbic acid, with activity closely related to the number and position of hydroxyl groups on their aromatic rings (Vitale et al., 2020).

Regarding cardiovascular health, phenolic compounds may act through multiple mechanisms. Their antioxidant activity mitigates oxidative stress, which contributes to endothelial dysfunction and low-density lipoprotein (LDL) oxidation, early events in atherogenesis. Additionally, phenolics inhibit inflammatory signaling pathways (e.g., nuclear factor κB (NF-κB), mitogen-activated protein kinases (MAPKs)) and reduce pro-inflammatory cytokines such as tumor necrosis factor α (TNF-α) and interleukin 6 (IL-6), thereby attenuating vascular inflammation. Although most studies to date are limited to *in vitro* models or in animal experiments, the accumulating evidence indicates that marine fungal phenolics could serve as nutraceuticals or lead compounds for the prevention and management of CVDs (Nagahawatta et al., 2024).

Overall, these lipid-derived metabolites frequently demonstrate health-promoting properties. Phospholipids can inhibit platelet-activating factor biosynthesis, exerting both anti-inflammatory and antithrombotic effects, while phenolics and carotenoids provide potent antioxidant, anti-inflammatory, and even anti-aging benefits through their actions on diverse molecular signaling cascades (Vitale et al., 2020). Representative terpenoids, polyketides, and other cardioprotective metabolites are presented in Figs. 1 and 2, and also summarized in Table 3.

3. Marine Fungal Polysaccharides: Bioactivity, Structure, and Therapeutic Potential

Polysaccharides from marine fungi are emerging as significant bioactive agents. These high-molecular-weight carbohydrates include β-glucans, heteropolysaccharides, and exopolysaccharides (EPS), typically composed of monosaccharides such as glucose, mannose, and galactose. The biological activity of polysaccharides is

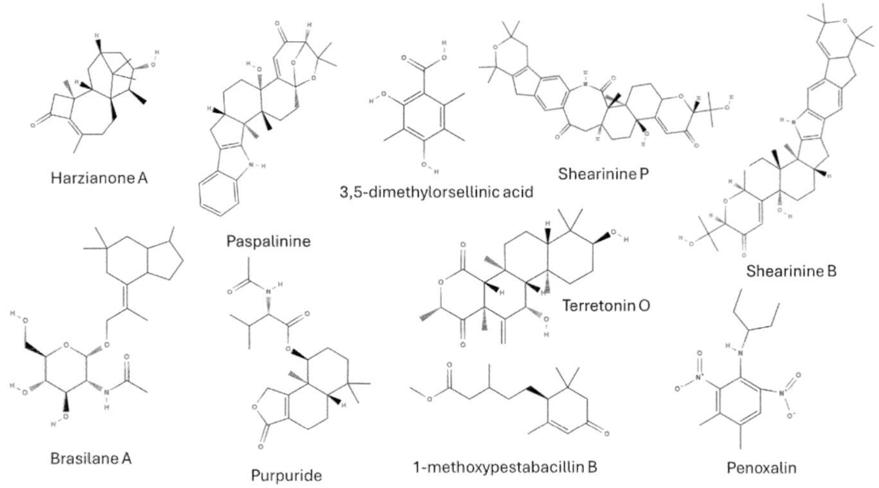

Fig. 1. The structures of the most prevalent terpenoids and polyketides in marine fungi species (Papikinou et al., 2024).

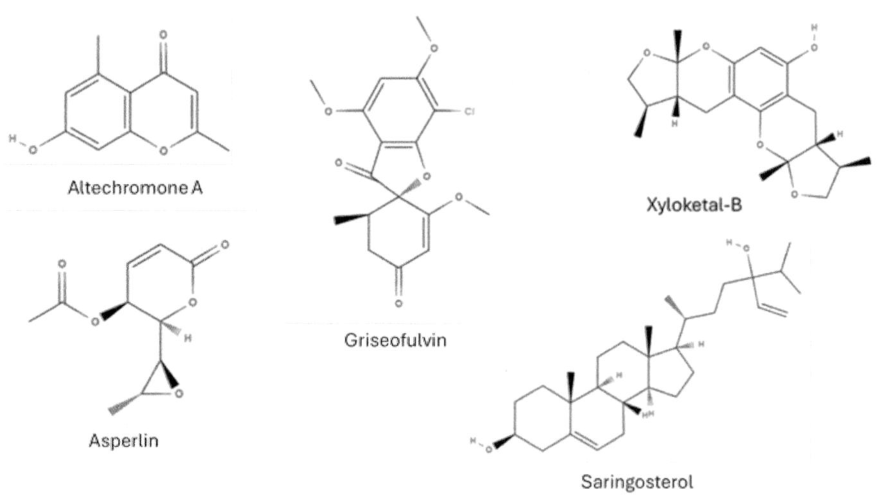

Fig. 2. The structures of the most prevalent bioactives with cardioprotective properties in marine fungi species (Papikinou et al., 2024).

strongly influenced by structural parameters such as their molecular weight (MW), degree of branching, sulphation, and other chemical modifications (Carrasqueira et al., 2025).

3.1 Antioxidant and Anti-Inflammatory Activities of Marine Fungi Polysaccharides

Marine fungal polysaccharides exhibit strong antioxidant activity, primarily through radical scavenging mechanisms (*in vitro*). More specifically, they neutralize

Table 3. Novel and common terpenoids and polyketides extracted from various strains of marine fungal species (Papikinou et al., 2024).

Bioactive Compounds	**Species**	**Compounds**	**Structures***
Monoterpenes	*Penicillium* sp. SCSIO 41691	1-methyl-12a,12b-epoxyarisugacin M	–
		1-methyl-4a,12b-epoxyarisugacin M	–
		2,3-dihydroxy-3,4a-epoxy-12a-dehydroxyisoterreulactone A	–
		2-hydroxy-12a-dehydroxyisoterreulactone A	–
		3′-demethoxyterritrems B′	–
		4a-hydroxyarisugacin P	–
		1-epi-arisugacin H	–
	Diaporthe sp. SYSU-MS4722	diaporterpene A-C	–
Sesquiterpenes	*Penicillium* sp. OPR23-FS02	purpuride	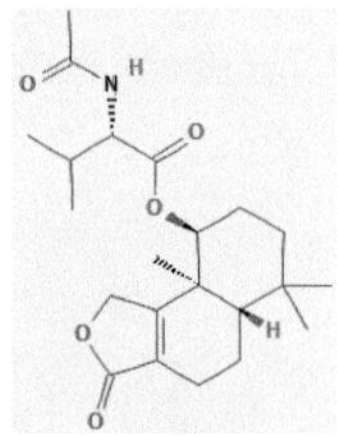
	Trichoderma lixii R22	brasilane A	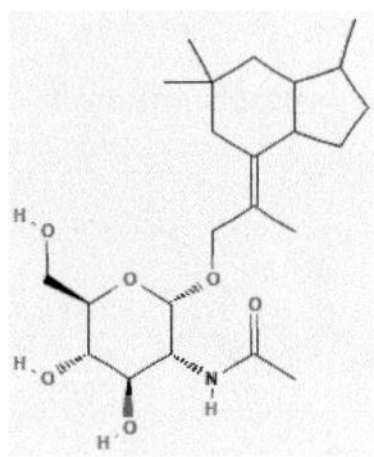
	Trichoderma hamatum Z36-7	5-hydroxyepicyclonerodiol oxide	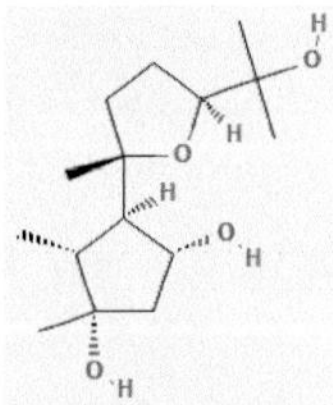
		4-hydroxyepicyclonerodiol oxide	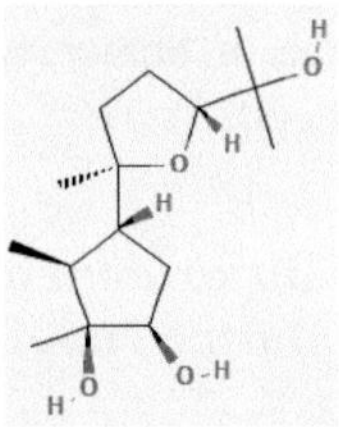

Table 3 contd. ...

...Table 3 contd.

Bioactive Compounds	Species	Compounds	Structures*
		trichodermol chlorohydrin	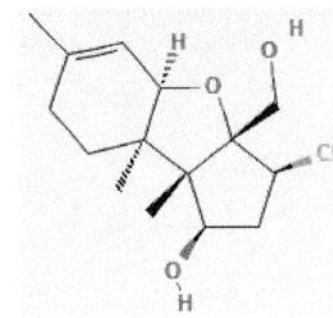
	Diaporthe sp. SCSIO 41011	1-methoxypestabacillin B	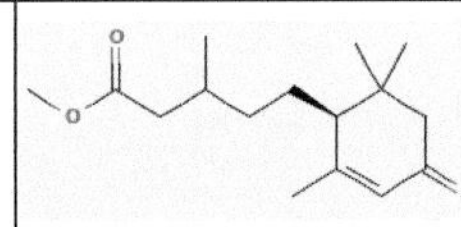
		11-Nor-8,9*R*-drimanediol	–
Diterpenes	*Penicillium sp.* OPR23-FS02	penioxalicin	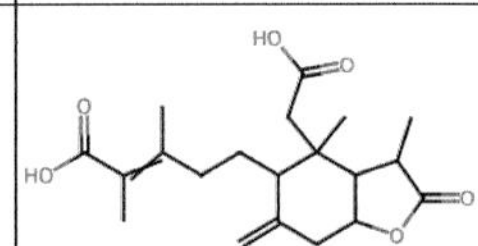
	Trichoderma lixii R22	harzianone A	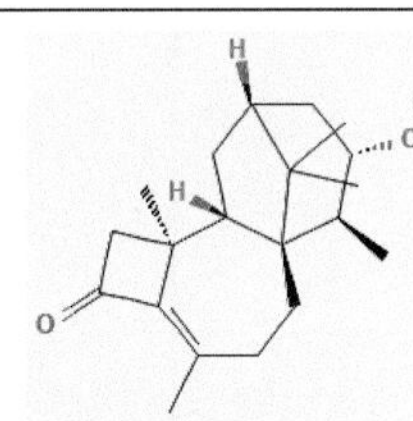
	Penicillium sp. KFD28	7-hydroxypaxilline-13-ene	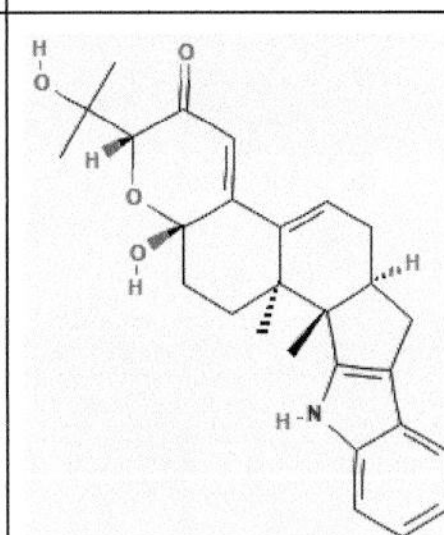
		paspalinine	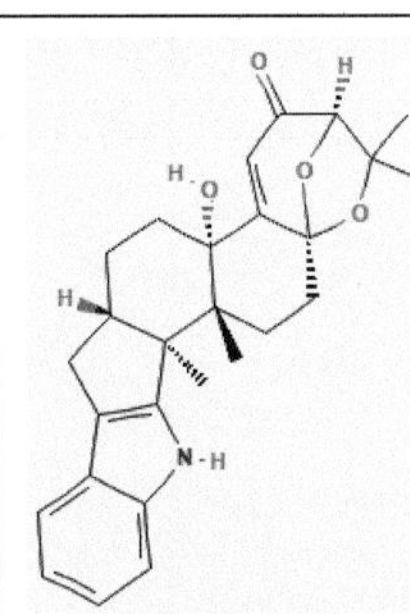

Table 3 contd. ...

...*Table 3 contd.*

Bioactive Compounds	**Species**	**Compounds**	**Structures***
		paspaline B	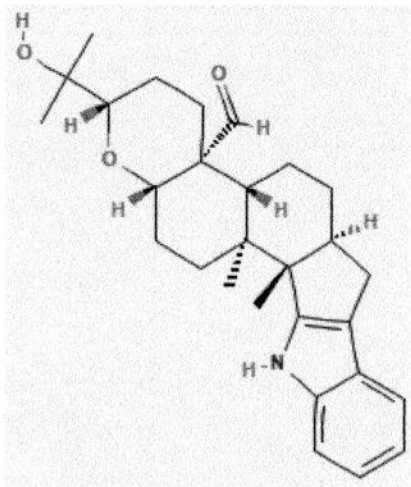
		pyrapaxilline	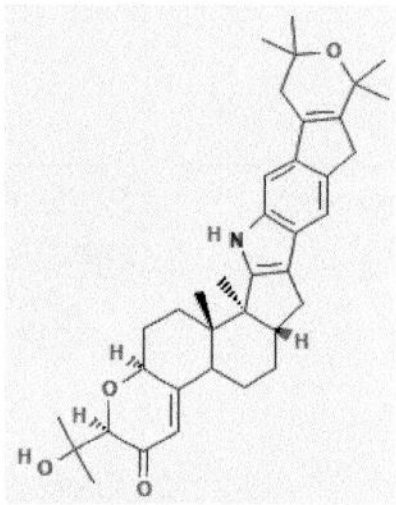
		shearinine B	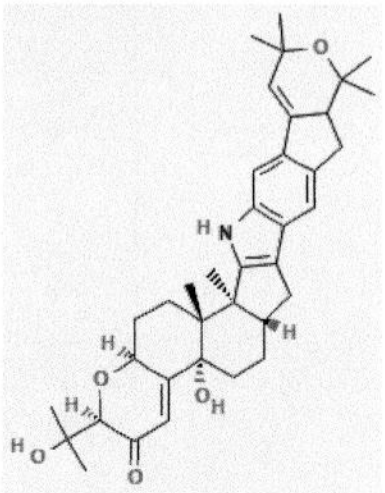
		shearinine P	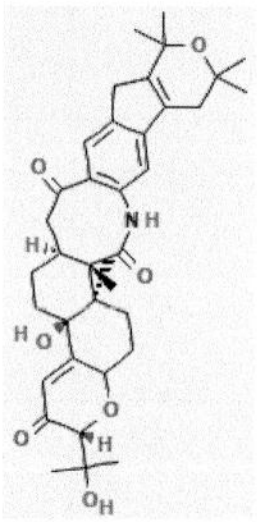
		3-deoxo-4b-deoxypaxilline	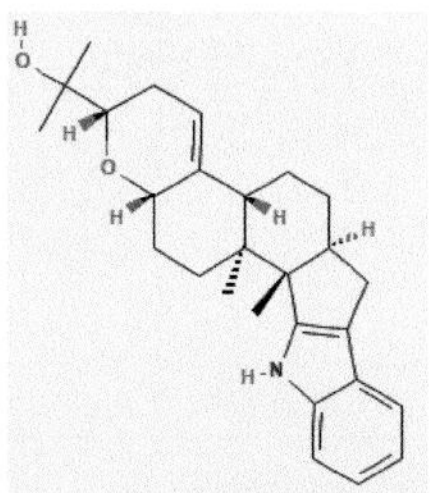

Table 3 contd. ...

...Table 3 contd.

Bioactive Compounds	**Species**	**Compounds**	**Structures***
	Peroneutypa sp. M16	solanapyrones	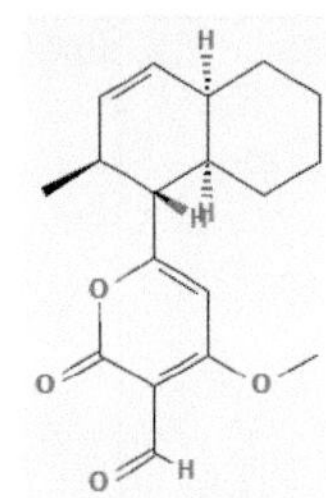 i.e.,
Meroterpenes	*Aspergillus* sp. CSYZ-1	3,5-dimethylorsellinic acid	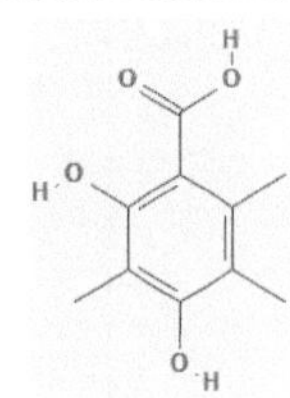
	Penicillium sp. A18	penimeroterpenoids A-C	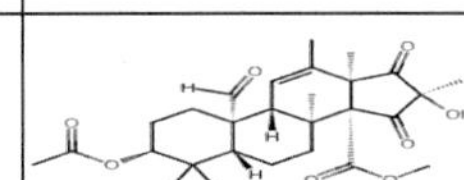 penimeroterpenoid A 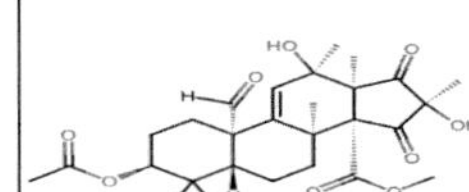 penimeroterpenoid B 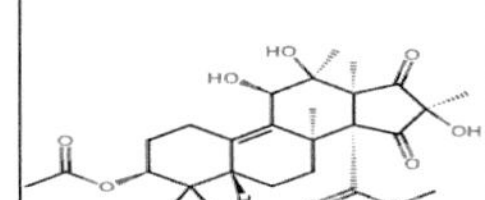 penimeroterpenoid C
	Diaporthe sp. SCSIO 41011	chrodrimanins A/B/E/F/G/H	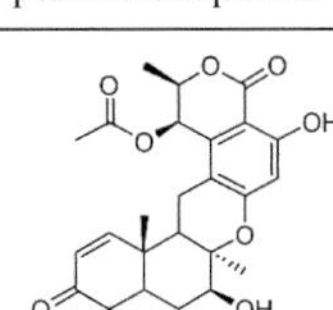 i.e.,

Table 3 contd. ...

...Table 3 contd.

Bioactive Compounds	**Species**	**Compounds**	**Structures***
	Stemphylium sp. FJJ006	tricycloalterfurenes E-G	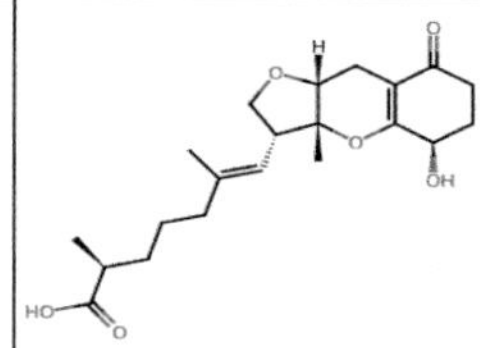 tricycloalterfurene E 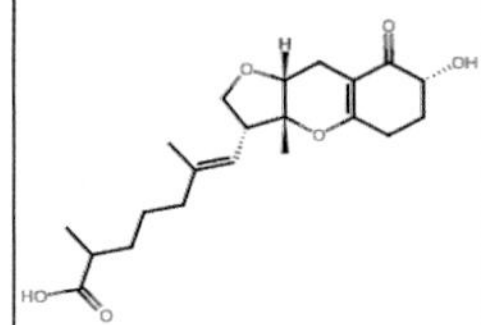 tricycloalterfurene F 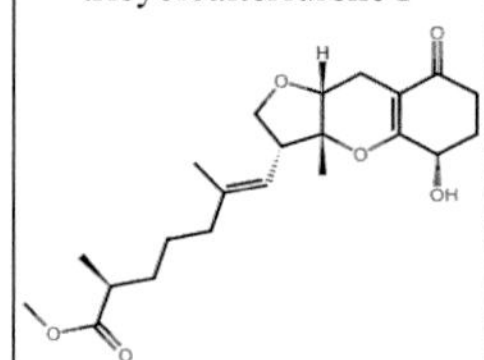 tricycloalterfurene G
	Aspergillus terreus GZU-31-1	aspermeroterpene A-C	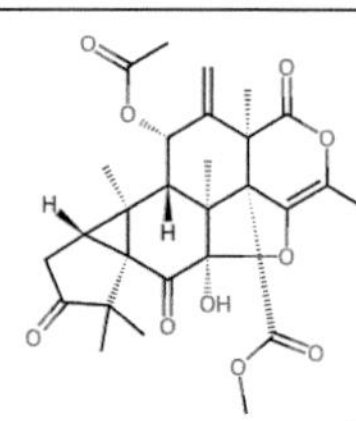 aspermeroterpene A 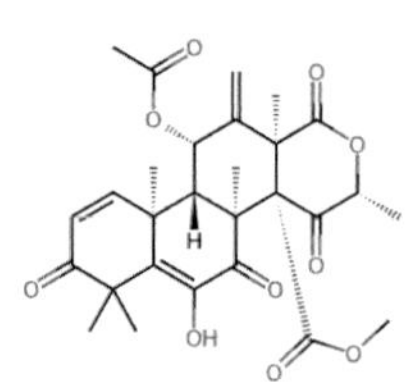 aspermeroterpene B 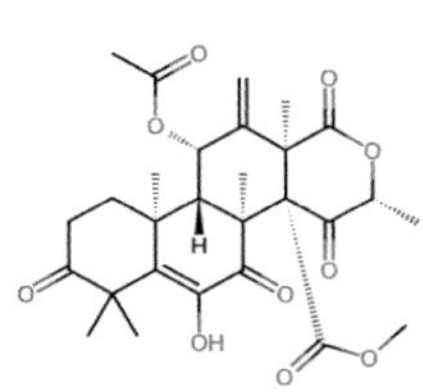 aspermeroterpene B

Table 3 contd. ...

...*Table 3 contd.*

Bioactive Compounds	**Species**	**Compounds**	**Structures***
		furanasperterpenes A and B	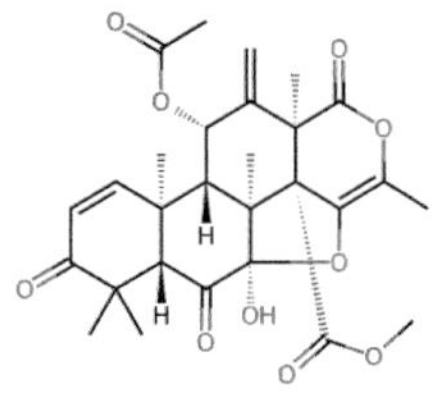 furanasperterpene A 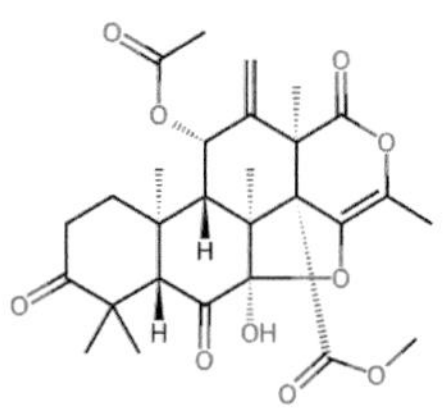 furanasperterpene B
		11-acetoxy-terretonin E	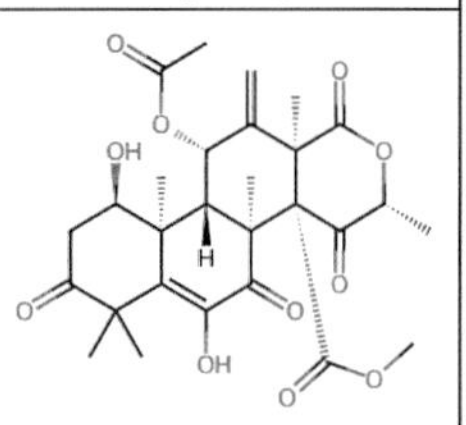
	Aspergillus terreus LGO13	terretonin O	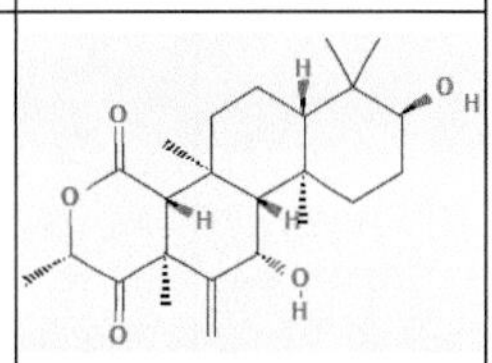
	Penicillium bilaiae MA-267 & *Penicillium chermesinum* EN-480	chermebilaenes A and B	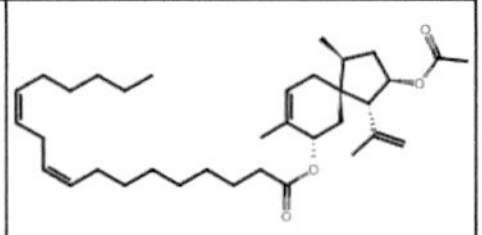 chermebilaene A 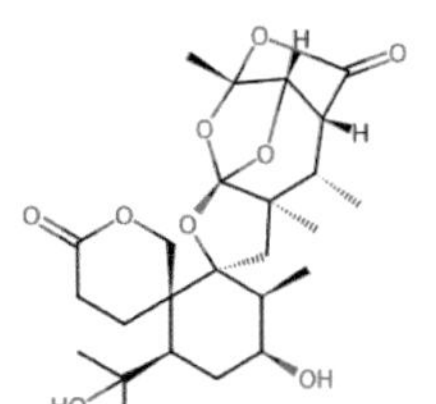 chermebilaene B

Table 3 contd. ...

...Table 3 contd.

Bioactive Compounds	**Species**	**Compounds**	**Structures***
	Aspergillus sp. ZYH026	asperaustins A-C	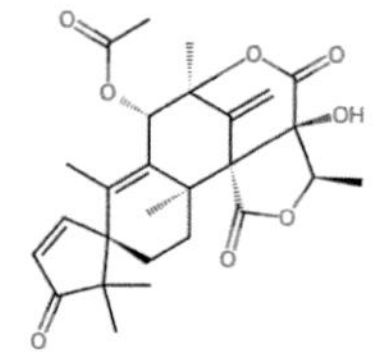 asperaustin A 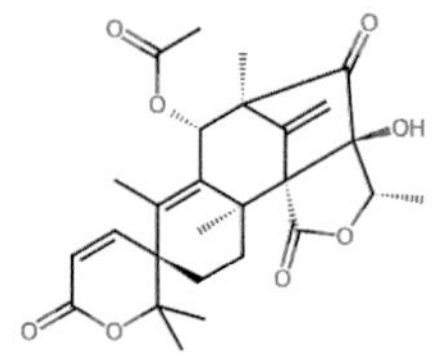 asperaustin B 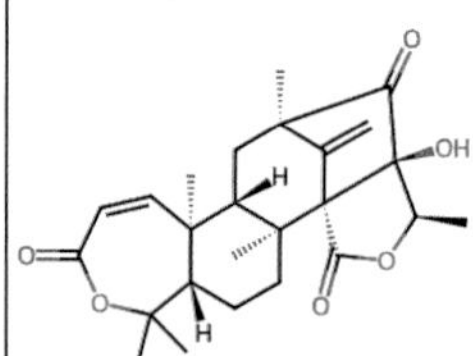 asperaustin C
Polyketides	*Penicillium* sp. OPR23-FS02	*trans*-3,4-dihydro-3,4,8-trihydroxynaphthalen-1(2*H*)-one	–
		chloromonilinic acid B	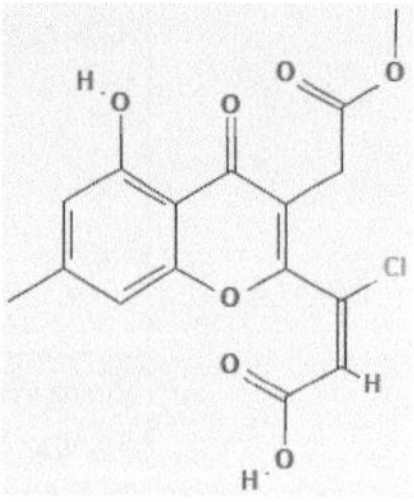
		chrysophanol	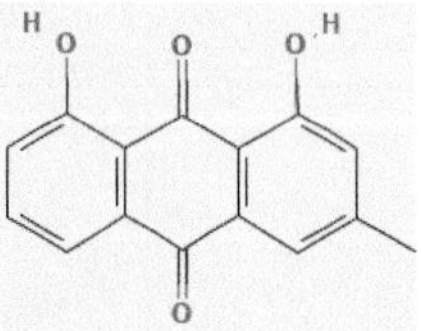
		emodin	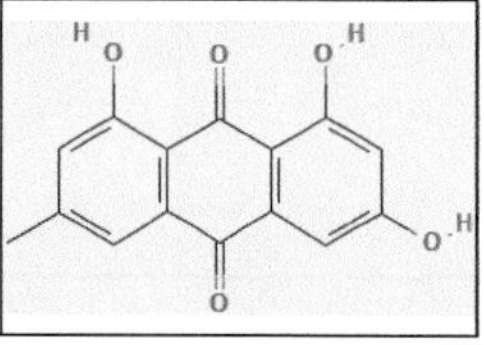

Table 3 contd. ...

...Table 3 contd.

Bioactive Compounds	**Species**	**Compounds**	**Structures***
		ω-hydroxyemodin (citreorosein)	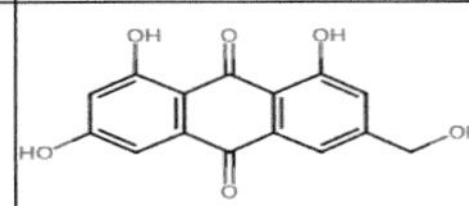
		questin	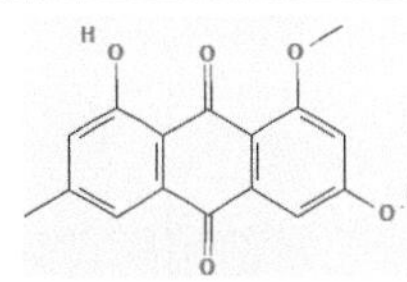
		endocrocin	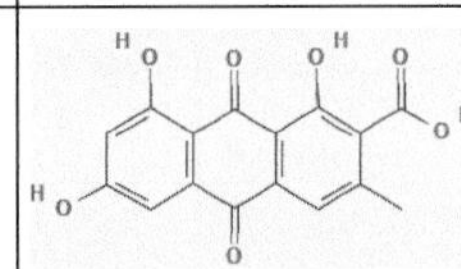
		ergochrome F	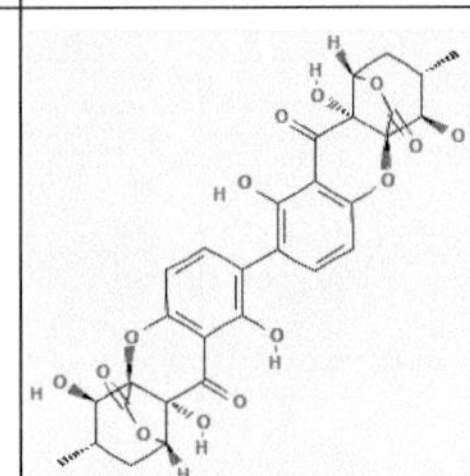
	Diaporthe sp. SYSU-MS4722	acropyrone	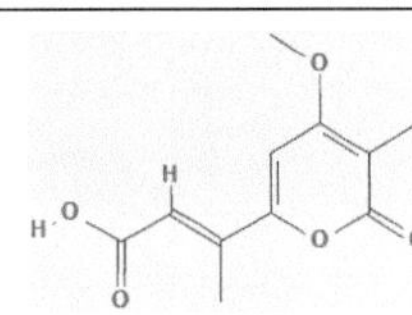
		nectriapyrone	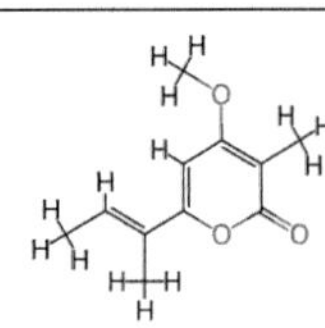
		monodictyphenone	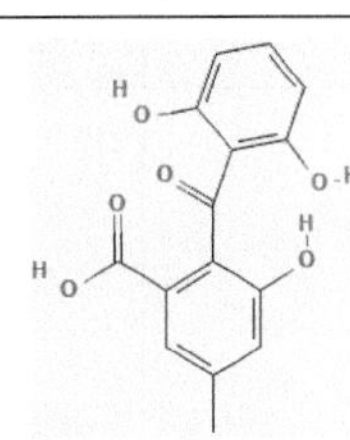
		2,2',6'-trihydroxy-4-methyl-6-methoxy-acyl-diphenylmethanone	–

Table 3 contd. ...

...Table 3 contd.

Bioactive Compounds	Species	Compounds	Structures*
		pestacin	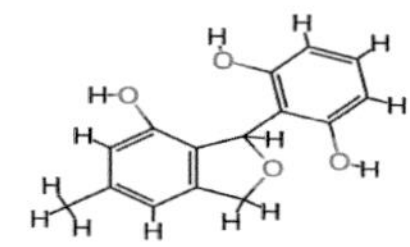
		3-(2,6-dihydroxyphenyl)-4-hydroxy-6-methyl-isobenzofuran-1(3*H*)-one	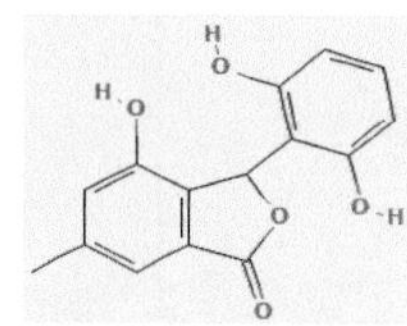
		3,4-dihydro-6,8-dihydroxy-3-methylisocoumarin	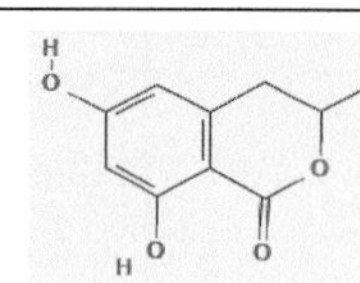
		2,5-dimethyl-7-hydroxychromone	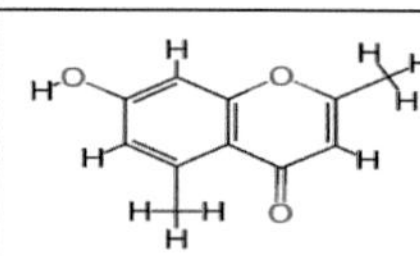
		4-hydroxyphenethyl-2-(4-hydroxyphenyl) acetate	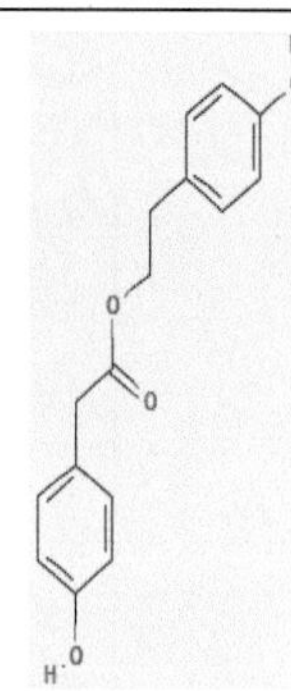
Anthraquinones	*Aspergillus unguis* 158SC-067 & *A. Flocculosus* 01NT-1.1.5	averantin	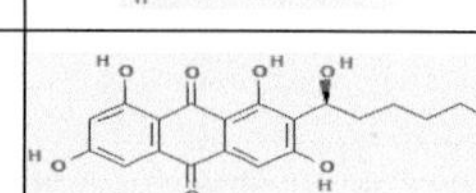
		7-chloroaverantin	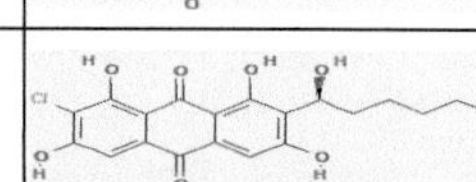
		1′-*O*-methylaverantin	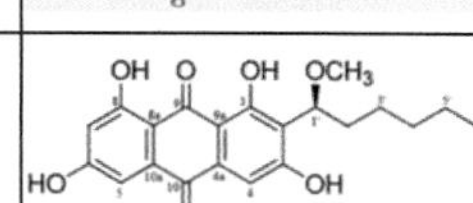

Table 3 contd. ...

...Table 3 contd.

Bioactive Compounds	**Species**	**Compounds**	**Structures***
Bianthraquinones	*Stemphylium sp.* FJJ006	alterporriol Z1– Z3	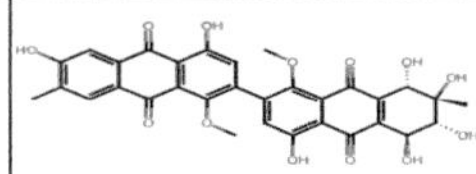 alterporriol Z1 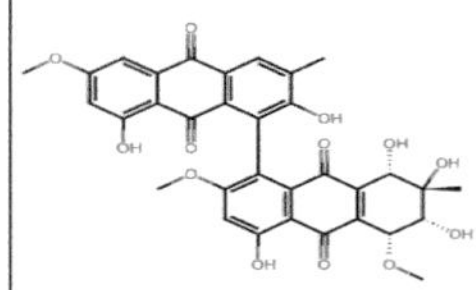 alterporriol Z3
Phenols	*Aspergillus unguis* 158SC-067/ *A. Flocculosus* 01NT-1.1.5	grifolin B	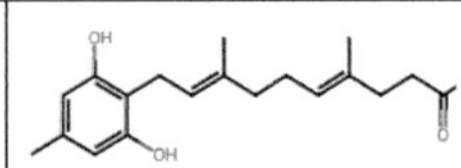
		12-hydroxyhomovalencic acid	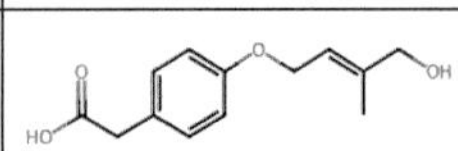

DPPH, hydroxyl, superoxide, and 2,2'-azino-bis (3-ethylbenzothiazoline-6-sulfonic acid) ABTS. For instance, an extracellular polysaccharide (YSS) from *Aspergillus terreus*, mainly composed of mannose and galactose ((MW) ~18.6 kDa), exhibited potent DPPH scavenging with a half-maximal effective concentration (EC_{50}) of ~2.8 mg/mL. Similarly, *Fusarium oxysporum* produced a galactofuranose-rich EPS (Fw-1; MW ~61.2 kDa) that demonstrated EC_{50} values of ~1.1 mg/mL for hydroxyl radical scavenging and ~2.0 mg/mL for superoxide radicals. These results compare favorably with other fungal polysaccharides such as AWP from *Aspergillus versicolor* (EC_{50} ~4.0 mg/mL) (Wang et al., 2013).

In addition to their antioxidant effects, marine fungal polysaccharides also modulate inflammatory responses. Although many studies on marine polysaccharides originate from algae, fungi-derived β-glucans and related polymers have demonstrated the ability to regulate cytokine release and modulate signaling pathways such as NF-κB and MAPKs, thereby suppressing pro-inflammatory enzymes. Structural features like branching and sulphation play a decisive role in determining their potency (Elbandy, 2023).

3.2 *Antitumor, Immunomodulatory, and Enzyme-Inhibitory Effects*

Marine fungal polysaccharides also exhibit antitumor and immunomodulatory activities *in vivo* and *in vitro*. For example, water-soluble polysaccharides from submerged mycelium of *Flammulina velutipes* inhibited murine leukemia P388 *in vivo*, with purified fractions showing greater tumor suppression compared to crude extracts (Krasnopolskaya et al., 2016). Polysaccharides from *Ganoderma lucidum*, including xylomannan and fucogalactan, revealed strong antitumor activity *in vivo*, where daily oral administration (2 mg/kg) resulted in notable tumor growth inhibition (Krasnopolskaya et al., 2015). Beyond anticancer effects, marine fungal polysaccharides also demonstrate enzyme-inhibitory activities relevant to metabolic disorders. Certain EPS fractions inhibit diabetes-related enzymes such as α-glucosidase and PTP1B. However, in fungi, the most potent enzyme inhibitors often derive from non-polysaccharide secondary metabolites, while polysaccharide-based inhibition is more extensively reported in algae and other marine organisms. However, the antioxidant and immunomodulatory properties of fungal polysaccharides support their potential in addressing metabolic syndrome and related conditions (Abdala-Díaz et al., 2024). Table 4 presents some novel and common polysaccharides found in marine fungi.

Table 4. Common and novel polysaccharides derived from various marine fungi strains (*Glc stands for "Galactose", Man for "Mannose", Glc for "Glucose", Xyl for "Xylose", GlcA for "Glucuronic acid", Rha for "Rhamnose", and Ara for "Arabinose").

Polysaccharides	Marine Fungi Species	MW of Compounds (Da)	Monosaccharide Composition*	References
YSS	*Aspergillus terreus*	1.86×10^4	Glc/Man (8.6:1.0)	Wang et al. (2013)
Fw-1	*Fusarium oxysporum*	6.12×10^4	Gal/Glc/Man (1.33:1.33:1.00)	Chen et al. (2015)
AWP	*Aspergillus versicolor*	5×10^7	Glc/Man (8.6:1.0)	Chen et al. (2013)
EPS	*Flintiella sanguinaria*	1.5×10^6	Xyl/Gal/GlcA/Rha/Glc/Ara (47:21:14:10:6:2)	Gaignard et al. (2018)
AS2-1	*Alternaria* sp.	2.74×10^4	Man/Glc/Gal (1.00:0.67:0.35)	Gaignard et al. (2018)

4. Bioactive Peptides from Marine Fungi: Emerging Therapeutic Potential

Marine fungi represent an underexplored yet highly promising source of bioactive peptides with wide-ranging health applications. These peptides are generally small, structurally diverse molecules, often produced as secondary metabolites that enable fungi to adapt and survive under extreme marine conditions. Their reported bioactivities include antimicrobial, anticancer, antiviral, antioxidant, and immunomodulatory effects, making them attractive candidates for pharmaceutical and nutraceutical development (Hafez Ghoran et al., 2023).

4.1 Antimicrobial and Antiviral Activities

Several marine-derived fungal peptides exhibit broad-spectrum antimicrobial activity against both Gram-positive and Gram-negative bacteria, including multidrug-resistant strains, and are therefore classified as antimicrobial peptides (AMPs). Importantly, many of these peptides also inhibit pathogenic fungi, viruses, and parasites, with a lower likelihood of inducing resistance. Most AMPs are ribosomally synthesized and undergo post-translational modifications, while others are non-ribosomal peptides (NRPs) produced by multi-enzyme complexes such as non-ribosomal peptide synthases (NRPSs). Their mechanisms of action vary: some disrupt microbial membranes through amphipathic or cationic structures, whereas others interfere with essential processes such as protein synthesis, enzyme function, or biofilm formation. In the antiviral context, peptides from marine fungi have shown inhibitory activity against viral enzymes and the suppression of replication in RNA viruses, highlighting their potential as lead compounds for novel antiviral therapies (Hafez Ghoran et al., 2023).

4.2 *Anticancer Properties*

Marine fungal peptides also display potent anticancer activity. Notable examples include depsipeptides and diketopiperazines (DKPs) produced by marine *Aspergillus* and *Fusarium* strains. These peptides can induce apoptosis, inhibit angiogenesis, and arrest the cell cycle in various tumor models (El-Kashef et al., 2023). Mechanistically, some peptides act by modulating key signaling pathways such as the phosphatidylinositol 3-kinase (PI3K)/Akt and MAPK cascades, while others promote the generation of ROS, leading to programmed cell death. Significantly, several peptides demonstrate preferential toxicity toward tumor cells over normal cells, underscoring their therapeutic selectivity and reducing potential side effects (Shahidi and Saeid, 2025).

4.3 *Antioxidant and Immunomodulatory Functions*

Beyond antimicrobial and anticancer effects, marine fungal peptides demonstrate antioxidant activity by scavenging ROS, inhibiting lipid peroxidation, and regulating oxidative stress pathways. These properties suggest protective roles against chronic diseases linked to oxidative damage, such as CVDs and neurodegenerative disorders. In addition, immunomodulatory peptides derived from marine fungi can either stimulate or suppress immune responses. For example, certain peptides enhance macrophage activation and cytokine production, whereas others dampen pro-inflammatory signaling, offering potential applications in managing chronic inflammation, autoimmune disorders, and immune-related pathologies (Youssef et al., 2019).

5. Marine Fungi Bioactives with Cardioprotective Potential

CVDs remain one of the leading global health burdens. Current therapeutic approaches, although effective, are often associated with adverse effects such as hypotension, bradycardia, arrhythmias, and electrolyte imbalances. In recent years, natural bioactives from plants, microbes, and marine organisms have attracted vast attention as alternative or complementary strategies. Among these, marine fungi represent a valuable source of novel metabolites with diverse pharmacological activities. Compounds such as ω-3 acid ethyl esters, xyloketal B (Xyl-B), asperlin, and saringosterol have shown promising cardioprotective effects. Notably, Sajaroketide A, altechromone A, and griseofulvin, polyketides isolated from fungal strains KMM4718 and KMM4747 (associated with the sea urchin *Scaphechinus mirabilis*), exhibited significant protection in an *in vitro* infectious myocarditis model using H9c2 cardiomyocytes co-cultured with *Staphylococcus aureus* (Leshchenko et al., 2023).

5.1 Marine Fungal Extracts and Hypertension

Hypertension is a major risk factor for stroke, ischemic heart disease, dementia, chronic kidney disease, and other CVDs (Xiang et al., 2023). Global data from 2019 indicate that 32% of women and 34% of men aged 30–79 suffer from hypertension (Roth et al., 2020). Marine bioactives, including chitooligosaccharide derivatives (COS) and phlorotannins, are recognized as natural angiotensin-converting enzyme (ACE) inhibitors and have been developed as nutraceuticals to mitigate the side effects of synthetic drugs (Badawy et al., 2022; Roth et al., 2020; Woodward, 2019).

Xyl-B, a marine fungal metabolite, demonstrated antihypertensive activity in a two-kidney, two-clip (2K2C) rat model. Daily administration of 20 mg/kg for 12 weeks resulted in a significant reduction in both systolic and diastolic pressures. *In vitro*, pretreatment with Xyl-B (20 μM) inhibited phenylephrine-induced vascular contractions in both endothelium-intact and endothelium-denuded aortic rings, indicating a vasorelaxant effect mediated by both endothelium-dependent and -independent mechanisms (Zhao et al., 2018).

5.2 Marine Fungal Extracts and Atherosclerosis

Atherosclerosis, a chronic and progressive vascular disorder, is primarily driven by hyperlipidemia and LDL-cholesterol accumulation, leading to endothelial injury and inflammation (Festa et al., 2020; Soehnlein and Libby, 2021). Marine-derived metabolites offer safer and potentially more effective alternatives than conventional therapies (Patil et al., 2018). Platelet-activating factor (PAF) plays a pivotal role by triggering platelet aggregation and inflammatory cascades through the PAF/PAF-R axis. Marine fungal compounds such as exserolides enhanced [^{3}H]-cholesterol efflux, reduced oxidized-LDL uptake in macrophages, lowered total cholesterol and triglycerides in HepG2 cells, and suppressed CD36 expression. These effects were mediated through liver X receptor α (LXRα)-adenosine triphosphate-binding cassette A1 (ABCA1) and peroxisome proliferator-activated receptor α (PPARα) signaling pathways, highlighting their potential in mitigating atherogenesis and their remarkable immunomodulatory potential (Zhang et al., 2024).

5.3 Marine Fungal Extracts and Ischemic Heart Disease

Ischemic heart disease (IHD), caused by impaired coronary blood flow and endothelial dysfunction, remains the leading cause of morbidity and mortality worldwide (Khan et al., 2020). In 2019, it accounted for 197 million cases and 9.14 million deaths (Wu et al., 2023). Interestingly, marine fungal alkaloids represent promising agents for IHD management. *Penicillium expansum* Y32 yielded three novel and nine known alkaloids, including fumiquinazoline Q (4), cottoquinazoline A (5), and prelapatin B (6), that promoted angiogenesis, suggesting therapeutic potential for ischemic conditions (Fan et al., 2015). Similarly, deep-sea fungi such as *Chaetomium globosum* YP-106 and *Neopestalotiopsis* sp. HN-1-6 produce azaphilone and chloro-azaphilone derivatives with dose-dependent pro-angiogenic effects in zebrafish models (Fan et al., 2023; Feng et al., 2024).

5.4 *Antithrombotic and Anticoagulant Activity of Marine Fungal Extracts*

Thrombosis significantly contributes to CVD-related mortality, including myocardial infarction, IHD, and peripheral vascular disease. Marine fungal metabolites exhibit notable antithrombotic potential. For instance, isaridin E suppressed adenosine diphosphate (ADP)-induced platelet activation and ATP release without affecting collagen- or thrombin-induced aggregation, indicating selective inhibition of ADP-mediated pathways. Importantly, it revealed strong antiplatelet and antithrombotic effects without altering coagulation parameters or platelet counts (Pan et al., 2022).

Other promising compounds include FGFC1, a pyranoindolone alkaloid originating from *Stachybotrys longispora* FG216, which showed fibrinolytic, anti-inflammatory, and antioxidant action. FGFC1 enhanced clot vulnerability to fibrinolysis by modulating fibrinopeptide B release and plasminogen activation (Gao et al., 2023; Lin et al., 2022). Similarly, versiase, a bifunctional fibrinolytic enzyme from *Aspergillus versicolor* ZLH-1, efficiently dissolved clots with minimal hemolysis and no cytotoxicity, making it a safe candidate for thrombolytic therapy (Zhao et al., 2022). Additionally, asperlin from *A. versicolor* LZD4403 promoted cholesterol efflux and inhibited foam cell formation, further supporting anti-atherosclerotic potential (Zhou et al., 2017).

5.5 *Antioxidant and Anticoagulant Activity of Marine Fungal Extracts*

Oxidative stress, driven by an imbalance between free radicals and antioxidants, is central to CVD progression through LDL oxidation and endothelial dysfunction. Marine fungi are a rich source of antioxidant metabolites. For example, *Aspergillus fumigatus* WA7S6, isolated from a sponge, produced compounds such as methyl ester, dehydromevalonic lactone, 9-tetradecynoic acid, and 11-hexadecynoic acid, which demonstrated marked antioxidant, antimicrobial, and anticancer properties. Its crude extract showed potent radical scavenging activity (half-maximal inhibitory concentration (IC_{50}) of 21.35 μg/mL) and cytotoxicity against HeLa and MCF cell lines (Alharbi et al., 2024; Hassan et al., 2024). Furthermore, Xyl-B also reduced oxidative stress by scavenging ROS and upregulating endogenous antioxidant enzymes, thereby protecting vascular endothelium and improving vascular function (Su et al., 2014). Similarly, an *Aspergillus terreus* metabolite preserved cardiomyocyte integrity under oxidative stress, while gliorosein, a *Lopadostoma pouzarii* polyketide, exerted radical scavenging and cardioprotective effects under hypoxic conditions (Fan et al., 2023; Trinh et al., 2022).

5.6 Anti-Inflammatory and Anticoagulant Activity of Marine Fungal Extracts

Inflammation is a major contributor to endothelial dysfunction and CVD pathogenesis. Marine fungi, particularly *Aspergillus* and *Penicillium* species, produce a wide array of anti-inflammatory metabolites, including alkaloids, terpenoids, polyketides, and peptides (Gao et al., 2022). Xyl-B was shown to downregulate pro-inflammatory cytokines and adhesion molecules, thus protecting endothelial integrity and preventing plaque formation (Su et al., 2014). Likewise, *A. terreus* EGF7-0-1 produced aspergteroid G, which inhibited nucleotide-binding leucine-rich pyrin domain-containing protein 3 (NLRP3) inflammasome activation via the glycogen synthase kinase-3 β (GSK-3β)/NLRP3 pathway, protecting cardiomyocytes from inflammation-induced damage (Fan et al., 2023). The most representative structures of these compounds are summarized in Table 5, while relevant preclinical and clinical studies are presented in Table 6. An overview of the cardioprotective and broader health-promoting properties of marine fungal metabolites is illustrated in Fig. 3.

6. Other Health-Promoting Properties of Marine Fungal Bioactives

6.1 Antidiabetic Effects of Marine Fungal Extracts

Marine fungi are emerging as valuable sources of antidiabetic agents. EPS PJ1-1 significantly improved glucose homeostasis in a Type 2 diabetes mellitus (T2DM) mouse model, reducing fasting blood glucose by up to 42% in a dose-dependent manner, while also enhancing glucose tolerance and insulin sensitivity (Shao et al., 2023).

Scedosporium apiospermum F41-1 produced fumiquinazoline alkaloids with insulin-sensitizing activity, notably scequinadoline D (EC_{50} 0.27 μM), which activated PPARγ and promoted adipocyte differentiation, highlighting its therapeutic potential in T2DM management (Li et al., 2020). Likewise, *Aspergillus flavipes* HN4-13 yielded flavipesolides A-C and other metabolites with strong α-glucosidase inhibitory activity, surpassing standard drugs such as acarbose (IC_{50} 101 μM), while displaying low cytotoxicity (Wang et al., 2016). *Paraconiothyrium brasiliense* HDN15-135 produced brasilterpenes A-E, of which A and C reduced hyperglycemia in zebrafish by improving insulin sensitivity and suppressing gluconeogenesis, with brasilterpene C outperforming rosiglitazone (Wang et al., 2022). Other deep-sea fungal metabolites, including cladosporol C, tenellone F, and circumdatin G, inhibited key metabolic enzymes (α-amylase,

Table 5. Structures of compounds detected in the referenced marine fungi presenting a cardioprotective and health-promoting role (Papikinou et al., 2024).

Compounds	Structures*
(4) fumiquinazoline Q	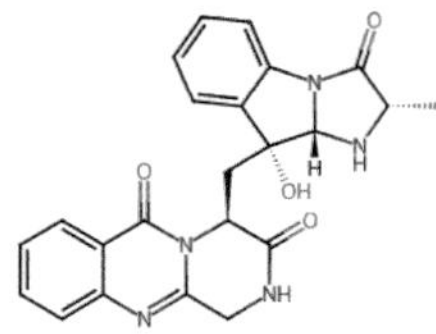
(5) cottoquinazoline A	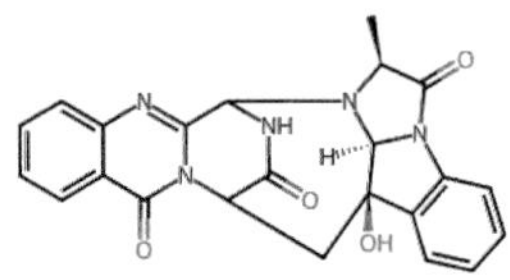
(6) prelapatin B	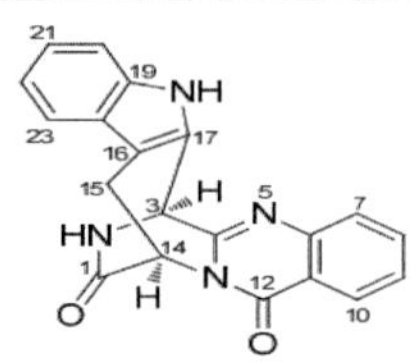 6
(8) protuboxepin E	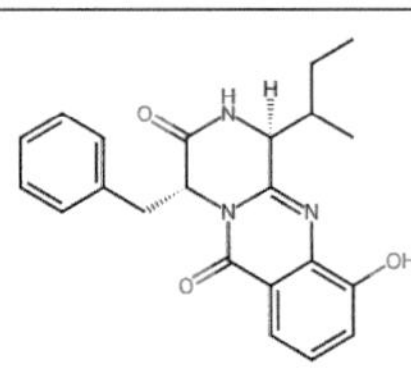
(9) protuboxepin A	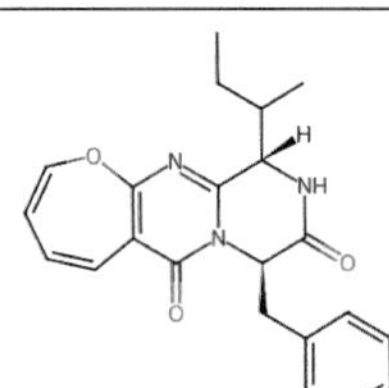
(10) protuboxepin B	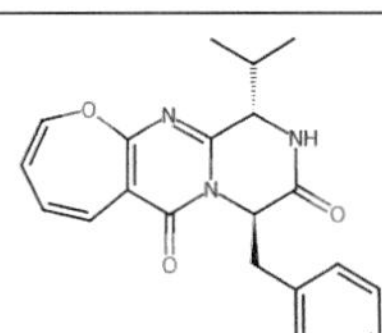
(11) chaetoglobosin C	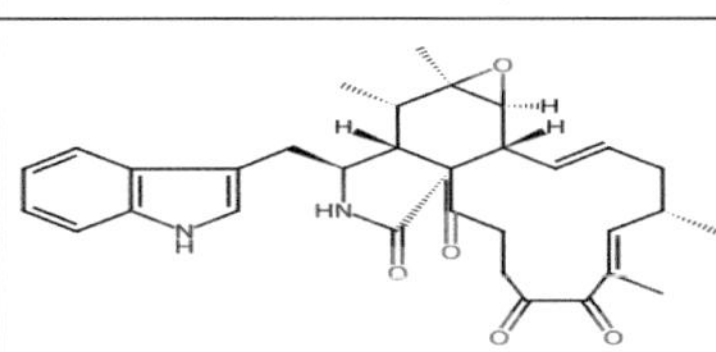

Table 5 contd. ...

...Table 5 contd.

Compounds	Structures*
(12) penochalasin E	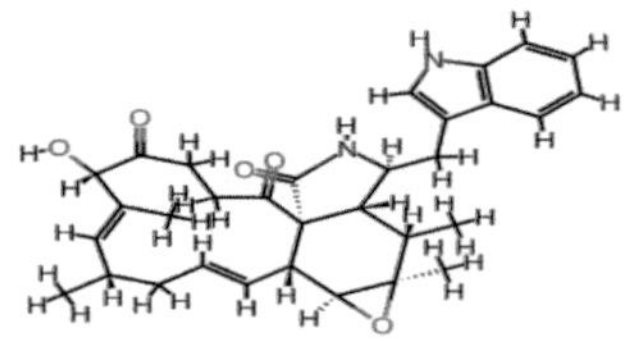
isaridin E	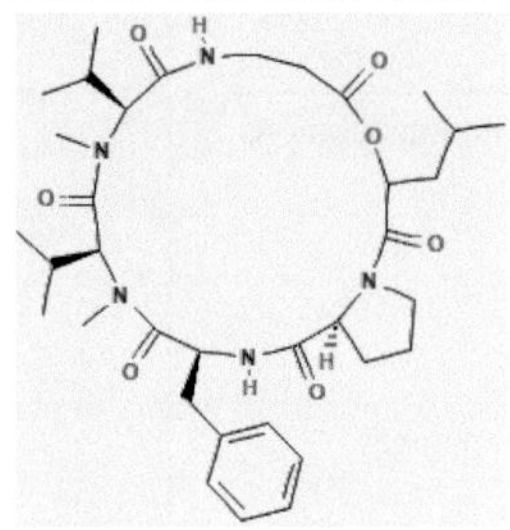
methyl ester	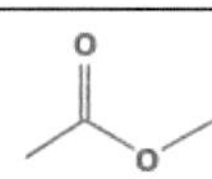
dehydromevalonic lactone	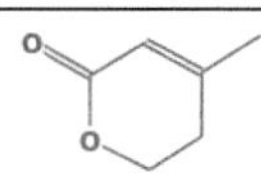
9-tetradecynoic acid	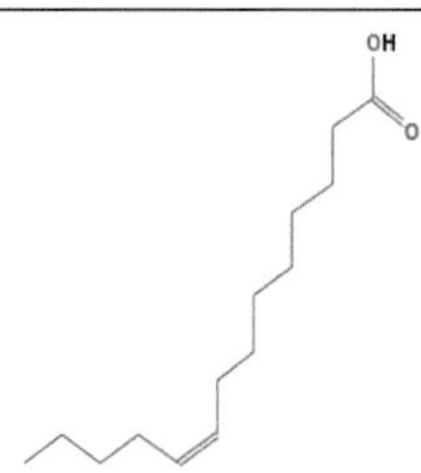
11-hexadecynoic acid	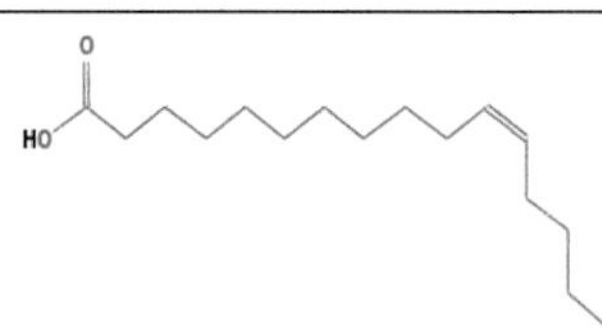
asperlin	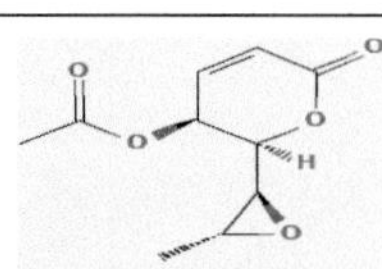
aspergteroid G	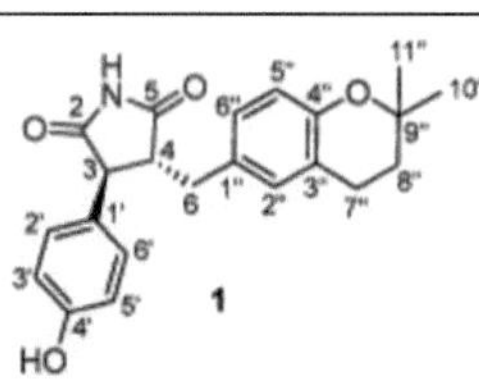 1

Table 6. Studies, interventions, and clinical trials regarding the benefits of bioactive extracts derived from marine fungi against cardiovascular diseases (Papikinou et al., 2024).

Hypothesis-Intervention	**Study Design-/ Parameters Examined**	**Main Findings**	**Structures**	**Specific Benefits/Other Benefits Observed-Mechanisms of Action(s)**	**Year of Study (2014–2024)**
Lopadostoma pouzarii strain 168CLC-57.3 was isolated from an unidentified marine sponge and cultured for low–molecular–weight metabolite isolation	Fungal strain 168CLC-57.3 was isolated from a sponge at Cu Lao Cham, Vietnam Assays tested DPPH, cell viability, and cardioprotective activities	The isolated compounds, namely, lopouazone A and B, showed weak cytotoxicity against PC-3 cells and H9c2 cardiomyocytes; gliorosein exhibited cardioprotective effects	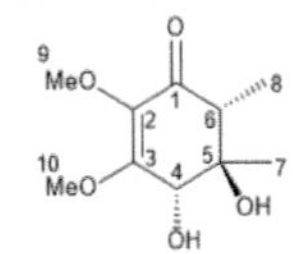 lopouazone A 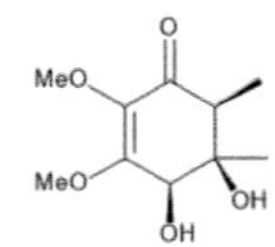 lopouazone B	Gliorosein's anti-radical effect in the DPPH test was statistically significant but not highly potent, and may not directly correlate with intracellular ROS action due to differing free radical concentrations	2022
We report the isolation of 11 polyketides from the natural complex of the sea urchin-associated fungi *P. sajarovii* KMM4718 and *A. protuberus* KMM4747	The fungal complex, likely from the sea urchin *S. mirabilis* (Sea of Japan), includes co-cultured *P. sajarovii* and *A. protuberus* Procedures involved DNA extraction, cultivation, urease inhibition assay, and antimicrobial activity tests	Compounds 1 and 2, namely sajaroketide A and sajaroketide B, were analyzed; some showed cardioprotective effects against *S. aureus*-induced myocarditis	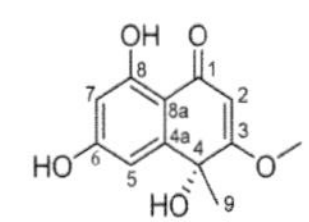 sajaroketide A 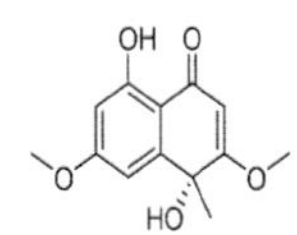 sajaroketide B	Altechromone A increased cell viability by up to 39.9% Compounds 2, 6, and 11 showed no significant effect	2023

Table 6 contd. ...

...Table 6 contd.

Hypothesis-Intervention	**Study Design-/ Parameters Examined**	**Main Findings**	**Structures**	**Specific Benefits/Other Benefits Observed-Mechanisms of Action(s)**	**Year of Study (2014–2024)**
Analyzed xyloketal B (XKB) and rat Cyp3a2 interactions via homology modelling, molecular docking, and oral XKB's effects on hepatic Cyp3a	A rat Cyp3a2 homology model was created; XKB's effects were assessed using high-performance liquid chromatography (HPLC) and luminescence	XKB inhibited Cyp3a-mediated midazolam metabolism in rats, affecting hepatic Cyp3a activity and suggesting possible CYP3A4/Cyp3a2 interactions	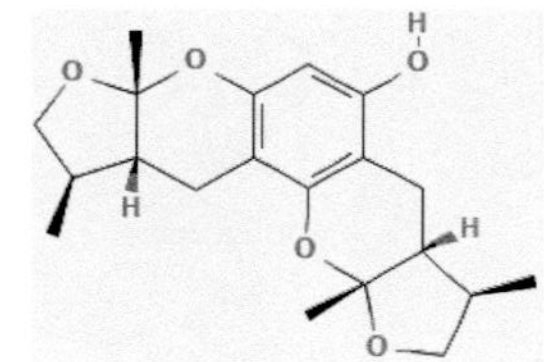 xyloketal B	XKB interacted with the rat Cyp3a2 model via hydrogen bonds Oral XKB increased midazolam's area under the curve (AUC) and the maximum achieved concentration of a drug (C_{max}) while decreasing oral clearance (CL/F), showing similar inhibition to ketoconazole	2014
The FGFC1's effect on fibrin clot structure, lysis, and plasminogen activation using various assays was studied	FGFC1, isolated from *Stachybotrys longispora* FG216 with > 98% purity, was extracted from fermentation samples using 2-butanone and ethyl acetate	FGFC1, a marine alkaloid, promoted fibrin dissolution *in vitro* and effectively treated pulmonary thrombosis in rats	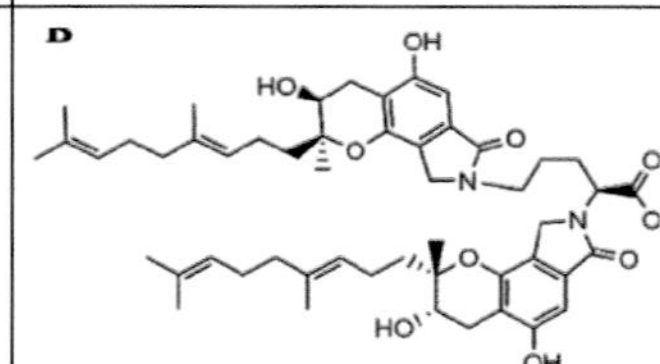 FGFC1	This study showed that FGFC1 had strong fibrinolytic activity in plasma, by reducing clot aggregation and increasing lysis sensitivity It provided a basis for future clinical research on FGFC1 in thrombotic diseases	2023
Versiase, a fibrinolytic enzyme from *Aspergillus versicolor*, showed anticoagulant potential	*Aspergillus versicolor* ZLH-1, isolated from *Callyspongia* sp., was cultured, and its fibrinolytic enzyme was tested for its activity and toxicity	Versiase's anticoagulant activity increased with concentration, exceeding 185 s at 100 μg/mL It showed thrombolytic activity *in vitro*	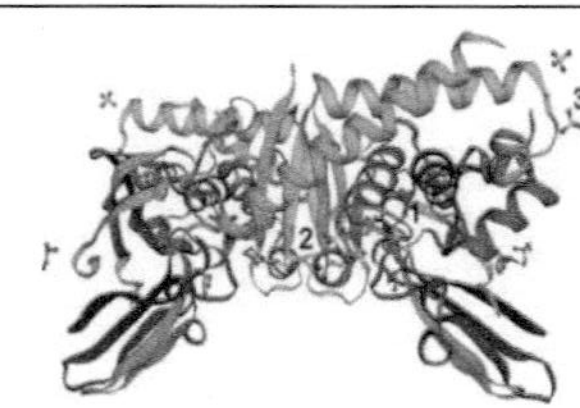 Versiase	Versiase effectively dissolved blood clots with thrombolytic and anticoagulant properties and was safe for use, showing low hemolysis and minimal impact on HUVEC proliferation	2022

Isaridin E effects on arterial thrombus and the PI3K/Akt pathway were studied in an appropriate mouse model	Mice were treated with isaridin E or clopidogrel before tail transection to measure bleeding time Hematologic and coagulation parameters were assessed from blood samples after treatment	Isaridin E inhibited ADP-induced platelet activation and aggregation in a concentration-dependent manner, without affecting collagen or thrombin-induced aggregation	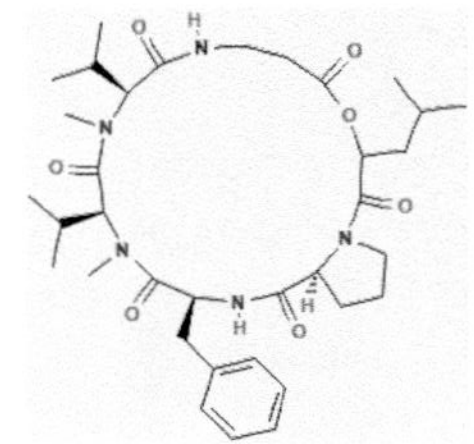 isaridin E	It is safe at 400 μM, with no significant effects on bleeding time or coagulation, making it a promising candidate for treating thrombotic diseases	2022
This study aimed to explore *Setosphaeria* sp.-derived exserolides (I, J, E, F) lower lipids by enhancing reverse cholesterol transport (RCT) *in vitro*	Procedure: Preparation of lipoproteins, cell viability assay, cholesterol efflux assay, determination of intracellular TC and TG levels, and immunoblotting and molecular docking	Exserolide J reduced ox-LDL accumulation and cholesterol efflux via LXRα-ABCA1 and PPARα pathways, decreased CD36 uptake in macrophages, and lowered total cholesterol (TC) and triglyceride content (TG) in HepG2 cells Exserolides I, J, and E enhanced PPARα levels; J showed the strongest LXRα interaction, while E had the weakest	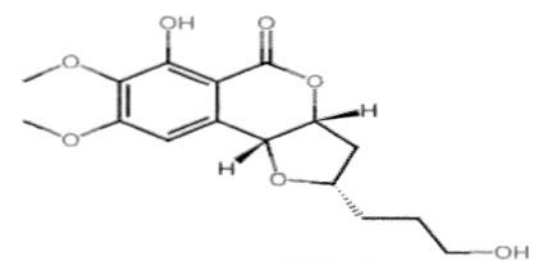 exserolide I 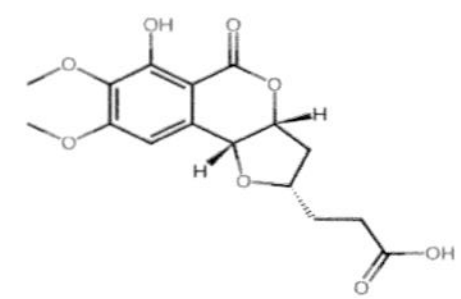 exserolide J 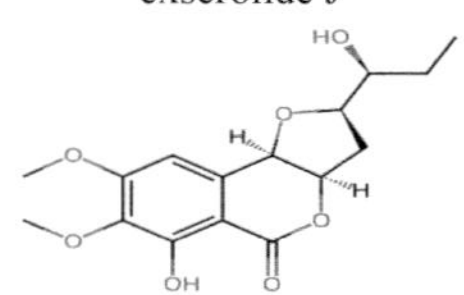 exserolide E 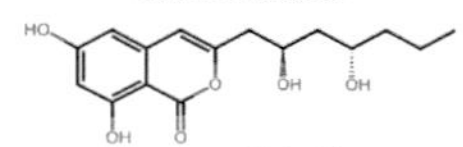 exserolide F	Hyperlipidemia, especially LDL cholesterol, is a key risk factor for atherosclerosis and cardiovascular diseases. Exserolides reduced ox-LDL accumulation and boosted [^{3}H]-cholesterol efflux in RAW264.7 macrophages	2024

Table 6 contd. ...

...Table 6 contd.

Hypothesis-Intervention	Study Design-/ Parameters Examined	Main Findings	Structures	Specific Benefits/Other Benefits Observed-Mechanisms of Action(s)	Year of Study (2014–2024)
Searching for new drugs to combat CVDs is crucial. Penicillium expansum Y32, isolated from Indian Ocean seawater, may provide lead compounds	Key experiments included extraction, purification, zebrafish heart rate, vasculogenesis, and antiangiogenic vessel growth assays	Compounds 4, 6, 8, and 10, identified by spectroscopy and quantum Electronic Circular Dichroism (ECD), reduced bradycardia and enhanced vasculogenesis in zebrafish, showing potential for cardiovascular disease	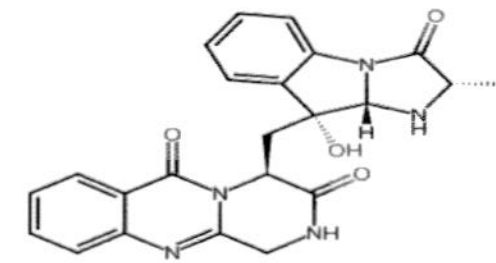 fumiquinazoline Q (4) 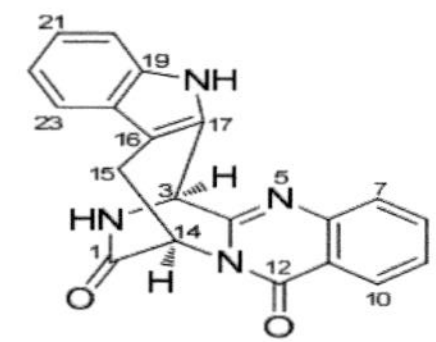 6 prelapatin B (6) 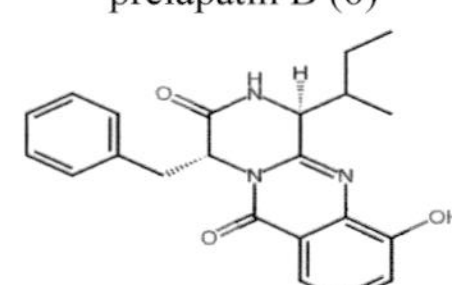 protuboxepin E (8) 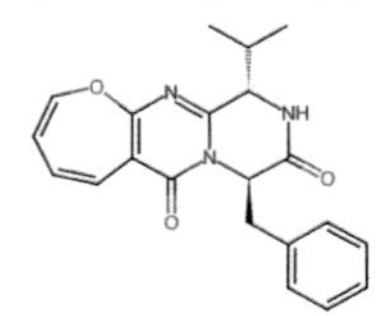 protuboxepin B (10)	Compounds 4, 6, 8, and 10, namely fumiquinazoline Q, prelapatin B, protuboxepin E, and protuboxepin B, respectively, promoted vessel growth; others had limited or no activity	2015

In this search for proangiogenic metabolites, *Chaetomium globosum* YP-106 was isolated from hadal zone seawater	*C. globosum* YP-106 was isolated from Yap Trench seawater, cultured in 50 flasks at 28°C, and extracted after fermentation Spectral data and pro-angiogenic activity were then assessed	Chaetoviridin L (1) and four analogues (2–5) were isolated from *Chaetomium globosum* YP-106 Compounds 1, 2, and 5, namely chaetoviridin L, chaetomugilin A, and chaephilone D, respectively, promoted vascular growth in zebrafish models, while 3 and 4, namely chaetoviridin E and chaetomugilin O, did not. Compounds 1, 2, and 5 are promising for CVD treatment	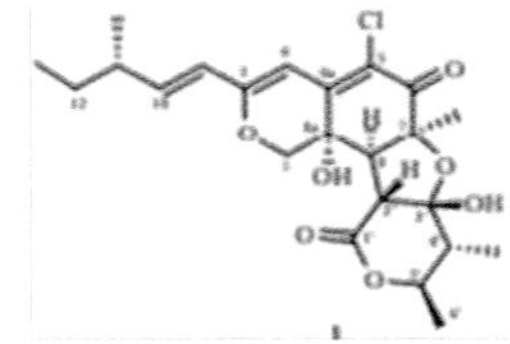 chaetoviridin L (1) 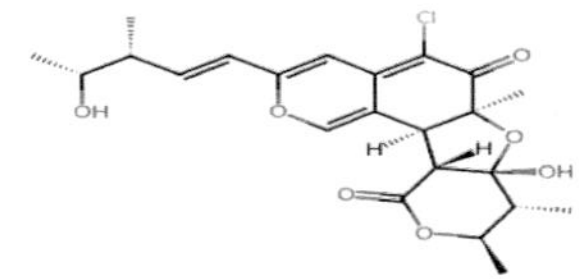 chaetomugilin A (2) 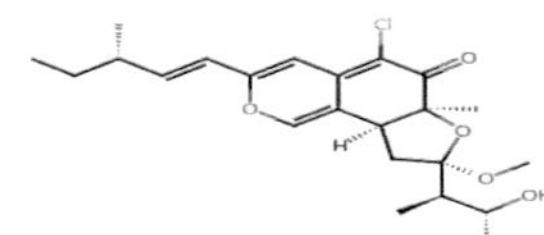 chaephilone D (5) 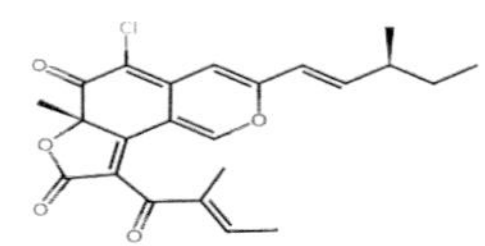 chaetoviridin E (3) 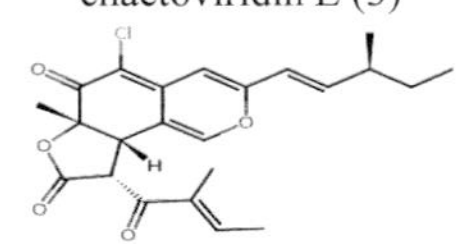 chaetomugilin O (4)	Azaphilones, from deep-sea fungi, may protect against photodamage and may display potential uses as food colorants and UV-proof products	2023

Table 6 contd. ...

...*Table 6 contd.*

Hypothesis-Intervention	Study Design-/ Parameters Examined	Main Findings	Structures	Specific Benefits/Other Benefits Observed-Mechanisms of Action(s)	Year of Study (2014–2024)
Four new compounds were isolated from Neopestalotiopsis sp. HN-1-6. Azaphilones (1–3, 5–7) were screened for proangiogenic activity in zebrafish	A fungal strain from Qinzhou was tested for cytotoxicity, antibacterial, and proangiogenic effects using Tg zebrafish embryos	From Neopestalotiopsis sp., three new azaphilones and one phenylpropanoid showed proangiogenic activity and no cytotoxicity.	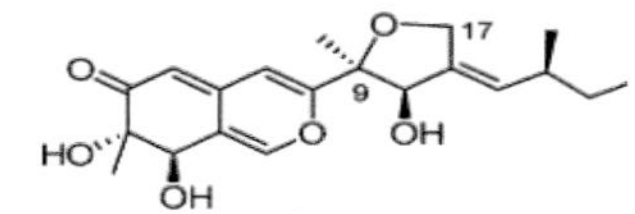 **3** pestaphilone J 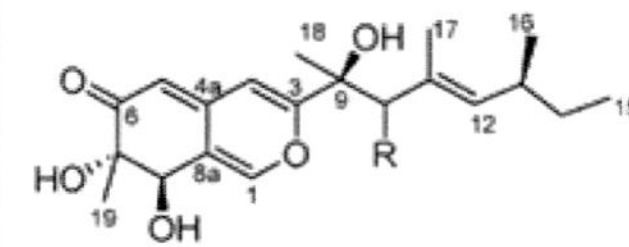 pestaphilone G 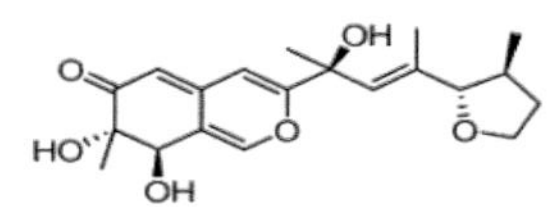 **7** pestaphilone I	Compounds 3, 5, and 7, namely pestaphilones J, G, and I, demonstrated remarkable proangiogenic activities	2024
Asperlin was isolated from *Aspergillus versicolor* LZD4403. Its anti-inflammatory and atheropreventive activities were confirmed in high-fat diet (HFD)-fed ApoE−/− mice	*Aspergillus versicolor* LZD4403 was isolated from a gorgonian and analyzed for anti-inflammatory and lipid effects	Asperlin enhanced cholesterol efflux via PPARγ-ABCA1/G1 and shifted macrophage polarization to M2, indicating potential for atherosclerosis prevention and treatment	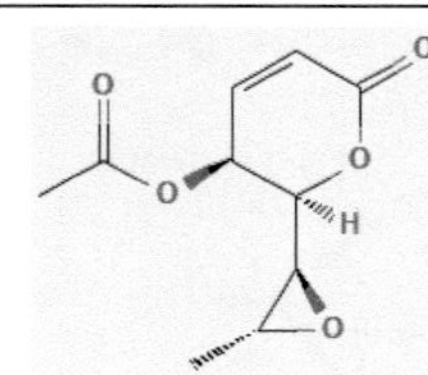 asperlin	Asperlin (1–10 μM) effectively suppressed LPS-induced foam cell formation, comparable to simvastatin, by shifting macrophage polarization from M1 to M2	2017

Xyl-B, isolated from *Xylaria* sp., was evaluated for hypertension effects using a 2-kidney, 2-clip rat model	Xyl-B, isolated and characterized by HPLC and 1D-nuclear magnetic resonance spectroscopy (D-NMR), was tested in 2K2C rats, with various vascular and cellular assays	This study showed that Xyl-B reduced blood pressure in 2K2C hypertensive rats and induced endothelium-dependent relaxation in aortic rings	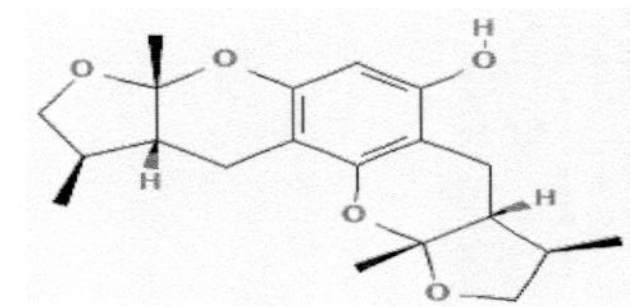 xyloketal B	A study limitation is the unverified nitric oxide–soluble guanylate cyclase–cyclic guanosine monophosphate (NO-sGC/cGMP) pathway and Ca^{2+} signaling mechanisms in vascular smooth muscle cells (VSMCs) Caution is needed in interpreting Xyl-B's antihypertensive effects	2018
Two new 4,5-disubstituted asperg eroids and four known 4,5-disubstituted analogs were isolated from soft-coral-associated symbiotic and epiphytic fungus *Aspergillus terreus* EGF7-0-1	Strain EGF7-0-1 was isolated from soft coral in the South China Sea and identified as *Aspergillus terreus* Isolation, Quantum Chemical Calculations, and Western Blot assays were conducted	The results suggested that compound 1, otherwise asperg eroid G, may reduce the inflammatory response and protect cardiomyocytes through the GSK-3β/NLRP3 pathway	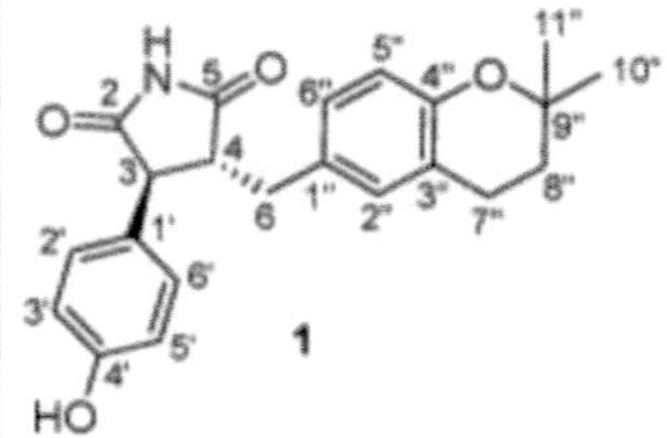 aspergeroid G	The expression of caspase-3 and Bax was regulated, the antiapoptotic gene Bcl-2 was upregulated, the antioxidant capacity of cells was improved, and apoptosis was inhibited	2023

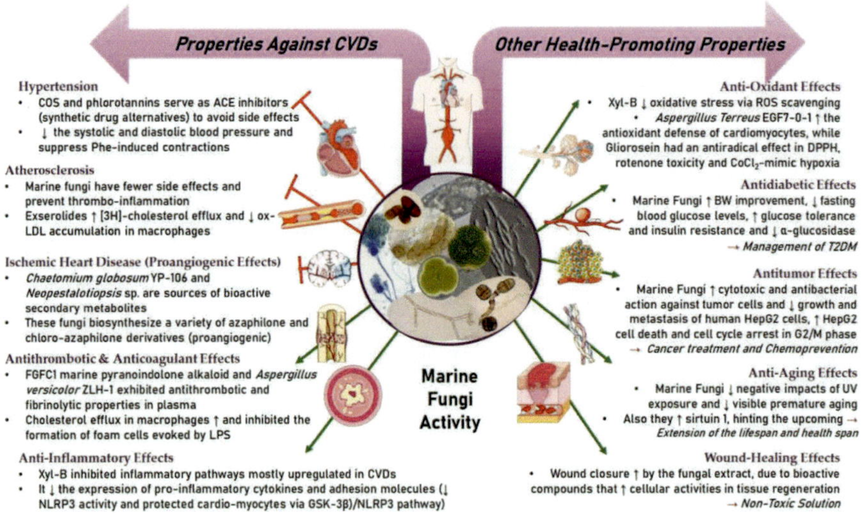

Fig. 3. A summary of the cardioprotective benefits and other health-promoting properties of marine fungi and their bioactives (Papikinou et al., 2024).

α-glucosidase, PTP-1B), further supporting their antidiabetic relevance (Alanzi et al., 2023).

6.2 *Antitumor Effects of Marine Fungal Extracts*

Marine fungi also represent promising sources of anticancer compounds. *Penicillium* sp. OUCMDZ-1435 produced meleagrine and oxaline, which suppressed HepG2 hepatocellular carcinoma cell proliferation, induced apoptosis, and caused G2/M cell cycle arrest (He et al., 2023). Danthron, a hydroxy-anthraquinone, inhibited angiogenesis and exhibited both antioxidant and anticancer effects (Vidal et al., 2023). From *Lopadostoma pouzarii*, novel polyketides such as lopouzanones A and B showed low cytotoxicity toward normal cells while exhibiting activity against prostate cancer cells (Trinh et al., 2022). Additional fungal metabolites, including compound XYA-2 and oxirapentyn A, induced apoptosis and inhibited migration in pancreatic cancer and glioma cells, respectively, with oxirapentyn A demonstrating a remarkable blood–brain barrier permeability activity (Guan et al., 2024; Jiang et al., 2023).

6.3 *Anti-Aging Effects of Marine Fungal Extracts*

β-Glucan, abundant in marine yeasts, is a potent antioxidant and anti-aging compound with applications in skincare. It enhances collagen synthesis, improves moisture retention, and protects against ultraviolet (UV)-induced oxidative stress, thereby reducing wrinkles and restoring skin elasticity. Further studies have identified *Scopalina hapalia* as a promising source of additional anti-aging bioactives (Sarkar et al., 2024).

6.4 Wound-Healing Effects of Marine Fungal Extracts

Melanin isolated from the halotolerant black yeast *Hortaea werneckii* AS1 was evaluated for its wound healing value using a fibroblast scratch assay. Although it did not accelerate closure beyond natural healing, it showed no cytotoxic effects and preserved fibroblast function. Its strong antioxidant and antimicrobial activities support its application in wound care by reducing infection risk and oxidative stress, hence indirectly promoting effective tissue regeneration (Elsayis et al., 2022). Collectively, marine fungi produce diverse metabolites with therapeutic potential against oxidative stress- and inflammation-related disorders. The structures of marine fungi beneficial against chronic diseases are summarized in Table 7, whereas indicative studies and clinical trials supporting their anti-inflammatory and antioxidant roles are listed in Table 8. The compounds included in these interventions are depicted in Fig. 4.

7. Limitations, Challenges, and Future Perspectives

Marine fungi represent an emerging yet underexplored reservoir of bioactive macromolecules, particularly polysaccharides and peptides, with broad therapeutic potential spanning antioxidant, anti-inflammatory, immunomodulatory, antitumor, and cardiometabolic applications. Despite their promising bioactivities, several challenges must be addressed before these compounds can be fully translated into clinical and industrial use (Geng et al., 2024).

For marine fungal polysaccharides, major obstacles include low natural yields, complex purification procedures, and insufficient structural characterization. Determining precise branching patterns, glycosidic linkages, molecular weights, and sulphation degrees remains essential for establishing clear structure-activity relationships (SARs). Moreover, standardized extraction and characterization protocols are urgently needed to ensure reproducibility and comparability across studies, as variability in fungal strains, culture conditions, and analytical methods greatly affects biological outcomes (Geng et al., 2024). Another key limitation is the lack of *in vivo* and clinical validation. While many polysaccharides demonstrate potent effects *in vitro*, their bioavailability, metabolism, pharmacokinetics, and safety remain largely underexplored. Addressing these gaps via animal models, preclinical pharmacology, and early translational studies will be vital to advance these biopolymers toward therapeutic use (Geng et al., 2024).

Future research should focus on the integration of synthetic biology and metabolic engineering to enhance yields and enable tailored structural modifications that improve potency and selectivity. The combination of genomic, transcriptomic, and metabolomic profiling could facilitate the discovery of novel polysaccharide biosynthetic gene clusters. Moreover, machine learning and computational modeling may aid in predicting biological activity from structural data, expediting the identification of high-value agents. When coupled with nanotechnology and advanced bioprocessing, such approaches may accelerate the application of marine fungal polysaccharides in drug delivery, tissue engineering, and regenerative medicine (Laurienzo, 2010).

Table 7. Structures of certain marine fungi-derived compounds associated with chronic disorders (Papikinou et al., 2024).

Compounds	**Structures**
exopolysaccharide PJ1-1	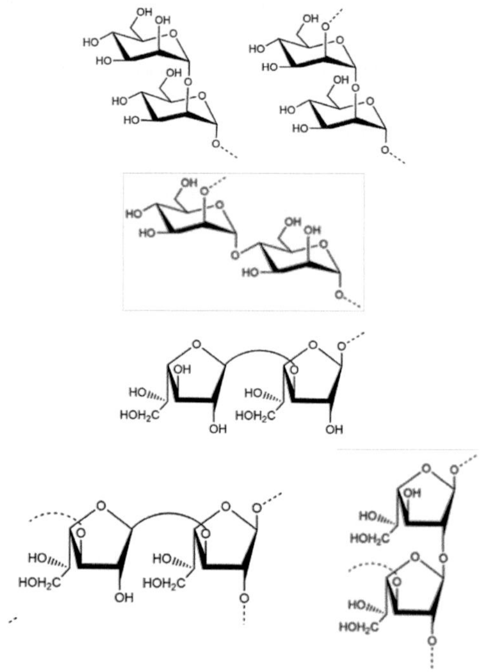 (Possible structures of the main repeating disaccharides in PJ1-1)
scequinadoline A, B, D & E (compounds 7-10, most important and effective fumiquinazolines)	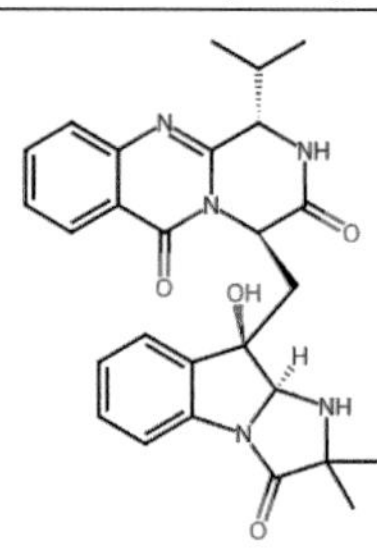 scequinadoline A 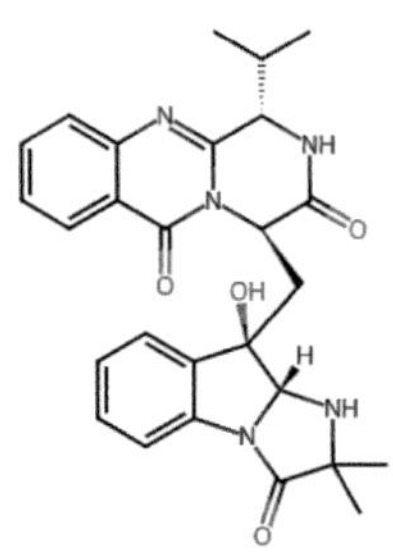 scequinadoline B

Table 7 contd. ...

...Table 7 contd.

Compounds	Structures
	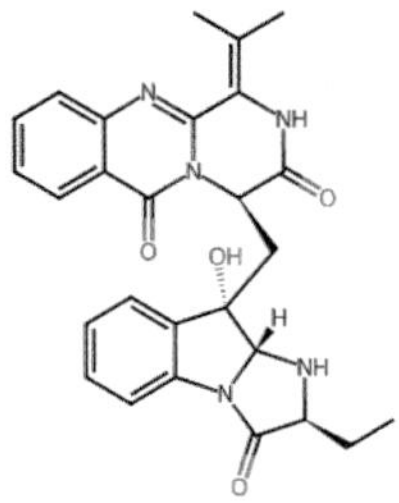 scequinadoline D 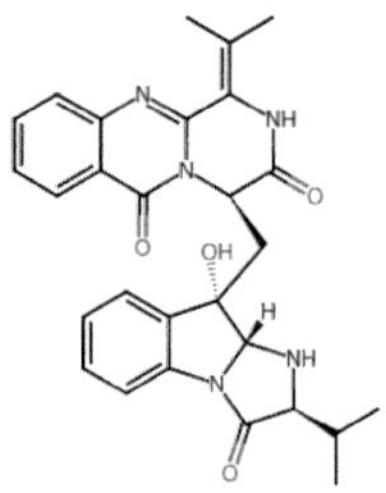 scequinadoline E
flavipesolides A-C	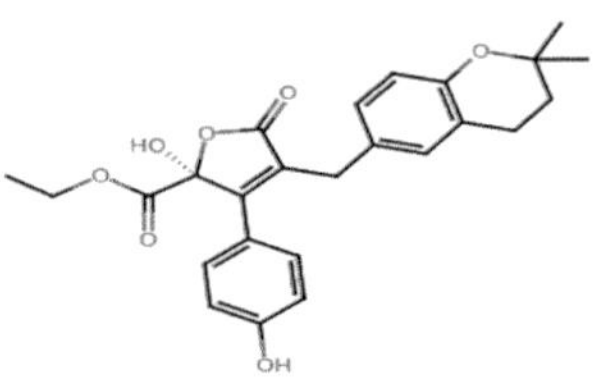 flavipesolide A 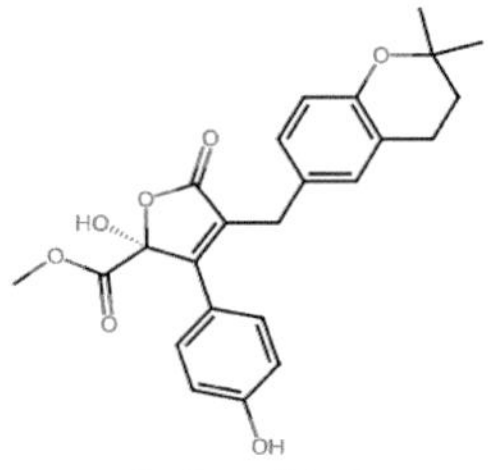 flavipesolide B 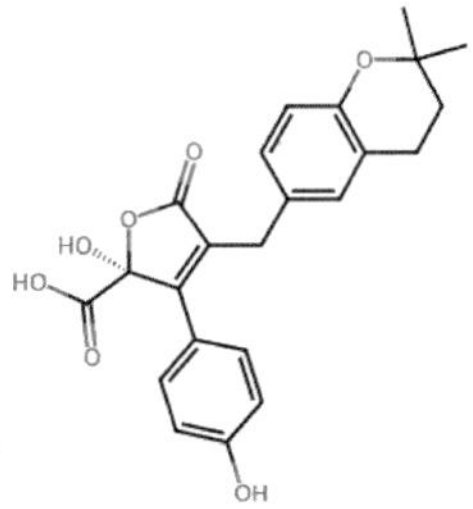 flavipesolide C

Table 7 contd. ...

...Table 7 contd.

Compounds	**Structures**
5–[(3,4–dihydro–2,2–dimethyl–2*H*–1–benzopyran–6–yl) methyl]–3–hydroxy4–(4–hydroxyphenyl)–2(5*H*) furanone (compound 4) & aspernolide (compound 5)	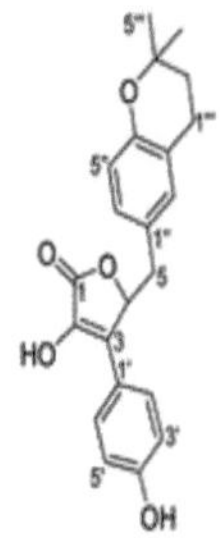 5–[(3,4–dihydro–2,2–dimethyl–2*H*–1–benzopyran–6–yl) methyl]–3–hydroxy4–(4–hydroxyphenyl)–2(5*H*) furanone (4) 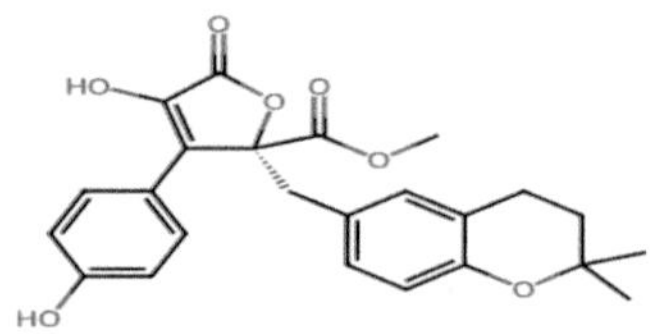 aspernolide A (5)
emodin	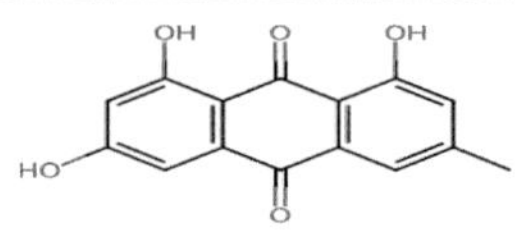
geodin hydrate (8) methyl dichloroasterrate (9) & monomethylosoic acid (10)	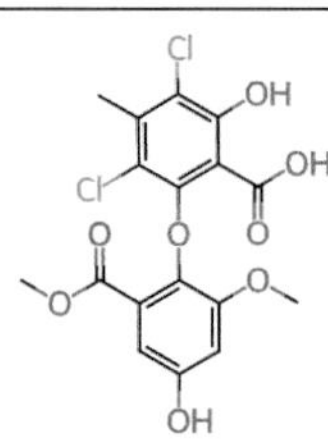 geodin hydrate (8) 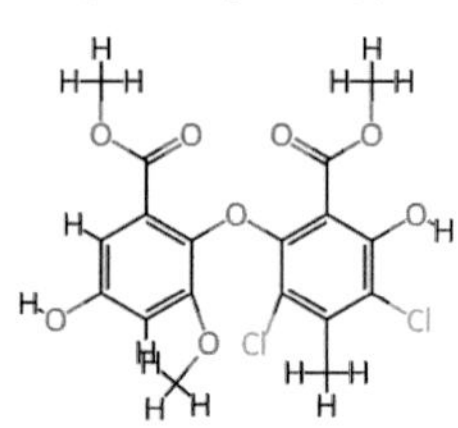 methyl dichloroasterrate (9) 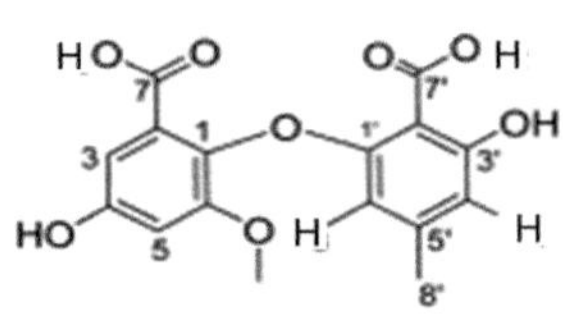 monomethylosoic acid (10)

Table 7 contd. ...

...Table 7 contd.

Compounds	Structures
epicoccolide B (13)	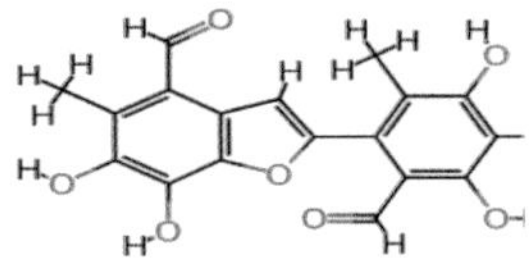
brasilterpenes A and C	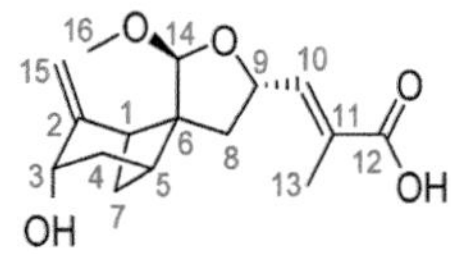 brasilterpene A (1) 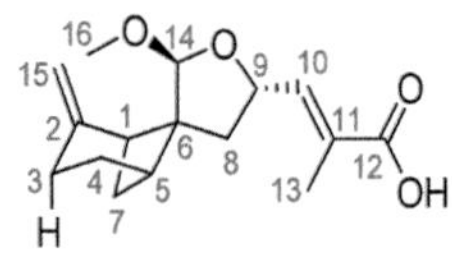 brasilterpene C (3)
cladosporol C	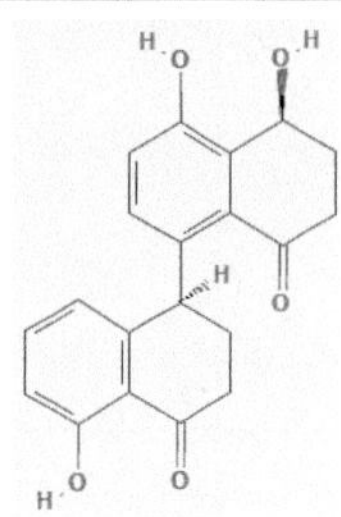
tenellone F	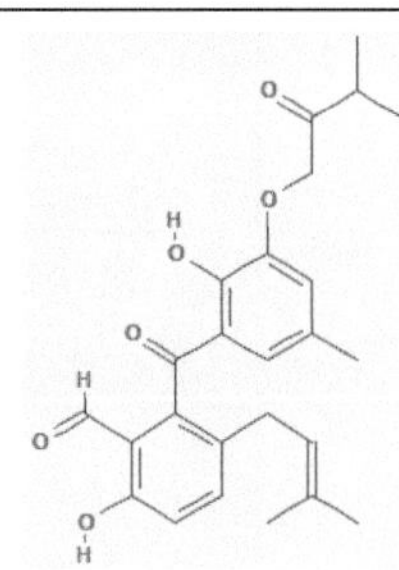
ozazino-cyclo-(2,3-dihydroxyl-trp-tyr)	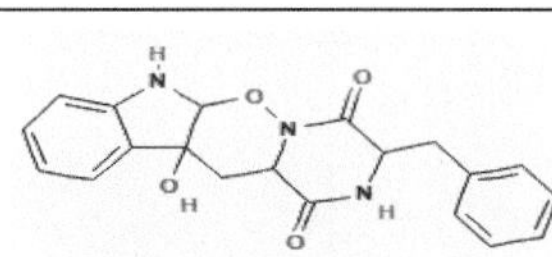
penicillactam	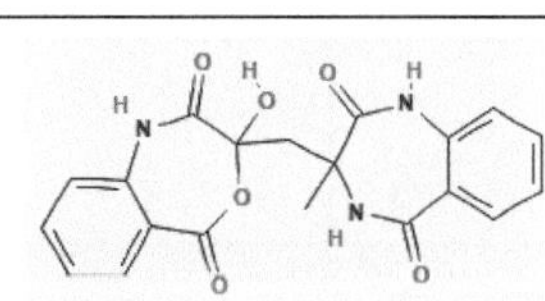

Table 7 contd. ...

...Table 7 contd.

Compounds	Structures
circumdatin G	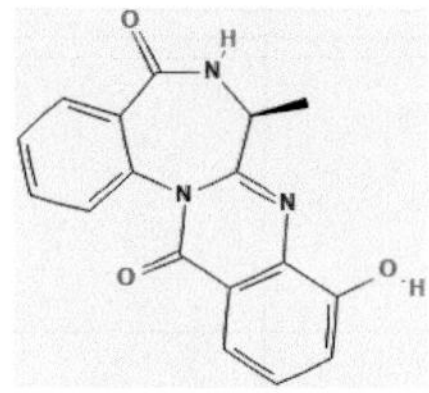
meleagrine	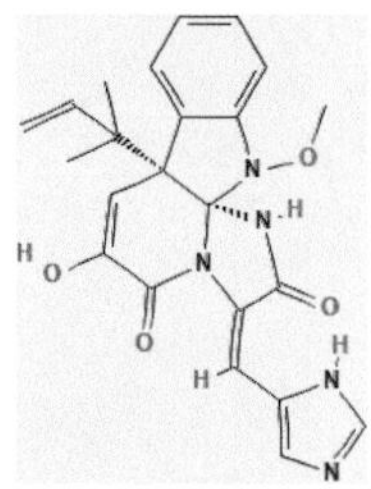
oxaline	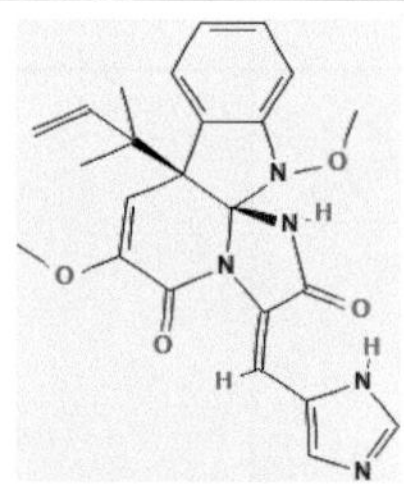
danthron	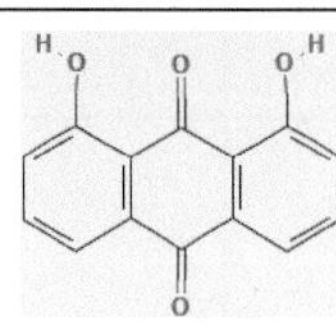
lopouzanones A and B	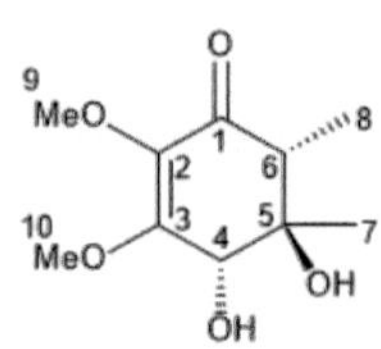 lopouazone A (1) 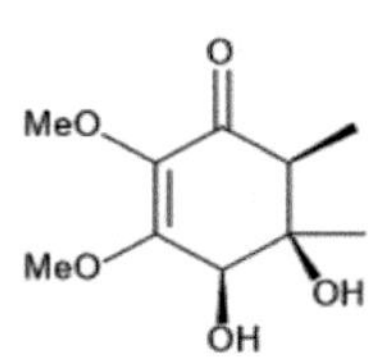 lopouazone B (2)

Table 7 contd. ...

...Table 7 contd.

Compounds	Structures
1-O-acetyl dendrodochol B	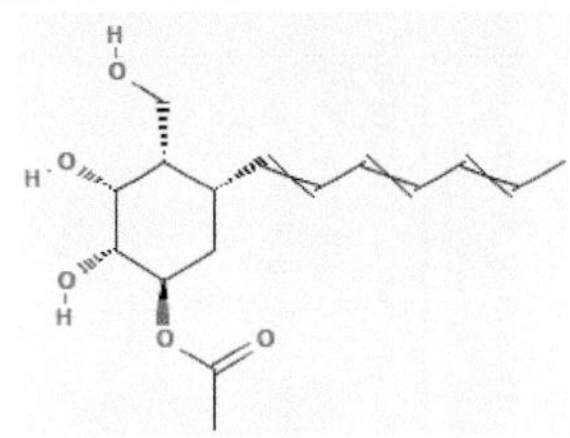
oxirapentyn A	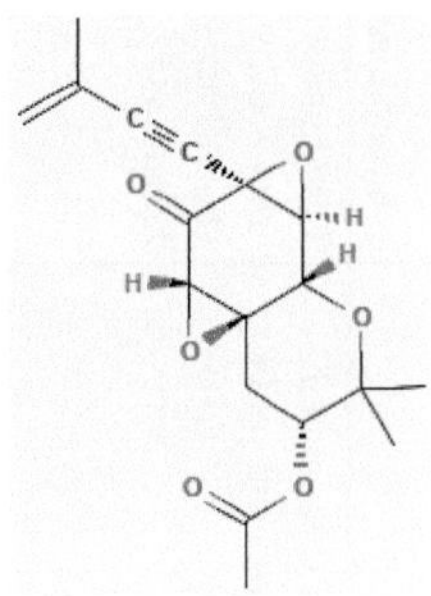

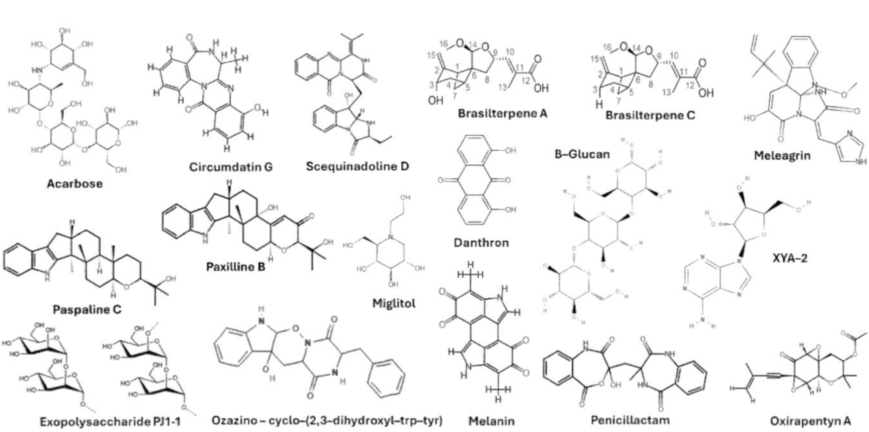

Fig. 4. Marine fungi-derived bioactive compounds and extracts used in the interventions and clinical trials, which are presented in Table 8 (Papikinou et al., 2024).

Similarly, marine fungal peptides hold exceptional potential as next-generation bioactives, yet their discovery and scalable production remain limited. Many of these peptides are synthesized in trace amounts and require genome mining, heterologous expression systems, or total chemical synthesis to achieve feasible yields. Their pharmacokinetics, stability, and toxicity profiles must be rigorously investigated before clinical translation (Hafez Ghoran et al., 2023). Emerging approaches, like artificial intelligence (AI)-assisted peptide design, *in silico* docking, and multi-omics-guided discovery pipelines, can notably enhance peptide

Table 8. Studies, interventions, and clinical trials of the anti-diabetic, anticancer, anti-aging, and wound healing effects of bioactive extracts derived from marine fungi (Papikinou et al., 2024).

Hypothesis-Intervention	**Study Design-/Parameters Examined**	**Main Findings**	**Specific Benefits/Other Benefits Observed - Mechanisms of Action(s)**	**Year of Study**
α-Glucosidase inhibitors from *Aspergillus flavipes* HN4-13 are introduced	*Aspergillus flavipes* HN4-13 was isolated from Lianyungang sediment by fermentation and extraction, and assays	α-Glucosidase inhibitors like acarbose and miglitol treat diabetes but cause side effects	Natural products from *Aspergillus* fungi show anti-tobacco mosaic virus (TMV), DPPH scavenging, α-glucosidase inhibition, and antiviral activities	2016
This study isolated three new and 12 known fumiquinazoline alkaloids from *Scedosporium apiospermum* F41-1 and assessed their antidiabetic potential	*Scedosporium apiospermum* F41-1 was isolated from *Lobophytum crassum* in Hainan, China Fermentation, Extraction, 3T3-L1 Differentiation, Triglyceride Assay, Oil Red O Staining, quantitative polymerase chain reaction (qPCR), procedures were held	Results suggested that scequinadoline D (9) targeted adipocytes and could treat T2DM effectively	Fumiquinazolines and related alkaloids from *Scedosporium apiospermum* F41-1 showed antiviral activity against hepatitis C	2020
Sixteen indole diterpenoids, including five novel ones, were tested for their cytotoxic and antimicrobial activities	The strain produced a crude extract, which was purified and tested for antimicrobial activity using disk diffusion	Paspaline (C–D)- and paxilline (B–D)-type indole diterpenoids rarely exhibit antimicrobial activity but may have variable cytotoxic effects across different cancer cell lines	Five new and eleven known indole diterpenoids from *Penicillium brefeldianum* WZW-F-69 were observed, highlighting the potential of marine fungi for discovering novel bioactive compounds	2022
This manuscript presents danthron as a new angiogenesis inhibitor, with antiangiogenic, antitumor, and antioxidant properties	Cell culture included human umbilical vein endothelial cells (HUVECs), treated with danthron or controls. Chick cell adhesion molecule (CAM) assays and cell cycle analysis followed	Danthron, a hydroxy anthraquinone from a marine fungus, inhibited angiogenesis and tumor cell proliferation by inducing apoptosis	The link between danthron's antiangiogenic and antioxidant activities warrants further study, supporting its potential as an anti-cancer drug	2023

In this study, a homogeneous exopolysaccharide, PJ1-1, was obtained from *Penicillium janthinellum* N29's fermented broth	C57BL/6J mice were used; *P. janthinellum* N29 derived from *Acanthus ilicifolius* Fermentation, composition, and analysis methods, as well as *in vivo* experiments and fasting insulin assays, were held	PJ1-1 enhanced glucose metabolism, improved insulin efficiency, and lipid metabolism in T2DM mice, suggesting its potential as an anti-diabetic agent	PJ1-1 improved lipid metabolism in T2DM mice by lowering TG, TC, LDL-C, and increasing high-density lipoprotein (HDL)-C	2023
Five bergamotane sesquiterpenoid derivatives, brasilterpenes, were isolated from *Paraconiothyrium brasiliense* HDN15-135	The fungus was isolated from the Indian Ocean sediment Procedures held: fermentation, NMR, glucose assays, toxicity evaluation, hypoglycemic mechanism studies	Brasilterpenes A (1) and C (3) lowered blood glucose in hyperglycemic zebrafish by enhancing insulin sensitivity	The hypoglycemic activity of compounds 1 and 3 is linked to the S configuration at C-14 Compound 3, with a hydroxyl group at C-3, is more active	2022
β-Glucan methyl ester compounds were checked for their applications in the cosmetics field	Assays used for antioxidant, metal-chelating, reducing power, and antibacterial activities B-glucan was applied in cosmetics and used for almond facial scrub preparation	β-Glucan improved skin firmness, collagen production, and has antioxidant, antibacterial effects	Carboxymethylated β-glucan showed greater antibacterial activity against *S. aureus* than the original compound	2024
Structure-based virtual screening (SBVC) of various deep-sea fungi derived metabolites was performed	Retrieve and prepare protein/ligand structures, virtual screening, molecular docking, molecular dynamics (MD) simulations, residue interaction analysis, and binding free-energy determination were performed	*Cladosporol C, tenellone F*, ozazino-cyclo-(2,3-dihydroxyl-trp-tyr), penicillactam, and circumdatin G were identified as potential inhibitors of key diabetes-related enzymes	Inhibiting α-amylase, α-glucosidase, and hexokinase II (HK-II) delayed sugar absorption, while protein tyrosine phosphatase 1B (PTP-1B) impaired insulin signaling	2023
Marine fungi are valuable for antitumor drug development	Annexin FITC/PI assay was used for compound analysis and apoptosis detection L-tryptophan and L-histidine are key precursors in Meleagrin biosynthesis	Challenges in the chemical synthesis of bioactive compounds led us to the optimization of Meleagrin's fermentation The initial pH significantly affected the yield: pH 3.0 was optimal for a simpler composition and higher Meleagrin production	These compounds inhibited HepG2 cell proliferation and metastasis, induced apoptosis, and caused cell cycle arrest in G2/M phase	2023

Table 8 contd. ...

...Table 8 contd.

Hypothesis-Intervention	**Study Design-/Parameters Examined**	**Main Findings**	**Specific Benefits/Other Benefits Observed - Mechanisms of Action(s)**	**Year of Study**
The current study attempted to examine the various valuable medical and environmental utilizations of this beneficial pigment	The purified melanin was evaluated for antioxidant activity, cytotoxicity, *in vitro* anti-cancer effects, wound healing, antimicrobial activity, and environmental impact	The pigment showed an antibacterial activity toward *S. aureus* and *A. hydrophila* It can further be applied as an antimicrobial agent against fatal pathogenic strains affecting the aquaculture industry	These findings showed that melanin at a concentration of 100 μg/ml did not influence cell migration	2022
The impact of XYA-2, a nitrogenated azaphilon previously reported from a deep-sea-derived fungus, on the progression of pancreatic cancer cells was evaluated	Procedures conducted: Cell cycle arrest assay, apoptosis assay, Western blot assay, wound healing assay, transwell migration assay	XYA-2 exhibited significant effects on various cellular processes in pancreatic cancer Furthermore, it influenced the transition of cancer cells during the cell cycle, leading to increased apoptosis at higher concentrations	The results from the wound healing assay underscored the concentration-dependent inhibitory effects of XYA-2 on the migratory capacity of MIA-PACA2 cells	2024
In this research on ascidian-derived fungi in the South China Sea, it was discovered that a culture extract of the fungus *Amphichorda felina* SYSU-MS7908 exhibited moderate cytotoxicity against U87-MG human glioma cells	Procedures held: X-ray crystallographic analysis, antiproliferative activity by MTT assay, prediction of blood–brain barrier permeation *in Silico*, cell migration by wound-healing assay, cell invasion by transwell assay, and cell apoptosis assay by flow cytometry	In an anti-glioma assay, oxirapentyn A (7) effectively inhibited the proliferation, migration, and invasion of glioma cells and induced their apoptosis Furthermore, an *in silico* analysis suggested that oxirapentyn A has the potential to penetrate the blood–brain barrier	These findings highlighted the potential of oxirapentyn A as a candidate for the development of novel anti-glioma drugs	2023

identification and optimization. Plus, peptide-nanocarrier conjugates or smart delivery systems could improve bioavailability and tissue targeting, leading to natural, effective, and safe cosmetic production (Di Guardo et al., 2025; He and Jia, 2022; Kim, 2025).

In summary, the future of marine fungal research lies in a multidisciplinary convergence of marine biotechnology, synthetic biology, and, importantly, bioinformatics. By leveraging these technologies, researchers can unlock the full potential of marine fungal polysaccharides and peptides as clinically relevant, sustainable, and functionally tunable therapeutics for chronic and degenerative diseases.

8. Conclusions

Marine natural products have emerged as a prolific source of structurally diverse compounds with potent biological activities. Many of these molecules exhibit efficacy against a wide spectrum of diseases, often acting through novel mechanisms of action and, in some cases, representing the most powerful inhibitors within their pharmacological classes. Recent advances in sampling technologies, nanomole-scale structure elucidation, genome sequencing, and biosynthetic pathway mining have greatly accelerated the pace of marine drug discovery. Furthermore, longstanding supply challenges are increasingly being overcome through total synthesis, microbial fermentation, and molecular biology approaches, paving the way for environmentally sustainable and economically feasible drug production.

The extreme and fluctuating conditions of marine ecosystems drive the evolution of unique metabolic pathways, resulting in a vast repertoire of secondary metabolites with therapeutic potential. Extracts and purified compounds from marine fungi, algae, and other marine organisms have revealed an impressive chemical diversity encompassing polysaccharides, terpenoids, phenolics, sterols, carotenoids, alkaloids, and fatty acids, many of which display antioxidant, anti-inflammatory, antitumor, antidiabetic, and cardioprotective properties. Despite this wealth of molecular diversity, marine fungi remain markedly underexplored, and only a limited number of their metabolites have advanced to clinical evaluation.

Looking ahead, future research should emphasize the responsible, sustainable, and ethically guided development of marine bioactives to combat the escalating global burden of chronic and degenerative burden of chronic diseases. Integration of genomic, metabolomic, and synthetic biology approaches with computational and AI-driven drug discovery will be crucial for accelerating the identification and optimization of novel compounds. With their remarkable structural novelty, therapeutic versatility, and feasibility for scalable production, marine-derived metabolites represent one of the most promising frontiers in modern pharmaceutical science. Continued exploration of marine fungal bioactives and other disease-modifying agents is poised to yield innovative, effective, and safer solutions to unmet medical challenges.

Author Contributions

Conceptualization, A.T.; methodology, A.T.; software, All Authors; validation, A.T.; investigation, All Authors; data curation, All Authors; writing—original draft preparation, M.-A.P., K.P., and A.T.; writing—review and editing, T.A., A.O., C.A., and A.T.; visualization, A.T.; supervision, A.T. All authors have read and agreed to the published version of the manuscript.

Funding

This research received no external funding.

Acknowledgments

The authors would like to thank the School of Chemistry, Faculty of Sciences of the Democritus University of Thrace, and the Fisheries Research Institute at Kavala, Greece, for their continuous support.

Conflicts of Interest

The authors declare no conflicts of interest.

References

Abdala-Díaz, R.T., Casas-Arrojo, V., Castro-Varela, P., Riquelme, C., Carrillo, P. et al. (2024). Immunomodulatory, antioxidant, and potential anticancer activity of the polysaccharides of the fungus *Fomitiporia chilensis*. Molecules, 29: 3628. https://doi.org/10.3390/molecules29153628.

Alanzi, A.R., Parvez, M.K., Alqahtani, M.J. and Al-Dosari, M.S. (2023). Deep-sea fungal metabolites as potential inhibitors of glucose-regulatory enzymes: *in silico* structure–activity analysis. Saudi Pharm. J. https://doi.org/10.1016/j.jsps.2023.101776.

Alharbi, N.K., Azeez, Z.F., Alhussain, H.M., Shahlol, A.M.A., Albureikan, M.O.I. et al. (2024). Tapping the biosynthetic potential of marine *Bacillus licheniformis* LHG166, a prolific sulphated exopolysaccharide producer: Structural insights, bio-prospecting its antioxidant, antifungal, antibacterial, and anti-biofilm potency as a novel anti-infective lead. Front. Microbiol., 15: 1385493. https://doi.org/10.3389/fmicb.2024.1385493.

Astudillo, A.M., Balboa, M.A. and Balsinde, J. (2023). Compartmentalized regulation of lipid signaling in oxidative stress and inflammation: Plasmalogens, oxidized lipids, and ferroptosis as new paradigms of bioactive lipid research. Prog. Lipid Res., 89: 101207. https://doi.org/10.1016/j.plipres.2022.101207.

Badawy, M.A.E.M.D., Naing, L., Johar, S., Ong, S., Rahman, H.A. et al. (2022). Evaluation of cardiovascular diseases risk calculators for CVDs prevention and management: Scoping review. BMC Public Health, 22: 1742. https://doi.org/10.1186/s12889-022-13944-w.

Calder, P.C. (2015). Functional Roles of fatty acids and their effects on human health. J. Parenter. Enteral. Nutr., 39: https://doi.org/10.1177/0148607115595980.

Carrasqueira, J., Bernardino, S., Bernardino, R. and Afonso, C. (2025). Marine-derived polysaccharides and their potential health benefits in nutraceutical applications. Mar. Drugs, 23: 60. https://doi.org/10.3390/md23020060.

Chen, Yin, Mao, W., Gao, Y., Teng, X., Zhu, W. et al. (2013). Structural elucidation of an extracellular polysaccharide produced by the marine fungus *Aspergillus versicolor*. Carbohydr. Polym., 93: 478–483. https://doi.org/10.1016/j.carbpol.2012.12.047.

Chen, Y.L., Mao, W.J., Tao, H.W., Zhu, W.M., Yan, M.X. et al. (2015). Preparation and characterization of a novel extracellular polysaccharide with antioxidant activity, from the mangrove-associated fungus *Fusarium oxysporum*. Mar. Biotechnol., 17: 219–228. https://doi.org/10.1007/s10126-015-9611-6.

Deshmukh, S.K., Prakash, V. and Ranjan, N. (2018). Marine fungi: A source of potential anticancer compounds. Front. Microbiol., 8: 2536. https://doi.org/10.3389/fmicb.2017.02536.

Deshmukh, S.K., Takahashi, J.A. and Saxena, S. (2024). Fungi Bioactive Metabolites: Integration of Pharmaceutical Applications. Springer Nature Singapore, Singapore. https://doi.org/10.1007/978-981-99-5696-8.

Di Guardo, A., Trovato, F., Cantisani, C., Dattola, A., Nisticò, S.P. et al. (2025). Artificial intelligence in cosmetic formulation: Predictive modeling for safety, tolerability, and regulatory perspectives. Cosmetics, 12: 157. https://doi.org/10.3390/cosmetics12040157.

Elbandy, M. (2023). Anti-inflammatory effects of marine bioactive compounds and their potential as functional food ingredients in the prevention and treatment of neuroinflammatory disorders. Molecules, 28(1): 2. https://doi.org/10.3390/molecules28010002.

El-Kashef, D.H., Obidake, D.D., Schiedlauske, K., Deipenbrock, A., Scharf, S. et al. (2023). Indole diketopiperazine alkaloids from the marine sediment-derived fungus *Aspergillus chevalieri* against pancreatic ductal adenocarcinoma. Mar. Drugs, 22(5). https://doi.org/10.3390/md22010005.

Elsayis, A., Hassan, S.W.M., Ghanem, K.M. and Khairy, H. (2022). Suggested sustainable medical and environmental uses of melanin pigment from halotolerant black yeast *Hortaea werneckii* AS1. Front. Microbiol., 13. https://doi.org/10.3389/fmicb.2022.871394.

Fan, H., Wang, L., Zhang, Z.K., Wu, P.P., He, Y.P. et al. (2023). Bioactive Aspergteroids G–J from soft-coral-associated symbiotic and epiphytic fungus *Aspergillus terreus* EGF7-0-1. Bioeng., 10: 805. https://doi.org/10.3390/bioengineering10070805.

Fan, Y., Jiang, C., Li, P., Wang, C. and Chen, H. (2023). A new chloro-azaphilone derivative with pro-angiogenesis activity from the hadal trench-derived fungus Chaetomium globosum YP-106. J. Oceanol. Limnol., 41: 1145–1151. https://doi.org/10.1007/s00343-022-2017-1.

Fan, Y.Q., Li, P.H., Chao, Y.X., Chen, H., Du, N. et al. (2015). Alkaloids with cardiovascular effects from the marine-derived fungus *Penicillium expansum* Y32. Mar. Drugs, 13(10): 6489–6504. https://doi.org/10.3390/md13106489.

Feng, T., Wu, R., Wang, Y., Wang, P., Zhou, L. et al. (2024). Proangiogenic azaphilones from the marine-derived fungus *Neopestalotiopsis* sp. HN-1-6. Mar. Drugs, 22(6): 241. https://doi.org/10.3390/md22060241.

Festa, M., Sansone, C., Brunet, C., Crocetta, F., Di Paola, L. et al. (2020). Cardiovascular active peptides of marine origin with ACE inhibitory activities: Potential role as anti-hypertensive drugs and in prevention of SARS-CoV-2 infection. IJMS, 21: 8364. https://doi.org/10.3390/ijms21218364.

Gaignard, C., Macao, V., Gardarin, C., Rihouey, C., Picton, L. et al. (2018). The red microalga Flintiella sanguinaria as a new exopolysaccharide producer. J. Appl. Phycol., 30: 2803–2814. https://doi.org/10.1007/s10811-018-1389-2.

Gao, C., Bao, B., Bao, C. and Wu, W. (2023). Fungi fibrinolytic compound 1 plays a core role in modulating fibrinolysis, altering plasma clot structure, and promoting susceptibility to lysis. Pharm., 15(9): 2320. https://doi.org/10.3390/pharmaceutics15092320.

Gao, C., Tang, S., Zhang, H., Zhang, H., Zhang, T. et al. (2022). A Novel marine pyran-isoindolone compound enhances fibrin lysis mediated by single-chain urokinase-type plasminogen activator. Mar. Drugs, 20(8): 495. https://doi.org/10.3390/md20080495.

Geng, H., Chen, M., Guo, C., Wang, W. and Chen, D. (2024). Marine polysaccharides: Biological activities and applications in drug delivery systems. Carbohydr. Res., 538: 109071. https://doi.org/10.1016/j.carres.2024.109071.

González-Becerra, K., Ramos-Lopez, O., Barrón-Cabrera, E., Riezu-Boj, J.I., Milagro, F.I. et al. (2019). Fatty acids, epigenetic mechanisms and chronic diseases: A systematic review. Lipids Health Dis., 18: 178. https://doi.org/10.1186/s12944-019-1120-6.

Guan, X., Li, Y., Guan, Xiaodan, Fan, L. and Ying, J. (2024). XYA-2: A marine-derived compound targeting apoptosis and multiple signaling pathways in pancreatic cancer. PeerJ., 12: e16805. https://doi.org/10.7717/peerj.16805.

Hafez Ghoran, S., Taktaz, F., Sousa, E., Fernandes, C. and Kijjoa, A. (2023). Peptides from marine-derived fungi: Chemistry and biological activities . Mar. Drugs, 21(10): 510. https://doi.org/10.3390/md21100510.

Hasan, S., Ansari, M.I., Ahmad, A. and Mishra, M. (2015). Major bioactive metabolites from marine fungi: A review. Bioinformation, 11(4): 176–181. https://doi.org/10.6026/97320630011176.

Hassan, M.G., Elmezain, W.A., Baraka, D.M., AboElmaaty, S.A., Elhassanein, A. et al. (2024). Anti-cancer and anti oxidant bioactive metabolites from *Aspergillus fumigatus* WA7S6 isolated from marine sources: *in vitro* and *in silico* studies. Microorganisms, 12(1): 127. https://doi.org/10.3390/microorganisms12010127.

He, J. and Jia, Y. (2022). Application of omics technologies in dermatological research and skin management. J. Cosmet. Dermatol., 21: 451–460. https://doi.org/10.1111/jocd.14100.

He, X., Jin, Y., Kong, F., Yang, L., Zhu, M. et al. (2023). Discovery, antitumor activity, and fermentation optimization of roquefortines from *Penicillium* sp. OUCMDZ-1435. Molecules, 28(7): 3180. https://doi.org/10.3390/molecules28073180.

Jiang, M., Wu, Q., Guo, H., Lu, X., Chen, S. et al. (2023). Shikimate-derived meroterpenoids from the ascidian-derived fungus *Amphichorda felina* SYSU-MS7908 and their anti-glioma activity. J. Nat. Prod., 86(12): 2651–2660. https://doi.org/10.1021/acs.jnatprod.3c00664.

Jones, E.B.G., Pang, K.L., Abdel-Wahab, M.A., Scholz, B., Hyde, K.D. et al. (2019). An online resource for marine fungi. Fungal Diver., 96: 347–433. https://doi.org/10.1007/s13225-019-00426-5.

Khan, M.A., Hashim, M.J., Mustafa, H., Baniyas, M.Y., Al Suwaidi, S.K.B.M. et al. (2020). Global epidemiology of ischemic heart disease: Results from the global burden of disease study. Cureus, 12(7): e9349. https://doi.org/10.7759/cureus.9349.

Kim, J. (2025). Smart nanocarriers in cosmeceuticals through advanced delivery systems. Biomimetics, 10(4): 217. https://doi.org/10.3390/biomimetics10040217.

Krasnopolskaya, L.M., Shuktueva, M.I., Avtonomova, A.V., Yarina, M.S., Dzhavakhyan, B.R. et al. (2016). Antitumor and antioxidant properties of water-soluble polysaccharides from submerged mycelium of *Flammulina velutipes*. Antibiot. Khimioter., 61(11-12): 16–20.

Krasnopolskaya, L.M., Yarina, M.S., Avtonomova, A.V., Usov, A.I., Isakova, E.B. et al. (2015). Antitumor activity of polysaccharides from *Ganoderma lucidum* mycelium: *in vivo* comparative study. Antibiot. Khimioter., 60: 29–34.

Laurienzo, P. (2010). Marine polysaccharides in pharmaceutical applications: An overview. Mar. Drugs, 8: 2435–2465. https://doi.org/10.3390/md8092435.

Leshchenko, E.V., Berdyshev, D.V., Yurchenko, E.A., Antonov, A.S., Borkunov, G.V. et al. (2023). Bioactive polyketides from the natural complex of the sea urchin-associated fungi *Penicillium sajarovii* KMM 4718 and *Aspergillus protuberus* KMM 4747. Int. J. Mol. Sci., 24(23): 16568. https://doi.org/10.3390/ijms242316568.

Leyton, A., Flores, L., Mäki-Arvela, P., Lienqueo, M.E. and Shene, C. (2019). *Macrocystis pyrifera* source of nutrients for the production of carotenoids by a marine yeast *Rhodotorula mucilaginosa*. J. Appl. Microbiol., 127: 1069–1079. https://doi.org/10.1111/jam.14362.

Li, C.J., Chen, P.N., Li, H.J., Mahmud, T., Wu, D.L. et al. (2020). Potential antidiabetic fumiquinazoline alkaloids from the marine-derived fungus *Scedosporium apiospermum* F41-1. J. Nat. Prod., 83(4): 1082–1091. https://doi.org/10.1021/acs.jnatprod.9b01096.

Lin, W., Li, H., Wu, Z., Su, J., Zhang, Z. et al. (2022). Paspalines C–D and Paxillines B–D: New indole diterpenoids from *Penicillium brefeldianum* WZW-F-69. Mar. Drugs, 20(11): 684. https://doi.org/10.3390/md20110684.

Mapelli-Brahm, P., Gómez-Villegas, P., Gonda, M.L., León-Vaz, A., León, R. et al. (2023). Microalgae, seaweeds and aquatic bacteria, archaea, and yeasts: Sources of carotenoids with potential antioxidant and anti-inflammatory health-promoting actions in the sustainability era. Mar. Drugs, 21: 340. https://doi.org/10.3390/md21060340.

Nagahawatta, D.P., Liyanage, N.M., Jayawardena, T.U. and Jeon, Y.J. (2024). Marine polyphenols in cardiovascular health: Unraveling structure–activity relationships, mechanisms, and therapeutic implications. IJMS, 25: 8419. https://doi.org/10.3390/ijms25158419.

Nash, K.L., Van Putten, I., Alexander, K.A., Bettiol, S., Cvitanovic, C. et al. (2022). Oceans and society: Feedbacks between ocean and human health. Rev. Fish Biol. Fisheries, 32: 161–187. https://doi.org/10.1007/s11160-021-09669-5.

Nguyen, V.D. and Pham, T.T., (2022). *Penicillium vietnamense* sp. nov., the first novel marine fungi species described from Vietnam with a unique conidiophore structure and molecular phylogeny of *Penicillium* section *Charlesia*. Mycobiology, 50: 155–165. https://doi.org/10.1080/12298093.2022.2068750.

Ninkuu, V., Zhang, L., Yan, J., Fu, Z., Yang, T. et al. (2021). Biochemistry of terpenes and recent advances in plant protection. Int. J. Mol. Sci., 22: 5710. https://doi.org/10.3390/ijms22115710.

Pan, N., Li, Z.C., Li, Z.H., Chen, S.H., Jiang, M.H. et al. (2022). Antiplatelet and antithrombotic effects of Isaridin E isolated from the marine-derived fungus via downregulating the PI3K/Akt signaling pathway. Mar. Drugs, 20(1): 23. https://doi.org/10.3390/md20010023.

Papikinou, M.A., Pavlidis, K., Cholidis, P., Kranas, D., Adamantidi, T. et al. (2024). Marine fungi bioactives with anti-inflammatory, antithrombotic, and antioxidant health-promoting properties against inflammation-related chronic diseases. Mar. Drugs, 22: 520. https://doi.org/10.3390/md22110520.

Patil, N.P., Le, V., Sligar, A.D., Mei, L., Chavarria, D. et al. (2018). Algal polysaccharides as therapeutic agents for atherosclerosis. Front. Cardiovasc. Med., 5: 153. https://doi.org/10.3389/fcvm.2018.00153.

Phukhamsakda, C., Nilsson, R.H., Bhunjun, C.S., De Farias, A.R.G., Sun, Y.R. et al. (2022). The numbers of fungi: Contributions from traditional taxonomic studies and challenges of metabarcoding. Fungal Diver., 114 : 327–386. https://doi.org/10.1007/s13225-022-00502-3.

Poli, A., Bovio, E., Ranieri, L., Varese, G.C. and Prigione, V. (2020). News from the Sea: A new genus and seven new species in the pleosporalean families roussoellaceae and thyridariaceae. Divers., 12: 144. https://doi.org/10.3390/d12040144.

Roth, G.A., Mensah, G.A., Johnson, C.O., Addolorato, G., Ammirati, E. et al. (2020). Global burden of cardiovascular diseases and risk factors, 1990–2019. J. Am. Coll. Cardiol., 76: 2982–3021. https://doi.org/10.1016/j.jacc.2020.11.010.

Sarkar, A., Philip, A.M., Thakker, D.P. and Bhaskara Rao, K.V. (2024). Extraction and characterization of beta-glucan methyl esters derived from UV-mutated marine yeast and their applications in cosmetics. Thalassas, 40: 721–733. https://doi.org/10.1007/s41208-024-00661-x.

Shahidi, F. and Saeid, A. (2025). Bioactivity of marine-derived peptides and proteins: A review. Mar. Drugs, 23: 157. https://doi.org/10.3390/md23040157.

Shao, Z., Tian, Y., Liu, S., Chu, X. and Mao, W. (2023). Anti-Diabetic activity of a novel exopolysaccharide produced by the mangrove endophytic fungus *Penicillium janthinellum* N29. Mar. Drugs, 21(5): 270. https://doi.org/10.3390/md21050270.

Shi, L. and Tu, B.P. (2015). Acetyl-CoA and the regulation of metabolism: Mechanisms and consequences. Curr. Opin. Cell Biol., 33: 125–131. https://doi.org/10.1016/j.ceb.2015.02.003.

Shin, H.J. (2020). Natural products from marine fungi. Mar. Drugs, 18: 230. https://doi.org/10.3390/md18050230.

Silber, J., Kramer, A., Labes, A. and Tasdemir, D. (2016). From discovery to production: Biotechnology of marine fungi for the production of new antibiotics. Mar. Drugs, 14: 137. https://doi.org/10.3390/md14070137.

Soehnlein, O. and Libby, P. (2021). Targeting inflammation in atherosclerosis: From experimental insights to the clinic. Nat. Rev. Drug. Discov., 20: 589–610. https://doi.org/10.1038/s41573-021-00198-1.

Sogin, M.L., Morrison, H.G., Huber, J.A., Welch, D.M., Huse, S.M. et al. (2006). Microbial diversity in the deep sea and the underexplored rare biosphere. Proc. Natl. Acad. Sci. USA, 103: 12115–12120. https://doi.org/10.1073/pnas.0605127103.

Song, R., Hu, M., Qin, X., Qiu, L., Wang, P. et al. (2023). The roles of lipid metabolism in the pathogenesis of chronic diseases in the elderly. Nutrients, 15: 3433. https://doi.org/10.3390/nu15153433.

Su, J., Chang, C., Xiang, Q., Zhou, Z.-W., Luo, R. et al. (2014). Xyloketal B, a marine compound, acts on a network of molecular proteins, and regulates the activity and expression of rat cytochrome p450

3a: A bioinformatic and animal study. Drug Des, Devel. Ther., 2014(8): 2555–2602. https://doi.org/10.2147/DDDT.S73476.

Trinh, P.T.H., Yurchenko, A.N., Khmel, O.O., Dieu, T.V.T., Ngoc, N.T.D. et al. (2022). Cytoprotective polyketides from sponge-derived fungus *Lopadostoma pouzarii*. Molecules, 27(21): 7650. https://doi.org/10.3390/molecules27217650.

Tsoupras, A., Adamantidi, T., Finos, M.A., Philippopoulos, A., Detopoulou, P. et al. (2024). Re-assessing the role of platelet activating factor and its inflammatory signaling and inhibitors in cancer and anti-cancer strategies. Front. Biosci., 29: 345. https://doi.org/10.31083/j.fbl2910345.

Tsoupras, A., Brummell, C., Kealy, C., Vitkaitis, K., Redfern, S. et al. (2022). Cardio-protective properties and health benefits of fish lipid bioactives: The effects of thermal processing. Mar. Drugs, 20: 187. https://doi.org/10.3390/md20030187.

Tsoupras, A., Lordan, R. and Zabetakis, I. (2018). Inflammation, not cholesterol, is a cause of chronic disease. Nutrients, 10: 604. https://doi.org/10.3390/nu10050604.

Uras, I.S., Karsli, B., Konuklugil, B., Ocsoy, I. and Demirbas, A. (2023). Organic–inorganic nanocomposites of *Aspergillus terreus* extract and its compounds with antimicrobial properties. Sustainability, 15: 4638. https://doi.org/10.3390/su15054638.

Vargas-Gastélum, L. and Riquelme, M. (2020). The mycobiota of the deep sea: What omics can offer. Life, 10: 292. https://doi.org/10.3390/life10110292.

Vidal, I., Torres-Vargas, J.A., Sánchez, J.M., Trigal, M., García-Caballero, M. et al. (2023). Danthron, an anthraquinone isolated from a marine fungus, is a new inhibitor of angiogenesis exhibiting interesting antitumor and antioxidant properties. Antioxidants, 12(5): 1101. https://doi.org/10.3390/antiox12051101.

Vitale, G.A., Coppola, D., Esposito, F.P., Buonocore, C., Ausuri, J. et al. (2020). Antioxidant molecules from marine fungi: Methodologies and perspectives. Antioxidants, 9(12): 1183. https://doi.org/10.3390/antiox9121183.

Wang, C., Guo, L., Hao, J., Wang, L. and Zhu, W. (2016). α-Glucosidase inhibitors from the marine-derived fungus *Aspergillus flavipes* HN4-13. J. Nat. Prod., 79(11): 2977–2981. https://doi.org/10.1021/acs.jnatprod.6b00766.

Wang, C., Mao, W., Chen, Z., Zhu, W., Chen, Y. et al. (2013). Purification, structural characterization and antioxidant property of an extracellular polysaccharide from *Aspergillus terreus*. Process Biochem., 48: 1395–1401. https://doi.org/10.1016/j.procbio.2013.06.029.

Wang, W., Shi, Y., Liu, Y., Zhang, Y., Wu, J. et al. (2022). Brasilterpenes A–E, bergamotane sesquiterpenoid derivatives with hypoglycemic activity from the deep sea-derived fungus *Paraconiothyrium brasiliense* HDN15-135. Mar. Drugs, 20(5): 338. https://doi.org/10.3390/md20050338.

Wijayawardene, N.N., Boonyuen, N., Ranaweera, C.B., de Zoysa, H.K.S., Padmathilake, R.E. et al. (2023). OMICS and other advanced technologies in mycological applications. J. Fungi (Basel), 9: 688. https://doi.org/10.3390/jof9060688.

Woodward, M. (2019). Cardiovascular disease and the female disadvantage. IJERPH, 16: 1165. https://doi.org/10.3390/ijerph16071165.

Wu, P., Yu, S., Wang, J., Zou, S., Yao, D.S. et al. (2023). Global burden, trends, and inequalities of ischemic heart disease among young adults from 1990 to 2019: A population-based study. Front. Cardiovasc. Med., 10: 1274663. https://doi.org/10.3389/fcvm.2023.1274663.

Xiang, Z., Han, M. and Zhang, H. (2023). Erratum to: Nanomaterials based flexible devices for monitoring and treatment of cardiovascular diseases (CVDs). Nano Res., 16: 8051–8051. https://doi.org/10.1007/s12274-022-5342-y.

Youssef, F.S., Ashour, M.L., Singab, A.N.B. and Wink, M. (2019). A Comprehensive review of bioactive peptides from marine fungi and their biological significance. Mar. Drugs, 17: 559. https://doi.org/10.3390/md17100559.

Zhang, Y., Wang, X., Liu, T., Zhang, Z.Y., Song, W.G. et al. (2024). Exserolide J ameliorates lipid accumulation *in vitro* by regulating liver X receptor alpha and peroxisome proliferator-activated receptor alpha proteins. Heliyon, 10: e31861. https://doi.org/10.1016/j.heliyon.2024.e31861.

Zhang, Z.F., Liu, F., Liu, L.R., Li, M., Cai, L. et al. (2024). Culturing the uncultured marine fungi in the omics age: Opportunities and challenges. Fungal Biol. Rev., 48: 100353. https://doi.org/10.1016/j.fbr.2023.100353.

Zhao, L., Li, J., Huang, X., Wang, G.H., Lv, X. et al. (2018). Xyloketal B exerts antihypertensive effect in renovascular hypertensive rats via the NO-sGC-cGMP pathway and calcium signaling. Acta Pharmacol. Sin., 39: 875–884. https://doi.org/10.1038/aps.2018.12.

Zhao, L., Lin, X., Fu, J., Zhang, J., Tang, W. and He, Z. (2022). A novel Bi-functional fibrinolytic enzyme with anticoagulant and thrombolytic activities from a marine-derived fungus *Aspergillus versicolor* ZLH-1. Mar. Drugs, 20(6): 356. https://doi.org/10.3390/md20060356.

Zhao, Y., Guo, L., Xia, Y., Zhuang, X. and Chu, W. (2019). Isolation, identification of carotenoid-producing *Rhodotorula* sp. from marine environment and optimization for carotenoid production. Mar. Drugs, 17: 161. https://doi.org/10.3390/md17030161.

Zhou, Y., Chen, R., Liu, D., Wu, C., Guo, P. et al. (2017). Asperlin inhibits LPS-evoked foam cell formation and prevents atherosclerosis in ApoE−/− mice. Mar. Drugs, 15(11): 358. https://doi.org/10.3390/md15110358.

8

Potential of Marine-Associated Fungi for Biotechnology, Medicine, and Ecosystem Health

Vinit Kumar

Marine fungi represent a fascinating biological frontier where environmental stresses have engineered unprecedented biochemical innovations. These microorganisms, which inhabit environments from oxygen-limited deep-sea sediments to temperature-rich hydrothermal vents of more than 100°C, have evolved new metabolic architectures that defy conventional notions of fungal biochemistry. Recent whole-genome sequencing suggests that marine fungi contain 60–67 biosynthetic gene clusters per species, 98% of which are cryptic under laboratory conditions, orders of magnitude greater than those on land in chemical space. This reservoir has already yielded game-changing therapeutics like plinabulin (Phase III trials), turbinmicin (active against pandrug-resistant *Candida auris*), and 16-epiascomylactam B (MIC: 0.20–0.80 μg/mL against multidrug-resistant pathogens). Outside of medicine, marine fungi are keystone species of ocean biogeochemical function, providing quantifiable contributions to the biological carbon pump and nitrogen-cycling mechanisms hitherto reserved for prokaryotes. Additional stress due to climate change provides the topic with added significance: in a background of mounting warming and ocean acidification, stress-response processes of these organisms increasingly

State Key Laboratory Breeding Base of Green Pesticide and Agricultural Bioengineering, Guizhou University, Guizhou, China.
Email: vinitk56@gmail.com

DOI: 10.1201/9781003768708-11

determine ecosystem stability and the development of climate-resilient biotechnologies. Synergism of AI-gene mining with culture-independent technologies has increased discovery rates 100-fold and positioned marine mycology to address humanity's most significant challenges, ranging from the antimicrobial resistance pandemic costing 1.27 million lives annually to green carbon sequestration and utilization technologies. This synthesis argues that marine fungi are not only untapped biodiversity but also natural biological solutions already optimized by evolution for global-scale challenges.

1. Introduction

Marine-associated fungi represent an evolutionary frontier, a kingdom of life that has conquered Earth's most challenging aquatic ecosystems (Bahram and Netherwa, 2022). Thriving under conditions of immense pressure, extreme thermal gradients near hydrothermal vents, and fluctuating salinity, these organisms have been forged by environmental forces fundamentally different from those on land. This intense selective pressure has driven the evolution of unique metabolic architectures and novel biochemical pathways, creating a vast reservoir of untapped biological potential (Dick, 2019). Recent genomic analyses confirm this, revealing that marine fungal species harbor an average of 60–67 biosynthetic gene clusters (BGCs), the genetic blueprints for natural products, a density that dwarfs that of their terrestrial relatives (Han et al., 2024).

A significant gap between this genomic wealth of promise and our ability to understand and tap into it. A potential up to 98% of all BGCs is generally silent or "cryptic" under normal laboratory cultivation. This is reinforced by a "cultivation paradox" where less than 5% of marine fungi are accessible for study using standard techniques, hence their ecological function and chemical potential are largely unrevealed (Oppong-Danquah et al., 2020). The distance between what we currently know in their DNA and what we can make or quantify today is the key research challenge in modern marine mycology.

This chapter presents a general overview of the novel developments bridging this gap. Through the integration of evidence from culture-independent genomics, artificial intelligence-supported biodiscovery, and functional ecology, we will explore the diverse promise of these amazing organisms. We will talk about their role as a source of next-generation drugs to combat the antimicrobial resistance pandemic, their use in industrial biotechnology through the production of robust enzymes, and their role as keystone species that enable ecosystem health and climate resilience. Lastly, this chapter concludes that marine fungi are not only a scientific interest, but a source of evolutionary solutions ready to address some of the science and societal problems of the 21st century.

2. Functional Definition of Marine Fungi

The definition of marine fungi itself has radically shifted, reflecting advances in molecular ecology, culture-independent methodologies, and systems biology approaches. Traditional definitions, beginning with Barghoorn and Linder (1944), depended on only one criterion: physiological salt tolerance, growth in seawater. The prevailing Kohlmeyer and Kohlmeyer (1979) paradigm set the obligate-facultative dichotomy: obligate marine fungi as being restricted to growth and sporulation in marine environments alone, as opposed to facultative marine terrestrial or freshwater fungi with marine-colonizing potential (Cunliffe, 2023). However, even this binary model has been unable to fully capture the reality of marine fungal ecology. Culture-independent molecular surveys show that a significant percentage of fungi found in the marine environment are taxonomically linked to non-marine taxa but have active gene expression and environmental correlations suggestive of actual marine activity and not passive dispersal (Richards et al., 2015; Amend et al., 2019). Phylogenetic analyses demonstrate that marine lineages have evolved from terrestrial ancestors through multiple independent transitions, creating a continuum of marine adaptation rather than discrete categories (Ren and Wang, 2022). Contemporary challenges with traditional definitions include: (1) taxonomic bias toward culturable species representing less than 5% of environmental diversity, (2) genomic plasticity enabling rapid ecological transitions between terrestrial and marine lifestyles, (3) functional redundancy where morphologically distinct taxa perform similar ecological roles, and (4) temporal variability in habitat occupancy reflecting life cycle complexity and environmental fluctuations.

2.1 *Proposed Integrative Definition*

Based on a comprehensive analysis of marine fungal ecology, evolution, and functional diversity, we propose an integrative functional definition that transcends traditional taxonomic boundaries:

"*Marine fungi comprise a functionally diverse ecological assemblage of organisms that demonstrate quantifiable ecological integration within marine ecosystems through one or more of the following criteria*:

(1) Active metabolic participation in marine biogeochemical cycles evidenced by measurable enzyme activity, substrate processing, or biomarker production.

(2) Reproductive success under marine conditions, including spore production, germination, or vegetative propagation in seawater environments.

(3) Evolutionary adaptation by genomic imprint of marine-specific gene regulation, osmotic stress response systems, or specialized metabolic pathways.

(4) Ecological function by describing symbiotic relationships, pathogenic interactions, or mutualistic associations with marine organisms.

(5) Environmental response patterns by statistical correlations with marine abiotic variables and temporal persistence in marine habitats outside passive dispersal timescales.

(6) Marine-diagnostic molecular signatures, including marine-specific biosynthetic gene clusters (BGCs) and characteristic secondary metabolite profiles that distinguish them from their terrestrial counterparts (Han et al., 2024; Shao et al., 2020).

(7) Functional integration within marine microbial communities as indicated by network centrality analysis in marine food webs and detectable ecological interactions with other marine microorganisms (Chen et al., 2024).

(8) Quantifiable contribution to marine ecosystem services, for example, carbon sequestration, nutrient cycling, organic matter decomposition, or marine habitat formation processes (Breyer et al., 2022; Lechtenfeld et al., 2015; Naselli-Flores and Padisák, 2023; Ng et al., 2025; Perkins et al., 2023).

(9) Adaptive responses to marine climate conditions, i.e., tolerance and physiological optimization along temperature, pH, salinity, and oxygen gradients characteristic for marine environments (Ahumada-Rudolph et al., 2018; Kumar et al., 2021; Lobo et al., 2024).

(10) Integrated omics signatures comprising genomic, transcriptomic, proteomic, and metabolomic information showing marine specialization and concerted metabolic adaptation at multiple biological levels (Breyer et al., 2022; Zhang et al., 2022a).

This definition recognizes marine fungi as ecological specialists, rather than taxonomic entities, with functional integration being emphasized over morphology. It encompasses the whole spectrum from extremely specialized obligate marine species with unique morphological specializations (appendaged spores, pressure-resistant enzymes) to ecologically versatile facultative species with active populations in both terrestrial and marine environments. Significantly, it requires evidence of active ecological participation and not merely detection, distinguishing functionally integrated marine fungi from inactive propagules or DNA remnants.

The revised framework addresses important gaps in previous definitions by being inclusive of quantifiable molecular markers, network-based ecological metrics, and multi-omics integration approaches that reflect the state of systems biology. Unlike the Pang et al. (2016) definition, which emphasized broad physiological criteria, this framework provides exact, measurable parameters for the identification of marine fungi that can be standardized across research laboratories as well as environmental monitoring initiatives.

Justification for this integrative approach stems from several key observations:

(1) Molecular evidence reveals active transcription of marine-specific genes in supposedly "terrestrial" taxa found in marine environments.

(2) Experimental studies demonstrate rapid evolutionary adaptation to marine conditions in laboratory settings.

(3) Ecological data show that fungal community composition correlates more strongly with environmental function than taxonomic identity.

(4) Genomic analyses reveal that marine adaptation involves regulatory evolution rather than wholesale genomic restructuring, enabling rapid ecological transitions.

(5) Multi-omics studies demonstrate coordinated molecular responses across transcriptomic, proteomic, and metabolomic levels that distinguish marine-adapted from terrestrial fungi.

This functional framework encompasses over 2,149 documented marine fungal species across eleven phyla, while acknowledging that environmental DNA surveys suggest > 50,000 additional cryptic species awaiting discovery. The definition provides operational criteria for distinguishing marine fungi from terrestrial contaminants while recognizing the ecological continuum that characterizes fungal adaptation to aquatic environments. Furthermore, the inclusion of ecosystem services and climate adaptation criteria positions this definition to address contemporary challenges in marine conservation, climate change mitigation, and sustainable biotechnology development.

3. Diversity of Marine Fungal Habitats

Marine fungal habitats exhibit extraordinary diversity across spatial and temporal scales. Deep-sea sediments harbor thermophilic fungi adapted to geothermal heating, with species like *Calcarisporium* sp., KF525 containing 60 unique BGCs and encoding novel secondary metabolites (Velez et al., 2022). Hydrothermal vent communities support extremophilic fungi that utilize sulfur compounds and heavy metals in their metabolism, producing compounds with unprecedented metal-binding capabilities. Coral reef ecosystems host complex fungal microbiomes that influence coral immunity through antimicrobial metabolite production and nutrient cycling (Zeng et al., 2021a). Mangrove forests support the highest documented marine fungal diversity, with endophytic communities producing over 300 characterized bioactive compounds, including the clinical candidate chevalone E (Fig. 1).

The historical neglect of marine mycology stems from fundamental methodological limitations (Cadamuro et al., 2021; Britto Martins de Oliveira et al., 2025). Early cultivation-dependent studies captured less than 1% of environmental fungal diversity, creating a profound "cultivation bias" that underestimated both species richness and metabolic potential (Schultz et al., 2023). The past decade has witnessed a methodological revolution through environmental DNA sequencing, single-cell genomics, and culture-independent metabolomics that has revealed marine fungi as quantitatively significant contributors to oceanic carbon cycling, nitrogen fixation, and biogeochemical processes previously attributed exclusively to bacteria and archaea.

This biochemical sophistication positions marine fungi as prime resources for addressing contemporary global challenges. The emerging crisis of antimicrobial resistance, with pandrug-resistant *Candida auris* being a worldwide pandemic, requires new chemicals with unprecedented mechanisms of action. Curbing climate change requires new carbon sequestration processes and stress-resistance biological systems. Sustainable biotechnology requires robust enzyme systems functioning in industrial conditions without sacrificing the environment. Marine fungi, which have evolved the solutions over geological timeframes, are nature's R&D (research and development) department for planet-scale challenges.

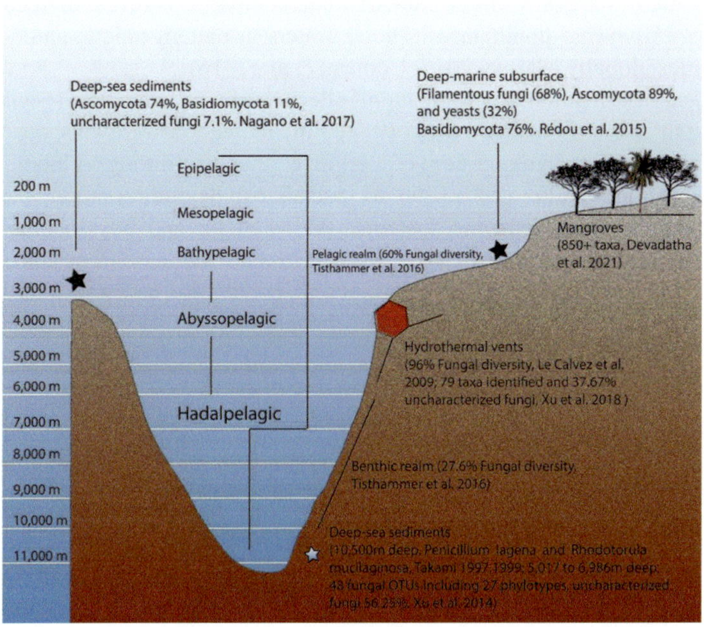

Fig. 1. Distribution of marine fungal diversity across ocean depth zones, showing sampling locations and fungal community composition percentages from deep-sea sediments, hydrothermal vents, mangroves, and subsurface environments. Ocean zones are labeled from epipelagic (surface) to hadalpelagic (deepest trenches), with fungal diversity data from multiple studies (2009–2021) (Multiple sources have been mentioned in the figure; background inspired from Raghukumar, 2017).

3.1 Taxonomic and Ecological Diversity

Marine fungal phylogenetic diversity includes all the major fungal phyla, dispelling previous assumptions of aquatic vs. terrestrial specialization (Naranjo-Ortiz and Gabaldón, 2019). Environmental DNA surveys suggest Ascomycota as the most abundant fungal phylum in marine ecosystems, particularly among culturable isolates. *Basidiomycota* typically comprise 10–30% of marine fungal sequences, while *Chytridiomycota* exhibit surprising abundance in polar marine environments, often emerging as the dominant phylum in Arctic and Antarctic waters, highlighting their unique ecological specialization in cold marine niches (Richards et al., 2015; Comeau et al., 2016; Gao et al., 2024). However, this apparent taxonomic simplicity masks extraordinary cryptic diversity: single mangrove trees harbor > 200 distinct fungal OTUs, deep-sea sediment cores contain > 400 phylotypes, and coastal metabarcoding studies reveal that 60% of marine fungal sequences represent potentially novel species.

3.2 Biogeographic Patterns and Habitat Specialization

Latitudinal diversity gradients in marine fungi follow complex patterns distinct from terrestrial systems. Tropical mangrove ecosystems support the highest documented diversity, with up to 38% of all known marine fungi originating from coastal regions. The Atlantic-Pacific fungal divide shows minimal overlap between ocean basins for many species, while Arctic-Antarctic comparisons reveal high endemism with *Chytridiomycota* dominance in polar waters, a pattern unique among fungal communities globally (Sarma, 2012; Comeau et al., 2016).

Vertical stratification in oceanic fungi reflects profound ecological specialization across depth zones. Epipelagic fungi (0–200 m) include fast-growing opportunists that process phytoplankton-derived organic matter during bloom cycles. Mesopelagic communities (200–1000 m) specialize in degrading sinking particulate matter through unique enzyme systems. Bathypelagic and abyssopelagic fungi (1000–6000 m) demonstrate extreme pressure adaptation with novel protein variants, while hadalpelagic taxa (> 6000 m) represent potentially endemic lineages with extraordinary biochemical innovations (Bochdansky et al., 2017; Li et al., 2019).

Habitat-specific adaptations reveal remarkable evolutionary convergence. Hydrothermal vent fungi express thermostable enzyme systems and heavy metal tolerance mechanisms, hypersaline lagoon species like *Hortaea werneckii* utilize specialized osmotic regulation pathways, and deep-sea sediment communities show unique substrate specificity for chemosynthetic bacterial necromass versus surface-derived organic matter (Pang et al., 2020; Plemenitaš, 2021; Zeng et al., 2021b).

3.3 Morphological and Physiological Adaptations

Marine fungi exhibit specialized morphological features that enable thriving in aquatic environments. Spore appendages represent the most distinctive adaptation: ribbon-like, filamentous, or polar appendages facilitate buoyancy control, substrate attachment, and dispersal in three-dimensional water columns (Gladfelter et al., 2019). Deliquescing asci in *Halosphaeriaceae* enable passive spore release in aquatic environments, while mucilaginous sheaths provide protection against osmotic stress and enhance particle entrapment. Physiological adaptations to marine conditions involve sophisticated molecular mechanisms (Susan et al., 1997). Osmotic regulation employs specialized transporters, compatible solute accumulation, and cell wall modifications that maintain turgor pressure under variable salinity. Pressure adaptation includes modified membrane compositions, pressure-resistant enzyme variants, and altered protein folding pathways. Temperature tolerance ranges from psychrophilic Antarctic species growing at – 2°C to thermophilic vent taxa surviving > 100°C (Plemenitaš et al., 2014).

3.4 Ecological Strategies and Functional Diversity

Marine fungi employ diversified ecological tactics that reflect aquatic adaptation. Endophytic marine fungi exhibit intimate tissue associations with hosts, which compose extremely modular communities wherein host phylogeny determines composition rather than geography. The creatures form defense chemicals to protect plants from illness as they acquire specialized nutrients that are not obtainable in free-living taxa (El-Bondkly et al., 2021; Zhu et al., 2023).

Epiphytic lifestyles encompass surface colonization with an impressive dispersal ability. Phylogeographic research demonstrates cosmopolitan species with identical genotypes across ocean basins, which implies active oceanic dispersal over passive transport. These fungi develop specialized attachment structures and hydrophobic spore coatings that facilitate prolonged survival in seawater (Fernández-Valero et al., 2023). Saprobic marine fungi act as ecosystem engineers, breaking down complex organic polymers that are recalcitrant to bacterial degradation. They exhibit unique carbohydrate-active enzyme inventories, including marine substrate-specific alginate lyases, fucoidan hydrolases, and chitin deacetylases. Stable isotope studies demonstrate these fungi contribute 15–25% of total organic matter processing in marine sediments despite representing $< 5\%$ of microbial biomass (Bonugli-Santos et al., 2015; Balabanova et al., 2018; Peng et al., 2024).

Symbiotic partnerships have evolved extraordinary complexity and functional integration. Coral-fungal associations involve fungi producing UV-protective mycosporine-like amino acids and antimicrobial compounds that enhance coral immunity (Roik et al., 2022; Mohamed et al., 2023). Sponge endosymbionts contribute up to 40% of host chemical defense metabolites. Algal partnerships include both parasitic and mutualistic relationships, with some fungi enhancing algal stress tolerance while others cause population crashes that influence marine food webs. Pathogenic marine fungi target diverse hosts across taxonomic groups (Taylor et al., 2007; Slaby et al., 2019). Coral diseases caused by *Aspergillus sydowii* and related taxa threaten reef ecosystem stability. Marine mammal infections by cosmopolitan genera like *Aspergillus*, *Candida*, and *Fusarium* suggest cross-ecosystem pathogen transmission. Fish and invertebrate pathogens, including *Lagenidium*, *Saprolegnia*, and *Exophiala* species, cause significant economic losses in aquaculture while serving as natural population regulators (Soler-Hurtado et al., 2016; Jones et al., 2019).

3.5 *Culture-Independent Revelations*

Next-generation sequencing approaches have revolutionized understanding of marine fungal diversity, revealing extensive "dark taxa" known only from environmental sequences. Time-series studies spanning 17 years demonstrate that marine fungal communities undergo predictable seasonal restructuring every 52 weeks, driven by temperature cycles and primary productivity patterns. Single-cell genomics has identified previously unknown lineages, including basal *Chytridiomycota* contributing significantly to carbon cycling and novel *Cryptomycota* parasitizing marine protists (Gautam et al., 2022; Naqvi et al., 2025). Functional gene analysis reveals metabolic capabilities absent from culture-based studies. Marine fungi express unique glycoside hydrolases for processing marine-specific polysaccharides, specialized transporters for trace metal acquisition, and novel secondary metabolite pathways triggered by marine environmental cues (Gonçalves et al., 2021; Peng et al., 2024). Metatranscriptomic studies show that marine fungi contribute significantly to oceanic enzyme activity, with implications for global biogeochemical cycling (Baltar et al., 2021; Gonçalves et al., 2021).

Table 1. Marine fungal biogeographical patterns and habitat specialization.

Habitat Type	Dominant Taxa	Unique Adaptation	Ecological Function	Estimated Species Richness	Sources
Tropical Mangroves	*Pestalotiopsis* and *Phyllosticta*	Appendaged spores and salt tolerance	Leaf litter decomposition	> 800 species	Liu et al. (2007); Sarma (2012)
Coral Reefs	*Aspergillus sydowii* and *Penicillium*	UV protection and antimicrobials	Pathogen/ commensal	> 200 species	Roik et al. (2022)
Deep-sea Sediments	*Emericella* and *Calcarisporium*	Pressure adaptation and psychrophily	Organic matter processing	> 400 OTUs	Li et al. (2016); Luo et al. (2020)
Hydrothermal Vents	Thermophilic Ascomycota	Heat shock proteins and metal tolerance	Extremophile metabolism	> 100 species	Xu et al. (2014); Tang et al. (2020)
Polar Waters	*Chytridiomycota* dominant	Antifreeze proteins and lipid modifications	Primary production cycling	> 300 OTUs	Kumar et al. (2021)
Hypersaline Lagoons	*Hortaea werneckii* and *Wallemia*	Osmotic regulation and melanin production	Halotolerant decomposition	> 50 species	Cantrell et al. (2006); Chen et al. (2020)
Pelagic Waters	*Malassezia*-like and planktonic fungi	Buoyancy controls and dispersal mechanisms	Carbon pump participation	> 500 OTUs	Taylor and Cunliffe (2016); Vass et al. (2022)

4. Bioactive Metabolites

Marine fungi represent Earth's most prolific producers of structurally complex secondary metabolites, with genomic analyses revealing biosynthetic capabilities that dwarf those of terrestrial counterparts (Kumar et al., 2018). Comprehensive surveys document 37,542 marine natural products by 2024, with marine fungi contributing > 7,000 unique structures representing 63.8% of all marine microbial metabolites (Carroll et al., 2020; Shin, 2020). This chemical diversity stems from evolutionary pressures unique to marine environments: intense interspecies competition, variable nutrient availability, osmotic stress, and chemical communication across three-dimensional water columns.

Polyketides dominate marine fungal secondary metabolite production, comprising 41.7% of newly discovered compounds and exhibiting extraordinary structural diversity. Recent discoveries exemplify this complexity: Harzianolides A–F from deep-sea *Trichoderma harzianum* ZN-4 feature unprecedented tetramic acid-polyketide hybrid structures with potent antifungal activity (MIC: 25−100 μg/mL against *Phytophthora theae*). Dicitrinols A–C from *Penicillium citrinum* TW132−59 represent the first citrinin derivatives with unique 6/5/7/5 tetracyclic cores, demonstrating remarkable antifungal potency (MIC: 4−16 μg/mL) that surpasses commercial fungicides (Zhou et al., 2025).

The penicisteckins, North Sea *Penicillium steckii* atropo-diastereomeric dimers, illustrate how the marine environment stimulates structural innovation. They exhibit excellent activity against methicillin-resistant *Staphylococcus aureus* (MIC: 4.0 μg/mL) through unprecedented membrane disruption mechanisms based on stereospecific lipid binding. X-ray crystallography reveals conformational flexibility to enable adaptation to different bacterial membrane compositions (Huang et al., 2025).

Alkaloids constitute 19.4% of new marine fungal compounds and demonstrate remarkable pharmacological versatility. The talarohydrazones A-D from Talaromyces amestolkiae feature phenylhydrazone scaffolds unknown in terrestrial fungi, showing broad-spectrum antibacterial activity through DNA gyrase inhibition. Most remarkably, 16-epiascomylactam B from deep-sea Microascus sp. SCSIO41821 exhibits extraordinary potency against bacterial pathogens (MIC: 0.20−0.80 μg/mL), approaching the activity levels of last-resort antibiotics while maintaining selectivity for pathogenic over commensal bacteria (Willems et al., 2020; Wu et al., 2024). Peptides and depsipeptides constitute 22.2% of marine fungal metabolites and are found to contain minimal mammalian toxicity with strong bioactivity. Simplicpeptaibs A-K are new peptaibiotic structures with distinctive amino acid sequences that display selective activity towards plant and fish diseases without harming beneficial microorganisms. These compounds exert membrane-active modes that are different from traditional antibiotics and are thus potential candidates to treat resistant pathogens (Youssef et al., 2019). Terpenoids from marine fungi (12.5% of novel compounds) have unique structural characteristics that are initiated by marine-specific biosynthetic modification.

Terpenoids from marine fungi (12.5% of new compounds) feature unique structural characteristics driven by marine-specific biosynthetic modifications. Sesterchaetins A-B from *Chaetomium globosum* SD-347 represent rare marine sesterterpenes with anti-*Vibrio* activity (MIC: 8.0−16 μg/mL), while trichoderenes A-D from *Trichoderma effusum* demonstrate broad antimicrobial activity through cell wall targeting mechanisms (Liu et al., 2024). Xanthones from marine sources exhibit superior bioactivity compared to terrestrial analogs. Dimeric xanthones from *Diaporthe* sp. SYSU-MS4722 show remarkable anti-inflammatory activity (IC50: 6.3−8.0 μM) that exceeds indomethacin by 5-fold, while ravenelin from *Setosphaeria rostrata* demonstrates significant anti-inflammatory effects (IC50: 6.27 μM) through selective COX-2 inhibition (Chen et al., 2022; Koopklang et al., 2024).

The true biosynthetic potential of marine fungi lies in their cryptic biosynthetic gene clusters, genomically encoded pathways that remain transcriptionally silent under laboratory conditions. Whole-genome sequencing of marine isolates reveals extraordinary genetic potential: *Calcarisporium* sp. KF525 harbors 60 BGCs (98% novel), while *Pestalotiopsis* sp. KF079 contains 67 BGCs (97% novel). RNA-seq analysis indicates that < 10% of these clusters are expressed under standard cultivation, suggesting vast untapped chemical diversity (Chen et al., 2022; Koopklang et al., 2024). Innovative BGC activation strategies are revolutionizing natural product discovery. OSMAC (One Strain Many Compounds) approaches systematically vary cultivation parameters, pH, salinity, temperature, and nutrients to trigger silent pathways. Chemical epigenetic modifications using HDAC inhibitors (trichostatin A, SAHA) and DNA methyltransferase inhibitors (5-azacytidine) activate chromatin-silenced BGCs. Co-cultivation strategies with marine bacteria or competing fungi mimic natural ecological triggers that activate defensive metabolite production (Romano et al., 2018). Heterologous expression systems enable production of complex marine fungal metabolites in tractable hosts. CRISPR-mediated BGC activation through targeted transcription factor overexpression has yielded novel compounds, including the alternapyrones, while synthetic biology approaches reconstruct marine fungal pathways in engineered *Saccharomyces cerevisiae* or *Aspergillus nidulans* for scalable production (Xu et al., 2022) (Table 2).

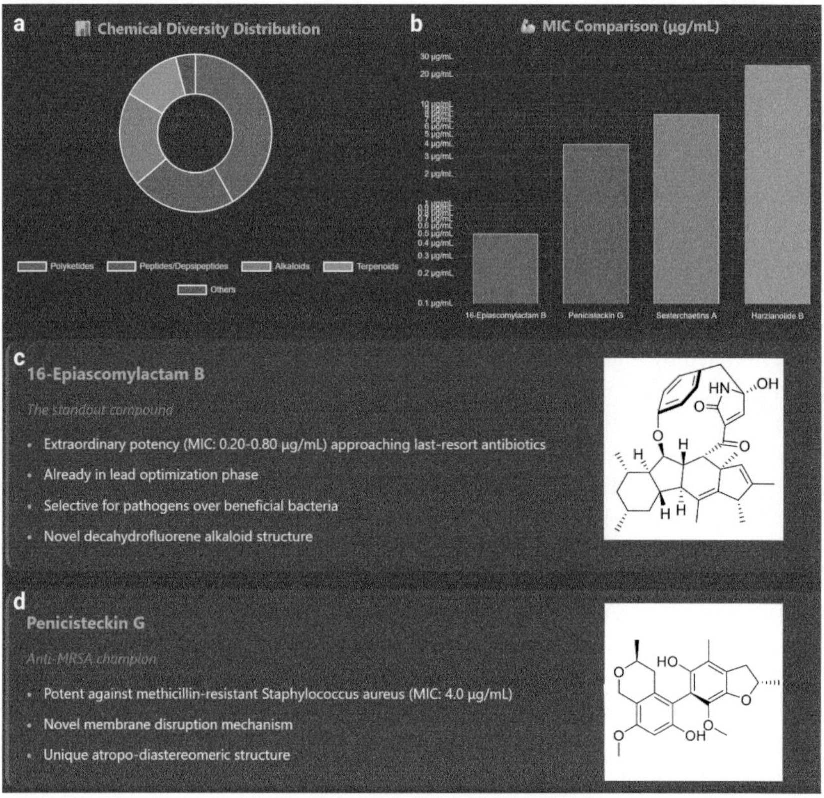

Fig. 2. Marine fungal bioactive metabolites: Chemical diversity and lead compound profiles.
(a) Chemical diversity distribution showing polyketides as the dominant class (41.7%), followed by peptides/depsipeptides (22.2%), alkaloids (19.4%), terpenoids (12.5%), and others (4.2%). (b) MIC comparison of lead compounds demonstrating exceptional potency of 16-epiascomylactam B (0.5 μg/mL average) compared to penicisteckin G (4.0 μg/mL), sesterchaetins A (8.0 μg/mL), and harzianolide B (25.0 μg/mL). (c) Structure and bioactivity profile of 16-epiascomylactam B, the standout compound with novel decahydrofluorene scaffold showing extraordinary antibacterial potency and pathogen selectivity. (d) Structure and bioactivity profile of penicisteckin G, featuring unique atropo-diastereomeric architecture with potent anti-MRSA activity through membrane disruption mechanisms.

Table 2. Marine fungal bioactive compounds with structural innovation and therapeutic potential.

Compound	Source Organism	Structure Class	Primary Activity	MIC/IC50	Mechanism	Clinical Status
Harzianolide B	*Trichoderma harzianum* ZN-4	Tetramic acid-PK hybrid	Antifungal	25 µg/mL	Cell wall synthesis	Preclinical
16-Epiascomylactam B	*Microascus* sp. SCSIO41821	Decahydrofluorene alkaloid	Antibacterial	0.20–0.80 µg/mL	Ribosome binding	Lead optimization
Penicisteckin G	*Penicillium steckii*	Atropo-diastereomeric dimer	Anti-MRSA	4.0 µg/mL	Membrane disruption	Preclinical
Sesterchaetins A	*Chaetomium globosum* SD-347	Sesterterpene	Anti-Vibrio	8.0 µg/mL	Cell wall targeting	Research phase
Phomoxanthone A	*Diaporthe* sp. SYSU-MS4722	Dimeric xanthone	Anti-inflammatory	7.5 µM	COX-2 selective	Preclinical
Turbinmicin	Marine microbiome	Polycyclic natural product	Antifungal	Active vs, *C. auris*	Vesicle trafficking	Phase I potential
Alternapyrone G	*Alternaria* sp. ZJ9-6B	Polyketide lactone	Anticancer	12.5 µM	Apoptosis induction	Research phase
Chevalone E	*Aspergillus milianensis* KUFA 0013	Meroterpenoid	MRSA synergist	12.5 µg/mL with oxacillin	β-lactamase modulation	Preclinical

5. Marine Fungi in Industrial Biotechnology

Marine fungi possess extraordinary enzymatic capabilities that have evolved under extreme conditions, making them invaluable resources for industrial biotechnology applications requiring robust, salt-tolerant, and thermostable biocatalysts. These organisms have developed enzyme systems that function optimally under conditions that denature terrestrial enzymes: high salinity (up to 35% NaCl), extreme temperatures (– 2°C to > 100°C), high pressure (> 1000 atmospheres), and alkaline pH (up to 10.5). Enzyme production represents the most immediately exploitable application of marine fungal biotechnology. Cellulases from marine fungi demonstrate superior salt tolerance and thermostability compared to terrestrial counterparts. *Aspergillus flavus* 9261 from South China Sea sediments produces cellulases with specific activities of 10.7482 U/mg that retain full activity in 3.5% NaCl solutions at 50°C, conditions that completely inactivate terrestrial cellulases. These enzymes show no activity loss after 24 hours at industrial operating conditions, making them ideal for biorefinery applications processing marine biomass.

Amylases from marine fungi exhibit remarkable thermostability and salt tolerance. Antarctic marine yeasts produce cold-active amylases with optimal activity at 15°C, enabling low-energy starch processing, while thermophilic vent fungi generate amylases stable at 85°C with half-lives exceeding 8 hours. Marine *Aspergillus* amylases maintain activity in 20% NaCl solutions, enabling direct processing of seawater-cultivated feedstocks without desalination.

Proteases from marine sources demonstrate exceptional alkaline stability and organic solvent tolerance. Deep-sea *Penicillium* species produce serine proteases that remain active in 70% ethanol solutions and pH > 11, making them suitable for biotransformation reactions in non-aqueous media (Li et al., 2025). These enzymes exhibit substrate specificities absent from terrestrial proteases, including novel endopeptidase activities that cleave marine-specific proteins like alginate-binding domains. Marine-derived laccases show superior thermostability, but the engineered variants from *Peniophora* sp. CBMAI 1063 with 32-fold improved thermostability and half-lives of 225 minutes at 75°C are actually from terrestrial basidiomycete PM1, not marine sources. The correct marine laccase data shows *Peniophora* sp. CBMAI 1063 (a marine isolate) with thermal stability ranging from 30 to 50°C (120 min) and $T50_{1}h$ = 56°C (Mainardi et al., 2018; Aza and Camarero, 2023). These enzymes achieve 93% decolorization of synthetic dyes within 2 hours under high-salt conditions, making them ideal for textile wastewater treatment and pulp bleaching applications.

Sustainable biofuel production through marine fungi-algae consortia represents a revolutionary approach to renewable energy. Bio-flocculation systems using *Nannochloropsis oceanica* and marine *Mortierella elongata* achieve 40% higher oil yields compared to monoculture algae while reducing harvesting costs by 50% (Du et al., 2018). The fungal partner produces natural flocculants that aggregate algal cells for efficient separation, while secreted lipases enhance oil extraction efficiency. Marine fungi enhance lipid accumulation in co-cultivated algae through several mechanisms: secretion of growth-promoting factors, stress-induced lipid synthesis, and nutrient cycling that maintains algal productivity. *Piriformospora indica* marine isolates increase algal biomass by 60% and lipid content by 40% through production of auxin-like compounds and phosphate-solubilizing enzymes (Bhatnagar et al., 2019).

Bioplastics and biodegradation applications address critical environmental challenges. Marine fungi demonstrate remarkable plastic-degrading capabilities: 42 of 68 tested marine strains degrade polyurethane, with *Parengyodontium album* achieving 0.044%/day degradation rates for UV-treated polyethylene (Steinbach et al., 2025). These fungi produce diverse plastic-degrading enzymes, including cutinases, esterases, laccases, and peroxidases that function effectively in seawater conditions. Novel bioplastic production utilizes marine fungal biomass and metabolites. Chitin and chitosan from marine fungi show superior properties for biodegradable plastic production: enhanced tensile strength, improved water resistance, and complete biodegradation in marine environments within 90 days. Marine *Aspergillus* species produce polyhydroxyalkanoates (PHAs) with unique monomer compositions that confer superior material properties (Oliveira et al., 2020). Marine biofilm applications span antifouling technologies and beneficial biofilm enhancement. Marine fungi produce both anti-biofilm compounds and biofilm-promoting factors, enabling precise control of surface colonization. Marine *Aspergillus terreus* isolates produce butyrolactones and other secondary metabolites that can regulate microbial interactions, while some marine-derived furanones from algae and bacteria inhibit quorum sensing in fouling bacteria (Barakat and Gohar, 2012).

Advanced biocatalysis exploits unique marine fungal enzyme specificities for pharmaceutical synthesis. Marine fungi catalyze stereospecific biotransformations unavailable through terrestrial organisms: hydroxylation of steroid compounds at novel positions, asymmetric reduction of ketones under high-salt conditions, and glycosylation reactions that proceed in seawater media (Virués-Segovia et al., 2023).

Biomineralization and materials science applications utilize marine fungi's natural mineral-processing capabilities. Deep-sea fungi participate in biomineralization of calcium carbonates, silicates, and metal sulfides through specialized proteins that nucleate crystal formation. These biological templates enable controlled synthesis of nanomaterials with precise size and morphology control. Environmental remediation capabilities position marine fungi as powerful bioremediation agents. Heavy metal sequestration proceeds through multiple mechanisms: biosorption via cell wall chitin and chitosan, intracellular accumulation in metallothionein-like proteins, and biotransformation to less toxic forms. Marine *Pleurotus* species tolerate and accumulate cobalt, copper, and nickel at concentrations up to 60 mg/L while maintaining metabolic activity (Alabssawy and Hashem, 2024).

Hydrocarbon degradation by marine fungi offers solutions for oil spill remediation. *Aspergillus sclerotiorum* achieves 99.7% pyrene degradation in 8 days and 76.6% benzo[a]pyrene degradation in 16 days through cytochrome P450 pathway systems enhanced by salt-tolerant enzyme variants. These fungi maintain degradative activity in the presence of dispersants and surfactants used in spill response (Vieira et al., 2018). Emerging xenobiotic degradation capabilities include the breakdown of pharmaceuticals, pesticides, and industrial chemicals in marine environments. Marine fungi express unique cytochrome P450 variants that metabolize compounds resistant to terrestrial microorganisms, including halogenated organic compounds and synthetic polymers (Khan et al., 2023; Virués-Segovia et al., 2023).

6. Therapeutic Applications

The global antimicrobial resistance crisis has reached alarming proportions, with multidrug-resistant pathogens causing 1.27 million deaths annually and threatening to return medicine to the pre-antibiotic era (Ayukekbong et al., 2017; Walsh et al., 2023). Marine fungi emerge as a critical resource for addressing this crisis through the production of structurally novel compounds with unprecedented mechanisms of action that circumvent existing resistance pathways. Urgent-threat pathogens represent the most immediate challenges (Pan et al., 2025). *Candida auris*, designated by WHO as a critical priority pathogen, has spread across 47 countries with mortality rates of 30–72% and resistance to multiple antifungal classes. The emergence of pandrug-resistant strains resistant to all available antifungals creates a desperate need for novel therapeutic approaches. Turbinmicin, discovered from marine fungal microbiomes, demonstrates potent activity against pandrug-resistant *C. auris* strains while showing excellent therapeutic tolerability in mouse infection models (De Gaetano et al., 2024; Zhang et al., 2020).

MRSA infections continue to challenge healthcare systems globally, with resistance rates reaching 42% in some regions. Marine fungal compounds offer novel anti-MRSA mechanisms: penicisteckins G-H from *Penicillium steckii* achieve MIC values of 4.0 μg/mL through membrane-disrupting mechanisms distinct from conventional antibiotics. 16-Epiascomylactam B from deep-sea *Microascus* sp. demonstrates exceptional potency (MIC: 0.20–0.80 μg/mL) against multiple MRSA strains through novel ribosomal binding mechanisms (Zhang et al., 2022b).

Carbapenem-resistant Enterobacterales (CRE) represent perhaps the greatest bacterial threat, with limited treatment options and high mortality rates. Marine fungi produce compounds that restore carbapenem sensitivity: chevalone E from *Aspergillus milianensis* shows synergistic activity with oxacillin against MRSA and similar mechanisms appear active against CRE strains through β-lactamase modulation. Novel mechanisms of action distinguish marine fungal compounds from terrestrial antibiotics (Smith et al., 2025). Traditional antibiotics target cell wall synthesis, protein synthesis, DNA replication, or membrane integrity. Marine fungal compounds exploit unique vulnerabilities: vesicle trafficking disruption (turbinmicin), quorum sensing inhibition (marine furanones), biofilm matrix degradation (alginate lyases), and metal homeostasis disruption (siderophore analogs) (Gadar and McCarthy, 2023).

Structure-activity relationships reveal how marine conditions drive antimicrobial innovation. High salinity selects for compounds with charged residues that maintain activity in ionic environments. Competition with marine bacteria favors molecules with novel scaffolds that resist enzymatic degradation. Three-dimensional aquatic diffusion requires compounds with optimal hydrophobic-hydrophilic balance for effective distribution (Durães et al., 2021; Xu et al., 2015). Anticancer drug development from marine fungi has yielded several clinical candidates with unique mechanisms. Plinabulin, a synthetic analog of marine-derived halimide, demonstrates dual activity as both a vascular disrupting agent and neutropenia protector. Phase III trials show comparable efficacy to pegfilgrastim in preventing chemotherapy-induced neutropenia while offering potential anticancer benefits through microtubule depolymerization (Gomes et al., 2015; Blayney et al., 2022). Salinosporamide A (Marizomib) represents another marine-inspired success, currently in Phase III trials for glioblastoma. This irreversible proteasome inhibitor uniquely crosses the blood-brain barrier and maintains activity against bortezomib-resistant cancers through 20S proteasome targeting mechanisms absent in terrestrial compounds (Gomes et al., 2015; Di et al., 2016). Apoptosis-inducing mechanisms employed by marine fungal compounds include both intrinsic and extrinsic pathway activation. Asperterpenoid A from marine *Aspergillus terreus* induces mitochondrial membrane depolarization and cytochrome c release, while penicilone activates death receptor pathways through TRAIL sensitization (Huang et al., 2013; Dalisay et al., 2024). These compounds demonstrate selectivity for cancer cells through targeting of metabolic vulnerabilities specific to transformed cells.

AI-enhanced drug discovery is revolutionizing marine fungal therapeutics development. Machine learning algorithms trained on marine natural product databases predict bioactivity with > 85% accuracy, enabling intelligent prioritization of synthesis targets. Deep learning models predict ADMET properties and toxicity profiles, reducing development timelines by identifying promising candidates early in the discovery process (Pereira and Aires-de-Sousa, 2018; Matsuzaka and Yashiro, 2022; Fu et al., 2024). Virtual screening approaches using marine fungal metabolite databases have identified over 860,000 potential antimicrobial peptides, with AI ranking systems prioritizing candidates most likely to overcome resistance mechanisms. Molecular docking studies reveal binding modes for marine compounds against resistant pathogen targets, guiding medicinal chemistry optimization (Santos-Júnior et al., 2024; Zhou et al., 2024). Computational challenges include data heterogeneity across marine environments, limited training datasets for marine-specific pharmacophores, and the complexity of predicting bioactivity in multi-target pathogen systems. Nevertheless, AI-driven approaches have accelerated hit identification by 100-fold compared to traditional screening methods.

The clinical pipeline is witnessing good activity in marine-derived therapeutics. Besides plinabulin and marizomib, another 21 marine-inspired molecules are in various phases of clinical development. Marine-fungi-derived therapeutics are a significantly untapped clinical pipeline compared to the general marine natural products field. Plinabulin (NPI-2358) is the only marine-fungi-derived molecule to date that has entered the clinic. This synthetic analog of halimide (phenylahistin), first isolated from marine *Aspergillus* sp. CNC139 collected from Philippine waters, is currently in Phase III clinical trials for the treatment and chemotherapy-induced neutropenia prevention of non-small cell lung cancer (Gomes et al., 2015). Plinabulin demonstrates dual vascular-disrupting and microtubule-destabilizing activities, and recent studies have identified additional therapeutic potential for drug-resistant epilepsy treatment (Gözcelioğlu, 2019; Copmans et al., 2022).

Cephalosporin C, while itself a legacy antibiotic from the marine fungus *Acremonium chrysogenum* that was isolated from Sardinian coastal sewage water in the 1940s, represents the greater legacy of marine fungi in drug discovery. The translational gap from clinic to use is particularly broad for marine fungi. Despite the fact that over 1,000 metabolites from marine fungi have been reported to possess a range of biological activities including anticancer, antibacterial, and antiviral, surprisingly few have progressed beyond preclinical development (Deshmukh et al., 2018; Chavhan et al., 2025). Recent discoveries from deep-sea marine fungi (2020–2022) have yielded 184 new bioactive compounds, yet clinical advancement remains limited (Wang et al., 2023). Promising preclinical candidates from marine fungi include various diketopiperazines, polyketides, and peptides with potent bioactivities, but most remain in early development phases, requiring significant optimization for clinical translation (Gomes et al., 2021).

The challenge lies in advancing these compounds through the rigorous requirements of clinical development, including selectivity between normal and cancer cells, activity against drug-resistant strains, and favorable safety profiles (Gomes et al., 2015). FGFC compounds (TMS-007/BIIB-131) from marine fungi have been commercialized for acute ischemic stroke treatment, demonstrating successful translation from discovery to clinical application (Jimenez et al., 2020; Cappello and Nieri, 2021; Saeed et al., 2021). Regulatory challenges to marine-derived therapeutics involve supply chain resilience, environmental impact analysis of collection technologies, and intellectual property concerns with naturally occurring compounds. Synthetic biology solutions increasingly address supply challenges with heterologous production systems, as exemplified with marine fungi where fermentation-based production has superiorities over harvesting macroorganisms. Future directions include combination therapies involving multiple marine compounds, individualized medicine approaches based on pathogen genomics, and design of marine-inspired drug delivery. Combination of marine natural products with existing drug development technologies promises continued therapeutic innovation.

7. Marine Fungi in Ecosystem and Climate Resilience

Marine fungi are key species in marine communities, and quantitative studies in recent years have established their functions in biogeochemical cycling comparable in scale and functional specificity to bacteria. Marine fungi are participants in complex ecological networks guaranteeing ocean well-being, acting as mediators of climate control, and enhancing mechanisms of resilience against human-manufactured environmental change. Their functions extend far beyond simple decomposition and encompass sophisticated regulatory processes in marine biogeochemistry and ecosystem stability.

7.1 Role in Biogeochemical Cycles

Marine fungal carbon cycling involves saprotrophic and primary decomposer functions, in addition to activity within the biological carbon pump. Stable isotope probing demonstrates that planktonic fungi play an active role within the assimilation of carbon within diatom extracellular polymeric substances (dEPS), and saprotrophic lineages of close Malassezia relatives are involved in 15–25% of pelagic carbon processing. This contradicts prevailing models that had apportioned marine carbon cycling to prokaryotes alone and reveals fungi as quantitatively significant participants in oceanic carbon processes. Depth-stratified functional specialization reveals sophisticated ecological partitioning within marine fungal assemblages. Surface waters (0–200 m) harbor rapidly growing opportunistic species with the ability to rapidly decay labile organic matter produced by phytoplankton blooms with turnover rates of 2–5 days during periods of peak productivity. Mesopelagic fungi (200–1000 m) are adapted to degrade recalcitrant marine polymers like chitin, alginate, and resistant polysaccharides inaccessible to degradation by bacteria with sinking particulate matter processing rates of 10–30% per month (Amend et al., 2019).

Subseafloor fungi in oceanic sediments contribute significantly to long-term carbon sequestration through specialized metabolic pathways. These organisms express unique carbohydrate-active enzymes (CAZymes), including GH7 cellobiohydrolases, CE6 acetylxylan esterases, and CBM48 chitin-binding modules that enable processing of marine-specific substrates unavailable to bacterial communities. Quantitative analysis reveals fungal biomass increases from 10^1–10^3 copies/mL in the water column to 10^4–10^6 copies/g in sediments, where they process 20–35% of total organic carbon burial, a previously unrecognized component of marine carbon sequestration (Orsi et al., 2022). Nitrogen cycling contributions by marine fungi have been significantly underestimated until recent isotope tracer studies. In oxygen minimum zones (OMZs), marine fungi contribute 18–22% of total N_2O production through denitrification pathways enhanced by specialized nitrate reductases adapted to low-oxygen conditions (Maeda et al., 2015; Otte et al., 2019; Peng et al., 2024).

Phosphorus cycling involves marine fungi in both mineralization and immobilization processes. Mycorrhizal-like associations between fungi and seagrass roots in Mediterranean *Posidonia oceanica* meadows enhance phosphate uptake by 40–60% compared to non-mycorrhizal plants through production of specialized phosphatases that liberate phosphate from organic compounds. These fungi maintain active phosphorus cycling in phosphate-limited marine environments where conventional bacterial mineralization is insufficient (Liu et al., 2021).

7.2 Symbiosis and Ecosystem Stability

Coral holobiont stability depends critically on fungal partners that contribute antimicrobial defense, nutrient cycling, and stress tolerance mechanisms. Healthy coral colonies harbor diverse fungal communities producing mycosporine-like amino acids (MAAs) that provide UV protection equivalent to SPF 30–50 sunscreen, while simultaneously secreting antimicrobial compounds that suppress pathogenic bacteria by 2–3 orders of magnitude. During thermal stress events, coral-associated fungi upregulate heat shock proteins and produce compatible solutes that enhance coral thermal tolerance by 2–4°C (Mohamed et al., 2023; Thompson et al., 2015).

Seagrass ecosystem engineering by endophytic fungi involves complex below-ground networks that enhance plant productivity and ecosystem stability. Root-associated fungal communities in seagrass meadows form ectomycorrhizal-like structures that increase nutrient acquisition surface area by 10–50 fold while producing plant growth hormones and stress-protective compounds. These partnerships enable seagrasses to colonize nutrient-poor sandy sediments and maintain productivity in fluctuating salinities (Seymour et al., 2018; Rajakaruna et al., 2024). Fungal-algal relationships vary from parasitism to mutualism with profound effects on marine primary productivity. Phytoplankton chytrid parasites of the sea can kill algal blooms within days through epidemic-sized infection, while endophytic fungi in macroalgae enhance host stress tolerance and produce defensive metabolites that repel herbivores. These interactions regulate primary production patterns and influence marine food web dynamics at multiple temporal and spatial scales (Sime-Ngando, 2012; Rasconi et al., 2014).

7.3 *Detoxification Services*

Hydrocarbon degradation capabilities of marine fungi provide essential ecosystem services for contaminated marine environments. Comprehensive metabolic studies reveal that *Aspergillus sclerotiorum* achieves complete degradation of pyrene (99.7% in 8 days) and substantial benzo[a]pyrene removal (76.6% in 16 days) through enhanced cytochrome P450 systems that maintain activity in the presence of dispersants, surfactants, and other spill-response chemicals (Passarini et al., 2011; Vieira et al., 2018). These degradation rates exceed those of specialized hydrocarbon-degrading bacteria under similar conditions. Heavy metal remediation proceeds through sophisticated fungal mechanisms adapted to marine chemistry. Biosorption processes utilize cell wall components, including chitin, chitosan, and melanins, that bind heavy metals with affinities of 50–200 mg/g dry weight (Fomina and Gadd, 2002; Parasnis et al., 2024). Marine *Pleurotus* species demonstrate active tolerance to cobalt, copper, and nickel concentrations up to 60 mg/L while accumulating metals in fungal biomass at concentrations 10–100 fold higher than ambient seawater, enabling effective bioremediation of contaminated sites.

Plastic degradation represents an emerging capability with profound environmental implications. Recent surveys identify 42 of 68 marine fungal strains capable of polyurethane degradation, with *Parengyodontium album* achieving measurable breakdown rates (0.044%/day) for UV-treated polyethylene, rates that translate to complete degradation of plastic films within 6–8 years under marine conditions (Zeghal et al., 2021; Vaksmaa et al., 2024). These fungi produce diverse plastic-degrading enzymes, including cutinases that cleave ester bonds in polyethylene terephthalate and laccases that oxidize aromatic polymer components.

7.4 *Adaptation to Climate Change*

Ocean acidification responses reveal complex interactions between marine fungi and changing ocean chemistry. Controlled experiments demonstrate that moderate acidification (pH 7.6–7.8, projected for 2100) can increase marine fungal abundance by almost an order of magnitude while altering community composition toward acid-tolerant taxa (O'Brien et al., 2016; Kumar et al., 2021; Yun et al., 2021). However, biodiversity-stability relationships show that high-diversity fungal assemblages reduce acidification impacts on vulnerable organisms by 50–90%, emphasizing the importance of fungal diversity conservation.

Temperature tolerance mechanisms vary significantly among marine fungal taxa with implications for climate warming responses. While some species show broad temperature tolerances (polar *Chytridiomycota* growing from –2°C to 15°C), many exhibit strict biogeographical restrictions with < 2°C overlap between thermal zones. Experimental evolution studies demonstrate that marine fungi can adapt to +3°C warming within 50–100 generations, but adaptation costs include reduced competitive ability and increased susceptibility to other stressors. Osmotic stress responses involve sophisticated molecular machinery, including the high osmolarity glycerol (HOG) signaling pathway. *Hortaea werneckii*, a model halotolerant marine fungus, employs stress-activated MAPK cascades that regulate glycerol synthesis (through HwGPD1A), ion transport, and cell wall modification in response to salinity fluctuations. Whole-genome analysis reveals expanded gene families for compatible solute synthesis (HwMET6, HwMET17) and osmotic stress response compared to terrestrial fungi, providing blueprints for engineering climate-resilient organisms (Kumar et al., 2021).

Ecosystem service valuation increasingly recognizes marine fungi as critical components of ocean health. Economic analysis estimates that marine fungal contributions to carbon sequestration, pollution remediation, and fisheries productivity provide ecosystem services valued at $2.1–4.8 billion annually in coastal zones alone. Biodiversity-function relationships demonstrate that fungal species loss reduces ecosystem stability by 25–40%, emphasizing the economic importance of marine fungal conservation. Climate change adaptation mechanisms in marine fungi reveal sophisticated molecular responses to environmental stressors. Ocean acidification studies demonstrate that marine fungi may benefit from decreasing pH, with some species showing increased proliferation under pH conditions projected for 2100 (pH 7.6–7.8). Moderate acidification can increase fungal abundance by almost an order of magnitude, potentially altering marine microbial community structure. Temperature tolerance varies significantly among marine fungal taxa, with implications for climate warming responses. While some species show broad temperature tolerances (4°C–35°C), many exhibit biogeographical restrictions to tropical, temperate, or polar waters with limited overlap between thermal zones (Kumar et al., 2021). Climate warming will likely drive poleward shifts in fungal community composition, with potential ecosystem consequences for carbon cycling and symbiotic relationships.

8. Challenges and Future Perspectives

Despite phenomenal promise, marine fungal research is faced with multi-dimensional challenges that limit biotechnological and therapeutic utilization. Such challenges include biological constraints, technological deficits, and regulatory intricacies that necessitate coordinated international efforts. Paradox of cultivation is the primary limitation with traditional laboratory media supporting growth of $< 5\%$ environmental marine fungal diversity. This poses the deepening bias against fast-growing generalist species at the cost of slow-growing specialist species attuned to oligotrophic marine conditions (Zhang et al., 2024). Environmental DNA surveys suggest global marine fungal diversity may be as high as 3.8 million species, yet only 2,149 species have been formally described and represent $< 0.1\%$ of estimated diversity.

Cryptic biosynthetic gene cluster expression renders this challenge frustrating, where even culturable organisms do not provide their full metabolic capacity. Genome mining studies reveal that marine fungi have 60–67 BGCs in each species, but $< 10\%$ are active at the transcriptional level in laboratory conditions (Kang and Kim, 2021). Such cryptic metabolism reflects advanced regulatory pathways evolved for marine environmental stimuli unattainable under ordinary environments. Viable but non-culturable (VBNC) conditions affect high rates of marine fungal populations, particularly in oligotrophic deep-sea biomes (Fulke et al., 2025). These organisms are metabolically active and viable but cannot be cultured using standard laboratory techniques, representing a unique survival strategy in response to harsh environmental conditions.

8.1 Advanced Solutions and Innovations

Microfluidics-based cultivation enables single-cell isolation under precisely controlled microscale conditions that better mimic marine environmental gradients. Extended cultivation approaches using months-long incubation at low temperatures (2–4°C) successfully isolate slow-growing deep-sea specialists, increasing isolation success rates by 300–500% (Xu et al., 2022).

Co-cultivation innovations exploit ecological interactions to activate silent pathways. Bacterial-fungal co-cultures mimic natural competitive relationships, triggering defensive metabolite production, while host-associated cultivation using marine plant or animal tissue provides physiological cues activating symbiosis-specific metabolic programs (Selegato and Castro-Gamboa, 2023).

Culture-independent genomic approaches increasingly circumvent cultivation limitations. Single-cell genomics enables genome assembly from individual uncultured cells, while metagenome-assembled genomes reconstruct complete fungal genomes from environmental samples using advanced binning algorithms (Garza and Dutilh, 2015; Mirete et al., 2025).

8.2 Synthetic Biology and Heterologous Expression

Synthetic biology solutions offer transformative routes through engineered biological systems. Heterologous expression platforms transfer marine fungal BGCs into tractable hosts like Saccharomyces cerevisiae, *Aspergillus nidulans*, or *Pichia pastoris* for compound production (Kjærbølling et al., 2019; Mózsik et al., 2022). CRISPR-mediated pathway reconstruction enables precise assembly of complex biosynthetic pathways in engineered chassis organisms, with success rates approaching 70–80% for well-characterized BGCs. AI-guided BGC prioritization using machine learning algorithms to predict which cryptic pathways are most likely to produce bioactive compounds, reducing screening time by 10-fold compared to random activation approaches (Sahayasheela et al., 2022).

8.3 Taxonomic Resolution and Database Development

Taxonomic resolution challenges stem from severe underrepresentation of marine fungi in reference databases. The UNITE database, the primary resource for fungal sequence identification, contains insufficient marine fungal sequences, rendering 60–80% of marine environmental OTUs unclassifiable even at the phylum level (Nilsson et al., 2019). *Chytridiomycota* and early-diverging lineages contribute to "fungal dark matter" due to primer bias in amplicon sequencing and incomplete phylogenetic frameworks (Grossart et al., 2016).

Advanced sequencing solutions address these limitations through full-length rRNA gene sequencing using PacBio long-read technology, phylogenomic analysis incorporating hundreds of conserved genes for robust phylogenetic placement, and multi-gene barcoding systems that integrate ITS, LSU, SSU, and protein-coding markers for enhanced taxonomic resolution (Krehenwinkel et al., 2019; Agustinho et al., 2024).

8.4 Regulatory Frameworks and International Cooperation

Legal barriers create significant challenges under the Nagoya Protocol on Access and Benefit-Sharing, with unclear sovereignty over marine genetic resources in Areas Beyond National Jurisdiction covering 64% of the ocean surface. Digital sequence information regulations threaten computational approaches to marine fungal research. Climate change impacts require urgent attention as ocean warming, acidification, and deoxygenation alter fungal community structure. Predictive modeling suggests 30–50% of current marine fungal habitat will experience conditions outside historical ranges by 2100, potentially causing massive community restructuring (O'Brien et al., 2016; Kumar et al., 2021).

9. Conclusion

Marine fungi represent one of the highest-level evolutionary solutions to hostile environments, having turned oceanic challenges into biochemical leaps on geological timescales. Evidence presented herein demonstrates such microbes to be biological chemical factories with unprecedented potential, harboring 6,067 biosynthetic gene clusters per species (98% cryptic), producing over 7,000 unique natural products, and delivering 15–25% of ocean carbon processing despite the representation by less than 5% of the marine microbial biomass.

9.1 Medical and Therapeutic Applications

The pipeline of the drug is very promising with clinical candidates already well advanced in trials. Plinabulin has dual anticancer and neutropenia protective mechanisms in phase III clinical trials, turbinmicin against pandrug-resistant *Candida auris*, and 16-epiascomylactam B has sub-micromolar activity against multidrug-resistant pathogens. With antimicrobial resistance causing 1.27 million deaths annually, marine fungi yield structurally unusual compounds with unprecedented mechanisms that avoid known resistance pathways altogether.

9.2 Industrial Biotechnology Revolution

Marine fungal enzymes exhibit groundbreaking industrial potential, working under extreme conditions (3.5% salinity, –2°C to > 100°C, > 1000 atmospheric pressure, pH >10) which denature their terrestrial counterparts. There are exhibited functionalities such as 40% increased biofuel yields by fungi-algae consortia, 42 of 68 strains with the degradation of polyurethane, and heavy metal sequestration at 100–1000-fold environmental concentrations for remediation processes.

9.3 Climate Adaptation and Environmental Solutions

These animals give vital answers to climate problems through enhanced coral thermal tolerance (2–4°C improvement through symbiotic relationships), oceanic production of N_2O (18–22%), and carbon burial processing in the deep sea (20–35%). Their sophisticated stress-response systems provide blueprints for the creation of climate-resistant biological systems and adaptation technologies of the next generation. The convergence of AI-gene genome mining (over 85% accuracy in predicting bioactivity), single-cell genomics, synthetic biology tools (70–80% success in reconstructing biosynthetic gene clusters), and culture-independent methods (revealing over 50,000 cryptic species) has transformed discovery rates by orders of magnitude. Convergence of technologies provides systematic access to hitherto inaccessible marine fungal chemical space.

9.4 Ecosystem Service Enhancement and Cultivation

Marine fungi offer irreplaceable ecosystem services that become more valuable under conditions of climate change. Their role in sequestering carbon, nutrient cycling, and stabilizing ecosystems creates opportunities for the development of nature-based climate solutions and ocean health maintenance. The cultivation paradox affects over

95% of marine fungal diversity, and cryptic biosynthetic potential is 10-fold higher than expressed metabolism. Current cultivation methods cannot induce the vast majority of biosynthetic gene clusters, excluding access to this biological bonanza.

9.5 Regulatory and Administrative Barriers

Regulatory structures trail decades behind scientific capacity, with Access and Benefit Sharing procedures acting as obstacles to non-commercial research. Global coordination is still incomplete despite the international nature of marine ecosystems and the problems they might solve.Even with technological advancements, significant knowledge gaps remain in activating the mechanisms of biosynthetic gene clusters, optimizing heterologous expression system designs, and translating laboratory breakthroughs to industrially relevant scales. Integrating culture-independent methods with functional studies remains ongoing methodological challenges.

9.6 Strategic Path Forward

Success requires unprecedented record coordination among scientific disciplines, regulatory policies, and national boundaries. The major priority actions include developing robust methods for the activation of biosynthetic gene clusters to access cryptic metabolic potential, embracing AI-accelerated discovery pipelines for compound discovery, and expanding deep-sea exploration in under-sampled regions, particularly hydrothermal vents, seamounts, and polar oceans. Equally important are policy and regulatory reforms that implement streamlined procedures for non-commercial marine research, create streamlined Access and Benefit Sharing procedures that facilitate scientific collaboration while protecting biodiversity rights, and shape international agreements on joint marine biotechnology research. To assist such activities includes augmenting worldwide systems for sharing resources and expertise, permeating artificial intelligence across species identification, predictive modeling, and laboratory automation systems, and aligning research activities with climate adaptation applications and sustainable biotechnology solutions as a priority.

Cumulatively, marine fungi have already solved the fundamental problems facing human civilization: antimicrobial resistance, industrial activity in extreme conditions, carbon sequestration, and ecosystem resilience in the face of environmental stress. These solutions have been perfected by evolution over billions of years, the issue is not one of discovery but one of access, understanding, and implementation of this vast biological knowledge base before planetary crises overwhelm human adaptive capacity. The confluence of the technological potential, urgent planetary issues, and unparalleled biological potential releases revolutionary promise at this transformative tipping point in marine mycology. These creatures that ruled Earth's most extreme environments on geological timescales hold perhaps the keys to human survival and planetary health in an era of unprecedented planetary change. Whether or not our species is able to tap into this evolutionary wisdom rests solely upon our capacity to organize scientific innovation, regulatory overhaul, and global collaboration on scales and timelines commensurate with the enormity of both the problems we confront and the solutions these incredible organisms have optimized. The unprecedented potential requires equally unprecedented scientific and policy action, the consequences for human civilization and the health of the planet have never been more at stake.

References

Agustinho, D.P., Fu, Y., Menon, V.K., Metcalf, G.A., Treangen, T.J. et al. (2024). Unveiling microbial diversity: Harnessing long-read sequencing technology. Nat. Methods, 21: 954–966. https://doi.org/10.1038/s41592-024-02262-1.

Ahumada-Rudolph, R., Novoa, V. and Becerra, J. (2018). Morphological response to salinity, temperature, and pH changes by marine fungus *Epicoccum nigrum*. Environ. Monit. Assess., 191: 35. https://doi.org/10.1007/s10661-018-7166-5.

Alabssawy, A.N. and Hashem, A.H. (2024). Bioremediation of hazardous heavy metals by marine microorganisms: A recent review. Arch Microbiol., 206: 103. https://doi.org/10.1007/s00203-023-03793-5.

Amend, A., Burgaud, G., Cunliffe, M., Edgcomb, V.P., Ettinger, C.L. et al. (2019). Fungi in the marine environment: Open questions and unsolved problems. mBio, 10: e01189-18. https://doi.org/10.1128/mBio.01189-18.

Ayukekbong, J.A., Ntemgwa, M. and Atabe, A.N. (2017). The threat of antimicrobial resistance in developing countries: Causes and control strategies. Antimicrob. Resist. Infect. Control, 6: 47. https://doi.org/10.1186/s13756-017-0208-x.

Aza, P. and Camarero, S. (2023). Fungal laccases: Fundamentals, engineering, and classification update. Biomolecules, 13: 1716. https://doi.org/10.3390/biom13121716.

Bahram, M. and Netherway, T. (2022). Fungi as mediators linking organisms and ecosystems. FEMS Microbiol. Rev., 46. https://doi.org/10.1093/femsre/fuab058.

Balabanova, L., Slepchenko, L., Son, O. and Tekutyeva, L. (2018). Biotechnology potential of marine fungi degrading plant and algae polymeric substrates. Front. Microbiol., 9. https://doi.org/10.3389/fmicb.2018.01527.

Baltar, F., Zhao, Z. and Herndl, G.J. (2021). Potential and expression of carbohydrate utilization by marine fungi in the global ocean. Microbiome, 9: 106. https://doi.org/10.1186/s40168-021-01063-4.

Barakat, K.M. and Gohar, Y.M. (2012). Antimicrobial agents produced by marine *Aspergillus terreus* var. *africanus* against some virulent fish pathogens. Indian J. Microbiol., 52: 366–372. https://doi.org/10.1007/s12088-012-0255-1.

Barghoorn, E.S. and Linder, D.H. (1944). Marine fungi: Their taxonomy and biology. Farlowia, 1: 395–467.

Bhatnagar, V.S., Bandyopadhyay, P., Rajacharya, G.H., Sarkar, S., Poluri, K.M. et al. (2019). Amelioration of biomass and lipid in marine alga by an endophytic fungus *Piriformospora indica*. Biotechnol. Biofuels, 12: 176. https://doi.org/10.1186/s13068-019-1516-6.

Blayney, D.W., Mohanlal, R., Adamchuk, H., Kirtbaya, D.V., Chen, M. et al. (2022). Efficacy of plinabulin vs pegfilgrastim for prevention of docetaxel-induced neutropenia in patients with solid tumors. JAMA Netw Open, 5: e2145446. https://doi.org/10.1001/jamanetworkopen.2021.45446.

Bochdansky, A.B., Clouse, M.A. and Herndl, G.J. (2017). Eukaryotic microbes, principally fungi and labyrinthulomycetes, dominate biomass on bathypelagic marine snow. ISME J., 11: 362–373. https://doi.org/10.1038/ismej.2016.113.

Bonugli-Santos, R.C., dos Santos Vasconcelos, M.R., Passarini, M.R., Vieira, G.A., Lopes, V.C. et al. (2015). Marine-derived fungi: Diversity of enzymes and biotechnological applications. Front. Microbiol., 6: 269. https://doi.org/10.3389/fmicb.2015.00269.

Breyer, E., Zhao, Z., Herndl, G.J. and Baltar, F. (2022). Global contribution of pelagic fungi to protein degradation in the ocean. Microbiome, 10: 143. https://doi.org/10.1186/s40168-022-01329-5.

Britto Martins de Oliveira, J., Corrêa Junior, D., Parente, C.E.T. and Frases, S. (2025). Fungi in mangrove: Ecological importance, climate change impacts, and the role in environmental remediation. Microorganisms, 13: 878. https://doi.org/10.3390/microorganisms13040878.

Cadamuro, R.D., da Silveira Bastos, I.M.A., Silva, I.T., da Cruz, A.C.C., Robl, D. et al. (2021). Bioactive compounds from mangrove endophytic fungus and their uses for microorganism control. J. Fungi, 7: 455. https://doi.org/10.3390/jof7060455.

Cantrell, S.A., Casillas-Martínez, L. and Molina, M. (2006). Characterization of fungi from hypersaline environments of solar salterns using morphological and molecular techniques. Mycol. Res., 110: 962–970. https://doi.org/10.1016/j.mycres.2006.06.005.

Carroll, A.R., Copp, B.R., Davis, R.A., Keyzers, R.A. and Prinsep, M.R. (2020). Marine natural products. Nat. Prod. Rep., 37: 175–223. https://doi.org/10.1039/C9NP00069K.

Chavhan, R.L., Jaybhaye, S.G., Hinge, V.R., Deshmukh, A.S., Shaikh, U.S. et al. (2025). Emerging applications of gene editing technologies for the development of climate-resilient crops. Front. Genome Ed., 7: 1524767. https://doi.org/10.3389/fgeed.2025.1524767.

Chen, S., Guo, H., Jiang, M., Wu, Q., Li, J. et al. (2022). Mono- and dimeric xanthones with anti-glioma and anti-inflammatory activities from the ascidian-derived fungus *Diaporthe* sp. SYSU-MS4722. Mar. Drugs, 20: 51. https://doi.org/10.3390/md20010051.

Chen, S., Xie, Z.X., Yan, K.Q., Chen, J.W., Li, D.X. et al. (2024). Functional vertical connectivity of microbial communities in the ocean. Sci. Adv., 10: eadj8184. https://doi.org/10.1126/sciadv.adj8184.

Chen, S., Xu, Y. and Helfant, L. (2020). Geographical isolation, buried depth, and physicochemical traits drive the variation of species diversity and prokaryotic community in three typical hypersaline environments. Microorganisms, 8: 120. https://doi.org/10.3390/microorganisms8010120.

Comeau, A.M., Vincent, W.F., Bernier, L. and Lovejoy, C. (2016). Novel chytrid lineages dominate fungal sequences in diverse marine and freshwater habitats. Sci. Rep., 6: 30120. https://doi.org/10.1038/srep30120.

Copmans, D., Kildgaard, S., Roux, E., Partoens, M., Steurs, G. et al. (2022). From the North Sea to drug repurposing, the antiseizure activity of halimide and plinabulin. Pharm., 15: 247. https://doi.org/10.3390/ph15020247.

Cunliffe, M. (2023). Who are the marine fungi? Environ. Microbiol., 25: 131–134. https://doi.org/10.1111/1462-2920.16240.

Dalisay, D.S., Tenebro, C.P., Sabido, E.M., Suarez, A.F.L., Paderog, M.J.V. et al. (2024). Marine-derived anticancer agents targeting apoptotic pathways: Exploring the depths for novel cancer therapies. Mar. Drugs, 22: 114. https://doi.org/10.3390/md22030114.

De Gaetano, S., Midiri, A., Mancuso, G., Avola, M.G. and Biondo, C. (2024). *Candida auris* outbreaks: Current status and future perspectives. Microorganisms, 12: 927. https://doi.org/10.3390/microorganisms12050927.

Deshmukh, S.K., Prakash, V. and Ranjan, N. (2018). Marine fungi: A source of potential anticancer compounds. Front. Microbiol., 8: 2536. https://doi.org/10.3389/fmicb.2017.02536.

Di, K., Lloyd, G.K., Abraham, V., MacLaren, A., Burrows, F.J. et al. (2016). Marizomib activity as a single agent in malignant gliomas: Ability to cross the blood-brain barrier. Neuro. Oncol., 18: 840–848. https://doi.org/10.1093/neuonc/nov299.

Dick, G.J. (2019). The microbiomes of deep-sea hydrothermal vents: Distributed globally, shaped locally. Nat. Rev. Microbiol., 17: 271–283. https://doi.org/10.1038/s41579-019-0160-2.

Du, Z.Y., Alvaro, J., Hyden, B., Zienkiewicz, K., Benning, N. et al. (2018). Enhancing oil production and harvest by combining the marine alga *Nannochloropsis oceanica* and the oleaginous fungus *Mortierella elongata*. Biotechnol. Biofuels, 11: 174. https://doi.org/10.1186/s13068-018-1172-2.

Durães, F., Szemerédi, N., Kumla, D., Pinto, M., Kijjoa, A. et al. (2021). Metabolites from marine-derived fungi as potential antimicrobial adjuvants. Mar. Drugs, 19: 475. https://doi.org/10.3390/md19090475.

El-Bondkly, E.A.M., El-Bondkly, A.A.M. and El-Bondkly, A.A.M. (2021). Marine endophytic fungal metabolites: A whole new world of pharmaceutical therapy exploration. Heliyon, 7: e06362. https://doi.org/10.1016/j.heliyon.2021.e06362.

Fernández-Valero, A.D., Reñé, A., Timoneda, N., Pou-Solà, N., Gordi, J. et al. (2023). The succession of epiphytic microalgae conditions fungal community composition: How chytrids respond to blooms of dinoflagellates. ISME Commun., 3. https://doi.org/10.1038/s43705-023-00304-x.

Fomina, M. and Gadd, G.M. (2003). Metal sorption by biomass of melanin-producing fungi grown in clay-containing medium. J. Chem. Technol. Biotechnol. https://doi.org/10.1002/jctb.736.

Fu, L., Shi, S., Yi, J., Wang, N., He, Y. et al. (2024). ADMETlab 3.0: An updated comprehensive online ADMET prediction platform enhanced with broader coverage, improved performance, API functionality, and decision support. Nucleic Acids Res., 52: W422–W431. https://doi.org/10.1093/nar/gkae236.

Fulke, A.B., Sharma, N. and Nadekar, J. (2025). Darkness to discovery: A comprehensive mini-review on culturable and non-culturable microbial diversity from deep sea. Microb. Ecol., 88: 77. https://doi.org/10.1007/s00248-025-02527-y.

Gadar, K. and McCarthy, R.R. (2023). Using next generation antimicrobials to target the mechanisms of infection. NPJ Antimicrob. Resist., 1: 11. https://doi.org/10.1038/s44259-023-00011-6.

Gao, M., Liu, B., Li, J., Deng, Y., Zhang, Y. et al. (2024). Diversity and distribution of fungi in the marine sediments of Zhanjiang Bay, China. J. Fungi, 10: 867. https://doi.org/10.3390/jof10120867.

Garza, D.R. and Dutilh, B.E. (2015). From cultured to uncultured genome sequences: Metagenomics and modeling microbial ecosystems. Cell Mol. Life Sci., 72: 4287–4308. https://doi.org/10.1007/s00018-015-2004-1.

Gautam, A.K., Verma, R.K., Avasthi, S., Sushma, Bohra, Y., Devadatha, B. et al. (2022). Current insight into traditional and modern methods in fungal diversity estimates. J. Fungi, 8: 226. https://doi.org/10.3390/jof8030226.

Gladfelter, A.S., James, T.Y. and Amend, A.S. (2019). Marine fungi. Curr. Biol., 29: R191–R195. https://doi.org/10.1016/j.cub.2019.02.009.

Gomes, N.G.M., Lefranc, F., Kijjoa, A. and Kiss, R. (2015). Can Some marine-derived fungal metabolites become actual anticancer agents? Mar. Drugs, 13: 3950–3991. https://doi.org/10.3390/md13063950.

Gomes, N.G.M., Madureira-Carvalho, Á., Dias-da-Silva, D., Valentao, P. and Andrade, P.B. (2021). Biosynthetic versatility of marine-derived fungi on the delivery of novel antibacterial agents against priority pathogens. Biomed. Pharmacother., 140: 111756. https://doi.org/10.1016/j.biopha.2021.111756.

Gonçalves, M.F., Hilário, S., Tacão, M., Van de Peer, Y., Alves, A. et al. (2021). Genome and metabolome MS-based mining of a marine strain of *Aspergillus affinis*. J. Fungi , 7: 1091. https://doi.org/10.3390/jof7121091.

Gözcelioğlu, B. (2019). Antioxidant and cytotoxic activity of three Turkish marine-derived fungi. Turk. J. Biochem., 44: 554–559. https://doi.org/10.1515/tjb-2019-0001.

Grossart, H.P., Wurzbacher, C., James, T.Y. and Kagami, M. (2016). Discovery of dark matter fungi in aquatic ecosystems demands a reappraisal of the phylogeny and ecology of zoosporic fungi. Fungal Ecol., 19: 28–38. https://doi.org/10.1016/j.funeco.2015.06.004.

Han, C., Song, A., He, Y., Yang, L., Chen, L. et al. (2024). Genome mining and biosynthetic pathways of marine-derived fungal bioactive natural products. Front. Microbiol., 15: 1520446. https://doi.org/10.3389/fmicb.2024.1520446.

Huang, X., Cai, J., Xiao, J., Zhou, X., Liu, Y. et al. (2025). Antimicrobial isochroman-derived atropo-diastereomeric dimers from *Penicillium Steckii* SCISO 41228. Chem. Biodivers., 22: e202402152. https://doi.org/10.1002/cbdv.202402152.

Huang, X., Huang, H., Li, H., Sun, X., Huang, H. et al. (2013). Asperterpenoid A, a new sesterterpenoid as an inhibitor of *Mycobacterium tuberculosis* protein tyrosine phosphatase B from the culture of *Aspergillus* sp. 16-5c. Org. Lett., 15: 721–723. https://doi.org/10.1021/ol303549c

Jones, E.G.G., Pang, K.L., Abdel-Wahab, M.A., Scholz, B., Hyde, K.D. et al. (2019). An online resource for marine fungi. Fungal Diver., 96: 347–433. https://doi.org/10.1007/s13225-019-00426-5.

Kang, H.S. and Kim, E.S. (2021). Recent advances in heterologous expression of natural product biosynthetic gene clusters in *Streptomyces* hosts. Curr. Opin. Biotechnol., 69: 118–127. https://doi.org/10.1016/j.copbio.2020.12.016.

Khan, M.F., Hof, C., Niemcová, P. and Murphy, C.D. (2023). Recent advances in fungal xenobiotic metabolism: Enzymes and applications. World J. Microbiol. Biotechnol., 39: 296. https://doi.org/10.1007/s11274-023-03737-7.

Kjærbølling, I., Mortensen, U.H., Vesth, T. and Andersen, M.R. (2019). Strategies to establish the link between biosynthetic gene clusters and secondary metabolites. Fungal Genet. Biol., 130: 107–121. https://doi.org/10.1016/j.fgb.2019.06.001.

Kohlmeyer, J. and Kohlmeyer, E. (1979). Marine Mycology: The Higher Fungi. Academic Press, New York.

Koopklang, K., Choodej, S., Hantanong, S., Intayot, R., Jungsuttiwong, S. et al. (2024). Anti-inflammatory properties of oxygenated isocoumarins and xanthone from Thai mangrove-associated endophytic fungus *Setosphaeria rostrata*. Molecules, 29: 603. https://doi.org/10.3390/molecules29030603.

Krehenwinkel, H., Pomerantz, A., Henderson, J.B., Kennedy, S.R., Lim, J.Y. et al. (2019). Nanopore sequencing of long ribosomal DNA amplicons enables portable and simple biodiversity assessments with high phylogenetic resolution across broad taxonomic scale. Giga Sci., 8 (5): giz006. https://doi.org/10.1093/gigascience/giz006.

Kumar, A., Sørensen, J.L., Hansen, F.T., Arvas, M., Syed, M.F. et al. (2018). Genome sequencing and analyses of two marine fungi from the North Sea unraveled a plethora of novel biosynthetic gene clusters. Sci. Rep., 8: 10187. https://doi.org/10.1038/s41598-018-28473-z.

Kumar, V., Sarma, V.V., Thambugala, K.M., Huang, J.J., Li, X.Y. et al. (2021). Ecology and evolution of marine fungi with their adaptation to climate change. Front. Microbiol., 12. https://doi.org/10.3389/fmicb.2021.719000.

Lechtenfeld, O.J., Hertkorn, N., Shen, Y., Witt, M. and Benner, R. (2015). Marine sequestration of carbon in bacterial metabolites. Nat. Commun., 6: 6711. https://doi.org/10.1038/ncomms7711.

Li, W., Wang, M., Burgaud, G., Yu, H. and Cai, L. (2019). Fungal community composition and potential depth-related driving factors impacting distribution pattern and trophic modes from epi- to abyssopelagic zones of the western Pacific Ocean. Microb. Ecol., 78: 820–831. https://doi.org/10.1007/s00248-019-01374-y.

Li, W., Wang, M.M., Wang, X.G., Cheng, X.L., Guo, J.J. et al. (2016). Fungal communities in sediments of subtropical Chinese seas as estimated by DNA metabarcoding. Sci. Rep., 6: 26528. https://doi.org/10.1038/srep26528.

Li, Z., Xing, Y., Liu, P., Liao, W. and Miao, L. (2025). Redox and solvent-stable alkaline serine protease from *Bacillus patagoniensis* DB-5: Heterologous expression, properties, and biotechnological applications. Front. Microbiol., 16: 1558419. https://doi.org/10.3389/fmicb.2025.1558419.

Liu, A.R., Wu, X.P. and Xu, T. (2007). Research advances in endophytic fungi of mangrove. Ying Yong Sheng Tai Xue Bao, 18: 912–918.

Liu, X., Xin, J., Sun, Y., Zhao, F., Niu, C. et al. (2024). Terpenoids from marine sources: A promising avenue for new antimicrobial drugs. Mar. Drugs, 22: 347. https://doi.org/10.3390/md22080347.

Liu, Y., Zhang, G., Luo, X., Hou, E., Zheng, M. et al. (2021). Mycorrhizal fungi and phosphatase involvement in rhizosphere phosphorus transformations improves plant nutrition during subtropical forest succession. Soil Biol. Biochem., 153: 108099. https://doi.org/10.1016/j.soilbio.2020.108099.

Lobo, L.Q., Izabel-Shen, D., Albertsson, J., Raymond, C., Gunnarsson, J.S. et al. (2024). Salinity and resource availability as drivers of Baltic benthic fungal diversity. Environ. DNA, 6: e526. https://doi.org/10.1002/edn3.526.

Luo, Y., Wei, X., Yang, S., Gao, Y.H. and Luo, Z.H. (2020). Fungal diversity in deep-sea sediments from the Magellan seamounts as revealed by a metabarcoding approach targeting the ITS2 regions. Mycology, 11: 214–229. https://doi.org/10.1080/21501203.2020.1799878.

Maeda, K., Spor, A., Edel-Hermann, V., Heraud, C., Breuil, M.C. et al. (2015). N2O production, a widespread trait in fungi. Sci. Rep., 5: 9697. https://doi.org/10.1038/srep09697.

Mainardi, P.H., Feitosa, V.A., de Paiva, L.B.B., Bonugli-Santos, R.C., Squina, F.M. et al. (2018). Laccase production in bioreactor scale under saline condition by the marine-derived basidiomycete Peniophora sp. CBMAI 1063. Fungal Biol., 122: 302–309. https://doi.org/10.1016/j.funbio.2018.01.009.

Matsuzaka, Y. and Yashiro, R. (2022). Applications of deep learning for drug discovery systems with BigData. BioMedInformatics, 2: 603–624. https://doi.org/10.3390/biomedinformatics2040039.

Mirete, S., Sánchez-Costa, M., Díaz-Rullo, J., González de Figueras, C., Martínez-Rodríguez, P. et al. (2025). Metagenome-Assembled Genomes (MAGs): Advances, challenges, and ecological insights. Microorganisms, 13: 985. https://doi.org/10.3390/microorganisms13050985.

Mohamed, A.R., Ochsenkühn, M.A., Kazlak, A.M., Moustafa, A. and Amin, S.A. (2023). The coral microbiome: Towards an understanding of the molecular mechanisms of coral–microbiota interactions. FEMS Microbiol. Rev., 47: fuad005. https://doi.org/10.1093/femsre/fuad005.

Mózsik, L., Iacovelli, R., Bovenberg, R.A.L. and Driessen, A.J.M. (2022). Transcriptional activation of biosynthetic gene clusters in filamentous fungi. Front. Bioeng. Biotechnol., 10: 901037. https://doi.org/10.3389/fbioe.2022.901037.

Naqvi, S.A.H., Abbas, A., Hasnain, A., Bilal, Z., Hakim, F. et al. (2025). Advancing fungal phylogenetics: Integrating modern sequencing, dark taxa discovery, and machine learning. Arch. Microbiol., 207: 192. https://doi.org/10.1007/s00203-025-04392-2.

Naranjo-Ortiz, M.A. and Gabaldón, T. (2019). Fungal evolution: Major ecological adaptations and evolutionary transitions. Biol. Rev. Camb. Philos. Soc., 94: 1443–1476. https://doi.org/10.1111/brv.12510.

Naselli-Flores, L. and Padisák, J. (2023). Ecosystem services provided by marine and freshwater phytoplankton. Hydrobiologia, 850: 2691–2706. https://doi.org/10.1007/s10750-022-04795-y.

Ng, M.S., Soon, N., Chin, M.Y., Ho, S.K., Drescher, L. et al. (2025). Fungi promote cross-domain interactions even in deep anoxic mangrove sediments. Environ. Microbiome, 20: 34. https://doi.org/10.1186/s40793-025-00686-6.

Nilsson, R.H., Larsson, K.H., Taylor, A.F.S., Bengtsson-Palme, J., Jeppesen, T.S. et al. (2019). The UNITE database for molecular identification of fungi: Handling dark taxa and parallel taxonomic classifications. Nucleic Acids Res., 47: D259–D264. https://doi.org/10.1093/nar/gky1022.

O'Brien, P.A., Morrow, K.M., Willis, B.L. and Bourne, D.G. (2016a). Implications of ocean acidification for marine microorganisms from the free-living to the host-associated. Front. Mar. Sci., 3. https://doi.org/10.3389/fmars.2016.00047.

Oliveira, J., Belchior, A., Da Silva, V.D., Rotter, A., Petrovski, Ž. et al. (2020). Marine environmental plastic pollution: Mitigation by microorganism degradation and recycling valorization. Front. Mar. Sci., 7. https://doi.org/10.3389/fmars.2020.567126.

Oppong-Danquah, E., Budnicka, P., Blümel, M. and Tasdemir, D. (2020). Design of fungal co-cultivation based on comparative metabolomics and bioactivity for discovery of marine fungal agrochemicals. Mar. Drugs, 18: 73. https://doi.org/10.3390/md18020073.

Orsi, W.D., Vuillemin, A., Coskun, Ö.K., Rodriguez, P., Oertel, Y. et al. (2022). Carbon assimilating fungi from surface ocean to subseafloor revealed by coupled phylogenetic and stable isotope analysis. ISME J., 16: 1245–1261. https://doi.org/10.1038/s41396-021-01169-5.

Otte, J.M., Blackwell, N., Ruser, R., Kappler, A., Kleindienst, S. et al. (2019). N_2O formation by nitrite-induced (chemo)denitrification in coastal marine sediment. Sci. Rep., 9: 10691. https://doi.org/10.1038/s41598-019-47172-x.

Pan, C., Hassan, S.S.U., Muhammad, I. and Jin, H. (2025). Marine fungi as a goldmine for novel antibiotics: A 2024 perspective. Front. Mar. Sci., 11. https://doi.org/10.3389/fmars.2024.1538136.

Pang, K.L., Chiang, M.W.L., Guo, S.Y., Shih, C.Y., Dahms, H.U. et al. (2020). Growth study under combined effects of temperature, pH and salinity and transcriptome analysis revealed adaptations of *Aspergillus terreus* NTOU4989 to the extreme conditions at Kueishan Island Hydrothermal Vent Field, Taiwan. PLoS One, 15: e0233621. https://doi.org/10.1371/journal.pone.0233621.

Parasnis, M.S., Deng, E., Yuan, M., Lin, H., Kordas, K. et al. (2024). Heavy metal remediation by dry mycelium membranes: Approaches to sustainable lead remediation in water. Langmuir, 40: 6317–6329. https://doi.org/10.1021/acs.langmuir.3c03811.

Passarini, M.R.Z., Rodrigues, M.V.N., da Silva, M. and Sette, L.D. (2011). Marine-derived filamentous fungi and their potential application for polycyclic aromatic hydrocarbon bioremediation. Mar. Pollut. Bull., 62: 364–370. https://doi.org/10.1016/j.marpolbul.2010.10.003.

Peng, X., Amend, A.S., Baltar, F., Blanco-Bercial, L., Breyer, E. et al. (2024). Planktonic marine fungi: A review. J. Geophys. Res.-Biogeosciences, 129: e2023JG007887. https://doi.org/10.1029/2023JG007887.

Pereira, F. and Aires-de-Sousa, J. (2018). Computational methodologies in the exploration of marine natural product leads. Mar. Drugs, 16: 236. https://doi.org/10.3390/md16070236.

Perkins, A.K., Rose, A.L., Grossart, H.P., Schulz, K.G., Neubauer, D. et al. (2023). Fungi increases kelp (*Ecklonia radiata*) remineralisation and dissolved organic carbon, alkalinity, and dimethylsulfoniopropionate (DMSP) production. Sci. Total Environ., 905: 166957. https://doi.org/10.1016/j.scitotenv.2023.166957.

Plemenitaš, A. (2021). Sensing and responding to hypersaline conditions and the HOG signal transduction pathway in fungi isolated from hypersaline environments: *Hortaea werneckii* and *Wallemia ichthyophaga*. J. Fungi, 7: 988. https://doi.org/10.3390/jof7110988.

Plemenitaš, A., Lenassi, M., Konte, T., Kejžar, A., Zajc, J. et al. (2014). Adaptation to high salt concentrations in halotolerant/halophilic fungi: A molecular perspective. Front. Microbiol., 5. https://doi.org/10.3389/fmicb.2014.00199.

Raghukumar, S. (2017). Fungi in Coastal and Oceanic Marine Ecosystems: Marine Fungi. Springer, Cham. https://doi.org/10.1007/978-3-319-54304-8.

Rajakaruna, O., Wijayawardene, N.N., Udagedara, S., Jayasinghe, P.K., Gunasekara, S.S. et al. (2024). Exploring fungal diversity in seagrass ecosystems for pharmaceutical and ecological insights. J. Fungi, 10: 627. https://doi.org/10.3390/jof10090627.

Rasconi, S., Grami, B., Niquil, N., Jobard, M. and Sime-Ngando, T. (2014) Parasitic chytrids sustain zooplankton growth during inedible algal bloom. Front. Microbiol., 5: 229. https://doi.org/10.3389/fmicb.2014.00229.

Ren, M. and Wang, J. (2022). Phylogenetic divergence and adaptation of Nitrososphaeria across lake depths and freshwater ecosystems. ISME J., 16: 1491–1501. https://doi.org/10.1038/s41396-022-01199-7.

Richards, T.A., Leonard, G., Mahé, F., Del Campo, J., Romac, S. et al. (2015). Molecular diversity and distribution of marine fungi across 130 European environmental samples. Proc. R. Soc. Lond B Biol. Sci. https://doi.org/10.1098/rspb.2015.2243.

Roik, A., Reverter, M. and Pogoreutz, C. (2022). A roadmap to understanding diversity and function of coral reef-associated fungi. FEMS Microbiol. Rev., 46: fuac028. https://doi.org/10.1093/femsre/fuac028.

Romano, S., Jackson, S.A., Patry, S. and Dobson, A.D.W. (2018). Extending the "One Strain Many Compounds" (OSMAC) principle to marine microorganisms. Mar. Drugs, 16: 244. https://doi.org/10.3390/md16070244.

Saeed, A.F.U.H., Su, J. and Ouyang, S. (2021). Marine-derived drugs: Recent advances in cancer therapy and immune signaling. Biomedicine & Pharmacotherapy, 134: 111091. https://doi.org/10.1016/j.biopha.2020.111091.

Sahayasheela, V.J., Lankadasari, M.B., Dan, V.M., Dastager, S.G., Pandian, G.N. et al. (2022). Artificial intelligence in microbial natural product drug discovery: Current and emerging role. Nat. Prod. Rep., 39: 2215–2230. https://doi.org/10.1039/d2np00035k.

Santos-Júnior, C.D., Torres, M.D., Duan, Y., Del Río, Á.R. and Schmidt, T.S. (2024). Discovery of antimicrobial peptides in the global microbiome with machine learning. Cell, 187: 3761–3778.e16. https://doi.org/10.1016/j.cell.2024.05.013.

Sarma, V.V. (2012). Diversity and distribution of marine fungi on *Rhizophora* spp. in mangroves. pp. 243–275. *In*: Raghukumar, C. (ed.). Biology of Marine Fungi. Springer, Berlin, Heidelberg.

Schultz, J., Modolon, F., Peixoto, R.S. and Rosado, A.S. (2023). Shedding light on the composition of extreme microbial dark matter: Alternative approaches for culturing extremophiles. Front. Microbiol., 14: 1167718. https://doi.org/10.3389/fmicb.2023.1167718.

Selegato, D.M. and Castro-Gamboa, I. (2023). Enhancing chemical and biological diversity by co-cultivation. Front. Microbiol., 14: 1117559. https://doi.org/10.3389/fmicb.2023.1117559.

Seymour, J.R., Laverock, B., Nielsen, D.A., Trevathan-Tackett, S.M. and Macreadie, P.I. (2018). The microbiology of seagrasses. pp. 343–392. *In*: Larkum, A.W.D., Kendrick, G.A. and Ralph, P.J. (eds.). Seagrasses of Australia: Structure, Ecology, and Conservation. Springer International Publishing, Cham.

Shao, M., Sun, C., Liu, X., Wang, X., Li, W. et al. (2020). Upregulation of a marine fungal biosynthetic gene cluster by an endobacterial symbiont. Commun. Biol., 3: 527. https://doi.org/10.1038/s42003-020-01239-y.

Shin, H.J. (2020). Natural products from marine fungi. Mar. Drugs, 18: 230. https://doi.org/10.3390/md18050230.

Sime-Ngando, T. (2012). Phytoplankton chytridiomycosis: Fungal parasites of phytoplankton and their imprints on the food web dynamics. Front. Microbiol., 3: 361. https://doi.org/10.3389/fmicb.2012.00361.

Slaby, B.M., Franke, A., Rix, L., Pita, L., Bayer, K. et al. (2019). Marine sponge holobionts in health and disease. pp. 81–104. *In*: Li, Z. (ed.). Symbiotic Microbiomes of Coral Reefs Sponges and Corals. Springer Netherlands, Dordrecht.

Smith, H.Z., Hollingshead, C.M. and Kendall, B. (2025). Carbapenem-resistant enterobacterales. In: StatPearls. StatPearls Publishing, Treasure Island, Florida, USA.

Soler-Hurtado, M.M., Sandoval-Sierra, J.V., Machordom, A. and Diéguez-Uribeondo, J. (2016). *Aspergillus sydowii* and other potential fungal pathogens in gorgonian octocorals of the Ecuadorian Pacific. PLoS One, 11: e0165992. https://doi.org/10.1371/journal.pone.0165992.

Steinbach, R.M., Whitner, S. and Amend, A.S. (2025). Marine fungi degrade plastic and can be conditioned to do it faster. Mycologia, 117: 1–8. https://doi.org/10.1080/00275514.2024.2422598.

Susan, J.R., Jones, E.B.G. and Stephen, T.M. (1997). Ultrastructural observations on *Nimbospora bipolaris* (Halosphaeriaceae, Ascomycetes). Philos. Trans. R. Soc. Lond., B, Biol. Sci., 339: 483–489. https://doi.org/10.1098/rstb.1993.0043.

Tang, X., Yu, L., Xu, W., Zhang, X., Xu, X. et al. (2020). Fungal diversity of deep-sea sediments in Mid-Oceanic Ridge area of the East Pacific and the South Indian Oceans. Bot. Mar., 63: 183–196. https://doi.org/10.1515/bot-2018-0112.

Taylor, J.D. and Cunliffe, M. (2016). Multi-year assessment of coastal planktonic fungi reveals environmental drivers of diversity and abundance. ISME J., 10: 2118–2128. https://doi.org/10.1038/ismej.2016.24.

Taylor, M.W., Radax, R., Steger, D. and Wagner, M. (2007). Sponge-associated microorganisms: Evolution, ecology, and biotechnological potential. Microbiol. Mol. Biol. Rev., 71: 295–347. https://doi.org/10.1128/MMBR.00040-06.

Thompson, J.R., Rivera, H.E., Closek, C.J. and Medina, M. (2015). Microbes in the coral holobiont: Partners through evolution, development, and ecological interactions. Front. Cell Infect. Microbiol., 4: 176. https://doi.org/10.3389/fcimb.2014.00176.

Vaksmaa, A., Vielfaure, H., Polerecky, L., Kienhuis, M.V.M., Van Der Meer, M.T.J. et al. (2024). Biodegradation of polyethylene by the marine fungus *Parengyodontium album*. Sci. Total Environ., 934: 172819. https://doi.org/10.1016/j.scitotenv.2024.172819.

Vass, M., Eriksson, K., Carlsson-Graner, U., Wikner, J. and Andersson, A. (2022). Co-occurrences enhance our understanding of aquatic fungal metacommunity assembly and reveal potential host–parasite interactions. FEMS Microbiol. Ecol., 98: fiac120. https://doi.org/10.1093/femsec/fiac120.

Velez, P., Salcedo, D.L., Espinosa-Asuar, L., Gasca-Pineda, J., Hernandez-Monroy, A. et al. (2022). Fungal diversity in sediments from deep-sea extreme ecosystems: Insights into low- and high-temperature hydrothermal vents, and an oxygen minimum zone in the Southern Gulf of California, Mexico. Front. Mar. Sci., 9. https://doi.org/10.3389/fmars.2022.802634.

Vieira, G.A., Magrini, M.J., Bonugli-Santos, R.C., Rodrigues, M.V. and Sette, L.D. (2018). Polycyclic aromatic hydrocarbons degradation by marine-derived basidiomycetes: Optimization of the degradation process. Braz. J. Microbiol., 49: 749–756. https://doi.org/10.1016/j.bjm.2018.04.007.

Virués-Segovia, J.R., Muñoz-Mira, S., Durán-Patrón, R. and Aleu, J. (2023). Marine-derived fungi as biocatalysts. Front. Microbiol., 14: 1125639. https://doi.org/10.3389/fmicb.2023.1125639.

Walsh, T.R., Gales, A.C., Laxminarayan, R. and Dodd, P.C. (2023). Antimicrobial resistance: Addressing a global threat to humanity. PLOS Medicine, 20: e1004264. https://doi.org/10.1371/journal.pmed.1004264.

Wang, Z., Qader, M., Wang, Y., Kong, F., Wang, Q. et al. (2023) Progress in the discovery of new bioactive substances from deep-sea associated fungi during 2020–2022. Front. Mar. Sci., 10. https://doi.org/10.3389/fmars.2023.1232891.

Willems, T., De Mol, M.L., De Bruycker, A., De Maeseneire, S.L. and Soetaert, W.K. (2020). Alkaloids from marine fungi: Promising antimicrobials. Antibiotics, 9: 340. https://doi.org/10.3390/antibiotics9060340.

Wu, J., Wang, W., Yang, Y., Shah, M., Peng, J. et al. (2024). Phenylhydrazone alkaloids from the deep-sea cold seep derived fungus *Talaromyces amestolkiae* HDN21-0307. J. Nat. Prod., 87: 1407–1415. https://doi.org/10.1021/acs.jnatprod.4c00132.

Xu, L., Meng, W., Cao, C., Wang, J., Shan, W. et al. (2015). Antibacterial and antifungal compounds from marine fungi. Mar. Drugs, 13: 3479–3513. https://doi.org/10.3390/md13063479.

Xu, W., Pang, K.L. and Luo, Z.H. (2014). High fungal diversity and abundance recovered in the deep-sea sediments of the Pacific Ocean. Microb. Ecol., 68: 688–698. https://doi.org/10.1007/s00248-014-0448-8.

Xu, Y., Du, X., Yu, X., Jiang, Q., Zheng, K. et al. (2022). Recent advances in the heterologous expression of biosynthetic gene clusters for marine natural products. Mar. Drugs, 20: 341. https://doi.org/10.3390/md20060341.

Youssef, F.S., Ashour, M.L., Singab, A.N.B. and Wink, M. (2019). A comprehensive review of bioactive peptides from marine fungi and their biological significance. Mar. Drugs, 17: 559. https://doi.org/10.3390/md17100559.

Yun, M.S., Sun, J., Lovejoy, C. and Lee, S.H. (2021). Editorial: Microbial response to a rapidly changing marine environment: Global warming and ocean acidification. Front. Microbiol., 12: 731732. https://doi.org/10.3389/fmicb.2021.731732.

Zeghal, E., Vaksmaa, A., Vielfaure, H., Boekhout, T. and Niemann, H. (2021). The potential role of marine fungi in plastic degradation: A review. Front. Mar. Sci., 8. https://doi.org/10.3389/fmars.2021.738877.

Zeng, X., Alain, K. and Shao, Z. (2021a). Microorganisms from deep-sea hydrothermal vents. Mar. Life Sci. Technol., 3: 204–230. https://doi.org/10.1007/s42995-020-00086-4.

Zeng, X., Alain, K. and Shao, Z. (2021b). Microorganisms from deep-sea hydrothermal vents. Mar. Life Sci. Technol., 3: 204–230. https://doi.org/10.1007/s42995-020-00086-4.

Zhang, F., Zhao, M., Braun, D.R., Ericksen, S.S., Piotrowski, J.S. et al. (2020). A marine microbiome antifungal targets urgent-threat drug-resistant fungi. Science, 370: 974–978. https://doi.org/10.1126/science.abd6919.

Zhang, H., Gu, B., Zhou, Y., Ma, X., Liu, T. et al. (2022a). Multi-omics profiling reveals resource allocation and acclimation strategies to temperature changes in a marine dinoflagellate. Appl. Environ. Microbiol., 88: e01213−1222. https://doi.org/10.1128/aem.01213-22.

Zhang, R., Wang, H., Chen, B., Dai, H., Sun, J. et al. (2022b). Discovery of Anti-MRSA secondary metabolites from a marine-derived fungus *Aspergillus fumigatus*. Mar. Drugs, 20: 302. https://doi.org/10.3390/md20050302.

Zhang, Z.F., Liu, F., Liu, L.R., Li, M., Cai, L. et al. (2024). Culturing the uncultured marine fungi in the omics age: Opportunities and challenges. Fungal Biol. Rev., 48: 100353. https://doi.org/10.1016/j.fbr.2023.100353.

Zhou, C., Pan, J., Ge, Y., Wu, X. and Wu, B. (2025) New polyketides from *Trichoderma harzianum* and their antifungal activity. Chem. Biodivers., 22: e202500140. https://doi.org/10.1002/cbdv.202500140.

Zhou, G., Rusnac, D.V., Park, H., Canzani, D., Nguyen, H.M. et al. (2024). An artificial intelligence accelerated virtual screening platform for drug discovery. Nat. Commun., 15: 7761. https://doi.org/10.1038/s41467-024-52061-7.

Zhu, C., Lin, Y., Wang, Z., Luo, W., Zhang, Y. et al. (2023). Community assembly and network structure of epiphytic and endophytic phyllosphere fungi in a subtropical mangrove ecosystem. Front. Microbiol., 14: 1147285. https://doi.org/10.3389/fmicb.2023.1147285.

9

Marine Yeasts

Hidden Microbial Treasures of the Ocean

Ashly Augustine,[1] *Nisha Alex,*[1] *Nimsi K.A.*[2]
and *K. Manjusha*[1,*]

Marine yeasts, the unicellular eukaryotic members of either phylum Ascomycota or Basidiomycota of Kingdom Fungi, are microbial communities that adapt to thrive in saline as well as extreme marine conditions. Over the past few years, researchers have uncovered various biotechnological potentials of these unicellular forms. They seem to inhabit a wide variety of marine environments, such as seawater, deep-sea sediments, hydrothermal vents, mangroves and estuaries, as well as in association with different marine flora and fauna. In comparison to their terrestrial counterparts, these marine inhabitants were reported to exhibit distinct physiological, biochemical, and ecological characteristics. Their diversity, distribution, and abundance are greatly influenced by complex marine factors, including salinity, pressure, temperature, nutrient availability, and oceanic currents. *Candida*, *Rhodotorula*, *Debaryomyces*, *Cryptococcus*, *Aureobasidium*, *Metschnikowia*, and *Saccharomyces* comprise the most commonly isolated yeast genera. Their survival in fluctuating marine conditions depends on an impressive set of physiological and molecular adaptations, including changes in gene expression profiles, enzyme function, and membrane fluidity. These distinctive adaptations qualify them for enormous biotechnological applications in the fields of food,

[1] Department of Marine Biosciences, Faculty of Ocean Science and Technology, Kerala University of Fisheries and Ocean Studies, Kochi, Kerala, India.
[2] Department of Biosciences, MES College, Marampally, Aluva, Kerala, India.
* Corresponding author: manjusha.k@kufos.ac.in

DOI: 10.1201/9781003768708-12

pharmaceuticals, bioremediation, and biofuels. The capacity to produce a wide variety of bioactive molecules (antimicrobials and antioxidants), enzymes (lipases and amylases), biopolymers, pigments, polysaccharides, and biosurfactants makes the marine yeasts an ideal candidate suited for industrial application. *Candida* (18%), *Aureobasidium* (12%), and *Rhodotorula* (11%) are the marine yeasts that find wide industrial applications. However, genera such as *Yarrowia*, *Geotrichum*, *Pichia*, and *Wickerhamomyces* have so far only made a minor contribution and could hold the key to future potential applications in the industry. Despite their environmental and economic importance, marine yeasts are not as well understood as terrestrial yeasts. This calls for in-depth taxonomic, genomic, and functional research. Apprehending the full potential of marine yeasts in terms of its diversity and bioactivity, may lead to the advancements in innovative solutions for pressing problems in the industry, environmental sustainability, and the health sector.

1. Introduction

Yeasts are microscopic fungi that live in a variety of habitats on Earth, including "extreme environments", which are uninhabitable to most life forms. Marine yeasts are a diverse group of extremophilic or extremotolerant unicellular fungi that inhabit saline environments and can grow better on medium prepared using seawater rather than freshwater (Chi et al., 2012).

The evolutionary history of the first simple unicellular fungi in aquatic environments dates back more than ~ 1000 million years ago (Ma); however, yeast in terrestrial environments (grouped as facultative marine yeast) have been reported to have evolved only over ~ 600 Ma (Barbee and Taylor, 2010). Obligatory or indigenous, marine yeasts are those that are restricted to marine environments (Kohlmeyer and Kohlmeyer, 1959; Kutty and Philip, 2008). However, no sufficient evidence has been found to explain the indispensability of seawater for obligate marine yeasts. It has been almost two centuries since studies on yeast diversity began and most of these studies have been focused on terrestrial yeasts (Boekhout, 2005; Lachance, 2006; Kurtzman et al., 2015; Péter et al., 2017). *Torula* sp. and *Mycoderma* sp. were the first red and white marine yeasts to be isolated from the Atlantic Ocean by Bernhard Fisher in 1894 (Fisher, 1894). Following this report, there have been studies from around the world that documented the isolation of marine yeasts from a variety of marine samples, such as saltwater, marine sediments, seaweeds, fish, marine mammals, and even seabirds (Hunter, 1920; Bhat et al., 1955a, 1955b; Suehiro, 1960; van Uden and Fell, 1968; Yu et al., 2023; Mapelli-Brahm et al., 2023). ZoBell and Feltham (1934) have reported the presence of yeasts from both coastal and open ocean environments. Additionally, Russian microbiologists documented the quantitative distribution of yeasts in marine ecosystems, including the Black Sea, Okhotsk Sea, Pacific Ocean, and Arctic Sea (Rukina and Novozhilova, 1952; Kriss, et al. 1952; Kriss, 1955; Novozhilova, 1955).

The observation of high yeast densities in the North Sea (Meyers et al., 1967) suggested the possible presence of native marine yeasts in these areas. Similarly,

Kriss and Rukina (1949) reported high yeast densities during plankton blooms in the Black Sea and the Pacific Ocean, suggesting a potential ecological association between phytoplankton activity and yeast proliferation. From this observation it is clear that the organic matter released during phytoplankton blooms can influence yeast populations, or they may interact positively or opportunistically with phytoplankton communities (Diver et al., 2024). Marine yeasts are broadly distributed across various habitats, *viz*., seawater, sediments, mangroves, hydrothermal vents, and marine invertebrates (Nimsi et al., 2023a) and play a crucial role in oceanic biogeochemical cycles (Starmer et al., 2011). Marine yeasts have an ability to thrive in highly saline environments, utilizing osmoregulatory mechanisms and oxygen-rich but nutrient-limited marine environments by their oxidative metabolism (Hohmann, 2002; Segal et al., 2022). They are one of the important components of marine ecosystems and a valuable resource for industrial and environmental applications (Zaky et al., 2014; Chi et al., 2009; Sarkar et al., 2010). Despite the importance of marine yeasts, our knowledge of their diversity, adaptation mechanisms, and bioprospecting potential remains limited. Against this backdrop, a review of over 60 peer-refereed studies conducted between 1971–2024 highlights the diversity and widespread presence of yeasts in marine environments, while also noting significant gaps in regional data.

2. Taxonomic and Ecological Diversity

Yeasts (unicellular eukaryotic fungi) are distinguished in having a defined nucleus, mitochondria, and a chitinous cell wall, making them structurally more complex than prokaryotic organisms like bacteria (Lipke and Ovalle, 1998). Taxonomically, yeasts do not form a single phylogenetic grouping and span multiple fungal lineages, and display considerable morphological and reproductive diversity. Kurtzman et al. (2011) have defined Yeasts as ascomycetes or basidiomycetes fungi, that reproduce asexually through budding or fission and possess sexual states that are not encapsulated within fruiting bodies. These features clearly differentiate yeasts from other filamentous fungi.

The exact classification and demarcation of yeasts within the fungal phyla Ascomycota and Basidiomycota still remains an area of continuing research and taxonomic ambiguity; despite the considerable advances in fungal taxonomy. In the phylum **Ascomycota,** yeasts are primarily grouped into Saccharomycotina and Taphrinomycotina. These two phyla basically differ in their cell division mechanisms and ecological roles. Within **Basidiomycota** yeasts are grouped into the three major subdivisions, i.e., Ustilaginomycotina, Pucciniomycotina, and Agaricomycotina (Li et al., 2021). This extensive distribution of yeasts among the diverse fungal taxa reflects the evolutionary convergence of this unicellular form and their ecological adaptability, particularly in marine environments.

Some of the yeast genera like *Candida*, *Rhodotorula*, *Debaryomyces*, *Cryptococcus*, *Aureobasidium*, *Metschnikowia*, and *Saccharomyces* are frequently reported from diverse marine habitats. It reflects their adaptability and ecological significance in saline environments. *Candida tropicalis* and *Candida albicans* have been repeatedly isolated from seawater and marine sediments, whereas *Rhodotorula mucilaginosa*, from mangrove ecosystems and coastal waters (Kutty and Philip, 2008). *Debaryomyces hansenii*, a halotolerant yeast species, has been isolated from marine fish and seaweed surfaces, and *C. laurentii* and *C. albidus*, have been recovered from deep-sea sediments and polar marine environments. *Aureobasidium pullulans*, a melanized yeast, is commonly found in marine biofilms and intertidal zones, whereas *Metschnikowia zobellii*, in marine invertebrates (Gao et al., 2013). Interestingly, *Saccharomyces cerevisiae*, traditionally associated with terrestrial and anthropogenic environments, has also been detected in marine ecosystems—likely introduced via human activities and runoff.

Despite these findings, our current understanding of marine yeast diversity remains incomplete. Traditional culture-dependent approaches are limited in scope, as it is estimated that only 0.01–0.1% of marine microbes can be cultured under standard laboratory conditions. This underscores the vast unexplored marine yeasts diversity that exists in marine systems.

3. Identification

According to MycoBank, over 2,000 yeast species have been isolated, identified, and classified. Many of these species originate from extreme environments (Segal et al., 2022). Several sequencing-free techniques are available for molecular identification, including polymerase chain reaction (PCR) with primers and hybridization probes, random amplified polymorphic DNA (RAPD), amplified fragment length polymorphisms (AFLPs), restriction fragment length polymorphisms (RFLPs), DNA fingerprinting, real-time PCR, and MALDI-ToF mass spectrometry (Kurtzman, 2015; Turhan et al., 2017). However, the current gold standard for rapid identification and phylogenetic assignment relies on PCR amplification of target sequences such as the 18S rRNA gene and the internal transcribed spacer (ITS) located between the small subunit (SSU) and large subunit (LSU) of ribosomal RNA genes (Kurtzman et al., 2011). Historically, yeast identification was based on sequence analysis of the D1/D2 domains of the LSU, followed by homolog matching to databases (e.g., GenBank) and phylogenetic placement (Fell et al., 2000).

Whole genome shotgun (WGS) sequencing is now being increasingly applied in microbiome research, including that of marine yeasts, for taxonomic resolution, gene annotation, and metabolic pathway elucidation. However, WGS analysis often requires substantial investment in bioinformatics infrastructure, skilled personnel, and robust analytical pipelines (Mardis, 2010). WGS offers a powerful tool for the identification and functional characterisation of yeasts as well as metagenomics offers direct analysis of environmental DNA and RNA, thereby overcoming the limitations of culture-dependent approaches (Purahong et al., 2019; Zeng et al., 2023). The majority of the studies that have employed a metagenomic approach to explore fungal diversity in mangroves (Haldar and Nazareth, 2019;

Zuo et al., 2022) have sidelined the diversity of yeasts. Insights into fungal diversity associated with mangroves have been provided in the works of De Souza Sebastian et al. (2013), Chi et al. (2019), and Purahong et al. (2019). Next-generation sequencing (NGS) could enable large-scale comparative genomic studies of diverse species (Metzker, 2009).

4. Ecology, Distribution, and Habitats

The biotic and abiotic components of an ecosystem interact, this facilitates the energy flow and nutrient cycling that are necessary to sustain life (Chapin et al., 2011). Compared to other ecosystems, marine ecosystems are considered the most expansive and biologically productive (Kutty, 2010). There exists a wide range of habitats with marine ecosystems, like the open waters, mangroves, estuaries, deep-sea trenches, coral reefs, etc. Some marine yeast species are cosmopolitan in distribution, while others exhibit endemism. Their distribution is influenced by climatic variations, oceanic currents, dispersal mechanisms, interactions with the marine vegetation and fauna, and distinct ecological niches (Nawaz et al., 2022; Ren et al., 2023).

Yeasts unlike bacteria, cyanobacteria, microalgae, and filamentous fungi have not been extensively documented in marine environments. Though marine yeasts have been recognised as a vital component of the marine microbial community, they remain underexplored. They play an indispensable role in biogeochemical cycles and degrade a wide range of organic substrates, including recalcitrant xenobiotics. Their ecological significance is matched by emerging interest in biotechnological exploitation.

Until the 20th century, the majority of the yeast diversity studies have been from the developed countries of the Northern Hemisphere (Kurtzman et al., 2015). However, the 21st century has noted a broader geographic engagement, with more studies being reported from Asian nations (China, India, and Thailand), as well as from South American countries (Brazil and Argentina). Marine yeast research in Africa remains limited, with sporadic studies from South Africa, Egypt, and Nigeria. A more widespread assessment of diversity to tackle this gap in our understanding has been enabled by the expansion of accessible molecular and cultivation technologies.

The yeast genera commonly isolated from marine habitats include *Candida* (39%), *Debaryomyces* (25%), *Rhodotorula* (23%), *Saccharomyces* (19%), *Cryptococcus* (18%), *Kluyveromyces* (17%), and *Geotrichum* (10%). In addition to these prominent taxa, numerous other yeast genera have been identified across different continents. Figure 1 provides a graphical representation of the continent-wise distribution of marine yeast genera, illustrating a clear trend: Asia accounts for the highest diversity, while Europe records the lowest. Although studies have been conducted in the Arctic region, there is a conspicuous absence of recent reports from Antarctica, most likely due to the logistical and technical challenges of sampling in these remote environments.

4.1 Sea Water

The Pacific Ocean has been home to species such as *Candida*, *Torulopsis*, *Trichosporon*, and *Endomycopsis*, as documented by Van Uden (1967), Van Uden and Zobell (1962), and Van Uden and Fell (1963). Yeasts like *Leucosporidium*, *Rhodosporidium*, and *Sympodiomyces* were observed in the Indo-Pacific region (Fell 1976). The presence of *Torulaspora*, *Dekkera*, and *Candida* species in Rendaiji, Shizouka Prefecture, was noted by Urano et al. (1998), while the Southern Ocean yielded *Blastobotrys parvus* (Fell and Statzell, 1971).

Different *Candida* spp. have been reported from varied marine environments; for example, Candida *maris* from Pacific seawater (Gadanho and Sampaio, 2005), *Candida rhizophoriensis* (Burgaud et al., 2011) and *Candida sharkiensis* (Fell et al., 2010) from Florida Everglades, and *Candida torressi* (Sampaio, 2005) from Torres Strait, Australia. *Metchnikowia australis* was found in the South Shetland Islands, Antarctica (Fell, 1961; Lachance, 2011a). *Torula*, *Mycoderma*, *Rhodotorula*, *Saccharomyces*, and *Cryptococcus* were some of the yeast species documented from the Atlantic Ocean, Biscayne Bay, China Sea, and the Gulf Stream (Phaff and Fell, 1970; Roth et al., 1962; Fell and Hunter, 1968; Tian et al., 2021).

4.2 Marine Sediment

Diverse yeast genera, *viz*., *Rhodotorula*, *Debaryomyces*, *Torulopsis*, *Cryptococcus*, *Candida*, *Trichosporon*, *Hansenula*, and *Saccharomyces* were identified, from Biscayne Bay and the Bahamas, (Fell et al., 1973). *Pichia* and *Kluyveromyces* were found in marshes of the Louisiana coast (Ahearn and Meyers, 1972), while *Rhodotorula* was noted from Florida (Lazarus and Koburger, 1974). *Candida*, *Cryptococcus*, *Rhodotorula*, *Torulopsis*, and *Trichosporon* were isolated from coastal Massachusetts and the Bahamas (MacGillivray and Shiaris, 1993; Abe et al., 2001). The deep-sea trenches were reported to harbour genera such as *Kluyveromyces*, *Rhodotorula*, *Debaryomyces*, *Cryptococcus*, *Candida*, and *Pichia* (Abe et al., 2001). The Arabian Sea and the Bay of Bengal were noted for their rich yeast diversity comprising *Candida*, *Cryptococcus*, *Debaryomyces*, *Pichia*, and *Rhodotorula* (Lakshmi, 2005). Zang et al. (2012) reported *Aureobasidium*, *Debaryomyces*, *Rhodotorula*, *Sporidiobolus*, and *Yarrowia* from Anatartica.

4.3 Mangrove Ecosystems

India's mangroves from the East Coast have been studied at length for yeast diversity, these studies report the presence of *Barnettozyma alicaria*, *Saccharomyces cerevisiae*, and *Kuraishia capsulate* from East Coast (Manivannan and Kathiresan, 2009; Kathiresan et al., 2011; Saravanan et al., 2013), whereas *Candida parapsilopsis* and *Hortaea werneckii* have been recorded from the West Coast (Prabhakaran and Gupta, 1990; Nazarath and Gonsalves, 2015; Kachiprath et al., 2019). Prasannakumar et al. (2020) created a DNA barcode library of marine yeasts and this was the first such effort from the Indian coast. Kumbalam and Puthuvype mangroves, from central

Kerala, were studied by Nimsi et al. (2024b), and they isolated 215 manglicolous yeasts.

Mangrove yeasts have been found to be highly diverse. With several novel species being reported from this ecosystem like *Heterocephalacria mucosa* and *Kluyveromyces* from Thailand (Am-In et al., 2008; Kunthiphun et al., 2018), and *Vishniacozyma taiwanica* from Taiwan (Chang et al., 2019). *Lachancea meyersii* and *Kwoniella mangroviensis* have been documented from the mangroves of Philippines, Bahamas, Rio de Janeiro, and China (Fell et al., 2004; Statzell-Tallman et al., 2008). *Candida* and *Issatchenkia*, have been identified as faecal indicators in polluted estuaries (De Almeida, 2005). From decaying mangrove vegetation, *Kazachstania unispora* and *Aureobasidium pullulans* have been isolated in Rio de Janeiro (Hagler et al., 1993) and *Kluyveromyces* species in Hainan, China (Buzdar et al., 2011). *Debaryomyces* and *Rhodotorula* were isolated from the *Rhizophora mangle* seedlings (Newell, 1976).

4.4 Estuaries

Candida and *Rhodotorula* are the dominant genera reported from the Suwannee estuary (Lazarus and Koburger, 1974) and Swedish estuary (Norkrans, 1966a). *Debaryomyces*, *Saccharomyces*, *Endomycopsis*, *Torulopsis*, and *Pichia* were isolated from the estuary in the west coast of Taiwan indicating that this location harboured diverse groups of yeasts (Cheng and Lin, 1977). Apart from the cosmopolitan genera like *Candida* and *Rhodotorula*, other genera like *Debaryomyces*, *Hanseniaspora*, and *Torulopsis* were recorded from the estuary of Rio de Janeiro (Hagler et al., 1993).

4.5 Algae, Corals, and Seaweeds

Rhodosporidium was noted to be associated with the planktonic communities of the Pacific Ocean (Fell et al., 1973). Suehiro (1960) observed that algae harboured the *Candida*, *Torulopsis*, *Trichosporon*, and *Endomycopsis*. *Metschnikowia* was identified as a dominant yeast species in Southern California's giant kelp ecosystems (Van Uden and Castelo-Branco, 1963). *Sporobolomyces* was isolated from marsh plants in England (Pugh and Lindsey, 1975); *Metschnikowia*, *Candida*, *Pichia*, *Kluyveromyces*, and *Torulopsis* were isolated and identified from both algae and corals (Van Uden and Zobell, 1962). *Candida taylori* was found in coral reefs (Statzell-Tallman et al., 2010; Gadanho and Sampaio, 2005). These observations clearly indicate that yeasts interact with the phytoplanktons and vegetation in the marine ecosystem, and also that the flora influences yeast diversity.

4.6 Invertebrates

Shrimp eggs and sponges were found to harbour *Debaryomyces*, *Torulopsis*, and *Trichosporon* (Siepmann and Hohnk, 1962). While, *Penaeus setiferus* harboured *Trichosporon*, *Rhodotorula*, *Candida*, and *Pichia* (Phaff et al., 1952),

Penaeus schmitti contained *Candida*, *Geotrichum*, *Trichosporon*, *Debaryomyces*, *Rhodotorula*, *Kloeckera*, *Cryptococcus*, and *Hansenula* (Pagnocca, 1989) and the brine shrimp (*Artemia salina*) hosted *Metschnikowia* (Fize et al., 1970). *Rhodotorula* and *Metschnikowia* were isolated from the amphipod *Podocerus brasiliensis* (Fell and Hunter, 1968), and the marine copepod *Calanus* (Siepmann and Hohnk, 1962) respectively. *Candida*, *Torulaspora*, and *Lachancea* were associated with several insect species (Morais et al., 1992; Fell et al., 2004). *Malasseziales* was isolated from the mussels from the Fiji Basin (Van Dover et al., 2007). Crabs and bivalve mollusks were found to host *Candida*, *Debaryomyces*, *Issatchenkia*, *Kloeckera*, *Kluyveromyces*, *Pichia*, *Saccharomyces*, *Torulaspora*, *Exophiala*, *Fonsecaea*, *Rhodotorula*, and *Torulopsis* (Araujo et al., 1995; Vicente et al., 2012). Additionally, the spiny lobster (*Panulirus argus*) and conch (*Strombus gigas*) were found to harbour *Candida*, *Rhodotorula*, and *Cryptococcus* (Volz et al., 1974). *Metchnikowia bicuspidata* was identified as a pathogen of marine invertebrates (Fell et al., 2004). These findings exemplify the incredible variety of yeasts that are associated with marine invertebrates and that both positive and negative interaction exists between yeasts and their invertebrate host.

4.7 Fishes, Birds, and Mammals

Saccharomyces, *Debaryomyces*, and *Rhodotorula* were isolated from the gastrointestinal tract of rainbow trout (Fell and van Uden, 1963). *Candida* was found to be cosmopolitan in distribution in fishes, sea birds, and mammals, as this yeast genera was isolated from bird excreta (Andlid et al., 1995), stomach of Dolphins and Porpoises (Van Uden and Castelo-Branco, 1963), Chilean seabass (Valderrama et al., 2021), the beluga fish (*Huso huso*) (Fami et al., 2021), killer whales, harbour seals, dolphins, and sea lions (Ka-Lai et al., 2024). Ka-Lai et al. (2024) noted that *Candida* spp. are the normal mycobiota of marine mammals. Yeasts like *Geotrichum*, *Cryptococcus*, *Rhodotorula*, *Trichosporon*, *Candida*, and *Debaryomyces*, were found in various fish species (Tartor et al., 2018; Fami et al., 2021). Both *Metchnikowiazobellii* and *Rhodosporidium* were observed in fish guts in the Pacific Ocean (Lachance, 2011b; Van Uden and Castelo-Branco, 1961; Reinoso et al., 2023). These findings underscore the ubiquity of yeast species across a broad range of marine and aquatic organisms, indicating their importance in marine microbiomes.

4.8 Hydrothermal Vents

A complex mix of yeasts, including *Pichia*, *Cryptococcus*, *Rhodospordium*, *Rhodotorula*, *Dioszegia*, *Clavispora*, *Sporidiobolus*, *Aremonium*, *Geomyces*, *Aspergillus*, and *Fomitopsis*, were isolated from the South Mid-Atlantic Ridge hydrothermal site (Xu et al., 2017; 2018). *Candida atmospherica*, *Candida oceani*, and *Candida spencermartinsiae* were reported from the hydrothermal vents (Gadanho et al., 2003; Van Uden and ZoBell, 1962; Fell et al., 2010).

Table 1. Studies marine yeast in different geographic regions.

Study Station/Location	Identified Yeast	References
Sea Water		
Pacific Ocean	*Candida, Torulopsis, Trichosporon,* and *Endomycopsis*	Van Uden (1967); Van Uden and Zobell (1962); Van Uden and Fell (1963)
Rendaiji, Shizouka Prefecture	*Torulaspora, Dekkera, Candida*	Urano et al. (1998)
Oceanic waters, Southern Ocean	*Blastobotrys parvus*	Fell and Statzell (1971); Fell (1976)
Florida Everglades	*Candida rhizophoriensis*	Burgaud et al. (2011)
Florida Everglades	*Candida sharkiensis*	Fell et al. (2010)
Seawater, South Shetland Islands, Antarctica	*Metchnikowia australis*	Fell (1961); Lachance (2011a)
Atlantic Ocean	Torula, Mycoderma	Phaff and Fell (1970)
Biscayne Bay	Candida, Rhodotorula	Roth et al. (1962)
Gulf Stream, Bahamas	*Candida, Rhodotorula, Cryptococcus, Debaryomyces*, and black yeasts	Fell and Hunter (1968)
China Sea	*Saccharomyces*	Tian et al. (2021)
Arabian Gulf surrounding Qatar	*Candida, Pichia*, and *Meyerozyma*	Fotedar et al. (2022)
Marine Sediment		
Biscayne Bay, Bahamas	*Rhodotorula, Debaryomyces, Torulopsis, Cryptococcus, Candida, Trichosporon, Hansenula*, and *Saccharomyces*	Fell et al. (1973)
Marshes, Louisiana coast	*Pichia, Kluyveromyces*	Ahearn and Meyers (1972)
Florida	*Rhodotorula*	Lazarus and Koburger (1974)
Bahamas	*Cryptococcus*	Abe et al. (2001)
Coastal Massachusetts	*Candida, Cryptococcus, Rhodotorula, Torulopsis,* and *Trichosporon*	MacGillivray and Shiaris (1993)
Japan Trench	*Cryptococcus*	Abe et al. (2001)
Arabian Sea and Bay of Bengal, India	*Candida, Rhodotorula, Cryptococcus, Debaryomyces*, and *Pichia*	Lakshmi (2005)
Antarctica	*Yarrowia, Debaryomyces, Rhodotorula, Rhodotorula, Sporidiobolus, Aureobasidium, Mrakia,* and *Guehomyces*	Zang et al. (2012)
Mangroves		
Chapora, Mandovi, and Zuari estuaries of Goa, India	*Candida, Debaryomyces, Geotrichum, Cryptococcus, Rhodosporidium, Kloeckera, Saccharomycete*, and *Trichosporon*	Da Costa and D'Souza (1979)
Mangalvan, Mangrove Ecosystem of Cochin Backwater, India	*Geotrichum, Lachancea*, and *Geotrichum*	Prabhakaran and Gupta (1990); Fell et al. (2004)

Table 1 contd. ...

...*Table 1 contd.*

Study Station/Location	Identified Yeast	References
Northern Bahamas Island	*Lachancea*	Fell et al. (2004)
Tagus Estuary, Portugal	*Candida, Clavispora, Rhodosporidium, Rhodotorula, Cryptococcus, Hansenula, Debaryomyces*, and *Pichia*	De Almeida (2005)
Ranong Province, Thailand	*Kluyveromyces*	Am-In et al. (2008)
Florida Everglades and Bahamas	*Kwoniella*	Fell (2008)
Phanganga Province, Thailand	*Candida*	Limtong et al. (2008)
Bahamas (Bimini and Highbourne Cay, Exuma, and Belize)	*Candida*	Statzell-Tallman et al. (2010)
Calatagan, Batangas, Philippines	*Candida, Pichia*	Simon Ramirez et al. (2010)
Tagus River estuary, Portugal	*Pichia, Candida, Debaryomyces*, and *Rhodotorula*	Coelho et al. (2010)
Shark River Slough of the Florida Everglades, USA	*Candida, Rhodotorula*, and *Cryptococcus*	Fell et al. (2011)
Ranong Province Thailand	*Kluyveromyces*	Am-In et al. (2011)
Rio de Janeiro, Brazil	*Kluyveromyces*	Araujo and Hagler (2011)
Southeast coast, India	*Candida, Debaryomyces, Geotrichum, Pichia, Rhodotorula, Cryptococcus*, and *Yarrowia*	Kathiresan and Saravanakumar (2011)
Fujian, Guangdong, and Hainan Provinces, China	*Candida, Saccharomycete, Trichosporon, Williopsis, Pichia, Issatchenkia, Kazachstania, Clavispora, Debaryomyces*, and *Geotrichum*	Chi et al. (2012)
Pichavaram mangrove forest, southeast coast, India	*Candida, Kuraishia*, and *Saccharomyces*	Senthilraja and Kathiresan (2015)
Mandovi estuary Goa, India	*Hortaea*	Nazareth and Gonsalves (2015)
Muthupet, India (Saviyarmunnai, Vavalthottam, Manakkattai, and Devvararul)	*Aureobasidium, Candida, Cryptococcus, Dioszella, Leucosporidilla Rhodotorula*, and *Saccharomyces*	Jayalakshmi and Umamaheshwari (2016)
Samutsakorn, Tart	*Prototheca, Saitozyma, Rhodotorula, Pseudozyma, Aureobasidium, Blastobotrys, Candida, Dipodascus, Kloeckera, Kluyveromyces, Kodamaea, Kuraishia, Metahyphopichia, Meyerozyma, Nakazawaea, Ogataea, Saccharomycopsis, Schwanniomyces, Torulaspora, Wickerhamomyces, Pseudozyma*, and *Rhodotorula*	Kunthiphun et al. (2018)

Table 1 contd. ...

...Table 1 contd.

Study Station/Location	Identified Yeast	References
Zhangzhou, Fujian Province, China	*Sporidiobolus*	Shen et al. (2019)
Malippuram, Ernakulam, India	*Candida*	Kachiprath et al. (2019)
Ranong province, Thailand	*Heterocephalacria*	Kunthiphun et al. (2019)
Coastlines of Thailand	*Saitozyma, Rhodotorula Cystobasidiomycetes, Sakaguchia Papiliotrema, Naganishia, Kwoniella Candida, Suhomyces, Wickerhamomyces, Kluyveromyces*, and *Pichia*	Hoondee et al. (2019)
Estuaries of Puerto Rico	*Candida, Saccharomyces*, and *Cryptococcus*	Rosario et al. (2019)
Surat, Gujarat in India	*Candida, Trichosporon*	Patel (2020)
Southeast coast (Kachchh, Gujarat) and West Bengal (Sundarbans), India	*Cystobasidiopsis, Slooffia, Udeniozyma, Colacogloea, Pichia, Candida, Debaryomyces*, and *Yamadazyma*	Prasannakumar et al. (2020)
Rio de Janeiro, Brazil	*Aureobasidium, Saccharomyces*	Hagler et al. (1993)
Godavari and Krishna deltas, east coast of India	*Candida*	Sarma and Vittal (2001)
Ranong Province, Thailand	*Candida, Kungkrabaensis*	Limtong et al. (2010)
Hainan, Guangdong, and Fujian Provinces, China	*Kluyveromyces*	Buzdar et al. (2011)
Fujian, Guangdong, and Hainan Provinces, China	*Aureobasidium, Clavispora Debaryomyces, Galactomyces Geotrichum, Issatchenkia Kazachstania, Kluyveromyces, Metschnikowia Kodamaea, Pichia, Rhodotorula Rhodosporidium, Saccharomyces Saturnispora, Trichosporon, Yarrowia*, and *Zygoascus*	Chi et al. (2012)
Shabi Creek, Gwadar, Pakistan	*Debaryomyces, Shizosaccharomyces, Candida, Debaryomyces*, and *Saccharomyces*	Ahmed (2019)
Ranong province, Thailand	*Heterocephalacria*	Kunthiphun et al. (2019)
Estuaries, Puerto Rico	*Meyerozyma, Kwoniella*, and *Jaminaea*	Rosario et al. (2019)
Red Sea Coast, Saudi Arabia	*Hortaea*	Hodhod et al. (2020)
Abaco Island, Bahamas	*Candida, Rhodotorula*, and *Cryptococcus*	Volz et al. (1974)
Sepetiba Bay, Rio de Janeiro, Brazil	*Candida, Geotrichum, Trichosporon, Debaryomyces, Rhodotorula, Kloeckera, Cryptococcus*, and *Hansenula*	Pagnocca (1989)
Rio de Janeiro, Brazil	*Candida, Torulaspora*	Morais et al. (1992)
Northern Bahamas Islands	*Lachancae*	Fell et al. (2004)
Coast of Northeastern region, Brazil	*Exophiala, Fonsecaea*	Vicente et al. (2012)

Table 1 contd. ...

...Table 1 contd.

Study Station/Location	Identified Yeast	References
Estuaries		
Estuary, Sweden	*Candida, Rhodotorula*	Norkrans (1966a)
Suwannee estuary, Florida	*Candida, Rhodotorula, Cryptococcus*, and *Hansenula*	Lazarus and Koburger (1974)
West coast estuary, Taiwan	*Saccharomyces, Torulopsis, Debaryomyces, Endomycopsis*, and *Pichia*	Cheng and Lin (1977)
Estuary, Rio de Janeiro, Brazil	*Candida, Rhodotorula, Debaryomyces, Hanseniaspora*, and *Torulopsis*	Hagler et al. (1993)
Weeds, Corals, and Algae		
Algae	*Candida, Torulopsis, Trichosporon*, and *Endomycopsis*	Suehiro (1960)
Algae and corals, Torres Strait region, Australia	*Metschnikowia, Candida, Pichia, Kluyveromyces, Torulopsis, Metschnikowia, Candida, Pichia, Kluyveromyces*, and *Torulopsis*	Van Uden and Zobell (1962)
Giant kelp, Southern California	*Metschnikowia*	van Uden and Castelo-Branco (1963)
Plankton, Pacific Ocean	*Rhodosporidium*	Fell et al. (1973)
Marsh plants, England	*Sporobolomyces*	Pugh and Lindsey (1975)
Invertebrates		
Shrimp, Mexico (*Penaeussetiferus)*	*Trichosporon, Rhodotorula, Candida*, and *Pichia*	Phaff et al. (1952)
Shrimp eggs and sponges, North Atlantic Ocean	*Debaryomyces, Torulopsis*, and *Trichosporon*	Siepmann and Hohnk (1962)
Amphipod (*Podocerus brasiliensis*) Atlantic Ocean	*Rhodotorula*	Fell and Hunter (1968)
Brine shrimp (*Artemia salina*), Western Asia	*Metschnikowia*	Fize et al. (1970)
Marine copepods (*Calanus*), Southern France	*Metschnikowia*	Siepmann and Hohnk (1962)
Drosophiloids, Rio de Janeiro, Brazil	*Candida, Torulaspora*	Morais et al. (1992)
Insects, Bahama Islands	*Lachancea*	Fell et al. (2004)
Crab (*Ucidescordatus*)	*Exophiala, Fonsecaea*	Vicente et al. (2012)
Spiny lobster (*Panulirus argus*), Conch (*Strombus gigas*), Bahama Islands	*Candida, Rhodotorula,* and *Cryptococcus*	Volz et al. (1974)
Shrimp (*Peneausschmitti)*	*Candida, Geotrichum, Trichosporon, Debaryomyces, Rhodotorula, Kloeckera, Cryptococcus*, and *Hansenula*	Pagnocca (1989)
Sponges in Sattahip Bay, Gulf of Thailand, South China Sea	*Cutaneotrichosporon, Cystobasidium, Naganishia, Papiliotrema, Rhodosporidiobolus, Rhodotorula, Trichosporon,* and *Vishniacozyma*	Kaewkrajay et al. (2021)

Table 1 contd. ...

...Table 1 contd.

Study Station/Location	Identified Yeast	References
Fishes, Birds, and Mammals		
Rainbow trout, Indian Ocean	*Debaryomyces, Saccharomyces*, and *Rhodotorula*	Fell and van Uden (1963)
Birds (excreta), Pacific and Atlantic	*Candida, Torulopsis*	Andlid et al. (1995)
Dolphin and porpoise (stomach), Indian Ocean	*Candida*	van Uden and Castelo-Branco (1963)
Genypterus chilensis (*G. chilensis*), Colorado Chile	*Rhodotorula, Candida, Cystobasidium*, and *Yamadazyma*	Valderrama et al. (2024)
Oreochromis, Clarias gariepinus, Dakahlia Governorate, Egypt	*Candida, Cryptococcus, Debaryomyces, Rhodotorula, Trichosporon,* and *Geotrichum*	Tartor et al. (2018)
Beluga (*Huso huso*) Fish	*Candida*	Fami et al. (2021)
mussels (*Bathymodiolusbrevior*), Fiji Basin	*Malasseziales*	Van Dover et al. (2007)
Shrimp eggs, North Atlantic Ocean	*Candida atlantica*	Seipmann and Hohnk (1962)
Pathogen of marine invertebrates, Bahamas Island	*Metchnikowia bicuspidata*	Fell et al. (2004)
Fish guts Pacific	*Metchnikowia zobellii*	Lachance (2011b)
Fish guts, Pacific	*Metchnikowia zobellii*	Van Uden and Castello-Branco (1961)
Porpoise (intestine), San Pedro, Santa Elena	*Rhodosporidium*	Reinoso et al. (2023)
Dolphin, killer whale, harbour seal, North Atlantic, and North Pacific oceans	*Candida*	Ka-Lai et al. (2024)
Hydrothermal Vents		
Hydrothermal site, South Mid-Atlantic Ridge	*Pichia, Cryptococcus, Rhodospordium, Rhodotorula, Dioszegia, Clavispora, Sporidiobolus, Aremonium, Geomyces*, and *Fomitopsis*	Xu et al. (2017, (2018)
Hydrothermal vent, Mid-Atlantic Ridge	*Candida oceani*	Van Uden and ZoBell (1962)
Hydrothermal vent, Coral reefs, Florida and Belize	*Candida spencermartinsiae*	Fell et al. (2010)

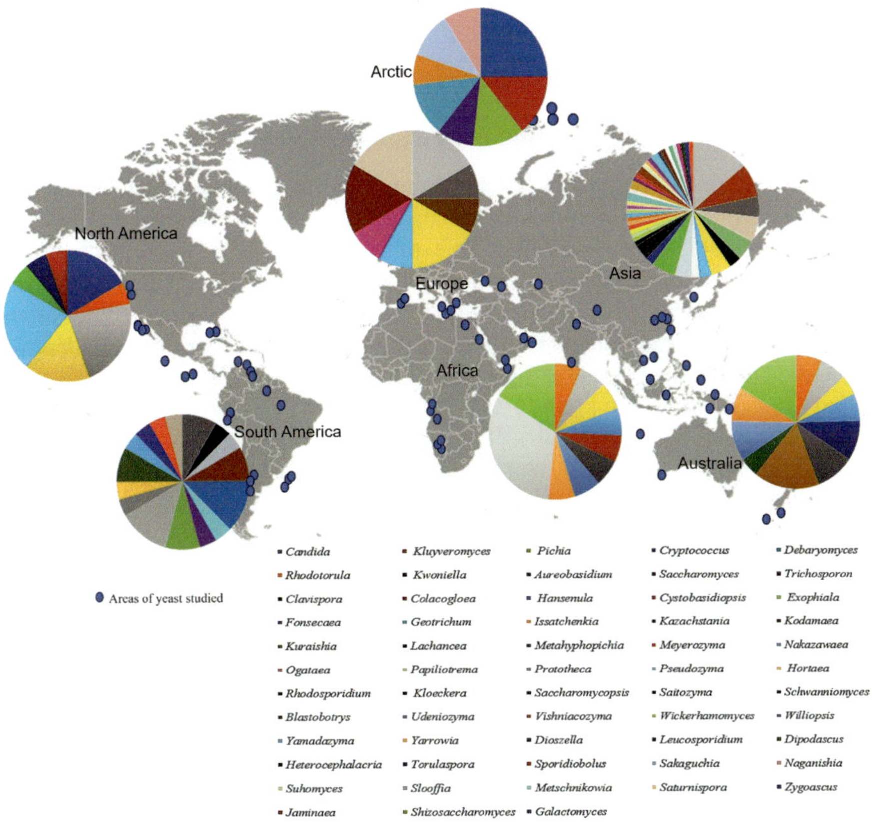

Fig. 1. The relative abundance of marine yeast genera.

The presence of yeasts in hydrothermal vents points out to their amazing adaptability to extremely harsh environments.

5. Marine Yeasts and their Adaptations

Marine yeasts reside in some of the most harsh and erratic environments on this Earth. Their habitats can range from frigid waters to high-pressure depths of the ocean to hypersaline lagoons. To survive and thrive in such challenging habitats, these organisms have evolved a suite of remarkable physiological and molecular adaptations (Table 2). Among the most studied are their mechanisms of osmotolerance and halotolerance, enabling them to maintain cellular integrity and function under high salinity conditions. Additionally, yeasts isolated from deep-sea environments exhibit adaptations to low temperatures and high hydrostatic pressure (HHP), including modifications in membrane fluidity, enzyme function, and gene expression profiles. These unique adaptations not only bring to light the ecological versatility of marine yeasts but also demonstrate their potential for biotechnological applications in stress-tolerant enzyme production, bioremediation, and industrial fermentation processes.

5.1 Adaptations to Hypersaline Conditions

Marine fungi, including yeasts isolated from hypersaline environments, deep-sea hydrothermal vents, and deep-sea sediments, exhibit diverse molecular and cellular mechanisms that facilitate survival under extreme osmotic stress. These mechanisms include stress-responsive gene expression, activation of the high-osmolarity glycerol (HOG) signalling pathway, melanization of the cell wall, regulated ion composition and accumulation, modifications in fatty acid biosynthetic enzymes, and adaptive changes in plasma membrane composition (Turk and Plemenitaš, 2002; Kogej et al., 2006; Romeo et al., 2020) (Table 2; Fig. 2).

Yeasts have been reported to be more osmotolerant than bacteria (Brown, 1978). *Debaryomyces hansenii* and *Saccharomyces rouxii* are notable for their high osmotolerance, capable of growing in media containing up to 4 M NaCl (Onishi, 1963; Norkrans, 1966), whereas *S. cerevisiae* is limited by NaCl concentrations exceeding 1.7 M (Onishi, 1963). In *D. hansenii*, exposure to high salt stress leads to decreased intracellular K^+ levels and increased Na^+ accumulation (Norkrans, 1968). However, intracellular salt alone is insufficient to counterbalance the external osmotic pressure; hence, compatible solutes like polyols accumulate in response to osmotic stress (Brown and Simpson, 1972; Gustafsson and Norkrans, 1976; Adler et al., 1985). Cells with elevated polyol content also demonstrate enhanced tolerance to sudden osmotic dehydration (Adler and Gustafsson, 1980).

In *D. hansenii*, glycerol is the primary osmolyte during exponential growth, while arabinitol becomes dominant in stationary-phase cells (Adler and Gustafsson, 1980). A direct correlation has been observed between intracellular glycerol concentration and environmental salinity (Adler et al., 1985; André et al., 1988). Glycerol levels can reach molar concentrations under high-salt conditions (Gustafsson and Norkrans, 1976). Its synthesis and catabolism are mediated by enzymes such as glycerol kinase and mitochondrial glycerol-3-phosphate dehydrogenase (Gancedo et al., 1968; Sprague and Cronan, 1977). In glycerol kinase-deficient yeasts, alternative pathways involving NAD^+-dependent glycerol dehydrogenase and dihydroxyacetone kinase are employed (Babel and Hofmann, 1982; May et al., 1982).

Among marine yeasts, *D. hansenii* is notable for its ability to accumulate high intracellular Na^+ without exhibiting Na^+ toxicity, showing similar tolerance to Na^+ and K^+ (Ross and Morris, 1962; Norkrans, 1966; Prista et al., 1997). Glycerol also functions as a compatible solute in a glycerol/Na^+ symporter, contributing to osmoregulation (Lages et al., 1999; Lucas et al., 1990). Effective osmotolerance in marine yeasts likely requires enhanced transporter activity and mechanisms for sodium efflux, potentially mediated by Na^+-ATPases or Na^+/H^+ antiporters (Fig. 2) (Ramos et al., 2011).

In *Saccharomyces cerevisiae*, osmotic stress is sensed by mitogen-activated protein kinase (MAPK) signalling pathways, particularly the HOG pathway (Fig. 2). Environmental stimuli are detected by membrane sensors, such as tyrosine kinase receptors, GPCRs, and histidine-aspartate phosphorelay systems—which activate a MAPK cascade involving sequential phosphorylation of MAPKKKs, MAPKKs, and MAPKs. Activated MAPKs translocate to the nucleus and modulate transcriptional responses to restore homeostasis (Plemenitaš et al., 2014).

The HOG pathway is highly conserved and functionally characterised in the halotolerant marine yeast *Hortaeawerneckii*, where it enables survival under fluctuating salinity, marine, and deep-sea conditions (Plemenitaš et al., 2014; Gunde-Cimerman et al., 2018; Romeo et al., 2020). Whole-genome sequencing of *H. werneckii* revealed upregulation of genes related to stress-activated MAPK cascades (GO:0051403), MAPKKK activity (GO:0004709), and cellular responses to osmotic, heat, and oxidative stress (GO:0071470, GO:0034605, GO:0034599), along with genes regulating the mitotic cell cycle (GO:0007346) (Romeo et al., 2020).

A critical component of this response is the activation of heat-shock proteins (HSPs) (Fig. 2), such as HSP88, mitochondrial HSP78, and HSP DnaJ. These proteins help maintain proteostasis and mitigate damage from osmotic and thermal

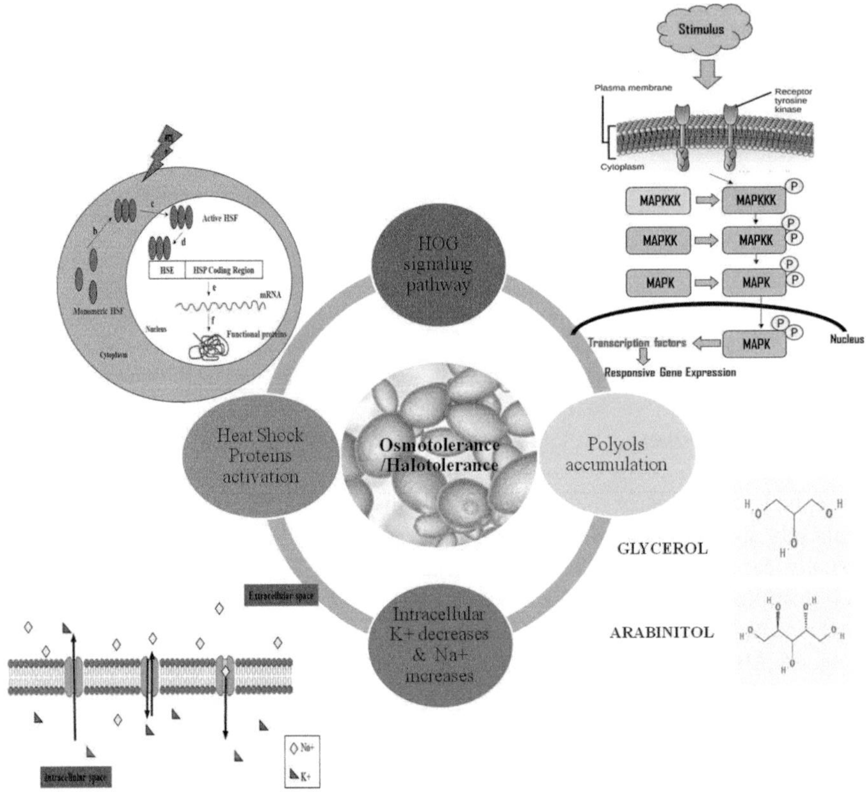

Fig. 2. Mechanism for adaptation to high saline conditions.

stress. HSP expression is regulated by heat-shock transcription factors (HSFs), particularly HSF1, which binds to heat-shock elements (HSEs) in the promoter regions of Hsp genes. In marine yeasts like *H. werneckii*, the heat shock response appears to be fine-tuned for hypersaline environments, with immediate modulation of pre-existing proteins and membrane transport systems, rather than relying solely on *de novo* protein synthesis (Kejžar et al., 2015; Romeo et al., 2020).

The multiple strategies adopted by yeasts such as osmosensing, regulation of ion transport, compatible solute accumulation, signal transduction, and expression of stress protein help them to continue their cellular function at same time, acclimatise to fluctuations in salinity. These traits form the foundation of their successful survival in these extreme environments as well as their imminent biotechnological exploitation.

5.2 *Adaptations to Cold and High-Pressure Conditions*

Though yeasts isolated from the deep-sea are most often psychrotolerant, they exhibit optimal growth at mesophilic conditions. Singh et al. (2010) in their study noted that, the isolates from deep-sea sediments showed more rapid growth at 30°C than at 5°C. Nagano et al. (2014) made the interesting observation that, when cultured at 3°C or 10°C, no significant difference in growth could be observed between deep-sea and terrestrial yeast strains. They also made a note that some terrestrial yeasts could survive under deep-sea settings such as low temperature (3°C) and high pressure (40 MPa). Despite these observations, marine yeasts, especially deep-sea isolates demonstrate specific adaptations to high hydrostatic pressure and salinity.

Cultivating deep-sea yeasts under simulated natural conditions poses challenges, especially due to limited oxygen availability under high pressure and cold temperatures. Since most yeasts are aerobic, this oxygen limitation hampers long-term physiological studies. To address this, Hiraki et al. (2012) developed a specialised cultivation system with a continuous oxygen supply under high-pressure conditions.

At the molecular level, studies on the effects of high pressure in deep-sea yeasts are hindered by difficulties in performing genetic manipulations such as gene disruption, overexpression, and site-directed mutagenesis. *Saccharomyces cerevisiae*, a facultative anaerobe, is widely used as a model organism in high-pressure studies. One of the key adaptations of *S. cerevisiae* to high pressure and low temperature involves the uptake of the amino acid tryptophan, mediated by the tryptophan permeases Tat1 and Tat2. Cell growth under such extreme conditions is directly influenced by tryptophan availability. Experimental wild-type strains typically carry nutrient auxotrophic markers (e.g., *ura3* (uracil), *ade2* (adenine), *his3* (histidine), *leu2* (leucine), *lys2* (lysine), or *trp1* (tryptophan)) for plasmid selection. Among these, the *trp1* mutation leads to growth deficiency at pressures of 15–25 MPa or at temperatures of 10–15°C, regardless of other auxotrophic markers. In contrast, tryptophan-prototrophic strains are capable of growth under these pressures and temperature ranges. Supplementation with excess tryptophan, insertion of the *TRP1* gene, or overexpression of *TAT1* or *TAT2* enhances tryptophan

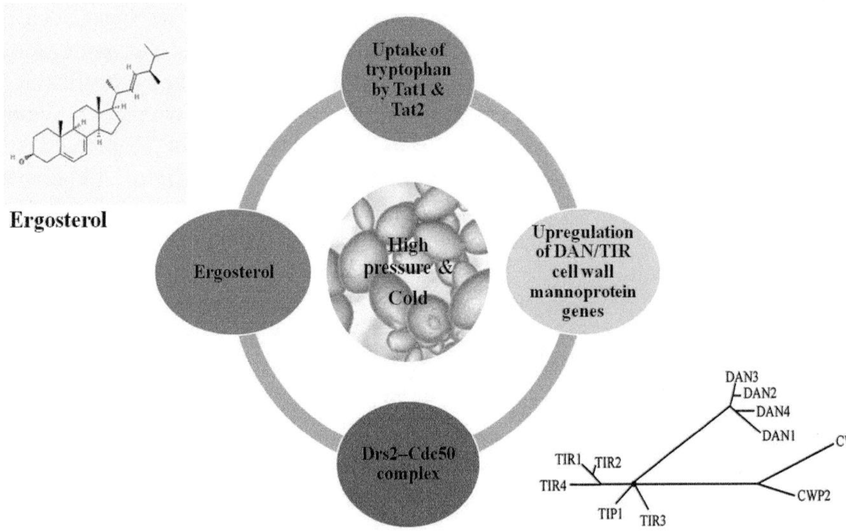

Fig. 3. Mechanism for adaptation to high pressure and cold conditions.

uptake and enables *trp1* mutants to grow under high-pressure or low-temperature conditions (Fig. 3) (Abe and Horikoshi, 2000).

The physicochemical properties of the plasma membrane also play a crucial role in yeast adaptation to high pressure and low temperature. These conditions increase the order of lipid bilayers, restricting the rotational motion of acyl chains. Kaneshina et al. (1998) demonstrated that with increasing pressure up to 100 MPa, the main phase transition temperature (gel to liquid crystalline, Lβ/La) of stearoyloleylphosphatidylcholine (SOPC) and dioleylphosphatidylcholine (DOPC) membranes rises by 18.1°C and 23.3°C, respectively. Consequently, tryptophan uptake becomes sensitive to decreases in membrane fluidity under high-pressure or low-temperature conditions.

Transcriptomic responses further elucidate yeast adaptations. Exposure of tryptophan-prototrophic strains to 30 MPa at 25°C led to upregulation (more than twofold) of 366 genes and downregulation of 253 genes. The upregulated genes were associated with stress responses and the metabolism of carbohydrates, lipids, and amino acids (e.g., *INO1*, *OPI3*, *PST1*, *RTA1*, *SED1*, *PRM5*), as well as membrane structure biosynthesis (*POX1*) (Iwahashi et al., 2005). Similarly, exposure to 200 MPa for 30 minutes resulted in upregulation of stress defence and carbohydrate metabolism genes, while genes related to cell cycle progression and protein synthesis were downregulated (Fernandes et al., 2004).

Nagano et al. (2014) reported that under high-pressure and low-temperature conditions, out of 6,337 genes analysed, 561 were upregulated and 161 downregulated. Hypoxic stress in these conditions led to the upregulation of *DAN/TIR* family cell wall mannoprotein genes, contributing to a reinforced cell wall as a physical defence mechanism (Abe, 2007b; Abramova et al., 2001). In marine yeasts, the inability to form unsaturated lipid bonds under hypoxia further reduces membrane fluidity. It is hypothesised that membrane sensor(s) detect these fluidity changes and transduce intracellular signals that activate *DAN/TIR* gene transcription (Fig. 3) (Nagano et al., 2014). The genes involved in a broad range of cellular functions, such as amino acid biosynthesis, mitochondrial function, the actin cytoskeleton, membrane trafficking, transcription, ribosome biogenesis, chromatin structure, and unknown roles, support yeast cell growth under high pressure and low temperature. About 71 genes were required for growth at high pressure and 56 genes for low temperature (Abe and Minegishi, 2008).

Genes involved in diverse cellular functions—including amino acid biosynthesis, mitochondrial activity, actin cytoskeleton organization, membrane trafficking, transcription regulation, ribosome biogenesis, chromatin structure, and others with yet unknown roles—support yeast growth under high-pressure and low-temperature conditions. Abe and Minegishi (2008) identified 71 genes required for high-pressure tolerance and 56 genes for low-temperature tolerance. Amongst these, the Drs2–Cdc50 protein complex has been identified as critical for maintaining cellular activities under stress. *CDC50* is critical for trafficking of protein between the Golgi and vacuole/endosome as well as actin cytoskeleton integrity. *DRS2* codes for a P-type ATPase implicated in phospholipid asymmetry (Chen et al., 1999, 2006). Removal of either of these crucial genes (*CDC50* or *DRS2*) makes the yeasts cells vulnerable to the effects of low temperature (Misu et al., 2003) and high pressure (Abe and Minegishi, 2008), signifying that this complex is critical for sustaining the actin dynamics and intracellular trafficking in extreme environments.

Ergosterol, a major component of the plasma membrane of yeasts, has been identified to play a crucial role in its growth under low temperature and high pressure. The significance of ergosterols was revealed from the study of Abe and Minegishi (2008). They deleted genes like ERG24, ERG2, ERG6, and ERG3t that are involved in the late stages of ergosterol biosynthesis, and found that it resulted in the build up of abnormal sterols and poor growth. This pointed out that the precise structure of ergosterol is vital in the correct function and membrane protein localisation under extreme conditions.

5.3 Adaptations to High Hydrostatic Pressure

Regardless of their source of isolation (terrestrial or marine) yeasts demonstrate a fairly good tolerance to hydrostatic pressure. Despite this fact their physiological reactions under high pressure conditions still remains poorly understood (ZoBell and Johnson, 1949; Yamasato et al., 1974).

While a majority of deep-sea yeasts are piezotolerant and grow at pressures up to 20 MPa, most of them grow best at lower pressure ranges. Comparative research to evaluate the effect of hydrostatic pressure on deep-sea and terrestrial yeast strains showed that most deep-sea isolates have better growth at 20 MPa and 40 MPa compared to their terrestrial counterparts (Nagano et al., 2014). An interesting observation that was made was that growth of some deep-sea yeast strains is more negatively affected by hydrostatic pressure at 10°C than at 3°C. These findings are

Table 2. Adaptations of marine yeasts and their mechanisms.

Adaptation	Species	Mechanism	References
Osmotolerance/ Halotolerance		Gene expression, High osmolarity glycerol (HOG) signalling pathway, melanisation of the cell wall composition and accumulation of ions, enzymes involved in fatty acid modifications, and plasma membrane composition	Turk and Plemenitaš (2002); Kogej et al. (2006); Romeo et al. (2020)
	S. cerevisiae and *H. werneckii*	MAPK cascade called the HOG signalling pathway	Plemenitaš et al. (2014); Gunde-Cimerman et al. (2018); Romeo et al. (2020)
	H. werneckii	Heat Shock Proteins (Hsps) activation	Romeo et al. (2020)
	D. hansenii	intracellular K+ decrease intracellular Na+ increases	Norkrans (1968); Norkrans (1966); Ross and Morris (1962); Prista (1997)
		polyols (osmotically active solutes) accumulate intracellularly	Brown and Simpson (1972); Gustafsson and Norkrans (1976); Brown (1978); Adler et al. (1985)
Osmotolerance/ Halotolerance	*D. hansenii*	polyols (glycerol and arabinitol) produced and accumulated	Lages et al. (1999); Lucas et al. (1990); Adler and Gustafsson (1980)
		existence of sodium efflux process (extrusion of Na+ across the plasma membrane might be carried out via the function of Na+ -ATPase or Na+/H+ antiporters)	Ramos et al. (2011)
High pressure and cold	*S. cerevisiae* (Model organism)	uptake of tryptophan mediated by tryptophan permeases Tat1 and Tat2	Abe and Horikoshi (2000)
	S. cerevisiae (Model organism)	nutrient auxotrophic markers, such as ura3 (uracil), ade2 (adenine), his3 (histidine), leu2 (leucine), lys2 (lysine), or trp1 (tryptophan)	Abe and Horikoshi (2000)
	S. cerevisiae (Model organism)	upregulation of DAN/TIR cell wall mannoprotein genes allowing the cells to establish a physical defensive system on the cell wall structure	Abe (2007b)
	S. cerevisiae (Model organism)	genes involved in cellular functions, such as amino acid biosynthesis, mitochondrial function, the actin cytoskeleton, membrane trafficking, transcription, ribosome biogenesis, chromatin structure, and unknown roles	Nagano et al. (2014)
High hydrostatic pressure	*C. atlantica* and *P. guilliermondii*	unique kinds of filamentation: pseudo-hyphae (non-permanent)	Burgaud et al. (2015)
	C. viswanathii	true hyphae (non-permanent)	Burgaud et al. (2015)

in line with the observations of Singh et al. (2010) who too noted greater biomass production at 20 MPa and 5°C as compared to 20 MPa and 30°C, indicating a subtle interplay between pressure and temperature in determining microbial physiology. All these observations suggest that deep-sea yeasts have developed adaptive mechanisms to survive low temperatures under high pressure conditions.

Bringing about changes in morphology of the cell has been found to be another adaptation to high hydrostatic pressure (HHP) that marine deep-sea yeasts exhibit. The formation of reversible pseudo-hyphae in response to HHP has been observed in yeasts species like *Candida atlantica* and *Pichia guilliermondii*, which is non-permanent morphological response that can be advantageous for survival under extreme environments (Burgaud et al., 2015). Pseudo-hyphae are extensions of the cell with constrictions at septal junctions and are distinct from true hyphae. They are developed probably to increase the surface area for nutrient uptake or regulate cellular rigidity to counter pressure. A completely contrasting observation was made in the case of *Candida viswanathii*, that formed reversible true hyphae to overcome the stress of HHP (Burgaud et al., 2015). True hyphae differ from pseudo-hyphae in having continuous, parallel-sided walls, septation, and invasive growth capability, enabling more complex environmental interaction and nutrient uptake.

The transient nature of hyphal formation, depicts the morphological plasticity of deep-sea fungi in response to environmental stressors. The capability to switch from the single cell form to hyphal forms (true/pseudo hyphae) under different pressure conditions, points to the fact that this morphological adaptation is carefully regulated. This morphological plasticity plays a vital role in guaranteeing the survival of fungal species in high-pressure, deep-sea environments.

6. Bioprospecting Potential

Marine yeasts represent a largely untapped reservoir of biotechnological promise, thriving in diverse and often extreme marine habitats. Their unique physiological and metabolic adaptations confer the ability to produce a wide array of high-value bioactive compounds and industrially relevant metabolites. Recent advances in marine microbial bioprospecting have illuminated the multifaceted applications of marine yeasts, ranging from natural pigment and enzyme production to their utilization as nutritional supplements in food and aquaculture feed. Additionally, several marine yeast strains exhibit potent antimicrobial, antioxidant, and immunomodulatory properties, underscoring their potential in medicinal and pharmaceutical domains (Fig. 4, Table 3).

They have been identified as suitable agents in bioremediation due to their ability to break down both hydrocarbons and other pollutants in the environment. Moreover, marine yeasts have now emerged as ideal candidates for sustainable agriculture due to their plant growth promoting and phytopathogen control potentials. This part discusses the varied and increasing bioprospecting opportunities provided by marine yeasts and their applicability to industrial development and ecological sustainability.

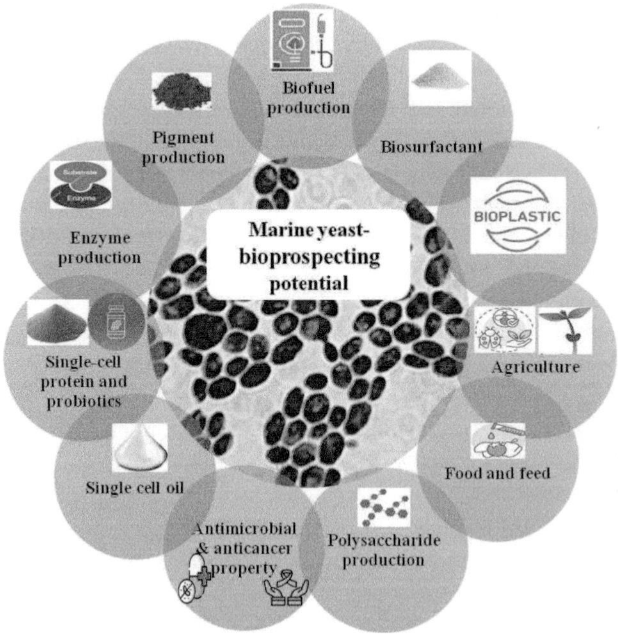

Fig. 4. Bioprospecting potentials of marine yeasts: An overview.

6.1 Pigment Production

6.1.1 Carotenoids

Carotenoids are naturally occurring pigments found in plants, animals, and microbes. Pigmented yeasts are one among the microbial sources. These pigments are lipid soluble and observed in varying colours like yellow, red, purple, and orange (Mukherjee et al., 2009). These pigment extracts exhibit antioxidant properties that defend cells via scavenging activity. Carotenoids are used in the food, animal health, pharmaceutical, and cosmetic industries. *Rhodotorula* sp., *Rhodosporidium* sp., *Sporobolomyces* sp., and *Xanthophylomyces* sp. are the carotenoid-producing marine yeast genera (Mannazzu et al., 2015).

The yeast *Rhodosporidium babjevae* isolated from the subarctic marine copepod *Calanus finmarchicus* was recognized as a source of β-carotene, torulene, and torularhodin (Sperstad et al., 2006). The carotenoid production was strongly influenced by culture conditions in this study. Muthezhilan et al. (2014) found that the pigment from *Rhodotorula* sp. (Amby109) isolated from marine sediment and water exhibited excellent antimicrobial activity. Their study indicated that the pigment from *Rhodotorula* sp. could be used in the food industry as an alternative to synthetic colourants. Leyton et al. (2019) and Zhao et al. (2019) also reported that *Rhodotorula* spp. were a good source of carotenoids and were suitable for commercialisation. Rekha et al. (2024) found that the carotenogenic mangrove yeast *Rhodotorula paludigena* VA 242 served as a colour-enhancing feed additive for ornamental fish Koi Carp. In a study by Nimsi et al. (2024c), *Rhodotorula mucilaginosa* PV 8 was found to be a potent carotenoid producer among 47 red-pigmented yeasts isolated from the mangrove ecosystem and the yield of carotenoid was found to be 194.78 μg/g underoptimised conditions. This carotenoid exhibited antioxidant and antibacterial activity against *Aeromonas* sp. and *Vibrio* sp.

Astaxanthin is a red-orange pigment and is included in the group of fat-soluble carotenoids known as xanthophylls. It is mainly found in microorganisms (bacteria, yeast, fungi, and microalgae) and marine animals (salmon and crustaceans). It is an antioxidant that protects cells from damage and has many health benefits, like anti-inflammatory and anti-apoptotic properties. The FDA and the European Union have approved astaxanthin as a food colourant or dye specially applied in animal or fish feeds. Mi (2011) reported that *Rhodotorula glutinis* YS-185 from the Pacific Ocean is a potent producer of astaxanthin and that optimisation increased astaxanthin yield. In a study by Ushakumari and Ramanujan (2013), they noticed that the astaxanthin extract demonstrated greater antimicrobial activity against both gram-negative and gram-positive bacteria when compared to the standard chloramphenicol.

6.1.2 Melanin

Melanin is a pigment that provides colour to skin, hair, and eyes. It protects the skin from UV radiation, has wound-healing properties, and has a role in primary immune response.

Melanin production from the marine environment has been reported mainly from the black yeast *Hortaea werneckii. H. werneckii* is a halophilic species often found in hypersaline environments like salt pans. Kejžar et al. (2013) found that melanin was crucial for growth and survival of the *Hortaea werneckii* in its natural hypersaline environment

Soundra Rani et al. (2013) reported that melanin from *H. werneckii* has great uses in agriculture, cosmetics, and pharmaceuticals. They noted a melanin pigment yield of 5.60 g/L under optimal conditions, and the extracted melanin showed strong antibacterial activity against harmful organisms. They also identified glucose as the best carbon source.

Elsayis et al. (2022) also studied melanin pigment extracted from *H. werneckii* AS1, which was taken from the solar saltern of Egypt. The extracted pigment had antioxidant properties and antimicrobial activity against certain bacterial and fungal strains. They also found that it was not toxic to the human skin fibroblast cell line (HSF). Wilson et al. (2024) observed that *H. werneckii* melanin (HwM) is eumelanin, with strong antioxidant and photoprotective activity, and that it was compatible with human epithelial cells.

Tyrosinase is an enzyme with multiple functions and has a major role in regulating melanin production. Amino acid tyrosine is hydroxylated to dihydroxyphenylalanine (DOPA), which is then oxidised to dopaquinone, facilitated by tyrosinase. Researchers have focused on these enzymes, which can be used in the cosmetics industry due to their whitening effects on the skin. Yu et al. (2023) isolated *Aureobasidium* sp. US-Sd3 from the marine sediment of Korea that exhibited tyrosinase inhibitory activity as well as antibacterial activity against *Bacillus subtilis*.

6.2 Riboflavin

Riboflavin (vitamin B2) is essential for all organisms; plants and microorganisms synthesise it. It is used commercially as a food colourant and in pharmaceuticals for its antibacterial properties.

Candida membranifaciens subsp. *flavinogenie*, isolated from seawater of China by Wang et al. (2008), was observed to produce a high amount of riboflavin in the medium containing a carbon source (galactose, maltose, sucrose, or xylose). Further optimisation studies of the same strain reported by Chi et al. (2008) using response surface methodology showed that the highest yield obtained was 22µg/ml within 54 hrs. Amornrattanapan (2012) reported a high riboflavin-yielding strain of *Candida tropicalis* MICBUU002 (254.22±5.51 g/ml) isolated from seawater samples of Thailand that was the best out of 47 yeast isolates.

6.3 Enzyme Production

Marine yeasts, as saprophytic organisms, possess an inherent ability to degrade a broad spectrum of organic substrates, making them a valuable source of hydrolytic enzymes with diverse industrial applications. Their adaptability to saline and often extreme environments contributes to the production of enzymes that are active and stable under varied physicochemical conditions. Several studies have highlighted the enzymatic potential of marine yeast strains isolated from distinct ecosystems. For instance, Duarte et al. (2013) reported lipase, protease, and xylanase activities in psychrotolerant yeast species such as *Candida glaebosa*, *Cryptococcus victoriae*, *Meyerozyma (Pichia) guilliermondii*, and *Rhodotorula* spp. from Antarctic marine and terrestrial samples.

Mangroves, a major biodiversity hotspot have emerged as a source of yeasts with a wide hydrolytic enzyme profile. *Starmerella etchellsii* (MY1) and *Candida parapsilopsis* (MY6) isolated from mangroves were found to be capable of producing lipase, protease, cellulase, glutaminase, and ligninase (Kachiprath et al., 2019). Among 482 yeast isolates from mangrove environments, Pothayi et al. (2022) observed that lipolytic activity was dominant (68%) among other hydrolytic potentials and that *Candida tropicalis* as suitable for industrial scale lipase enzyme production. Nimsi et al. (2024b) too in their work reiterated the dominance of lipase activity among manglicolous yeasts. From a broader perspective these findings highlight the enzymatic versatility of marine yeasts as well as their untapped prospectives as biofactories for enzymes suited for various industries.

6.3.1 Lipases

Among the different enzymes produced by marine yeasts, Lipases are the most widely researched and industrially significant enzyme. This enzyme catalyses triacylglycerol hydrolysis into free fatty acids and glycerol. Their applications cut across varied industries, ranging from food to pharmaceuticals, detergents, leather treatment, and biofuel production. The physiological tolerance of marine yeasts to high salinity, low temperature, and nutrient-poor conditions often leads to the secretion of lipolytic enzymes with a high degree of stability and activity. In a pioneering study, Wang et al. (2007a) isolated 427 yeast strains from Antarctic seawater and sediment samples, identifying nine strains—*Candida intermedia* YA01a, *Pichia guilliermondii* N12c, *Candida parapsilosis* 3eA2, *Lodderomyces elongisporus* YF12c, *Candida quercitrusa* JHSb, *Candida rugosa* Wl8, *Yarrowia lipolytica* N9a, *Rhodotorula mucilaginosa* L10-2, and *Aureobasidium pullulans* HN2.3—capable of producing cell-bound lipase. Both the crude and cell-bound lipases from these strains were capable of breaking down different natural oils, indicating that they were useful for industrial processing.

Highlighting the capability of yeasts found in mangroves to produce lipase; Pothayi et al. (2021) found that *Candida tropicalis* KGDP7 produced extracellular lipase that possessed traits suitable for biotechnological application Pothayi et al. (2024) also isolated and studied lipases from *Candida tropicalis* (VPY3) and *Diutina mesorugosa* (VPY42). Both showed strong lipase activity and potential for commercial use. Pothayi et al. (2022) and Nimsi et al. (2023b) have reported that 68% and 100% of strains respectively displayed lipase activity. These results support the statement that lipase activity is the most common enzymatic feature among marine yeast isolates. As diverse groups of marine yeasts (both pigmented and non-pigmented) are capable of producing lipase it suggests that marine environments contain a rich and underused source of this biocatalyst.

6.3.2 Proteases

Protease-producing marine yeasts have gained far more attention than yeasts from other sources, because the proteases they secrete extracellularly can work well under harsh environmental conditions. This makes them extremely appealing for use in pharmaceuticals, food processing, and waste treatment. Several strains from the ocean have been found to be strong producers of both alkaline and acid proteases. *Aureobasidium pullulans*, isolated from saltern sediments in China, was reported by Chi et al. (2007) to produce significant levels of alkaline protease (623.1 U/mg protein; 7.2 U/mL), with optimal activity at pH 9.0 and 45°C. Maximum enzyme production occurred when the strain was cultured in a medium containing soluble starch and sodium nitrate, incubated at 24.5°C for 30 hours. Expanding on this, Ni et al. (2008a) identified two protease-producing strains of *A. pullulans* (HN2-3 and N13d) from Antarctic salterns and seawater, characterising the enzymes as serine alkaline proteases with activity best suited to alkaline conditions. In more recent studies, Sarkar et al. (2024) isolated 14 yeast strains from mangrove sediments, among which *Candida orthopsilosis* AKB-1 exhibited the highest alkaline protease activity (978 IU/mL). This enzyme also displayed strong antibacterial activity, particularly against *Bacillus cereus*, suggesting dual utility in industrial and antimicrobial applications. On the other end of the pH spectrum, acid protease-producing marine yeasts also show biotechnological potential. *Metschnikowia reukaufii* W6b, isolated from sea sediments, was reported by Li et al. (2010) to produce a cell-bound acid protease suitable for applications in cheese production and food fermentation. Similarly, *Rhodotorula mucilaginosa* L7, obtained from Antarctic marine algae, demonstrated maximal acid protease activity in the temperature range of 45–55°C (Lario et al., 2015). These findings clearly highlight the industrial relevance of proteolytic enzymes from marine yeasts across a wide range of environmental and operational conditions.

6.3.3 Other Hydrolytic and Functional Enzymes

Beyond the widely studied lipases and proteases, marine yeasts also produce a variety of other industrially significant enzymes, including inulinase, pectinase, phytase, cellulase, amylase, glucanase, xylanase, and antioxidant enzymes like superoxide dismutase (SOD). Inulinase, which catalyses the hydrolysis of inulin to produce fructose, fructooligosaccharides, and bioethanol, is particularly valued in the food and beverage industry. Sheng et al. (2007) first reported inulinase production from the marine yeast *Cryptococcus aureus* G7a, which exhibited over 85 U/mL of enzyme activity under optimized conditions. Other strains such as *Pichia guilliermondii* (Gong et al., 2007; Gao et al., 2007b), *Yarrowia lipolytica*, and *Debaryomyces hansenii* have also been identified as high inulinase producers. *Cryptococcus* sp. SY3, isolated from marine fish, displayed 62.7 U/mL of activity within 48 hours, highlighting its potential for fructose syrup and bioethanol production (Bharathi et al., 2011). Similarly, *Aureobasidium* sp. P6, isolated from mangrove habitats, exhibited inulin-hydrolysing activity of 30.98 ± 0.8 U/mL (Jiang et al., 2016).

Other enzymes of note include pectinase, originally reported by Da Costa and D'Souza (1979) in various yeast genera from decaying mangrove vegetation, and phytase, which is critical in animal feed formulations. *Kodamaea ohmeri* BG3, isolated from marine fish, was optimised for phytase production via response surface methodology (Li et al., 2007b).

Aureobasidium pullulans 98 was reported to produce carboxymethyl cellulase (CMCase) at levels of 4.51 U/mg protein (Yanjun et al., 2015), while *A. pullulans* N13d demonstrated substantial amylase activity (58.5 units of protein in 56 hours) (Li et al., 2007a). Additional notable enzymes include Exo-β-1,3-glucanase from *Williopsis saturnus* WC91-2 (Peng et al., 2009), α-glucosidase from *Leucosporidium antarcticum* 171 (Turkiewicz et al., 2005), and endoxylanase from *Candida davisiana*, *Cryptococcus adeliensis*, and *Guehomyces pullulans* (Duarte et al., 2013).

Superoxide dismutase (SOD) is an antioxidant enzyme that finds use in both cosmetics and pharmaceutical industry. Researchers isolated a yeast strain *Debaryomyces hansenii* C-11, from the Pacific Ocean, that was found to be a strong SOD producer (Ramírez Orozco et al., 1998) making it useful for anti-inflammatory treatments, immune support, anti-aging therapies, and protection against oxidative stress.

The stability and diversity of marine yeast enzymes highlight their enormous possibility in ecologically sound industrial uses, calling for more research in this area.

6.4 Food and Feed

6.4.1 Single-Cell Protein and Probiotics

Single-cell protein (SCP) is a general term used to refer to microbial cells that are rich in protein content, grown and harvested for use as human food or animal feed. SCPs are derived from microorganisms, like algae, fungi, and bacteria. Due to their high protein content, ability to utilise low-cost substrates as carbon sources, as well as grow rapidly, fungi (mainly yeasts) and bacteria are found to be ideal candidates for SCP production (Sharif et al., 20121). They have been found to be ideal for addressing protein shortages and an alternative to expensive traditional protein sources like soy meal and fish meal.

Studies of workers like Rishipal and Philip (1998) and Gao et al. (2007) on marine yeasts *Candida* sp. and *Cryptococcus aureus* G7, respectively, have ascertained its suitability for SCP production clearly indicating the feasibility of marine yeasts as an apt source for industrial-scale, cost-effective protein production.

Another area, in which marine yeasts have gained attention, is as probiotic in aquaculture. Their ability to enhance nutrient assimilation, inhibit aquaculture pathogens, improve host immunity, and above all survive in saline environments has been the reason for their popularity. Moreover, their Generally Recognised as Safe (GRAS) status, better tolerance to acidic pH and bile salts have made them a better option than bacterial probiotics in aquaculture. These yeasts are also known to produce bioactive molecules, positively influence the gut microbiota of fishes as well as improve the quality of water of aquaculture systems. *Y. lipolytica* N-6 a marine isolate was found to act immunostimulant in Pacific Red Snapper. The strains *Y. lipolytica* D-1 and as well *Y. lipolytica* N-6 were found to enhance the antimicrobial activity and antioxidant activity and act as an immunostimulant against the aquaculture pathogen *V. parahaemolyticus* (Reyes-Becerril et al., 2021). Gilthead Sea Bream (*Sparus aurata*) when fed with a diet supplemented *D. hansenii* BCS004 greatly improved feed uptake, growth and immune response (Reyes-Becerril et al., 2021).

Overall, we can conclude that marine yeasts are as efficient as SCP and probiotics. They are especially useful in sustainable aquaculture practices. Adding them to feed and water treatment systems not only improves the health and productivity of farmed species but also supports sustainable aquaculture by lowering the need for antibiotics and synthetic additives.

6.4.2 Xylitol

Xylitol is a sugar alcohol with five carbon atoms that is found in small amounts in fruits and vegetables. Yeasts are among the dominant microorganisms capable of fermenting xylose and converting it into xylitol, which is commonly used as a sugar substitute (sweetener) in the food and pharmaceutical industries. Among the 21 marine fungi isolated by Kamat and co-workers (2013), only a single isolate, *Cyberlindnera (Williopsis) saturnus* (yeast) was found to synthesise xylitol from xylose and corn cob hydrolysate, a form of agricultural waste.

6.5 Single-Cell Oil

Single-cell oils (SCO), derived from oleaginous yeasts, are lipids that constitute more than 20% of the yeast's dry cell mass in the form of intracellular lipid droplets or triacylglycerols (TAGs). These microbial lipids have garnered significant attention due to their potential applications in medicine, nutrition, and biofuel production (López et al., 2022; Jape et al., 2014). Yeasts capable of producing SCO are considered promising alternatives for biodiesel production. The composition of SCO predominantly includes both polar lipids such as myristic (C14:0), palmitic (C16:0), stearic (C18:0), oleic (C18:1), linoleic (C18:2), and linolenic acid (C18:3), as well as non-polar lipids like triacylglycerols (C13 to C21) and steryl esters (Wissebach et al., 1969). Several marine yeasts have shown great promise as SCO producers. For instance, *Rhodotorula mucilaginosa* TJY15, isolated from the surface of marine fish, utilised cassava starch hydrolysate to achieve a lipid content of 52.9% through fed-batch fermentation (Li et al., 2010a, 2010b). The potential of *Yarrowia lipolytica* to be used for biodiesel production was proved by its trait to accumulate lipids from waste oils (Katre et al., 2012). In another study, the yeast strain *A. pullulans* var. *melanogenum* P1 isolated from mangroves was found to accumulate 66.3 g of oil per 100 g of dry cell mass (Wang et al., 2014).

The high lipid production from these marine yeasts highlights their potential for sustainable biofuel production. Liu et al. (2014) also reported *Aureobasidium pullulans* var. *melanogenum P5* can secrete large amounts of heavy oil into the medium containing 120 g/L of glucose and 0.1% corn steep liquor. Another study by Abdel-Wahab et al. (2023) on two oleaginous yeasts, *Hortaea werneckii* and *Rhodotorula mucilaginosa*, isolated from mangrove vegetation, also produced high lipid levels.

6.6 Bioethanol and Biofuel

Marine yeasts are also gaining attention as biofuel producers, particularly for bioethanol, due to their capacity to thrive in saline conditions and utilise diverse biomass substrates. Biofuels such as bioethanol, biodiesel, and biogas are sustainable and eco-friendly energy sources derived from agricultural and industrial waste, and they serve as alternatives to fossil fuels.

Kathiresan et al. (2011) investigated ten yeast strains from coastal mangrove sediments and identified *Pichia salicaria* as the most efficient, yielding 26.2 ± 8.9 g/L of bioethanol. Similarly, *Saccharomyces cerevisiae* C-19 and *Candida* spp. have been reported to ferment paper-shredder scrap and red algae, respectively, for bioethanol production (Obara et al., 2012; Khambhaty et al., 2012). A comparative study by Saravanakumar et al. (2013) found that a mangrove-derived strain of *S. cerevisiae* (JN387604) outperformed its terrestrial counterpart in ethanol yield. Zaky et al. (2018) evaluated *S. cerevisiae* AZ65, a marine yeast that exhibited efficient bioethanol production in seawater or saline media, outperforming the terrestrial strain *S. cerevisiae* NCYC2592. Exploring furthermore the potential for renewable energy generation using marine yeasts, Khajeh et al. (2025) most recently screened marine yeast strains and identified *Pichia* sp. BK10, *Pichia kudriavzevii* BK40 and BK50, *Pichia fermentans* MK20, and *Candida parapsilosis* BK30 as strong contenders for bioethanol production.

6.7 Antimicrobial and Anti-Cancer Activity

Killer yeasts are those that are capable of producing proteinaceous antimicrobial compounds referred to as killer toxins. They find use as biocontrol agents of phytopathogens in agriculture, antimicrobial agents of human and animal pathogens and in food preservation.

Antimicrobial potentials of a number of marine and manglicolous yeast strains have been recognised by several workers. *Metchnikowia bicuspidate*, the pathogenic yeast causing milky disease in crabs was significantly inhibited by marine yeasts *Aureobasidium pullulans* HN2.3, *Debaryomyces hansenii* hcx-1, *Pichia anomala* YF07b, *Pichia guilliermondii* GZ1, and *Williopsis saturnus* WC91-2 (Wang et al., 2008). The killer toxin of mangrove yeast isolate *K. siamensis* HN12-1 was successfully extracted and purified by Buzdar and co-workers (2011). This purified toxin with molecular weight 66.4 kDa exhibited greatest inhibitory action against *M. bicuspidate* WCY, a pathogenic yeast. The crude extract of *Hortaea werneckii,* an extremophilic black yeast isolated from the Red Sea mangrove-displayed antimicrobial activity against *Campylobacter jejuni*, methicillin-resistant *Staphylococcus aureus* (MRSA), and *Salmonella typhimurium* (Hodhod et al., 2020).

The screening studies carried out by Senthilraja and Kathiresan (2015) on yeasts strains isolated from coastal mangrove ecosystem revealed that, certain species like *Saccharomyces cerevisiae*, *Kuraishia capsulate* and *Candida albicans* produced biomolecules that possessed anti-cancer activity. These biomolecules from yeasts could lead to anti-cancer drugs in the future.

6.8 Polymalic Acid and Polyhydroxyalkanoates

A safe, water-soluble, biodegradable polymer made up of the monomeric units β-L-malic acid is called Polymalic acid (PMA). This can be synthesised either via microbial fermentation or purely chemically in the laboratory. This non-toxic polymer finds application as carrier material for drugs and surgical sutures in the medical field, as food additive in the food industry and in the manufacture of biodegradable plastics (Wang et al., 2015; Zeng et al., 2019). Strains of *Aureobasidium* sp. a black yeast that is found in hypersaline environments has been reported by workers (Ma et al., 2013, Wang et al., 2015) to be capable of producing large amounts of PMA. Wang et al. (2015) carried out batch fermentation for the production of Ca (2+)-poly malate (PMA) using the strain *Aureobasidium pullulans* var. *pullulans* MCW. They also noted that the purified PMA was composed of 1784 L-malic acid units.

Amongst the various bioplastics, Polyhydroxyalkanoates (PHAs) are a group of microbial polyesters, that are mostly extensively studied. Though bacteria have been identified as the major producer of these polyesters, a few marine yeasts too have been reported to have the capability to produce these biomolecules when nutrients are limited in the culture media. In media containing banana peel and chicken feather the yeast strain *P. kudriavzevii* VIT-NN02 produced PHA (Ojha and Das, 2020). A similar study reported the production of poly(3-hydroxybutyrate) by the marine yeast, *Pichia kudriavzevii* TSLS24 that showed excellent biodegradability (Thu NTT et al., 2023).

These studies indicate that marine yeasts are an excellent but undermined source of these bioplastics/biopolymers that could lead to development of green technologies in packaging and materials science. They offer biodegradable options to replace petroleum-based plastics and support the sustainable developmental goals.

6.9 Polysaccharides – Pullulan

Among polysaccharides of microbial origin, Pullulan, an exopolysaccharide, has been recognised as a product of high commercial value. They find use in the biomedical and pharmaceutical industry. Apart from this, they are extensively used in the food industry as packaging films that are impermeable to O_2, stabilising agents, adhesives, thickening agents, and as a low-calorie food ingredient.

The marine yeast *Aureobasidium pullulans* has exhibited potential in pullulan production. Chen et al. (2012) isolated *Aureobasidium pullulans* CJ001 from marine sediments and using this strain achieved a maximum yield of 26.13 g/L of pullulan in the media optimised using response surface methodology. Ma et al. (2014) screened more than 100 strains of *Aureobasidium* spp. from the mangrove ecosystem for the production of polysaccharides. Among them, *Aureobasidium pullulans* var. *melanogenum* P16 showed high productivity of exopolysaccharide (EPS) with a yield of 65.3 g/L under optimal conditions. This EPS was then purified and identified to be pullulan.

6.10 Nanoparticle Synthesis

Microorganisms mediated synthesis of nanoparticles has in recent years garnered the attention of researchers as a sustainable and an eco-friendly method. Marine yeasts, due to their distinctive metabolic potentials, have been identified as a promising agent in nanoparticle biosynthesis.

There are several reports that highlight the potentials of marine yeasts to reduce metal ions to form stable nanoparticles with antimicrobial activity and biomedical use. Agnihotri et al. (2009) in their study noted that the marine yeast *Yarrowia lipolytica* NCIM 3589 was capable of synthesising gold nanoparticles (AuNPs) from chloroauric acid ($HAuCl_4$).On screening for their nanoparticle synthesising ability among yeast strains isolated from mangrove sediments, Subramanian et al. (2010) observed that *Kuraishia capsulata* (formerly *Pichia capsulata*) was capable of synthesizing silver nanoparticle (AgNP). Similar studies carried out by Seshadri et al. (2011) and Dinesh et al. (2011) showed that *Rhodosporidium diobovatum* and *Candida* sp., VITDKGB were able to synthesise lead sulphide and silver nanoparticles, respectively. The silver nanoparticle synthesised by the psychrotrophic marine yeast *Yarrowia lipolytica* (NCYC 789) inhibited the biofilm formed by *Salmonella paratyphi*on glass and polystyrene surfaces.

6.11 Biocontrol Agents

Biocontrol agents can be defined as environmentally friendly and sustainable substitutes for chemical pesticides. They not only help in controlling weeds, pests, and plant diseases but also encourage plant growth. Microbes like viruses, bacteria, yeasts, fungi, and protozoa can act as biocontrol agents and function as integral components in integrated pest management strategies.

Yeasts from marine sources in particular, have received the attention of researchers in the field of agriculture for their antagonistic activity against phytopathogens. *Sporidiobolus pararoseus* ZMY-1, a novel yeast species isolated from the marine sediments of Zhangzhou mangroves displayed both *in vitro* and *in vivo* antagonistic activity against *Botrytis cinerea*, the causative agent of grey mould disease in strawberries. Its application was found to be effective in post-harvest disease management (Shen et al., 2019). Biocontrol potentials of manglicolous yeasts against fungal plant pathogens such as *Fusarium oxysporum*, *Phytophthora infestans*, *Sclerotium rolfsii*, and *Rhizoctonia solani*, were demonstrated in the study of Nimsi et al. (2024b). All these studies clearly highlight the scope of marine yeasts in the field of phytopathogen control.

6.12 Plant Growth-Promotion

Plant growth-promoting yeasts (PGPYs) are a promising group of unicellular fungi capable of promoting plant growth by solubilising nutrients, enhancing soil fertility, producing iron chelating molecules (siderophores), synthesising phytohormones, and enhancing tolerance to varied biotic and abiotic stress. As yeasts belong to the Recognised as Generally Regarded As Safe (GRAS) category, PGPYs in agriculture have been identified as sustainable substitutes to chemical fertilisers, and pesticides.

Though most of reported PGPY strains have been from terrestrial ecosystems, strains from the marine environment have also shown promising abilities in this regard. Marine samples, like mangrove sediments and seawater, have been a source of potent PGPY strains. For example, the Red Sea and bromeliad tank water isolates like *Rhodotorula*, *Kazhakistania*, and *Rhodosporidium* species exhibited quite a number of PGP traits (Marques et al., 2021; Nimsi et al., 2023b). In a greenhouse study, *Rhodotorula mucilaginosa* and *Candida orthopsilosis* isolated from the Red Sea considerably enhanced the growth of wheat and also provided resistance to *Fusarium oxysporum* (Abdel-Kareem et al., 2021). They also completely inhibited zearalenone, a harmful mycotoxin. The potential of *Wickerhamomyces anomalus* MSD1 isolated from sea weeds to act as biofertiliser and biostimulant was revealed from its genomic studies (Krishnan et al., 2020). The stimulation of growth of *Vigna radiata* seedlings on co-culturing with mangrove isolates *Cryptococcus* sp. and

Candida tropicalis was noted by Arya et al. (2023). Other mangrove isolates such as *Debaryomyces hansenii*, *Aureobasidium melanogenum*, and *Candida tropicalis* have exhibited multifarious growth-promoting traits, like production of ammonia, catalase, indole acetic acid, and siderophores, enhanced tolerance to abiotic stress and inhibition of phytopathogens (Nimsi et al., 2024b). These traits could be positively correlated with the enhanced growth of mangrove seedling observed during the pot studies, pointing to their use in mangrove restoration and conservation. Phosphate-solubilising ability has been reported in *Candida* sp. isolated from the rhizosphere of the mangrove ecosystem (Nurhariyati et al., 2020). *Aureobasidium pullulans* HN6.2, a marine isolate, was found to be capable of producing large quantities (1.1 mg/mL) of siderophores. These iron chelating compounds were responsible for antibacterial activity of *Aureobasidium pullulans* HN6.2 against bacterial pathogens *Vibrio* spp. (Wang et al., 2009a). In a later study Fusigen was identified as the active component by Wang and his team (2009b), whereas its physiological role was investigated by Chi et al. (2012). Murugappan et al. (2012) observed that yeast isolated *A. pullulans* from *Ulva fasciata* synthesised hydroxamate-type siderophores that too promoted plant growth. These findings highlight the huge possibility of employing marine yeasts in sustainable agriculture.

6.13 Bioremediation and Hydrocarbon Degradation

The ability of marine yeasts to break down hydrocarbons, produce biosurfactants, and detoxify heavy metals; make them suitable for restoring polluted marine and coastal ecosystems, particularly those affected by hydrocarbon and industrial contamination.

The first record of the hydrocarbon-degrading capabilities of marine yeasts comes from the work of Da Costa and D'Souza (1979) on yeast isolates from mangroves. They were successful in isolating several yeasts genera like *Candida*, *Cryptococcus*, *Debaryomyces*, *Geotrichum*, *Kloeckera*, *Rhodosporidium*, *Saccharomycete*, and *Trichosporon* metabolise different hydrocarbons, underscoring their role in the natural reduction of oil pollution in mangrove areas (Da Costa and D'Souza, 1979). Another notable example is *Yarrowia lipolytica* NCIM 3589, a tropical marine yeast reported by Oswal et al. (2002) and Jain et al. (2004), which could degrade hydrocarbons and triglycerides in marine environments contaminated with explosives. It transformed 2,4,6-trinitrotoluene (TNT) into 2,4-dinitrotoluene (DNT), making it a promising agent for cleaning up TNT-polluted marine areas. Recent studies have broadened the understanding of marine yeast strains that can degrade crude oil. Al-Enazi (2025) identified *Acrophialophora jodhpurensis* C3, *Pichia kudriavzevii* D1, *Hyphopichia khmerensis* U1, and *Candida tropicalis* W6 from coastal areas as alternative microbial agents for crude oil cleanup.

Apart from hydrocarbon degradation, yeasts can also accumulate metal ions. For instance, *Geotrichum* sp. CS-67, exhibited the ability to uptake heavy metal ions (Mi, 2011), indicating its potential in reducing metal pollution. These findings collectively affirm the ecological importance of marine yeasts in lowering environmental pollution and their potential use in sustainable cleanup technologies.

6.14 Biosurfactants

Microorganisms are capable of secreting certain surface-active biomolecules that are amphipathic; they are referred to as Biosurfactants. Being biological in origin they are biodegradable, non-toxic, and eco-friendly alternatives to chemical or synthetic surfactants. Biosurfactants find use not only in bioremediation but also in a large number of industries like cosmetics, food and feed, detergent, pharmaceuticals, agriculture, and medicine (Patel, 2020; Johnson et al., 2021). From among the different microbial sources of biosurfactants, marine yeasts have been identified as potent sources of mainly glycolipid biosurfactants. *Pseudozyma hubeiensis* SY62, an isolate from deep-sea cold-seep clams, Japan, was found to produce mannosylerythritol lipids (MEL) (Konishi et al., 2010). Whereas, from the Indian mangrove sediments, *Saccharomyces cerevisiae* VEMS 13, *Candida tropicalis* VEMS 27, *Starmerella bombicola* VEMS 39, and *Yarrowia lipolytica* VEMS 51 have been identified as promising strains for biosurfactant production (Thenmozhi et al., 2018). A novel glycolipid biosurfactant named cybersan (trigalactomargarate) from *Cyberlindnera saturnus* SBPN-27 was reported by Balan and co-workers (2019). Cybersan was found to be non-cytotoxic to mammalian fibroblast cells that possessed antibacterial activity. The isolates from Gujarat mangroves *Trichosporon asahii* and *Candida glabrata* were capable of producing glycolipid biosurfactants (Patel, 2020). Antioxidant and antimicrobial properties of the glycolipid biosurfactant from the red yeast *Rhodotorula paludigena* VA 242 isolated from Kerala mangroves was reported by Nimsi et al. (2022). Another isolate from mangroves of central Kerala, *Geotrichum candidum* PV 37 possessed antagonistic activity against multidrug-resistant Gram-negative bacteria and anticancer activity against breast cancer cell lines (Nimsi et al., 2024a). Strong hydrocarbon bio-emulsifying capability was demonstrated by the biosurfactant from *Y. lipolytica* LMS 24B, a marine zoanthids isolate (Barros de Oliveira et al., 2024).

All these studies point to the utility of marine yeasts biosurfactants in pharmaceutical and environmental fields.

7. Bioprospect Potential

Marine yeasts have demonstrated a wide spectrum of biotechnological applications due to their metabolic versatility, stress tolerance, and ability to produce a variety of valuable metabolites. Figure 5 reflects the relative contributions of different marine yeast genera to bioprospecting, based on literature and reported applications in industrial biotechnology, pharmaceuticals, bioremediation, and enzyme production.

7.1 Dominant Genera with High Bioprospecting Potential

***Candida* (18%)** emerges as the most prominent genus, reflecting its widespread distribution and extensive utility in bioprocessing, including the production of lipases, organic acids, and potential in hydrocarbon degradation. *Candida tropicalis* has been extensively studied for its bioremediation capabilities and tolerance to saline environments.

Table 3. Bioprospecting potentials of marine yeasts.

Yeast Strain	Sample	Country of Study	Product	References
Pigment Production				
Carotenoids				
Rhodosporidium babjevae	Marine copepod (*Calanus finmarchicus*)	Subarctic	Carotenoids	Sperstad et al. (2006)
Rhodotorula sp. (Amby109)	Water and sediment samples	India (Kelambakkam, East Coast of Tamil Nadu)	Food Colourants	Muthezhilan et al. (2014)
Rhodotorula sp. RY1801	Sediment samples	South Yellow Sea, China	Carotenoids	Zhao et al. (2019)
R. mucilaginosa	Seaweed	Chile	Carotenoids (ß-carotene), lycopene, and astaxanthin	Leyton et al. (2019)
Rhodotorula mucilaginosa PV 8	Mangrove (*Avicennia officinalis*)	India, Kerala	Carotene	Nimsi et al. (2024c)
Rhodotorula paludigena VA242	Valanthakkad Mangrove Islands	India (Valanthakkad, Kerala)	Carotenoids	Rekha et al. (2024)
Astaxanthin				
R. glutinis YS-185	Pacific Ocean	China	Astaxanthin	Mi (2011)
Yeast strain (name of strain not mentioned)	Marine sediments	Cochin, Kerala, India	Astaxanthin	Ushakumari and Ramanujan (2013)
Melanin				
Hortaeawerneckii	Solar salterns water samples	India (Pondicherry)	Melanin	Soundra Rani et al. (2013)
H. werneckii	Solar salterns	Slovenia	Melanin	Kejžar et al. (2013)
H. werneckii AS1	Sediment samples from solar saltern	Egypt	Melanin	Elsayis et al. (2022)
H. werneckii R23	Sediment samples (Arabian Sea and the Bay of Bengal)	India	Melanin	Wilson et al. (2024)

Table 3 contd. ...

...Table 3 contd.

Yeast Strain	**Sample**	**Country of Study**	**Product**	**References**
Tyrosinase				
Aureobasidium sp. (US-Sd3), *Geotrichum candidum* (GH-W2), *Metschnikowia bicuspidate* (BP-Up1), *Papiliotrema fonsecae* (HW-W2), *Rhodotorulas* p.TS1-6, *R. mucilaginosa* GSU-CS3, *Vishniacozyma carnescens* US-Sd1, *Vishniacozyma* sp. SM-Rc3, *Yamadazyma olivae* K2-6 and *Yarrowia lipolytica* HW-W3	Sediment, Seawater, *Ulva australis*, Seawater, Seawater, *Venerupis philippinarum*, Sediment, *Rumex crispus*, guts of *Pagrus major*, seawater	Korea	Antibacterial, tyrosinase inhibition, and antioxidant activity	Yu et al. (2023)
Riboflavin				
Candida membranifaciens subsp. *flavinogenie* W14-3	Seawater	China Eastern Sea	Riboflavin	Wang et al. (2008); Chi et al. (2008)
Candida tropicalis MICBUU002	Seawater samples	Thailand	Riboflavin	Amornrattanapan (2012)
Enzyme Production				
Candida davisiana, Cryptococcus adeliensis, and *Guehomyces pullulans*	Marine and terrestrial samples	Antarctic continent	Lipase, protease, and xylanase activity	Duarte et al. (2013)
C. tropicalis	Mangrove sediments	India	Hydrolytic Enzymes	Pothayi et al. (2022)
Lipase Production				
Aureobasidium pullulans HN2.3 *Candida intermedia* YA01a, *Candida parapsilosis* 3eA2, *C. quercitrusa* JHSb, *C. rugosa* W8, *Lodderomycese longisporus* YF12c, *Pichia guilliermondii* N12c, *R. mucilaginosa* L10-2 and *Yarrowia lipolytica* N9	Surface of *Sargassum pallidum,* Seawater, Gut of *Apostichopus japonicus,* Gut of *Pseudosciaenacrocea,* Salterns, Gut of *Nemipterusvirgatus,* Deep seawater, Surface of *Laminaria japonica,* Salterns	China (isolated from various seawater and sediments collected during Antarctic exploration in 2004)	Lipase	Wang et al. (2007a)
C. tropicalis KGDP 7	Sediment samples from the mangrove ecosystem	India (Kasaragod district of Kerala)	Lipase	Pothayi et al. (2021)

C. tropicalis	Sediment samples were collected from the mangroves of the 5 districts	India (Kerala-Kasaragod Dt, Kannur Dt, Kozhikode Dt, Malappuram Dt, Thrissur Dt)	Lipase	Pothayi et al. (2022)
C. tropicalis (VPY3) and *Diutinameso rugosa* (VPY42)	Mangrove sediments of the Kerala coast	India	Lipase	Pothayi et al. (2024)
C. parapsilopsis (MY6) and *Starmerella etchellsii* (formerly *C. etchellsii* (MY1))	Sediment samples from mangrove ecosystem	India (Ernakulam, Kerala)	Lipase, protease, cellulase, glutaminase, and ligninase	Kachiprath et al. (2019)
Protease Production				
Aureobasidium pullulans	Sediment of saltern in Qingdao	China	Protease	Chi et al. (2006)
A. pullulans	Marine-derived	China	Protease	Chi et al. (2009)
Metschnikowiareukaufii W6b	Sediment of the South Sea	China	Cell-bound acid protease	Li et al. (2010)
R. mucilaginosa	Marine and terrestrial samples	Antarctic continent	Protease	Duarte et al. (2013)
R. mucilaginosa L7	Marine alga	Antarctic	Extracellular acid protease	Lario et al. (2015)
Inulinase Production				
Cryptococcus aureus G7a	Sediment of the South Sea	China	Inulinase	Sheng et al. (2007)
Cryptococcus aureus G7a, *Debaryomyces hansenii* G7a1, *Pichia guilliermondii* OUC1, and *Yarrowialipolytica* OUC2	Seawater, sediments, mud of salterns, guts of the marine fish, and marine algae	China	Inulinase	Gao et al. (2007b)
Pichia guilliermondii	Marine environment	China	Inulinase	Gong et al. (2008)
Aureobasidium sp. P6	Mangrove ecosystems	China	Inulinase	Jiang et al. (2016)
Production of other enzymes				
Candida, *Cryptococcus*, *Debaryomyces*, *Geotrichum*, *Kloeckera*, *Rhodosporidium*, *Saccharomyces* and *Trichosporon*	Decaying vegetation in the rhizosphere of the mangrove plant (11 sites)	India (in the Chapora, Mandovi, and Zuari estuaries of Goa)	Pectinase	Da Costa and D'Souza (1979)

Table 3 contd. ...

...Table 3 contd.

Yeast Strain	**Sample**	**Country of Study**	**Product**	**References**
Kodamaeaohmeri BG3	Gut of marine fish (*Hexagrammesotakii*)	China	Phytases	Li et al. (2007b)
A. pullulans 98	Sea saltern of the Yellow Sea	China	Cellulase production	Yanjun et al. (2015)
A. pullulans 98 N13d	Isolated from deep-sea sediments of the Pacific Ocean	China	Extracellular amylase	Li et al. (2007a)
Williopsis saturnus WC91-2	Marine yeast	China	Exo-β-1,3 glucanase	Peng et al. (2009)
Leucosporidiumantaracticum 171	Seawater sample	Antarctica (Admiralty Bay, King George Island)	α-Glucosidases	Turkiewicz et al. (2005)
Candida davisiana, Cryptococcus adeliensis, and *Guehomyces pullulans*	Marine and terrestrial samples	Antarctic continent	Endoxylanase	Duarte et al. (2013)
Debaryomyces hansenii C-11	Marine environment	Mexico (Pacific Ocean, Baja California Sur)	Copper-zinc superoxide dismutase	Hernandez-Saavedra and Ochoa (1999)
Single-cell Protein and Probiotics				
Candida sp.	Coastal and offshore waters	India (Cochin)	Single-cell protein production	Rishipal and Philip (1998)
Cryptococcus aureus G7a	Sediment of the South Sea	China	Single-cell protein	Gao et al. (2007)
Yarrowia lipolytica N-6	hypersaline natural environment	Mexico	Probiotics	Alamillo et al. (2017)
Y. lipolytica D-1 and N-6 strains	Marine yeast	Mexico	Probiotics	Reyes-Becerril et al. (2021)
D. hansenii BCS004	Marine yeast	Mexico	Probiotics	Reyes-Becerril et al. (2021)
Xylitol				
Cyberlindnera (Williopsis) saturnus	Detritus-based mangrove wetlands	Indian west coast	Xylitol production	Kamat et al. (2013)

Single-cell Oil				
R. mucilaginosa TJY15a	Marine fish (surface)	China	SCO	Li et al. (2010)
R. mucilaginosa TJY15a	Marine fish	China	SCO	Li et al. (2010b)
Yarrowia lipolytica	Tropical marine yeast	India	SCO	Katre et al. (2012)
A. pullulans var.melanogenum P10	Mangrove ecosystems	China	SCO	Wang et al. (2014)
Aureobasidium pullulans var. *melanogenum* P5	Mangrove system	China	heavy oils	Liu et al. (2014)
H. werneckii and *R. mucilaginosa*	Mangroves	Saudi Arabia	SCO	Abdel-Wahab et al. (2023)
Bioethanol and Biofuel Production				
B. salicaria (formerly *Pichia salicaria*)	Sediments of the mangrove forests	India	Bioethanol	Kathiresan et al. (2011)
S. cerevisiae C-19	Coastal waters	Japan	Bioethanol	Obara et al. (2012)
Candida sp.	Seawater and sediment samples	India	Bioethanol	Khambhaty et al. (2012)
S. cerevisiae (JN387604)	Mangrove soil samples	India (Pichavaram)	Bioethanol	Saravanakumar et al. (2013)
S. cerevisiae AZ65	Marine samples	The UK, Egypt, and the USA	Bioethanol	Zaky et al. (2018)
Candida parapsilosis BK30, *Pichia* sp. BK10, *Pichia kudriavzevii* BK40, *Pichia kudriavzevii* BK50, and *Pichia fermentans* MK20	Water and sediments of the Bushehr coast	Iran	Bioethanol	Khajeh et al. (2025)
Antimicrobial and Anticancer Properties				
Killer Toxin Production				
A. pullulans HN2.3, *Debaryomyces hansenii* hcx-1, *Pichia anomala* YF07b, *Pichia guilliermondii* GZ1 and *Williopsis saturnus* WC91-2	Seawater, sediments, mud of salterns, guts of marine fish, and marine algae	China	Killer toxin production	Wang et al. (2008)

Table 3 contd. ...

...Table 3 contd.

Yeast Strain	**Sample**	**Country of Study**	**Product**	**References**
Kluyveromyces siamensis HN12-1	Mangrove ecosystem	China	Killer toxin	Buzdar et al. (2011)
H. werneckii	Decayed leaves of *Avicennia marina*	Saudi Arabia (Red Sea Coast)	Antimicrobial compounds	Hodhod et al. (2020)
R. minuta	Mangrove-derived marine yeasts	India	Immunostimulant	Subramanian et al. (2014)
S. cerevisiae	Coastal mangrove ecosystem	India (Tamil Nadu)	Anticancer agent	Senthilraja and Kathiresan (2015)
Poly (β-malic acid) (PMA)				
Aureobasidium sp. P6	Mangrove system	China	Poly (β-malic acid) (PMA)	Ma et al. (2013)
Aureobasidium sp. P6, A. *pullulans* var. *pullulans* MCW	Mangrove systems	South of China	Polymalate	Wang et al. (2015)
Polyhydroxyalkanoates (PHAs)				
Pichia kudriavzevii VIT-NN02	Marine seaweeds	Tamil Nadu, India	Polyhydroxyalkanoates (PHAs)	Ojha and Das (2020)
Pichia kudriavzevii TSLS24	Spratly Island	Vietnam	Polyhydroxyalkanoates (PHAs)	Thu NTT et al. (2023)
Polysaccharide - Pullulan production				
Aureobasidium pullulans C J001	Sea mud samples	Eastern China	Pullulan	Chen et al. (2012)
A. pullulans var. melanogenium P16	Mangrove system	China	Pullulan	Ma et al. (2014)
Metal nanoparticle synthesis				
Yarrowia lipolytica NCIM 3589	Tropical marine yeast	India (Mumbai)	Biosynthesis of gold nanoparticles	Agnihotri et al. (2009)
K. capsulate (formerly *Pichia capsulata*)	Mangrove sediment	India	Silver Nanoparticle synthesis	Subramanian et al. (2010)

Rhodosporidium diobovatum	Marine yeast	India	Lead sulfide nanoparticle	Seshadri et al. (2011)
Biocontrol agent				
*S. pararoseus*ZMY-1	Mangrove forest mud	China (Zhangzhou)	Biocontrol agent	Shen et al. (2019)
Plant growth promoters				
Wickerhamomyces anomalus MSD1	Seaweed	India	PGY (biofertilizer/ biostimulant)	Krishnan et al. (2020)
Candida orthopsilosis and *R. mucilaginosa*	Red Sea water	Egypt	Biological control agents and improvement in wheat growth	Abdel-Kareem et al. (2021)
Cryptococcus sp. and *Candida tropicalis*	Mangrove sediments	India (Kerala)	Plant growth-promoting yeasts	Arya et al. (2023)
A. melanogenum KV 35, *C. tropicalis* P 9, and *Debaryomyces hansenii* PV 23	Mangroves of Puthuvype and Kumbalam	India (Kerala)	Plant growth-promoting yeast	Nimsi et al. (2024b)
Phosphate Solubilizer				
Candida sp.	Rhizosphere soil in Mangrove Center	Indonesia (Jenu Tuban)	Phosphate solubilizer	Nurhariyati et al. (2020)
Siderophore				
Aureobasidium pullulans HN6.2	Marine environment China		Siderophore	Wang et al. (2009)
A. pullulans	*Ulva fasciata* (seaweed)	India (Gulf of Mannar)	Siderophore	Murugappan et al. (2012)
Bioremediation and Hydrocarbon Degradation				
Candida, *Cryptococcus*, *Debaryomyces*, *Geotrichum, Kloeckera*, *Rhodosporidium*, *Saccharomycete*, and *Trichosporon*	Decaying vegetation in the rhizosphere of the mangrove plant (11 sites)	India	Hydrocarbon degradation	Da Costa and D'Souza (1979)

Table 3 contd. ...

...Table 3 contd.

Yeast Strain	**Sample**	**Country of Study**	**Product**	**References**
Yarrowia lipolytica NCIM 3589	Tropical marine yeast	India (Mumbai)	Hydrocarbon degradation	Oswal et al. (2002)
Y. lipolytica NCIM 3589	Tropical marine yeast	India (Mumbai)	Bioremediation	Jain et al. (2004)
Geotrichum sp. CS-67	Polluted mangrove sediments	China	Bioremediation	Mi (2011)
Acrophialophora jodhpurensis C_3, *Candida tropicalis* W_6, *Hyphopichia khmerensis strain* U_1 and *Pichia kudriavzevii* D_1	Water and sediment of the coastal areas of the Arabian Gulf	Saudi Arabia	Bioremediation	Al-Enazi (2025)
Biosurfactant				
P. hubeiensis SY62	*Calyptogena soyoae* (deep-sea cold-seep clam)	Japan	Biosurfactant	Konishi et al. (2010)
C. tropicalis VEMS 27, *S. cerevisiae* VEMS, *Starmerella bombicola* VEMS 39, and *Y. lipolytica* VEMS 51	Mangrove sediment samples	India (Tamil Nadu)	Biosurfactant	Thenmozhi et al. (2018)
C. parapsilopsis (MY6)	Sediments of the mangrove forests	India	Bioremediation detergent additive	Kachiprath et al. (2019)
Cyberlindnera saturnus SBPN 27	Sea sediment	India (Tamil Nadu)	Biosurfactant producer (antibacterial activity)	Balan et al. (2019)
Candida glabrata and *Trichosporonasahii*	Mangrove areas	Gujarat in India	Biosurfactant	Patel (2020)
R. paludigena VA 242	Mangrove islands	India (Valanthakkad, Kerala)	Biosurfactant	Nimsi et al. (2022)
Geotrichum candidum PV 37	Mangrove tree species	India (Puthuvype (PV) Kerala)	Biosurfactant (antimicrobial and anticancer activity)	Nimsi et al. (2024a)
Y. lipolytica LMS 24B	Marine zoanthids	Brazil	Bio-emulsifying agent	Barros de Oliveira et al. (2024)

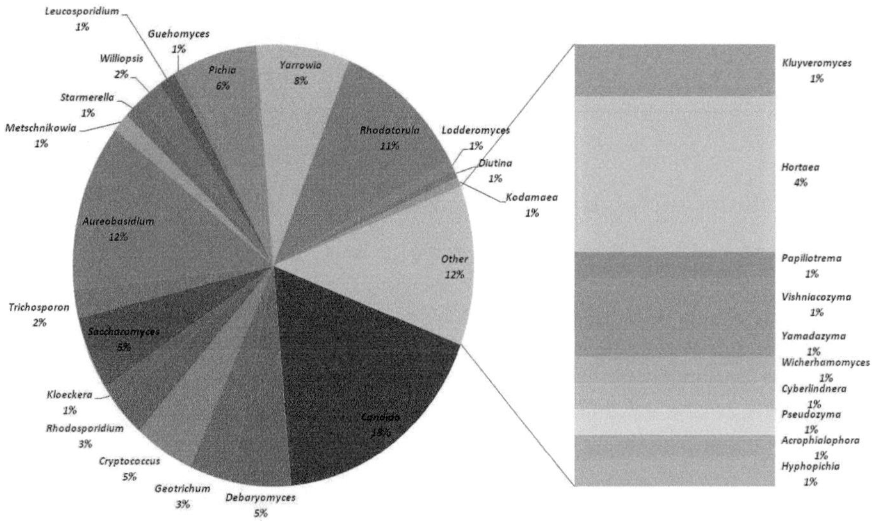

Fig. 5. Genus-wise bioprospecting potential of marine yeasts

***Aureobasidium* (12%)** ranks second, particularly *A. pullulans*, a black yeast known for its production of pullulan, enzymes (e.g., lipases, proteases), and potential in heavy metal biosorption and marine biofilm formation. Its melanin production also contributes to UV resistance and potential cosmetic applications.

***Rhodotorula* (11%)** is valued for its ability to synthesise carotenoids (e.g., torulene, β-carotene), lipids, and polyunsaturated fatty acids, making it a candidate for nutraceutical and biodiesel production. *R. mucilaginosa*, in particular, is noted for its prevalence in coastal ecosystems.

7.1.1 Genera with Moderate Contributions

***Yarrowia* (8%)**, particularly *Y. lipolytica* is recognised for lipid accumulation, production of lipases, and its use as a model organism in metabolic engineering for industrial biotechnology.

Pichia (6%), including *Pichia guilliermondii* is used for enzyme production, bioethanol, polyesters, antimicrobial properties, and bioremediation.

***Debaryomyces* (5%)**, especially *D. hansenii* is notable for its high salt tolerance and use in food fermentation, biosurfactant production, and biocontrol applications.

***Cryptococcus* (5%)**, includes species such as *Cryptococcus aureus* and *Cryptococcus adeliensis*, known for cold adaptation, extracellular polysaccharide production, and use in metal detoxification and dye degradation.

***Saccharomyces* (5%)**, primarily *S. cerevisiae*, although terrestrial in origin, is detected in marine environments and plays a central role in industrial fermentation, recombinant protein expression, and ethanol production.

***Hortaea* (4%) and *Geotrichum* (3%)** and *Metschnikowia* (1%) are represented for their role in enzyme production, biocontrol, and bioemulsifier synthesis.

***Trichosporon* (2%)**, *Willopsis* (2%), and *Leucosporidium* (1%) contribute to cold-adapted enzyme production and hydrocarbon degradation.

***Kluyveromyces* (1%)**, a killer toxin producer, has marine isolates with similar potential.

7.2 *Rare Biotechnologically Relevant Genera*

Several genera with a 1% contribution each, such as *Papiliotrema*, *Vishniacozyma*, *Yamadazyma*, *Wickerhamomyces*, *Cyberlindnera*, *Diutina*, *Pseudozyma*, *Kodamaea*, *Lodderomyces*, *Acrophialophora*, and *Hyphopichia*, represent emerging candidates in marine biotechnology. Though less frequently reported, these genera show promise in antimicrobial and biosurfactant production, plant growth-promoting activities, heavy metal bioremediation and enzyme production under extreme conditions.

The "Other" category (12%) indicates a long tail of rare or less studied genera that may harbour untapped biotechnological potential, especially with advancements in metagenomics and high-throughput screening.

8. Conclusion

Marine yeasts, a distinct group of eukaryotic microorganisms, form an integral component of marine microbial communities. Distributed across diverse marine habitats such as open oceans, coastal zones, estuarine waters, hydrothermal vents, polar seas, marine sediments, and in association with marine flora and fauna, these yeasts display remarkable ecological plasticity. Taxonomically, they span the phyla *Ascomycota* and *Basidiomycota*, with significant representation from genera such as *Candida*, *Debaryomyces*, *Saccharomyces*, *Cryptococcus*, *Rhodotorula*, *Aureobasidium*, and *Metschnikowia*. While traditional culturing techniques have provided initial insights, modern molecular, and metagenomic approaches have greatly expanded our understanding of their hidden diversity and ecological significance.

One of the most striking features of marine yeasts is their capacity to thrive under extreme and fluctuating environmental conditions. These include high salinity, variable temperatures (including polar cold and hydrothermal heat), high hydrostatic pressure, low nutrient availability, and intense UV radiation. Their survival is underpinned by several physiological and molecular adaptations:

In terms of biotechnological applications, marine yeasts represent an underutilised but promising frontier. The genera-wise bioprospecting analysis reveals that *Candida* (18%), *Aureobasidium* (12%), and *Rhodotorula* (11%) are among the most extensively explored for industrial potential. These yeasts are sources of bioactive compounds, cold-active enzymes, biosurfactants, pigments, antimicrobial metabolites, and polymers like pullulan. Though genera such as *Yarrowia*, *Geotrichum*, *Pichia*, and *Wickerhamomyces* contribute modestly to the overall profile, they too hold unique potential for lipid production, fermentation, and enzyme generation.

Moreover, the presence of numerous low-abundance or rarely studied genera such as *Papiliotrema*, *Cyberlindnera*, and *Vishniacozyma* points to a vast reservoir of unexplored functional traits. These genera, although less studied, hold promise for novel applications in enzyme production, antimicrobial activity, stress tolerance, and plant-microbe interactions. Their biotechnological relevance is expected to grow as tools like metagenomics, transcriptomics, and synthetic biology are increasingly applied to marine fungal research. These taxa may hold the key to novel industrial processes, especially in stress-resilient biotechnology applications such as marine biomass degradation, wastewater treatment, and climate-resilient agriculture.

In conclusion, marine yeasts embody a diverse and metabolically versatile group whose ecological roles and adaptive traits equip them to survive and flourish in some of the Earth's most extreme habitats. Their taxonomic spread, global distribution, biochemical capabilities, and resilience mechanisms together offer immense promise for sustainable biotechnological innovation. As we enter an era of intensified blue biotechnology research, marine yeasts are poised to emerge as key players in sectors like pharmaceuticals, agriculture, nutraceuticals, food and feed, energy, cosmetics biopolymers, biomedical and nanotechnology, houseload cleaning and personal care industry, and environmental bioremediation. Future research integrating omics technologies, bioprospecting, and systems biology will be crucial in unlocking their full potential.

References

Abdel-Kareem, M.M., Zohri, A.N.A. and Nasr, S.A.E.E. (2021). Novel marine yeast strains as plant growth-promoting agents improve defense in wheat (*Triticum aestivum*) against *Fusarium oxysporum*. J. Pl. Dis. Prot., 128: 973–988. https://doi.org/10.1007/s41348-021-00461-y.

Abdel-Wahab, M.A., Elgorban, A.M. and Bahkali, A.H. (2023). Single cell oil of oleaginous marine microbes from Saudi Arabian mangroves as a potential feedstock for biodiesel production. J. King Saud Univ. Sci., 35: 102615. https://doi.org/10.1016/j.jksus.2023.102615.

Abe, F. (2007). Induction of DAN/TIR yeast cell wall mannoprotein genes in response to high hydrostatic pressure and low temperature. FEBS Lett., 581(25): 4993–4998. https://doi.org/10.1016/j.febslet.2007.09.039.

Abe, F. and Horikoshi, K. (2000). Tryptophan permease gene TAT2 confers high-pressure growth in *Saccharomyces cerevisiae*. Mol. Cell. Biol., 20(21): 8093–8102. https://doi.org/10.1128/mcb.20.21.8093-8102.2000.

Abe, F. and Minegishi, H. (2008). Global screening of genes essential for growth in high-pressure and cold environments: Searching for basic adaptive strategies using a yeast deletion library. Genetics, 178(2): 851–872. https://doi.org/10.1534/genetics.107.083063.

Abe, F., Miura, T., Nagahama, T., Inoue, A., Usami, R. and Horikoshi, K. (2001). Isolation of a highly copper tolerant yeast, *Cryptococcus* sp., from the Japan Trench and the induction of superoxide dismutase activity by Cu 2+. Biotechnol. Lett., 23: 2027–2034. https://doi.org/10.1023/A:1013739232093.

Abramova, N., Sertil, O., Mehta, S. and Lowry, C.V. (2001). Reciprocal regulation of anaerobic and aerobic cell wall mannoprotein gene expression in *Saccharomyces cerevisiae*. J. Bacteriol., 183(9): 2881–2887. https://doi.org/10.1128/jb.183.9.2881-2887.2001.

Adler, L. and Gustafsson, L. (1980). Polyhydric alcohol production and intracellular amino acid pool in relation to halotolerance of the yeast *Debaryomyces hansenii*. Arch. Microbiol., 124: 123–130. https://doi.org/10.1007/BF00427716.

Adler, L., Blomberg, A. and Nilsson, A. (1985). Glycerol metabolism and osmoregulation in the salt-tolerant yeast *Debaryomyceshansenii*. J. Bacteriol., 162(1): 300–306. https://doi.org/10.1128/jb.162.1.300-306.1985.

Agnihotri, M., Joshi, S., Ravi Kumar, A., Zinjarde, S. and Kulkarni, S. (2009). Biosynthesis of gold nanoparticles by the tropical marine yeast *Yarrowia lipolytica* NCIM 3589. Mater. Lett., 63(15): 1231–1234. https://doi.org/10.1016/j.matlet.2009.03.017.

Ahearn, D.G. and Meyers, S.P. (1972). The role of fungi in the decomposition of hydrocarbons in the marine environment. Biodeter. Mat., 2: 12–18.

Ahmed, I., Haroon, A., Khan, M.N., Shah, S.R.U., Abbasi, M.A. and Buzdar, M.A. (2019). Occurrence and biodiversity of marine yeast in mangrove ecosystem of Shabi Creek, Gwadar-Pakistan. Pure Appl. Biol., 8(1): 680–687. https://doi.org/10.19045/bspab.2019.80008.

Alamillo, E., Reyes-Becerril, M., Cuesta, A. and Angulo, C. (2017). Marine yeast *Yarrowia lipolytica* improves the immune responses in Pacific red snapper (*Lutjanus peru*) leukocytes. Fish Shellfish Immunol., 70: 48–56. https://doi.org/10.1016/j.fsi.2017.08.036.

Al-Enazi, N.M. (2025). Isolation and characterization of crude oil degrading yeasts from the geo-marine sites of the Arabian Gulf. Geomicrobiol. J., 42(4): 324–331. https://doi.org/10.1080/01490451.2025.2462151.

Ambati, R.R., Phang, S.M., Ravi, S. and Aswathanarayana, R.G. (2014). Astaxanthin: Sources, extraction, stability, biological activities, and its commercial applications: A review. Mar. Drugs, 12(1): 128–152. https://doi.org/10.3390/md12010128.

Am-In, S., Limtong, S., Yongmanitchai, W. and Jindamorakot, S. (2011). *Candida andamanensis* sp. nov., *Candida laemsonensis* sp. nov. and *Candida ranongensis* sp. nov., anamorphic yeast species isolated from estuarine waters in a Thai mangrove forest. Int. J. Syst. Evol. Microbiol., 61(2): 454–461. https://doi.org/10.1099/ijs.0.022038-0.

Am-In, S., Yongmanitchai, W. and Limtong, S. (2008). *Kluyveromyces siamensis* sp. nov., an ascomycetous yeast isolated from water in a mangrove forest in Ranong Province, Thailand. FEMS Yeast Res., 8(5): 823–828. https://doi.org/10.1111/j.1567-1364.2008.00396.x.

Amornrattanapan, P. (2012). Riboflavin production by *Candida tropicalis* isolated from seawater. Sci. Res. Essays, 8: 43–47. https://doi.org/10.5897/SRE12.603.

Andlid, T., Juárez, R.V. and Gustafsson, L. (1995). Yeast colonizing the intestine of rainbow trout (*Salmo gairdneri*) and turbot (*Scophtalmus maximus*). Microb. Ecol., 30: 321–334. https://doi.org/10.1007/BF00171938.

André, L., Nilsson, A. and Adler, L. (1988). The role of glycerol in osmotolerance of the yeast *Debaryomyces hansenii*. Microbiol., 134(3): 669–677.

Anwar, A. and Saleemuddin, M. (1998). Alkaline protease: A review. Bioresour. Technol., 64(3): 175–183. https://doi.org/10.1016/S0960-8524(97)00182-X.

Arya, H., Nimsi, K.A. and Manjusha, K. (2023). *Cryptococcus* sp. K9 and *C. tropicalis* P9: Two plant growth-promoting yeasts (PGPY) from mangrove sediments. *In*: Recent Advances in Agricultural Sciences and Technology. Dilpreet Publishing House, Ariana Publishers & Distributors. New Delhi, pp. 655–668.

Babel, W. and Hofmann, K.H. (1982). The relation between the assimilation of methanol and glycerol in yeasts. Arch. Microbiol., 132: 179–184.

Balan, S.S., Kumar, C.G. and Jayalakshmi, S. (2019). Physicochemical, structural, and biological evaluation of Cybersan (trigalactomargarate), a new glycolipid biosurfactant produced by a marine yeast, *Cyberlindnerasaturnus* strain SBPN-27. Process. Biochem., 80: 71–180. https://doi.org/10.1016/j.procbio.2019.02.005.

Barros de Oliveira, V.P., Silva, J.R.M., Melo, V.M.M., Terceiro, P.S., de Oliveira, I.N. et al. (2024). Biosurfactants production by marine yeasts isolated from zoanthids and characterization of an emulsifier produced by *Yarrowia lipolytica LMS 24B*. Chemosphere, 355: 141807. https://doi.org/10.1016/j.chemosphere.2024.141807.

Bharathi, S., Saravanan, D., Radhakrishnan, M. and Balagurunathan, R. (2011). Bioprospecting of marine yeast with special reference to inulinase production. Int. J. Chem. Tech. Res., 3(3): 1514–1519.

Bhat, J.V. and Kachwalla, N. (1955a). Marine yeasts off the Indian coast. Proc. Ind. Acad. Sci., 41(1): 9–15.

Bhat, J.V., Kachwalla, N. and Mody, B.N. (1955b). Some aspects of the nutrition of marine yeasts and their growth. J. Sci. Ind. Res. C., 14: 24–27.

Brown, A.D. (1978). Compatible solutes and extreme water stress in eukaryotic micro-organisms. Adv. Microb. Physiol., 17: 181–242.

Brown, A.D. and Simpson, J.R. (1972). Water relations of sugar-tolerant yeasts: The role of intracellular polyols. Microbiol., 72(3): 589–591.

Buhagiar, R.W.M. and Barnett, J.A. (1971). The yeasts of strawberries. J. Appl. Bacteriol., 34(4): 727–739.

Burgaud, G., Hué, N.T.M., Arzur, D., Coton, M., Perrier-Cornet, J.M., Jebbar, M. and Barbier, G. (2015). Effects of hydrostatic pressure on yeasts isolated from deep-sea hydrothermal vents. Res. Microbiol., 166(9): 700–709. https://doi.org/10.1016/j.resmic.2015.07.005.

Buzdar, M.A., Chi, Z., Wang, Q., Hua, M.X. and Chi, Z.M. (2011). Production, purification, and characterization of a novel killer toxin from *Kluyveromyces siamensis* against a pathogenic yeast in crab. Appl. Microbiol. Biotechnol., 91: 1571–1579. https://doi.org/10.1007/s00253-011-3220-8.

Cao, K., Chen, J., Lu, X., Yao, Y., Huang, R. and Li, L. (2023). Matrine-producing endophytic fungus *Galactomyces candidum* TRP-7: Screening, identification, and fermentation conditions optimization for matrine production. Biotechnol. Lett., 45(1): 209–223. https://doi.org/10.1007/s10529-022-03280-z.

Chang, C.F., Huang, S.Y. and Lee, C.F. (2019). *Vishniacozyma changhuana* sp. nov., and *Vishniacozyma taiwanica* sp. nov., two novel yeast species isolated from mangrove forests in Taiwan. Int. J. Syst. Evol. Microbiol., 71(3): 004703. https://doi.org/10.1099/ijsem.0.004703.

Chen, C.Y., Ingram, M.F., Rosal, P.H. and Graham, T.R. (1999). Role for Drs2p, a P-type ATPase and potential aminophospholipid translocase, in yeast late Golgi function. J. Cell. Biol., 147(6): 1223–1236.

Chen, J., Wu, S. and Pan, S. (2012). Optimization of medium for pullulan production using a novel strain of *Aureobasidium pullulans* isolated from sea mud through response surface methodology. Carbohydr. Polym., 87(1): 771–774. https://doi.org/10.1016/j.carbpol.2011.08.062.

Chen, S., Wang, J., Muthusamy, B.P., Liu, K., Zare, S., Andersen, R.J. and Graham, T.R. (2006). Roles for the Drs2p–Cdc50p complex in protein transport and phosphatidylserine asymmetry of the yeast plasma membrane. Traffic, 7(11): 1503–1517. https://doi.org/10.1111/j.1600-0854.2006.00485.x.

Chi, W.C., Chen, W., He, C.C., Guo, S.Y., Cha, H.J. et al. (2019). A highly diverse fungal community associated with leaves of the mangrove plant *Acanthus ilicifolius var. xiamenensis* revealed by isolation and metabarcoding analyses. Peer J., 7: e7293. https://doi.org/10.7717/peerj.7293.

Chi, Z., Chi, Z., Zhang, T., Liu, G. and Wang, F. (2009). Production, characterization, and gene cloning of the extracellular enzymes from the marine-derived yeasts and their potential applications. Biotechnol. Adv., 27(2): 236–255. https://doi.org/10.1016/j.biotechadv.2009.01.002.

Chi, Z., Ma, C., Wang, P. and Li, H.F. (2006). Optimization of medium and cultivation conditions for alkaline protease production by the marine yeast *Aureobasidium pullulans*. Bioresour. Technol., 98(3): 534–538. https://doi.org/10.1016/j.biortech.2006.02.006.

Chi, Z., Wang, L., Ju, L. and Chi, Z. (2008). Optimisation of riboflavin production by the marine yeast *Candida membranifaciens* subsp. *flavinogenie* W14-3 using response surface methodology. Ann. Microbiol., 58(4): 677–681. https://doi.org/10.1007/BF03175574.

Chi, Z., Wang, X.X., Ma, Z.C. and Liu, G. (2012). The unique role of siderophore in marine-derived *Aureobasidium pullulans* HN6.2. Biometals, 25(2): 219–230. https://doi.org/10.1007/s10534-011-9499-1.

Chi, Z.M., Liu, T.T., Chi, Z., Liu, G.L. and Wang, Z.P. (2012). Occurrence and diversity of yeasts in the Mangrove ecosystems in Fujian, Guangdong, and Hainan Provinces of China. Ind. J. Microbiol., 52(3): 346–453. https://doi.org/10.1007/s12088-012-0251-5.

Chi, Z.M., Liu, G., Zhao, S., Li, J. and Peng, Y. (2010). Marine yeasts as biocontrol agents and producers of bio-products. Appl. Microbiol. Biotechnol., 86(5): 1227–1241. https://doi.org/10.1007/s00253-010-2483-9

Coelho, M.A., Almeida, J.M., Martins, I.M., Jorge da Silva, A. and Sampaio, J.P. (2010). The dynamics of the yeast community of the Tagus river estuary: Testing the hypothesis of the multiple origins of estuarine yeasts. Ant. van Leeuwenh., 98: 331–342. https://doi.org/10.1007/s10482-010-9445-1.

Da Costa, E. and D'Souza, J. (1979). Studies on estuarine yeasts: III. Hydrocarbon degraders. Mahasagar, 12: 155–161.

de Almeida, J.M. (2005). Yeast community survey in the Tagus estuary. FEMS Microbiol. Ecol., 53(2): 295–303.

de Araujo, F.V., Soares, C.A.G., Hagler, A.N. and Mendonça-Hagler, L.C. (1995). Ascomycetous yeast communities of marine invertebrates in a Southeast Brazilian mangrove ecosystem. Ant. van Leeuwenh., 68: 91–99.

de Souza Sebastianes, F.L., Romao-Dumaresq, A.S., Lacava, P.T., Harakava, R., Azevedo, J.L., De Melo, I.S. and Pizzirani-Kleiner, A.A. (2013). Species diversity of culturable endophytic fungi from Brazilian mangrove forests. Curr. Genet., 59: 153–166.

Diver, P., Ward, B.A. and Cunliffe, M. (2024). Physiological and morphological plasticity in response to nitrogen availability of a yeast widely distributed in the open ocean. FEMS Microbiol. Ecol., 100(5): fiae053.

Duarte, A.W.F., Dayo-Owoyemi, I., Nobre, F.S., Pagnocca, F.C., Chaud, L.C.S., Pessoa, A., Felipe, M. and Sette, L.D. (2013). Taxonomic assessment and enzymes production by yeasts isolated from marine and terrestrial Antarctic samples. Extremophiles, 17: 1023–1035.

Elsanhoty, R.M., Al-Turki, A.I. and El-Razik, M.M.A. (2017). Production of carotenoids from *Rhodotorula mucilaginosa* and their applications as colourant agent in sweet candy. J. Food Agri. Env., 15: 21–26.

Elsayis, A., Hassan, S.W.M., Ghanem, K.M. and Khairy, H. (2022). Suggested sustainable medical and environmental uses of melanin pigment from halotolerant black yeast Hortaea werneckii AS1. Front. Microbiol., 13: 871394. https://doi.org/10.3389/fmicb.2022.871394.

Fami Zaghrami, M., Hashemi, M. and Shokohi, T. (2021). Isolation and molecular identification of potential probiotic yeasts from the gastrointestinal tract of beluga fish. IJMCM, 11(1): 1471–1478.

Fell, J.W. (1976). Yeasts in oceanic regions. pp. 93–124. *In*: Jones, E.B.G. (ed.). Recent Advances in Aquatic Mycology. Wiley.

Fell, J.W. and Hunter, I.L. (1968). Isolation of heterothallic yeast strains of *Metschnikowia* Kamienski and their mating reactions with *Chlamydozyma* Wickerham spp. Ant. van Leeuwenh., 34: 365–376.

Fell, J.W. and Van Uden, N. (1963) Yeasts in marine environments. pp. 329–340. *In*: Charles C. Thomas (ed.). Symposium on Marine Microbiology. Springfield, USA.

Fell, J.W., Ahearn, D.G., Meyers, S.P. and Roth, Jr. F.J. (1960). Isolation of yeasts from Biscayne Bay, Florida and adjacent benthic areas 1, 2 Limnol. Oceanogr., 5(4): 366–371.

Fell, J.W., Boekhout, T., Fonseca, A., Scorzetti, G. and Statzell-Tallman, A. (2000). Biodiversity and systematics of basidiomycetous yeasts as determined by large-subunit rDNA D1/D2 domain sequence analysis. Int. J. Syst. Evol. Microbiol., 50(3): 1351–1371.

Fell, J.W., Hunter, I.L. and Tallman, A.S. (1973). Marine basidiomycetous yeasts (*Rhodosporidium* spp. n.) with tetrapolar and multiple allelic bipolar mating systems. Can. J. Microbiol., 19(5): 643–657.

Fell, J.W., Statzell-Tallman, A. and Kurtzman, C.P. (2004). *Lachancea meyersii* sp. nov., an ascosporogenous yeast from mangrove regions in the Bahama Islands. Stud. Mycol., 50: 359–363.

Fell, J.W., Statzell-Tallman, A., Scorzetti, G. and Gutiérrez, M.H. (2011). Five new species of yeasts from fresh water and marine habitats in the Florida Everglades. Ant. van Leeuwenh., 99: 533–549.

Fernandes, M.S. and Kerkar, S. (2017). Microorganisms as a source of tyrosinase inhibitors: A review. Ann. Microbiol., 67: 343–358.

Fernandes, P.M., Domitrovic, T., Kao, C.M. and Kurtenbach, E. (2004). Genomic expression pattern in *Saccharomyces cerevisiae* cells in response to high hydrostatic pressure. FEBS Lett., 556(1-3): 153–160.

Fischer, M. and Bacher, A. (2008). Biosynthesis of vitamin B2: Structure and mechanism of riboflavin synthase. Arch. Biochem. Biophy., 474(2): 252–265.

Fisher, B. (1894). Die Bakterien des Meersnach den Untersuchungen der Plankton-Expedition untergleichzeitigerBerucksichtigungeinigeralterer und neuererUntersuchungen. Ergebnisse Der Plankton-Expedition Der Humboldt-Stiftung, 4: 1–83.

Fize, A., Manier, J.F. and Maurand, J. (1970). Sur un casd'infestation du Copepode *Eurytemora velox* (Lillj) par unelevure du genre *Metschnikowia* (Kamienski). Ann. Parasitol. Hum. Comp., 45(3): 357–363.

Fotedar, R., Chatting, M., Kolecka, A., Zeyara, A., Al Malki, A., Kaul, R., Bukhari, S.J., Moaiti, M.A., Febbo, E.J., Boekhout, T. and Fell, J.W. (2022). Communities of culturable yeasts and yeast-like fungi in oligotrophic hypersaline coastal waters of the Arabian Gulf surrounding Qatar. Ant. van Leeuwenh., 115(5): 609–633.

Gancedo, C., Gancedo, M.J. and Sols, A. (1968). Glycerol metabolism in yeasts: Pathways of utilization and production. Eur. J. Biochem., 5: 165–172.

Gao, L., Chi, Z., Sheng, J., Li, J. and Wang, L. (2007a). Single-cell protein production from Jerusalem artichoke extract by a recently isolated marine yeast *Cryptococcus aureus* G7a and its nutritive analysis. Appl. Microbiol. Biotechnol., 77(4): 825–832. https://doi.org/10.1007/s00253-007-1210-7.

Gao, L., Chi, Z., Sheng, J., Wang, L., Li, J. and Gong, F. (2007b). Inulinase-producing marine yeasts: Evaluation of their diversity and inulin hydrolysis by their crude enzymes. Microb. Ecol., 54(4): 722–729. https://doi.org/10.1007/s00248-007-9231-4.

Gao, M., Su, R., Wang, K., Li, X. and Lu, W. (2013). Natural antifouling compounds produced by a novel fungus *Aureobasidium pullulans* HN isolated from marine biofilm. Mar. Pollut. Bull., 77(1-2): 172–176.

Gong, F., Sheng, J., Chi, Z. and Li, J. (2007). Inulinase production by a marine yeast *Pichia guilliermondii* and inulin hydrolysis by the crude inulinase. J. Ind. Microbiol. Biotechnol., 34(3): 179–185. https://doi.org/10.1007/s10295-006-0184-2.

Gunde-Cimerman, N., Plemenitaš, A. and Oren, A. (2018). Strategies of adaptation of microorganisms of the three domains of life to high salt concentrations. FEMS Microbiol., Rev. 42(3): 353–375.

Gustafsson, L. and Norkrans, B. (1976). On the mechanism of salt tolerance: Production of glycerol and heat during growth of *Debaryomyce shansenii*. Arch. Microbiol., 110: 177–183.

Haefner, S., Knietsch, A., Scholten, E., Braun, J., Lohscheidt, M. and Zelder, O. (2005). Biotechnological production and applications of phytases. Appl. Microbiol. Biotechnol., 68: 588–597.

Hagler, A.N., Rosa, C.A., Morais, P.B., Mendonça-Hagler, L.C., Franco, G.M., Araujo, F.V. and Soares, C.A. (1993). Yeasts and coliform bacteria of water accumulated in bromeliads of mangrove and sand dune ecosystems of southeast Brazil. Can. J. Microbiol., 39(10): 973–977.

Haldar, S. and Nazareth, S.W. (2019). Diversity of fungi from mangrove sediments of Goa, India, obtained by metagenomic analysis using Illumina sequencing. 3 Biotech., 9: 1–5.

Hasan, F., Shah, A.A. and Hameed, A. (2006). Industrial applications of microbial lipases. Enzyme Microb. Technol., 39: 235–251.

He, M., Xu, Y., Qiao, Y., Zhang, Z., Liang, J. et al. (2022). A novel yeast strain *Geotrichum* sp. CS-67 is capable of accumulating heavy metal ions. Ecotoxicol. Environ. Saf., 236: 113497. https://doi.org/10.1016/j.ecoenv.2022.113497.

Hernandez-Saavedra, N.Y. (1990). Levaduras Marinas Aisladasen la Costa Occidental de Baja California Sur. Mexico Thesis. UNAM-IztacalaMe´xico, 30 de Junio.

Hernandez-Saavedra, N.Y. and Ochoa, J.L. (1999). Copper-zi c superoxide dismutase from the marine yeast *Debaryomyces hansenii*. Yeast, 15: 6n57–668.

Hiraki, T., Sekiguchi, T., Kato, C., Hatada, Y., Maruyama, T., Abe, F. and Konishi, M. (2012). New type of pressurized cultivation method providing oxygen for piezotolerant yeast. J. Biosci. Bioeng., 113(2): 220–223.

Hodhod, M.S.E.D., Gaafar, A.R.Z., Alshameri, A., Qahtan, A.A., Noor, A. and Abdel-Wahab, M. (2020). Molecular characterization and bioactive potential of newly identified strains of the extremophilic black yeast *Hortaea werneckii* isolated from Red Sea mangrove. Biotechnol. Biotechnol. Equip., 34(1): 1288–1298.

Hoondee, P., Wattanagonniyom, T., Weeraphan, T., Tanasupawat, S. and Savarajara, A. (2019). Occurrence of oleaginous yeast from mangrove forest in Thailand. World J. Microbiol. Biotechnol., 35: 1–17.

Hunter, A.C. (1920). A pink yeast causing spoilage in oysters. US Dept. Agric. Bull., 819: 1–24.

Iwahashi, H., Odani, M., Ishidou, E. and Kitagawa, E. (2005). Adaptation of *Saccharomyces cerevisiae* to high hydrostatic pressure causing growth inhibition. FEBS Lett., 579(13): 2847–2852.

Jain, M.R., Zinjarde, S.S., Deobagkar, D.D. and Deobagkar, D.N. (2004). 2,4,6-Trinitrotoluene transformation by a tropical marine yeast, *Yarrowia lipolytica* NCIM 3589. Mar. Pollut. Bull., 49(9-10): 783–788. https://doi.org/10.1016/j.marpolbul.2004.06.007.

Jape, A., Harsulkar, A. and Sapre, V.R. (2014). Modified Sudan Black B staining method for rapid screening of oleaginous marine yeasts. Int. J. Curr. Microbiol. Appl. Sci., 3(9): 41–46.

Jayalakshmi, N. and Umamaheshwari, G. (2016). Isolation of marine yeast from Muthupet mangrove environment. Int. J. Pharm. Pharm. Res., 6: 200–205.

Jiang, H., Ma, Y., Chi, Z., Liu, G. and Wang, F. (2016) Production, purification, and gene cloning of a β-fructofuranosidase with a high inulin-hydrolyzing activity produced by a novel yeast *Aureobasidium* sp. P6 isolated from a mangrove ecosystem. Mar. Biotechnol., 18(5): 500–510. https://doi.org/10.1007/s10126-016-9712-x.

Johnson, P., Trybala, A., Starov, V. and Pinfield, V.J. (2021). Effect of synthetic surfactants on the environment and the potential for substitution by biosurfactants. Adv. Colloid. Interface. Sci., 288: 102340. https://doi.org/10.1016/j.cis.2020.102340.

Kachiprath, B., Solomon, S., Jayanath, G. and Sivasubramanian, K. (2019). Mangrove microflora is a potential source of hydrolytic enzymes for commercial applications. Indian J. Geo-Mar. Sci., 48(4): 678–684.

Kaewkrajay, C., Putchakarn, S. and Limtong, S. (2021). Cultivable yeasts associated with marine sponges in the Gulf of Thailand, South China Sea. Ant. van Leeuwenh., 114: 253–274.

Kalai, M., Becha, H. and Helali, K. (2024). Effect of artificial intelligence on economic growth in European countries: A symmetric and asymmetric cointegration based on linear and non-linear ARDL approach. Econ. Struct., 13: 22.

Kamat, S., Gaikwad, S., Ravi Kumar, A. and Deshpande, S. (2013). Xylitol production by *Cyberlindnera* (*Williopsis*) *saturnus*, a tropical mangrove yeast from xylose and corn cob hydrolysate. J. Appl. Microbiol., 115(6): 1357–1367. https://doi.org/10.1111/jam.12327.

Kaneshina, S., Ichimori, H., Hata, T. and Matsuki, H. (1998). Barotropic phase transitions of dioleoylphosphatidylcholine and stearoyl-oleoylphosphatidylcholine bilayer membranes. BBA – Biomembranes, 1374(1-2): 1–8.

Kathiresan, K., Saravanakumar, K. and Senthilraja, P. (2011a). Bio-ethanol production by marine yeasts isolated from coastal mangrove sediment. Int. Multidiscip. Res. J., 1(1): 19–24.

Kathiresan, K., Saravanakumar, K., Senthilraja, P. and Arockia, M.S. (2011b). Bio-ethanol production by marine yeasts isolated from coastal mangrove sediment. Int. Multidiscip. Res. J., 1: 19–24. https://doi.org/10.1016/j.jksus.2012.12.005.

Katre, G., Joshi, C., Khot, M., Zinjarde, S. and Ravikumar, A. (2012). Evaluation of single cell oil (SCO) from a tropical marine yeast *Yarrowia lipolytica* NCIM 3589 as a potential feed stock for biodiesel. AMB Express, 2: 36.

Kejžar, A., Cibic, M., Grøtli, M., Plemenitaš, A. and Lenassi, M. (2015). The unique characteristics of HOG pathway MAPKs in the extremely halotolerant *Hortaea werneckii*. FEMS Microbiol. Lett., 362(8): fnv046.

Kejžar, A., Gobec, S., Plemenitaš, A. and Lenassi, M. (2013). Melanin is crucial for growth of the black yeast *Hortaea werneckii* in its natural hypersaline environment. Fungal Biol., 117(5): 368–379. https://doi.org/10.1016/j.funbio.2013.03.006.

Khajeh, B., Zolgharnein, H. and Zamani, I. (2025). The potential of bioethanol production in marine yeasts and investigation of the optimal conditions of production in the selected isolates. Biologia, 80(4): 1085–1095. https://doi.org/10.1007/s11756-025-01891-x.

Khambhaty, Y., Upadhyay, D., Kriplani, Y., Joshi, N., Mody, K. and Gandhi, M.R. (2013). Bioethanol from macroalgal biomass: Utilization of marine yeast for production of the same. Bioenergy Res., 6: 188–195.

Kitano, H., Kataoka, K., Furukawa, K. and Hara, S. (2002). Specific expression and temperature dependent expression of the acid protease-encoding gene (PepA) in *Aspergillus oryzae* in solid-state culture (ricekoji). J. Biosci. Bioeng., 93: 563–567.

Kogej, T., Gostinčar, C., Volkmann, M., Gorbushina, A. A. and Gunde-Cimerman, N. (2006). Mycosporines in extremophilic fungi: Novel complementary osmolytes. Environ. Chem., 3(2): 105–110.

Kohlmeyer, J. and Kohlmeyer, E. (1959). Marine Mycology: The Higher Fungi. Academic Press, New York.

Konishi, M., Fukuoka, T., Nagahama, T., Morita, T., Imura, T. et al. (2010). Biosurfactant-producing yeast isolated from *Calyptogenasoyoae* (deep-sea cold-seep clam) in the deep sea. Biosci. Bioeng., 110(2): 169–175. https://doi.org/10.1016/j.jbiosc.2010.01.018.

Krishnan, S.R., Prabhakaran, N., Ragunath, K.S., Srinivasan, R., Ponni, K.K. et al. (2020). Unearthing the genes of plant-beneficial marine yeast: *Wickerhamomyces anomalus* strain MSD1. Int. J. Innov. Res. Sci. Eng. Technol., 10(9): 12492–12510. DOI:10.15680/IJIRSET.2021.1009032.

Kriss, A. and Rukina, E.A. (1949). Microbiology of the black sea. Mikrobiologia, 18: 141.

Kriss, A.E. (1955). Microbiological Research in the Region of the North Pole. California University, Scripps Institution of Oceanography.

Kriss, A.E., Rukina, E.A. and Tikhonenko, A.S. (1952). Distribution of yeast in the sea. Zhurnal Obshchei Biologii, 13(3): 232–242.

Kumar, C.G. and Takagi, H. (1999). Microbial alkaline protease: From a bioindustrial viewpoint. Biotechnol. Adv., 17(6): 561–594. https://doi.org/10.1016/S0734-9750(99)00038-3.

Kunthiphun, S., Chokreansukchai, P., Hondee, P., Tanasupawat, S. and Savarajara, A. (2018). Diversity and characterization of cultivable oleaginous yeasts isolated from mangrove forests. World J. Microbiol. Biotechnol., 34: 125. https://doi.org/10.1007/s11274-018-2507-7.

Kutty, S.N. and Philip, R. (2008). Marine yeasts: A review. Yeast, 25: 465–483. https:// doi.org/ 10.1002/ yea.1599.

Lages, F., Silva-Graça, M. and Lucas, C. (1999) Active glycerol uptake is a mechanism underlying halotolerance in yeasts: A study of 42 species. Microbiol., 145(9): 2577–2585.

Lakshmi, R. (2005). Marine Yeasts: Isolation, Characterization, and Screening for Probiotic Applications. MPhil Dissertation, Cochin University of Science and Technology, Kochi, India.

Lario, L.D., Chaud, L., das Graças Almeida, M., Converti, A., Sette, L.D. and Pessoa Jr, A. (2015). Production, purification, and characterization of an extracellular acid protease from the marine Antarctic yeast *Rhodotorula mucilaginosa* L7. Fungal. Biol., 119(11): 1129–1136. https://doi. org/10.1016/j.funbio.2015.08.012.

Lazarus, C.R. and Koburger, J.A. (1974). Identification of yeasts from the Suwannee River Florida estuary. Appl. Microbiol., 27(6): 1108–1111.

Leyton, A., Flores, L., Mäki-Arvela, P., Lienqueo, M.E. and Shene, C. (2019). *Macrocystis pyrifera* as a source of nutrients for the production of carotenoids by a marine yeast *Rhodotorula mucilaginosa*. J. Appl. Microbiol., 127(4): 1069–1079. https://doi.org/10.1111/jam.14242.

Li, H., Chi, Z., Wang, X. and Liu, G. (2007). Amylase production by the marine yeast *Aureobasidium pullulans* N13d. J. Ocean Univ. China, 6(1): 60–65. https://doi.org/10.1007/s11802-007-0060-3.

Li, J., Peng, Y., Wang, X. and Chi, Z. (2010). Optimum production and characterization of an acid protease from marine yeast *Metschnikowia reukaufii* W6b. J. Ocean Univ. China, 9(3): 359–364. https://doi. org/10.1007/s11802-010-1765-2.

Li, M., Liu, G.L., Chi, Z. and Chi, Z.M. (2010b). Single cell oil production from hydrolysate of cassava starch by marine-derived yeast *Rhodotorula mucilaginosa* TJY15a. Biomass Bioenergy, 34(1): 101–107. DOI: 10.1016/j.biombioe.2009.09.008.

Li, X., Chi, Z., Liu, Z., Li, Z., Wang, X. and Hirimuthugoda, N.Y. (2007b). Purification and characterization of extracellular phytase from a marine yeast *Kodamaea ohmeri* BG3. Mar. Biotechnol., 10: 190–197.

Li, X.Y., Liu, Z.Q. and Zhang, Z.M. (2008). Production of phytase by a marine yeast *Kodamaeaohmeri* BG3 in an oats medium: Optimization by response surface methodology. Bioresour. Technol., 99(14): 6386–6390.

Limtong, S. and Yongmanitchai, W. (2010). *Candida chanthaburiensis* sp. nov., *Candida kungkrabaensis* sp. nov. and *Candida suratensis* sp. nov., three novel yeast species from decaying plant materials submerged in water of mangrove forests. Ant. van Leeuwenh., 98: 379–388.

Limtong, S., Youngmanitchai, W., Kawasaki, H. and Seki, T. (2008). *Candida phangngensis* sp. nov., an anamorphic yeast species in the *Yarrowia* clade, isolated from water in mangrove forests in Phang-Nga Province, Thailand. Int. J. Syst. Evol. Microbiol., 58(2): 515–519.

Liu, Y.Y., Chi, Z., Wang, Z.P., Liu, G.L. and Chi, Z.M. (2014). Heavy oils, principally long-chain n-alkanes secreted by *Aureobasidium pullulans var. melanogenum* strain P5 isolated from mangrove system. J. Ind. Microbiol. Biotechnol., 41(12): 1329–1337. https://doi.org/10.1007/s10295-014-1484-6.

López, J.M.S., Vandeputte, M. and Van Bogaert, I.N.A. (2022). Oleaginous yeasts: Time to rethink the definition? Yeast, 39: 553–606.

Lucas, C., Da Costa, M. and Van Uden, N. (1990). Osmoregulatory active sodium-glycerol co-transport in the halotolerant yeast *Debaryomyces hansenii*. Yeast, 6(3): 187–191.

Ma, Y., Wang, G.Y., Liu, G.L., Wang, Z.P. and Chi, Z.M. (2013). Overproduction of poly (β-malic acid) (PMA) from glucose by a novel *Aureobasidium* sp. P6 strain isolated from a mangrove system. Appl. Microbiol. Biotechnol., 97(20): 8931–8939. https://doi.org/10.1007/s00253-013-5150-0.

Ma, Z.C., Fu, W.J., Liu, G.L., Wang, Z.P. and Chi, Z.M. (2014). High-level pullulan production by *Aureobasidium pullulans var. melanogenium* P16 isolated from a mangrove system. Appl. Microbiol. Biotechnol., 98(11): 4865–4873. https://doi.org/10.1007/s00253-014-5554-5.

MacGillivray, A.R. and Shiaris, M.P. (1993). Biotransformation of polycyclic aromatic hydrocarbons by yeasts isolated from coastal sediments. Appl. Environ. Microbiol., 59(5): 1613–1618.

Manivannan, S. and Kathiresan, K. (2009). Marine yeasts from rhizosphere soil of mangroves along the east coast of India: Conservation and management of mangroves in India. Kolkata: Zoological Survey of India, 269–274.

Mannazzu, S., Landolfo, T.L., Da Silva, P. and Buzzini, P. (2015). Red yeasts and carotenoid production: Outlining a future for non-conventional yeasts of biotechnological interest. World J. Microbiol. Biotechnol., 31(11): 1665–1673. https://doi.org/10.1007/s11274-015-1943-7.

Mapelli-Brahm, P., Gómez-Villegas, P., Gonda, M.L., León-Vaz, A., León, R. et al. (2023). Microalgae, seaweeds and aquatic bacteria, archaea, and yeasts: Sources of carotenoids with potential antioxidant and anti-inflammatory health-promoting actions in the sustainability era. Mar. Drugs, 21(6): 340.

Marques, A.R., Resende, A.A. and Gomes, F.C.O. (2021). Plant growth-promoting traits of yeasts isolated from the tank bromeliad *Vriesea minarum* L.B. Smith and the effectiveness of *Carlosrosaea vrieseae* for promoting bromeliad growth. Braz. J. Microbiol., 52(4): 1417–1429. https://doi.org/10.1007/s42770-021-00436-4.

Mata-Gómez, L.C., Montañez, J.C., Méndez-Zavala, A. and Aguilar, C.N. (2014). Biotechnological production of carotenoids by yeasts: An overview. Microb. Cell Fact., 13: 1–11. https://doi.org/10.1186/s12934-014-0011-4.

May, J.W., Marshall, J.H. and Sloan, J. (1982). Glycerol utilization by *Schizosaccharomyces pombe*: Phosphorylation of dihydroxyacetone by a specific kinase as the second step. Microbiol., 128(8): 1763–1766.

Menendez, E., Ramírez-Bahena, M.H., Fabryova, A., Igual, J.M., Benada, O., Mateos, P.F., Peix, A., Kolařík, M. and García-Fraile, P. (2015). *Pseudomonas coleopterorum* sp. nov., a cellulase-producing bacterium isolated from the bark beetle *Hylesinus fraxini*. Int. J. Sys. Evol. Microbiol., 65(9): 2920–2927. https://doi.org/10.1099/ijs.0.000344.

Mi, S. (2011). Identification of an astaxanthin-producing marine yeast strain YS-185 and optimization of its fermentation conditions. Environ. Sci. Biol. https://api.semanticscholar.org/CorpusID:88455865.

Misu, K., Fujimura-Kamada, K., Ueda, T., Nakano, A., Katoh, H. and Tanaka, K. (2003). Cdc50p, a conserved endosomal membrane protein, controls polarized growth in *Saccharomyces cerevisiae*. Mol. Biol. Cell., 14(2): 730–747.

Morais, P.B., Hagler, A.N. and Rosa, C.A. (1992). Yeasts associated with *Drosophila* in tropical forests of Rio de Janeiro, Brazil. Can. J. Microbiol., 38(11): 1150–1155.

Mukherjee, B., Mandal, S. and Banerjee, S. (2009). Turmeric as a carotenoid source on pigmentation and growth of fantail guppy, *Poecilia reticulata*. Proc. Zool. Soc., 62(2): 119–123. https://doi.org/10.1007/s12595-009-0033-5.

Murugappan, R., Karthikeyan, M., Aravinth, A. and Alamelu, M. (2012). Siderophore-mediated iron uptake promotes yeast-bacterial symbiosis. Appl. Biochem. Biotechnol., 168: 2170–2183.

Muthezhilan, R., Ragul, R. and Pushpam, R.L. (2014). Isolation, optimization, and extraction of microbial pigments from marine yeast *Rhodotorula* sp. (Amby109) as food colorants. Biosci. Biotechnol. Res. Asia, 11(1): 271–278.

Nagano, Y., Nagahama, T. and Abe, F. (2014). Cold-adapted yeasts in deep-sea environments. pp. 149–171. *In*: Buzzini, P. and Margesin, R. (eds.). Cold-adapted Yeasts. Springer, Berlin, Heidelberg. https://doi.org/10.1007/978-3-642-39681-6_7.

Nazareth, S. and Gonsalves, V. (2015). Occurrence of the halophilic black yeast *Hortaea werneckii* from brackish waters of mangroves. Kavaka, 39: 23–26.

Newell, S.Y. (1974). The Succession in the Mycoflora of Red Mangrove (*Rhizophora Mangle L.*) Seedlings. Doctoral dissertation, University of Miami, Coral Gables.

Ni, X.M., Chi, Z.M., Liu, Z.Q. and Yue, L.X. (2008a). Screening of protease-producing marine yeasts for the production of bioactive peptides. Acta Ocean. Sin., 27(1): 1–10.

Nimsi, K.A. and Manjusha, K. (2024a). Antibacterial and anticancer activity of glycolipid biosurfactant from manglicolous yeast *Geotrichum candidum* PV 37. Kavaka, 60(1): 9–20. https://doi.org/10.36460/Kavaka/60/1/2024/9-20.

Nimsi, K.A., Arya, H., Manjusha, K. and Kathiresan, K. (2024b). Multifarious plant growth-promoting traits of mangrove yeasts: Growth enhancement in mangrove seedlings (*Rhizophora mucronata*) for conservation. Arch. Microbiol., 206(1): 192. https://doi.org/10.1007/s00203-024-03913-9.

Nimsi, K.A., Manjusha, K. and Farzana, M.N. (2024c). Diversity, distribution, and bioprospecting potentials of carotenogenic yeast from the mangrove ecosystem. Arch. Microbiol., 206(1): 189. https://doi.org/10.1007/s00203-024-03879-8.

Nimsi, K.A., Manjusha, K., Hatha, A.A.M. and Kathiresan, K. (2023a). Diversity, distribution, and bioprospecting potentials of manglicolous yeasts: A review FEMS Micro. Eco., 99(5): fiad044.

Nimsi, K.A., Manjusha, K., Kathiresan, K. and Arya. H. (2023b). Plant growth-promoting yeasts (PGPY), the latest entrant for use in sustainable agriculture: A review. J. Appl. Microbiol., 134(1): lxac088. https://doi.org/10.1111/jam.16828.

Nimsi, K.A., Manjusha, K., Nefla, N., Kutty, S.N. and Aneymol, V.S. (2022). Biosurfactant from pigmented yeasts isolated from mangroves of Central Kerala and its application. Discourse Xaver Res. J., 10(1): 31–41.

Norkrans, B. (1966). Studies on marine occurring yeasts: Growth related to pH, NaCl concentration, and temperature. Arch. Microbiol., 54: 374–392.

Norkrans, B. (1966). On the occurrence of yeasts in an estuary off the Swedish westcoast. Svenska botaniskaforeningen Sven Bot. Tidskr., 60: 463–482.

Norkrans, B. (1968). Studies on marine occurring yeasts: Respiration, fermentation, and salt tolerance. Arch. Microbiol., 62: 358–372.

Novozhilova, M.I. (1955). The quantitative characteristics, species composition, and distribution of yeast like organisms in the Black Sea, the Sea of Okhotsk, and the Pacific Ocean. Tr. Inst. Mikrobiol Akad Nauk SSSR, 4: 155–195.

Nurhariyati, T., Dewi, P.A. and Widhiya, E.W. (2020). Potency of phosphate solubilizing yeast from Mangrove Center in Jenu, Tuban, Indonesia. Ecol. Environ. Conserv., 26(4): 1727–1730.

Obara, N., Ishida, M., Hamada-Sato, N. and Urano, N. (2012). Efficient bioethanol production from scrap paper shredder by a marine *Saccharomyces cerevisiae* derived C-19. Stud. Sci. Technol., 1: 127–132.

Ojha, N. and Das, N. (2020). Process optimization and characterization of polyhydroxyalkanoate copolymers produced by marine *Pichia kudriavzevii* VIT-NN02 using banana peels and chicken feather hydrolysate. Biocatalysis Agric. Biotechnol., 27: 101616. https://doi.org/10.1016/j.bcab.2020.101616.

Ônishi, H. (1963). Studies on osmophilic yeasts: Part xv. The effects of high concentrations of sodium chloride on polyalcohol production. Agric. Biological. Chem., 27(7): 543–547.

Oswal, N., Sarma, P.M., Zinjarde, S.S. and Pant, A. (2002). Palm oil mill effluent treatment by a tropical marine yeast. Bioresour Technol., 85: 35–37.

Pagnocca, F.G., Mendonça-Hagler, L.C. and Hagler, A.N. (1989). Yeasts associated with the white shrimp Penaeus schmitti, sediment, and water of Sepetiba Bay, Rio de Janeiro, Brasil. Yeast, 5: S479–Σ483.

Pandey, A., Szakacs, G., Soccol, C.R., Rodriguez-Leon, J.A. and Soccol, V.T. (2001). Production, purification, and properties of microbial phytases. Biores. Technol., 77(3): 203–214.

Patel, K. (2020). Screening of biosurfactant-producing yeasts isolated from the mangrove ecosystem of Surat region of Gujarat, India. Indian J. Sci. Technol., 13(19): 1927–1934. https://doi.org/10.17485/IJST/v13i19.204.

Peng, Y., Chi, Z.M., Wang, X.H. and Li, J. (2009). Purification and molecular characterization of exo-beta-1,3-glucanases from the marine yeast *Williopsis saturnus* WC91-2. Appl. Microbiol. Biotechnol., 85: 85–94.

Phaff, H.J. and Fell, J.W. (1970). Genus *Cryptococcus* Kutzing emend. Phaff et Spencer. pp. 1088–1145. *In*: Lodder, J. (ed.). The Yeasts: A Taxonomic Study. North Holland Publishing Co., Amsterdam.

Phaff, H.J., Mrak, E.M. and Williams, O.B. (1952). Yeasts isolated from shrimp. Mycologia, 44(4): 431–451.

Plemenitaš, A., Lenassi, M. and Konte, T. (2014). Adaptation to high salt concentrations in halotolerant/halophilic fungi: A molecular perspective. Front. Microbiol., 5: 199.

Pothayi, V. and Sebastian, C.D. (2021). Isolation, identification and screening of extracellular lipase producing yeasts from mangrove sediments of Kasaragod, North Kerala. Adv. Biores., 12(1): 81–89. https://doi.org/10.15515/abr.0976-4585.12.1.8189.

Pothayi, V. and Sebastian, C.D. (2022a). Yeast diversity in the mangrove sediments of North Kerala, India. Eur. J. Biol. https://doi.org/10.26650/EurJBiol.2022.1027475.

Pothayi, V., Athikkavil, F.M. and Devasia, S.C. (2024). Screening, purification and characterization of extracellular lipases from mangrove sediment yeasts. South Asian J. Exp. Bio., 14(1): 10–20. https://doi.org/10.38150/sajeb.

Pothayi, V., Kutty, S.N. and Devasia, S.C. (2022b). Effect of physico-chemical parameters in the production of hydrolytic enzymes from yeast *Candida tropicalis* isolated from the mangrove sediments of North Kerala, India. Jordan J. Biol. Sci., 15(2): 249–255. https://doi.org/10.54319/jjbs/150212.

Prabaharan, N. and Gupta, R. (1991). Yeasts from the sediment samples of the EEZ along the southwest coast of India. J. Mar. Biol. Assoc. India, 33(2): 455.

PrasannaKumar, C., Velmurugan, S., Subramanian, K., Pugazhvendan, S.R., Nagaraj, D.S. et al. (2020). Previously unrecorded distribution of marine sediments derived yeast isolates revealed by DNA barcoding. Biorxiv. https://doi.org/10.1101/2020.08.29.273490.

Prista, C., Almagro, A., Loureiro-Dias, M.C. and Ramos, J. (1997). Physiological basis for the high salt tolerance of *Debaryomyces hansenii*. Appl. Environ. Microbiol., 63(10): 4005–4009.

Pugh, G.J.F. and Lindsey, B.I. (1975). Studies of Sporobolomyces in a maritime habitat. Trans. Br. Mycol. Soc., 65(2): 201–209.

Orozco, M.R., Hernández-Saavedra, N.Y., Valle, F.A., González, B.A. and Ochoa, J.L. (1998). Cell yield and superoxide dismutase activity of the marine yeast *Debaryomyces hansenii* under different culture conditions. J. Mar. Biotechnol., 6(4): 255–259.

Ramos, J., Ariño, J. and Sychrova, H. (2011). Alkali–metal–cation influx and efflux systems in nonconventional yeast species. FEMS Microbiol. Lett., 317(1): 1–8.

Ravot, G., Masson, J.M. and Lefèvre, F. (2006). Applications of extremophiles: The industrial screening of extremophiles for valuable biomolecules. Methods Microbiol., 35: 785–813.

Reinoso, S., Gutiérrez, M.S., Domínguez-Borbor, C., Argüello-Guevara, W., Bohórquez-Cruz, M., Sonnenholzner, S., Nova-Baza, D., Mardones, C. and Navarrete, P. (2023). Selection of autochthonous yeasts isolated from the intestinal tracts of cobia fish (*Rachycentron canadum*) with probiotic potential. J. Fungi, 9(2): 274. https://doi.org/10.3390/jof9020274.

Rekha, R., Nimsi, K.A., Manjusha, K. and Sirajudheen, T.K. (2024). Marine yeast *Rhodotorula paludigena* VA 242: A pigment-enhancing feed additive for the ornamental fish Koi Carp. Aquac. Fish, 9: 66–70.

Reyes-Becerril, M., Alamillo, E. and Angulo, C. (2021). Probiotic and immunomodulatory activity of marine yeast *Yarrowialipolytica* strains and response against *Vibrio parahaemolyticus* in fish. Probiotics Antimicrob. Proteins, 13(5): 1292–1305. https://doi.org/10.1007/s12602-021-09769-5.

Reyes-Becerril, M., Angulo, C., Angulo, M. and Esteban, M.Á. (2021). Probiotic properties of *Debaryomyces hansenii* BCS004 and their immunostimulatory effect in supplemented diets for gilthead seabream (*Sparus aurata*). Aquac. Res., 52(2): 607–617. https://doi.org/10.1111/are.15123.

Romeo, O., Marchetta, A., Giosa, D., Giuffrè, L., Urzì, C. and De Leo, F. (2020). Whole genome sequencing and comparative genome analysis of the halotolerant deep sea black yeast *Hortaea werneckii*. Life, 10(10): 229.

Rong, Y., Zhang, L., Chi, Z. and Wang, X. (2015). A carboxymethyl cellulase from a marine yeast (*Aureobasidium pullulans 98*): Its purification, characterization, gene cloning, and carboxymethyl cellulose digestion. J. Ocean Univ. China, 14(6): 913–921. https://doi.org/10.1007/s11802-015-2574-4.

Rosario, C., Rivera Sotomayor, R. and Shelsy Torres, N. (2019). Survey of marine yeasts associated with mangroves in Puerto Rico. In: Proceedings of the 87th Annual Meeting of the Mycological Society of America Mayaguez: Department of Biology, University of Puerto Rico.

Ross, S.S. and Morris, E.O. (1962). Effect of sodium chloride on the growth of certain yeasts of marine origin. J. Sci. Food Agric., 13(9): 467–475.

Roth Jr, F.J., Ahearn, D.G., Fell, J.W., Meyers, S.P. and Meyer, S.A. (1962). Ecology and taxonomy of yeasts isolated from various marine substrates 1. Limnol Oceanogr., 7(2): 178–185.

Rukina, E.A. and Novozhilova, M.I. (1952). Species of yeasts isolated from various depths of Black Sea. Trudy Mikrobiologijasinstituts (Latvijas PSR Zinatnuakademija), 2: 150–156.

Saha, M., Sarkar, S., Sarkar, B., Sharma, B.K., Bhattacharjee, S. and Tribedi, P. (2016). Microbial siderophores and their potential applications: A review. Environ. Sci. Pollut. Res., 23(4): 3984–3999. https://doi.org/10.1007/s11356-015-4294-0.

Saravanakumar, K., Senthilraja, P. and Kandasamy, K. (2013). Bioethanol production by mangrove-derived marine yeast, *Saccharomyces cerevisiae*. J. King Saud Univ. Sci., 25(2): 121–127. https://doi.org/10.1016/j.jksus.2012.12.005.

Sarkar, A. and Rao, B. (2016). Marine yeast: A potential candidate for biotechnological applications—A review. Asian J. Microbiol. Biotechnol. Environ. Sci., 18(3): 627–634.

Sarkar, A., S.R., A., and K.V., B.R. (2024). Characterization of alkaline protease enzyme produced from marine yeast *Candida orthopsilosis* AKB-1 and its applications. Folia Microbiol. https://doi.org/10.1007/s12223-024-01216-6.

Sarma, V.V. and Vittal, B.P.R. (2001). Biodiversity of manglicolous fungi on selected plants in the Godavari and Krishna deltas, east coast of India. Fungal Divers., 6: 115–130.

Senthilraja, P. and Kathiresan, K. (2015). *In vitro* cytotoxicity MTT assay in Vero, Hep G2, and MCF-7 cell lines study of marine yeast. J. Appl. Pharm. Sci., 5(3): 080–084. https://doi.org/10.7324/JAPS.2015.50313.

Seshadri, S., Saranya, K. and Kowshik, M. (2011). Green synthesis of lead sulfide nanoparticles by the lead-resistant marine yeast, *Rhodosporidium diobovatum*. Biotechnol. Prog., 27(5): 1464–1469. https://doi.org/10.1002/btpr.651.

Shen, H., Wei, Y., Wang, X., Xu, C. and Shao, X. (2019). The marine yeast *Sporidiobolus pararoseus* ZMY-1 has antagonistic properties against *Botrytis cinerea in vitro* and in strawberry fruit Postharvest. Biol. Technol., 150: 1–8. https://doi.org/10.1016/j.postharvbio.2018.12.009.

Sheng, J., Chi, Z., Li, J., Gao, L. and Gong, F. (2007). Inulinase production by the marine yeast *Cryptococcus aureus* G7a and inulin hydrolysis by the crude inulinase. Proc. Biochem., 42(6): 805–811 https://doi.org/10.1016/j.procbio.2007.01.014.

Siepmann, R. and Höhnk, W. (1962). Über Hefen und einige Pilze (Fungi imp., Hyphales) aus dem Nordatlantik. Veröff Inst Meeresforsch Bremerh, 8: 79–97.

Singh, P. and Cameotra, S. (2004). Potential applications of microbial surfactants in biomedical sciences. Trends Biotechnol., 22(3): 142–146. https://doi.org/10.1016/j.tibtech.2004.01.010.

Singh, P., Raghukumar, C., Verma, P. and Shouche, Y. (2010). Phylogenetic diversity of culturable fungi from the deep-sea sediments of the Central Indian Basin and their growth characteristics. Fungal Divers., 40: 89–102.

Soundra Rani, M.H., Ramesh, T., Subramanian, J. and Kalaiselvam, M. (2013). Production and characterization of melanin pigment from halophilic black yeast *Hortaea werneckii*. J. Pharm. Res. Int., 2(8): 9–17.

Sperstad, S., Lutnæs, B.F., Stormo, S.K., Liaaen-Jensen, S. and Landfald, B. (2006). Torularhodin and torulene are the major contributors to the carotenoid pool of marine *Rhodosporidium babjevae* (Golubev). J. Ind. Microbiol. Biotechnol., 33(4): 269–273. https://doi.org/10.1007/s10295-005-0043-7.

Sprague, G.F. and Cronan, J.E. (1977). Isolation and characterization of *Saccharomyces cerevisiae* mutants defective in glycerol catabolism. J. Bacteriol., 129(3): 1335–1342.

Stahmann, K.P., Revuelta, J.L. and Seulberger, H. (2000). Three biotechnical processes using *Ashbya gossypii*, *Candida famata*, or *Bacillus subtilis* compete with chemical riboflavin production. Appl. Environ. Microbiol., 53(5): 509–516.

Statzell-Tallman, A., Belloch, C. and Fell, J.W. (2008). *Kwoniella mangroviensis* gen. nov., sp. nov. (Tremellales, Basidiomycota), a teleomorphic yeast from mangrove habitats in the Florida Everglades and Bahamas. FEMS Yeast Res., 8(1): 103–113.

Statzell-Tallman, A., Scorzetti, G. and Fell, J.W. (2010). *Candida spencermartinsiae* sp. nov., *Candida taylorii* sp. nov., and *Pseudozymaabaconensis* sp. nov., novel yeasts from mangrove and coral reef ecosystems. Int. J. Syst. Evol. Microbiol., 60: 1978–1984. https:// doi.org/ 10.1099/ ijs.0. 016857-0.

Subramanian, M., Alikunhi, N.M. and Kandasamy, K. (2010). In vitro synthesis of silver nanoparticles by marine yeasts from coastal mangrove sediment. J. Comput. Theor. Nanosci., 3(4): 428–433. https://doi.org/10.1166/asl.2010.1168.

Subramanian, M., Alikunhi, N.M. and Kandasamy, K. (2012). Immunostimulatory effect of mangrove-derived marine yeasts in Penaeus monodon. Aquac. Res., 44(4): 607–616. https://doi.org/10.1111/j.1365-2109.2012.03235.x.

Suehiro, S. (1960). Studies on the yeasts developing in the putrefied marine algae. Sci. Bull. Fac. Agric. Kyushu Univ., 17: 443–449.

Tartor, Y., Taha, M., Mahboub, H. and El Ghamery, M. (2018). Yeast species associated with diseased fish: Occurrence, identification, experimental challenges, and antifungal susceptibility testing. Aquacult., 488: 134–144.

Thenmozhi, M., Mani, P. and Boominathan, M. (2018). Screening and molecular identification of potential biosurfactant-producing estuarine yeasts isolated from mangrove sediment samples of Vellalar estuary, Porto Novo, India. Indian J. Pharma Innovation, 7(12): 208–213.

Thu, N.T.T., Hoang, L.H., Cuong, P.K., Viet-Linh, N., Nga, T.T.H., Kim, D.D., Leong, Y.K. and Nhi-Cong, L.T. (2023). Evaluation of polyhydroxy alkanoate (PHA) synthesis by *Pichia* sp. TSLS24 yeast isolated in Vietnam. Sci. Rep., 13(1): 3137. https://doi.org/10.1038/s41598-023-28220-z.

Tian, B.C., Liu, G.L., Chi, Z., Hu, Z. and Chi, Z.M. (2021). Occurrence and distribution of strains of *Saccharomyces cerevisiae* in China seas. J. Mar. Sci. Eng., 9(6): 590. https://doi.org/10.3390/jmse9060590.

Trivedi, N., Gupta, V., Kumar, M., Kumari, P., Reddy, C.R.K. and Jha, B. (2011). Solvent-tolerant marine bacterium *Bacillus aquimaris* secreting organic solvent-stable alkaline cellulase. Chemosphere, 83(5): 706–712. https://doi.org/10.1016/j.chemosphere.2011.02.006.

Turk, M. and Plemenitaš, A. (2002). The HOG pathway in the halophilic black yeast *Hortaea werneckii*: isolation of the HOG1 homolog gene and activation of HwHog1p. FEMS Microbiol. Lett., 216(2): 193–199.

Turkiewicz, M., Pazgier, M., Donachie, S.P. and Kalinowska, H. (2005). Invertase and α-glucosidase production by the endemic Antarctic marine yeast *Leucosporidium antarcticum*. Pol. Polar Res., 26: 125–136.

Umai, D., Kayalvizhi, R., Kumar, V. and Jacob, S. (2022). Xylitol: Bioproduction and applications: A review. Front. Sustain., 3: 826190. https://doi.org/10.3389/frsus.2022.826190.

Urano, N., Hirai, H., Ishida, M. and Kimura, S. (1998). Characterization of ethanol-producing marine yeasts isolated from coastal water. Fish Sci., 64(4): 633–637.

Ushakumari, U.N. and Ramanujan, R. (2013). Isolation of astaxanthin from marine yeast and study of its pharmacological activity. Int. Curr. Pharm. J., 2: 67–69.

Valderrama, B., Ruiz, J.J., Gutiérrez, M.S., Alveal, K., Caruffo, M., Oliva, M., Flores, H., Silva, A., Toro, M., Reyes-Jara, A. and Navarrete, P. (2021). Cultivable yeast microbiota from the marine fish species *Genypterus chilensis* and *Seriolella violacea*. J. Fungi, 7(7): 515. https://doi.org/10.3390/jof7070515.

Van Dover, C.L. (2014). Impacts of anthropogenic disturbances at deep-sea hydrothermal vent ecosystems: A review. Mar. Environ. Res., 102: 59–72.

Van Hamme, J.D., Singh, A. and Ward, O.P. (2006). Physiological aspects: Part 1 in a series of papers devoted to surfactants in microbiology and biotechnology. Biotechnol. Adv., 24(6): 604–620. https://doi.org/10.1016/j.biotechadv.2006.08.001.

Van Uden, N. (1967). Occurrence and origin of yeasts in estuaries. pp. 306–310. *In*: Lauff, G.H. (ed.). *Estuaries*. American Association for the Advancement of Science: Washington, DC.

Van Uden, N. and Castelo-Branco, R. (1963). Distribution and population densities of yeast species in Pacific water, air, animals, and kelp off southern California. Limnol. Oceanogr., 8: 323–329.

Van Uden, N. and Fell, J.W. (1968). Marine yeasts. Adv. Microbiol. Sea, 1: 167–201.

Van Uden, N. and Zobell, C.E. (1962). *Candida marina* nov. spec., *Torulopsis torresii* nov. spec. and *T. maris* nov. spec., three yeasts from the Torres Strait. Ant. van Leeuwenh., 28: 275–283.

Vicente, V.A., Orélis-Ribeiro, R., Najafzadeh, M.J., Sun, J., Guerra, R.S. et al. (2012). Black yeast-like fungi associated with Lethargic Crab Disease (LCD) in the mangrove-land crab, *Ucides cordatus* (Ocypodidae) Vet. Microbiol., 158(1-2): 109–122.

Vitale, G.A., Coppola, D., Palma Esposito, F., Buonocore, C., Ausuri, J. et al. (2020). Antioxidant molecules from marine fungi: Methodologies and perspectives. Antioxidant, 9(11): 1183. https://doi.org/10.3390/antiox9111183.

Volz, P.A., Jerger, D.E., Worzburger, A.J. and Hiser, J.L. (1974). A preliminary survey of yeasts isolated from marine habitats at Abaco Island, the Bahamas. Mycopathol. Mycol. Appl., 54: 313–316.

Wang, C.L., Li, Y., Xin, F.H., Liu, Y.Y. and Chi, Z.M. (2014). Evaluation of single cell oil from *Aureobasidium pullulans var. melanogenum* P10 isolated from mangrove ecosystems for biodiesel production. Process Biochem., 49(5): 725–731. https://doi.org/10.1016/j.procbio.2014.02.017.

Wang, L., Chi, Z., Wang, X., Ju, L., Chi, Z. and Guo, N. (2008). Isolation and characterization of *Candida membranifaciens* subsp. *Flavinogenie* W14-3, a novel riboflavin-producing marine yeast. Microbiol. Res., 163: 255–266.

Wang, L., Chi, Z., Wang, X., Liu, Z. and Li, J. (2007). Diversity of lipase-producing yeasts from marine environments and oil hydrolysis by their crude enzymes. Ann. Microbiol., 57(4): 495–501.

Wang, L., Yue, L., Chi, Z. and Wang, X. (2008). Marine killer yeasts active against a yeast strain pathogenic to crab *Portunus trituberculatus*. Dis. Aquat. Organ., 80(3): 211–218. https://doi.org/10.3354/dao01943.

Wang, W., Chi, Z., Liu, G., Buzdar, M.A., Chi, Z. and Gu, Q. (2009b). Chemical and biological characterization of siderophores produced by the marine-derived *Aureobasidium pullulans* HN6.2 and its antibacterial activity. Biomet., 22: 965–972.

Wang, W.L., Chi, Z.M., Chi, Z., Li, J. and Wang, X.H. (2009a). Siderophore production by the marine-derived *Aureobasidium pullulans* and its antimicrobial activity. Bioresour. Technol., 100: 2639–2641. https://doi.org/10.1016/j.biortech.2008.12.010.

Wang, Y.K., Chi, Z., Zhou, H.X., Liu, G.L. and Chi, Z.M. (2015). Enhanced production of Ca^{2+}-polymalate (PMA) with high molecular mass by *Aureobasidium pullulans var. pullulans* MCW. Microb. Cell. Fact., 14: 115. https://doi.org/10.1186/s12934-015-0296-3.

Wilson, W., Lowman, D. and Puthumana, J. (2024). Biocompatible melanin from the marine black yeast *Hortaea werneckii* R23 with antioxidant and photoprotection property. Braz. J. Microbiol., 55: 3171–3186. https://doi.org/10.1007/s42770-024-01483-y.

Winkelmann, G. (2002). Microbial siderophore-mediated transport. Biochem. Soc. Trans., 30: 691–696.

Wissebach, H. (1969). Pflanzen- und Tierfette (ausgenommen Milchfette): Vorkommen, Gewinnung, Zusammensetzung, Eigenschaften, Verwendung. *In*: pp. 9–180. Heimann, W. (ed.). Fette und Lipoide (Lipids). Springer.

Xu, W., Gong, L.F., Pang, K.L. and Luo, Z. H. (2018). Fungal diversity in deep-sea sediments of a hydrothermal vent system in the Southwest Indian Ridge. Deep Sea Res. I: Oceanogr. Res. Pap., 131: 16–26.

Yu, W.J., Chung, D., Bae, S.S., Kwon, Y.M., Cho, E.S. and Choi, G. (2023). Characterization and biological activities of yeasts isolated from marine environments. Microbiol. Res., 14: 1984–1999. https://doi.org/10.3390/microbiolres14040134.

Zaky, A.S., Greetham, D., Louis, E.J., Tucker, G.A. and Du, C. (2016). A new isolation and evaluation method for marine-derived yeast spp. with potential applications in industrial biotechnology. J. Microbiol. Biotechn., 26: 1891–1907. https://doi.org/10.4014/jmb.1605.05074.

Zeng, W., Zhang, B., Chen, G., Li, M. and Liang, Z. (2019). Efficient production of polymalic acid by a novel isolated *Aureobasidium pullulans* using metabolic intermediates and inhibitors. Appl. Biochem. Biotechnol., 187: 612–627. https://doi.org/10.1007/s12010-018-2825-0.

Zeng, X., Alain, K. and Shao, Z. (2021). Microorganisms from deep-sea hydrothermal vents. Mar. Life Sci. Technol., 3: 204–230.

Zhang, X., Hua, M., Song, C. and Chi, Z. (2012). Occurrence and diversity of marine yeasts in Antarctica environments. J. Ocean Univ. China, 11: 70–74.

Zhang, L. and Chi, Z.M. (2007). Screening and identification of a cellulase producing marine yeast and optimisation of medium and cultivation conditions for cellulase production. J. Ocean Univ. China, 37: 101–108.

Zhao, Y., Guo, L., Xia, Y., Zhuang, X. and Chu, W. (2019). Isolation, identification of carotenoid-producing *Rhodotorula* sp. from marine environment and optimization for carotenoid production. Mar. Drugs, 17(3): 161. https://doi.org/10.3390/md17030161.

ZoBell, C.E. and Feltham, C.B. (1934). Preliminary Studies on the Distribution and Characteristics of Marine Bacteria. Barkeley, California, University of California Press, pp. 279–295.

ZoBell, C.E. and Johnson, F.H. (1949). The influence of hydrostatic pressure on the growth and viability of terrestrial and marine bacteria. J. Bacteriol., 57(2): 179–189.

Zuo, Y., Zeng, R., Tian, C., Wang, J. and Qu, W. (2022). The importance of conditionally rare taxa for the assembly and interaction of fungal communities in mangrove sediments. Appl. Microbiol. Biotechnol., 106(9): 3787–3798.

10

Marine Endophytic Fungal Metabolites

An Underexploited Source of Novel Metabolites

B. Dinesh,[1] *Rushendran Rapuru,*[2] *Chandrashekhar G. Joshi*[1,*] and *Kandikere R. Sridhar*[3]

Marine endophytic fungi, residing asymptomatically within marine hosts such as seaweeds, mangroves, and sponges, are a gifted and underexplored source of bioactive leads against various diseases. This chapter aims to highlight their ecological significance, diversity, and unique adaptations to extreme marine environments. Further, this chapter focuses on the role of the key metabolic pathways in the biosynthesis of structurally diverse and pharmacologically potent compounds, including polyketides, peptides, alkaloids, and terpenoids. Special attention is given to silent gene clusters and strategies for their activation, as well as the influence of marine-specific conditions on gene expression. The pharmacological potential of metabolites from genera such as *Pestalotiopsis* and *Aspergillus* is explored, with a focus on their antimicrobial, anticancer, and neuroprotective activities. Moreover, the chapter addresses the

[1] Department of Biochemistry, Mangalore University, Mangalore 574 199, Karnataka, India.
[2] Department of Pharmacology, SRM College of Pharmacy, Faculty of Medicine and Health Sciences, SRM Institute of Science and Technology, Kattankulathur, Chengalpet-603203, Tamil Nadu, India.
[3] Central Research Laboratory, NITTE (Deemed to be University), KS Hegde Medical Academy (KSHEMA), Mangalore 575 018, Karnataka, India.
* Corresponding author: josheejoshee@gmail.com

DOI: 10.1201/9781003768708-13

deployment of genetic engineering approaches such as CRISPR-Cas systems and genome mining to overcome production bottlenecks and enhance yield. Despite significant promise, challenges related to cultivation, metabolite scalability, and regulatory concerns persist. The chapter concludes by bringing attention to the need for interdisciplinary collaboration and sustainable bioprospecting approaches, as well as the integration of AI and metabolomics for future drug discovery initiatives from marine endophytes.

1. Introduction

Endophytes are diverse group of microorganisms that reside within the tissues of plants or other living host tissues with no visible symptoms of disease (Fagorzi and Mengoni, 2022). The term "endophyte," derived from the Greek words *endon* (within) and *phyton* (plant), was originally used by de Bary to describe such internal symbionts (Lengrand et al., 2024). Endophytes cover an extensive range of microbial taxa, with bacteria, fungi, actinomycetes, and even viruses, and they are involved in a range of interactions with their hosts, ranging from neutral or commensal to parasitic (Khare et al., 2018) and mutualistic relationships (Gouda et al., 2016). Among these, endophytic fungi have received significant attention due to their widespread presence and potential functional roles within host organisms (Jha et al., 2023). Endophytic fungi colonize the different anatomical plant parts, such as foliage, stems, roots, and bark, in an asymptomatic manner that does not elicit disease responses in their host organisms (Waqar et al., 2024). Endophytic fungi are reported to be associated with almost all the plants identified and are increasingly being discovered to be effective against promoting host vigor, stress resistance, and ecological fitness (Chaudhary et al., 2022).

Marine environments, covering over 67% of the lithosphere, support some of the most diverse and biologically rich environments on the planet. Among these, marine endophytic fungi are a relatively poorly explored but ecologically important group (Gonçalves et al., 2022). Initially studied in terrestrial plants, endophytic fungi are now recognized as significant members of marine plant and animal microbiomes. The symbiotic association of endophytes with plants and animals is well known, including seaweeds, mangroves, seagrasses, and soft-bodied marine invertebrates (El-Bondkly et al., 2021). These associations are particularly important for their seemingly non-pathogenic nature and the putative mutualistic advantage they provide to the host (Jha et al., 2023). Marine endophytic fungi are especially suited to the demands of the marine environment: high salinity, tidal temperature fluctuations, alternating pressure, and constant biotic interaction (Gostinčar et al., 2022). In contrast to terrestrial hosts that are characterized as having hard defensive structures, most marine hosts possess softer tissues, and thus their endophytes tend to have beneficial or protective functions (Grabka et al., 2022). Marine macroalgae, or seaweeds, are dominant hosts, hosting complex microbial consortia on their surface (epiphytes) and within their tissue (endophytes) (Ren et al., 2022). Despite their ecological ubiquity and probable importance, marine endophytic fungi also open avenues for discovering novel natural products. These fungi represent a largely untapped reservoir of secondary metabolites compounds that may possess significant pharmaceutical and biotechnological potential (Pushparaj et al., 2023).

Marine endophytes can synthesize an array of low molecular weight secondary metabolites that are not directly involved in the primary life process of the organism, such as growth and reproduction. Instead, the substances mediate a variety of ecological functions, facilitating survival, competition, and interaction in the host habitat (Elhamouly et al., 2022). The variety of these metabolites covers an extensive collection of distinct chemicals with bioactive molecules, including alkaloids, steroids, terpenoids, flavonoids, quinones, xanthones, isocoumarins, peptides, phenolic acids, and saponins (El-Bondkly et al., 2021; Ortega et al., 2021; Singh and Kumar, 2023). The specific feature of secondary metabolites of marine origin, as opposed to those of terrestrial origin, is their specific chemical diversity, which is likely determined by the unique selective pressures of marine habitats (Chen et al., 2020). The high salinity, temperature fluctuations, and intense competition between species in marine habitats favor the evolution of novel biosynthetic pathways (dos Santos et al., 2022). As a result, marine endophytes often exhibit higher-level metabolic activities, producing compounds with novel structures and biological activities that are rarely seen in their terrestrial counterparts (El-Bondkly et al., 2021).

The biological activities of marine endophytic fungal metabolites cover an extensive range, including antimicrobial (Cadamuro et al., 2021), antiparasitic (Kaushik et al., 2014), antiprotozoal (Dembitsky et al., 2021), antituberculosis (Tapfuma et al., 2022), antiviral (Hawas et al., 2016), antidiabetic (Wu et al., 2018), anti-inflammatory–immunomodulatory (Quang et al., 2021), insecticidal (Chen et al., 2018), and anticancer (Sahoo et al., 2021) activities. Notably, certain marine endophytic fungi can synthesize phytochemicals that are chemically alike or similar to those of their host organisms, highlighting a deep symbiotic connection. This phenomenon presents a promising, sustainable alternative to sourcing bioactive compounds from endangered marine species or overexploited terrestrial plants (Alam et al., 2021). Against the background of intensifying demands for novel therapeutic compounds and growing concerns about synthetic antibiotic resistance and ecological responsibility, marine endophytic fungi stand as potential “chemical factories”. Their ability to synthesize structurally new and bioactive natural products places them at the forefront of drug discovery research (Kjer et al., 2010; Rateb and Ebel, 2011). Elucidation and further research on these marine symbionts may result in considerable breakthroughs in pharmaceuticals, biotechnology, and marine ecology. This chapter emphasizes various aspects pertaining to marine fungal cultivation, metabolites, biosynthesis, therapeutics, clinical applications, industrial avenues, and future avenues.

2. Marine Endophytic Fungal Diversity and Ecology

Marine endophytic fungi are a component of marine ecosystems, and they play a range of ecological roles in their host systems. In marine systems, the existence of such fungi is often defined by particular ecological niches and the process of their evolutionary development. Traditional mycological classifications differentiate between obligate marine fungal organisms, that require saline conditions for growth and reproduction, and facultative marine fungi, which have evolved from terrestrial or freshwater habitats but adapted to live and reproduce in marine systems (Pang et al., 2016). In practice, many marine-derived endophytes are facultative generalists; genera like *Penicillium* and *Aspergillus* are commonly isolated from land as well as the saline niche, while others, including certain *Halosphaeria* and *Corollospora* species, are considered true marine specialists (Kalaiselvam, 2015).

2.1 *Habitats and Hosts*

Marine endophytic fungi live in a highly diverse range of hosts, including macroalgae, angiosperms, and invertebrates (Fig. 1, Table 1). In the world of macroalgae (seaweeds), endophytic communities have been widely reported in many red, brown, and green algal species (Abdelrazek et al., 2024). For instance, Sahoo et al. (2021) reported 31 different fungal strains from these algal species, with the *Aspergillus* genus being dominant, followed by *Cladosporium*. Endophytes of algae must endure substantial fluctuations in moisture, salinity, and solar irradiance due to changes in tides (Sahoo et al., 2021). Studies carried out in the Bay of Bengal have, for example, listed several species of *Aspergillus*, *Chaetomium*, *Curvularia* endophytic in *Gracilaria* and *Sargassum* spp., and revealed a richness at the genus level comparable with the richness of terrestrial fungi (Noor et al., 2025).

Mangrove ecosystems and their halophytic vegetation also support abundant sources of fungal endophytes. Surveys carried out in Asia and South America have reported hundreds of fungal isolates from mangrove trees and ferns, mostly ascomycetes with a few basidiomycetes (Cadamuro et al., 2021). Zhu et al. (2024) reported that endophytes from two mangrove fern species (*Acrostichum* spp.) were composed of 75–81% Ascomycota and 15–19% Basidiomycota (Zhu et al., 2024).

Fig. 1. Niches for endophytic fungi in marine ecosystems.

Table 1. Diversity of endophytic fungi inhabiting marine flora and fauna.

Marine Habitat	Host	Isolated Endophytic Fungi	Reference(s)
Mangroves	*Rhizophora mucronata*	*Aspergillus* spp., *Cladosporium* spp., *Fusarium* spp., and *Penicillium* spp.	Zhou et al. (2018, 2022); Hamzah et al. (2018)
	Avicennia marina	*Alternaria* spp., *Phoma* spp., and *Trichoderma* spp.	Anjugam et al. (2019); Mukhopadhyay et al. (2024)
Marine sediment-associated plants	*Salicornia brachiata*	*Aspergillus terreus* and *Bipolaris* spp.	Khalmuratova et al. (2020); Valli et al. (2024)
Coral reefs	*Acropora cervicornis*	*Cladosporium* spp., *Fusarium solani*, and *Phoma* spp.	Abd El-Rahman et al. (2020); Klinges et al. (2023)
Seaweeds (green algae)	*Ulva lactuca*	*Aspergillus flavus* and *Curvularia lunata*	Rajulu et al. (2022); Abeygunawardane et al. (2025)
Seaweeds (brown algae)	*Sargassum wightii*	*Alternaria* spp., *Aspergillus* spp., and *Chaetomium* spp.	Suryanarayanan (2012); Rajulu et al. (2022)
Seaweeds (red algae)	*Gracilaria edulis*	*Mucor* spp., *Penicillium* spp., and *Rhizopus* spp.	Rajulu et al. (2022); Ariantari et al. (2024)
Seagrass	*Enhalus acoroides*	*Colletotrichum* spp., *Curvularia* spp., and *Phoma* spp.	Sakayaroj et al. (2010)
	Thalassia hemprichii	*Epicoccum* spp., *Alternaria* spp., and *Nigrospora* spp.	Supaphon et al. (2013); Qader et al. (2021)
Marine tunicates	*Didemnum molle*	*Alternaria* spp. and *Talaromyces* spp.	Nicoletti and Trincone (2016)
Marine invertebrate (sea cucumber)	*Holothuria edulis*	*Nigrospora oryzae* and *Phomopsis* spp.	Pivkin (2000)
Marine sponges	*Halichondria panicea*	*Aspergillus niger*, *Penicillium citrinum*, and *Talaromyces stipitatus*	HÖLLER et al. (2000)
	Xestospongia testudinaria	*Cladosporium* spp. and *Trichoderma harzianum*	Bakhtra et al. (2022)

Common tree hosts like *Avicennia* and *Rhizophora* also harbor diverse taxa such as *Cladosporium*, *Phomopsis*, and *Pestalotiopsis*, especially in the roots and stems. Although their exact functional roles are still being studied, these endophytes may contribute to improving salinity tolerance (Sridhar, 2018). Marine sponges (phylum Porifera) represent another prolific habitat for endophytic fungi. Sponges filter large volumes of seawater, thereby accumulating substantial microbial biomass up to 40–70% of their total tissue mass (Thomas et al., 2010). *Aspergillus*, *Penicillium*, *Cladosporium*, and *Fusarium* are commonly found in fungal isolates from sponges; many of these genera generate bioactive metabolites proved species-specific fungal communities in sponges: *Theonella* sp. hosted *Aspergillus versicolor* and *Penicillium chrysogenum*, while *Phakellia fusca* harbored *Penicillium lilacinum* and *Cladosporium* species. These results suggest that sponge identity and local microhabitats have a substantial role in shaping the fungal population dynamics (Proksch et al., 2002). Seagrasses, as true marine angiosperms, also support unique fungal endophytes. Ettinger et al. (2021) demonstrated a global ITS-based survey of *Zostera marina* by sampling leaves, roots, and rhizosphere sediments from 16 locations worldwide. Their results revealed distinct fungal communities between tissues and sites, indicating that geographic separation influences fungal composition. The core mycobiome of *Z. marina* consisted largely of unclassified Ascomycota and Chytridiomycota, suggesting that many associated taxa remain undescribed (Ettinger et al., 2021).

Corals and other cnidarians represent another substantial but poorly recognized niche for marine fungi. Endophytes have been found in coral mucosa, skeletons, and tissues, although their roles are less clear than those of bacterial or dinoflagellate symbionts (Roik et al., 2022). El-Bondkly et al. (2021) observed the presence of fungal endophytes across various marine hosts, including algae, seagrasses, mangroves, sponges, and corals. In healthy coral systems, yeast-like and filamentous fungi, mostly ascomycetes, seem to inhabit host structures without causing damage, though some species (e.g., *Candida*, *Fusarium*) can become pathogenic under stress conditions (El-Bondkly et al., 2021). These findings point to corals as reservoirs of distinct and potentially ecologically important fungal communities.

2.2 *Adaptations to Marine Conditions*

Marine endophytic fungi are exceptional owing to their endurance to the extreme environmental conditions. One major stressor is salinity; marine fungi must reduce osmotic pressure by accumulating compatible solutes like mannitol, glycerol, or trehalose and altering the lipid composition of their membranes, usually by increasing the percentage of unsaturated fatty acids (Ramesh et al., 2021; Papikinou et al., 2024). Some macroalgal endophytes can tolerate salt concentrations exceeding 30%, a testament to their halophilic or halotolerant nature (Corral et al., 2019). Another environmental problem, particularly in deep-sea environments, is hydrostatic pressure. Fungal cells change morphologically and biochemically when exposed to high pressure (Zhong et al., 2024). According to Deng et al. (2023), *Aspergillus sydowii*, which was isolated from the Mariana Trench, developed swollen, spherical hyphae in order to withstand pressures of 100 MPa. This adaptation is probably meant to lessen the stress that turgor causes on membranes. Additionally, in order to stabilize their cellular machinery in harsh environments, these deep-sea fungi up regulate stress-response proteins like chaperones and proteases (Deng et al., 2023). More specialization is required in deep ocean zones and polar regions due to their low temperatures. Psychrophilic fungi secrete cold-active enzymes that continue to function at temperatures close to freezing, change the lipid composition of their membranes, and produce antifreeze proteins (Dolashki et al., 2024). For instance, many Antarctic isolates have a considerable amount of polyunsaturated fatty acids besides secreting lipases as well as proteases that are adapted to the cold marine environment, allowing them to survive and perform ecological functions. These convergent adaptations highlight the ecological resilience and evolutionary versatility of marine endophytes by enabling them to colonize a broad range of habitats, from intertidal zones to the depths of the abyss.

2.3 Geographical Distribution and Sampling

Marine endophytic fungi can be found all over the world, living in pelagic and coastal environments ranging from tropical reefs to the Arctic. Regional environmental factors, ocean currents, and host range all influence their distribution (Sen et al., 2022). As an illustration, studies comparing Zostera marina populations in the Pacific and Atlantic have revealed different fungal assemblages, suggesting strong local environmental filtering and minimal transoceanic dispersal (Ettinger et al., 2021). Surveys of mangroves and seaweeds on various continents have also identified both region-specific taxa and globally prevalent genera (*Aspergillus*, *Penicillium*), indicating that marine fungal biogeography is both localized and cosmopolitan (Xu et al., 2018). The usual procedure for sampling these fungi is to gather healthy host tissues, such as thalli, leaves, roots, stems, or pieces of sponge, and then thoroughly sterilize the surface to get rid of epiphytic bacteria. Following that, the internal tissues are cultivated on selective media or marine agar, frequently with antibiotics added. Although many marine fungi are slow-growing or unculturable under normal conditions, culture-based techniques have produced thousands of fungal isolates from a variety of marine hosts (Elnaggar et al., 2016). Molecular methods are being used more often to overcome these constraints. Cryptic and uncultured taxa have been successfully discovered through amplicon sequencing of the fungal ITS region or 18S rRNA genes from host DNA (Lücking et al., 2020). In aquatic plant samples, Zheng et al. (2021) combined culture-based and sequencing techniques, isolating 1,689 strains from 154 genera through culturing and identifying 1,074 OTUs from 194 genera through sequencing; only 66 genera were shared between methods, highlighting the complementary nature of these approaches (Zheng et al., 2021a). The toolkit for isolating novel marine fungi is being further expanded by recent developments like single-cell microfluidics and culturomics (e.g., dilution-to-extinction, *in situ* diffusion chambers). Meta-transcriptomics, metagenomics, and environmental DNA (eDNA) provide strong, cultivation-independent methods for identifying uncommon species and evaluating their functional potential. These days, sampling procedures differ depending on the type of host; for endolithic fungi, sponge biopsies or coral rubble grinding are used (Kennedy et al., 2010). Finally, standardizing sterilization and processing procedures and integrating culture-based and molecular techniques are essential for capturing the entire range of marine fungal diversity. As these techniques continue to evolve and more regions and host taxa are studied, our understanding of marine endophyte ecology and function will continue to deepen (Joe Xavier Sneha et al., 2024).

3. Biosynthesis of Metabolites

Fungi are well known for their ability to generate a wide variety of secondary metabolites, which are substances that are vital to their ecological success but not necessary for their growth or reproduction. These metabolites have roles in symbiosis, signaling, defence, and competition (Fig. 2). Complex biosynthetic gene clusters (BGCs), which encode specialized enzymes, coordinate their production,

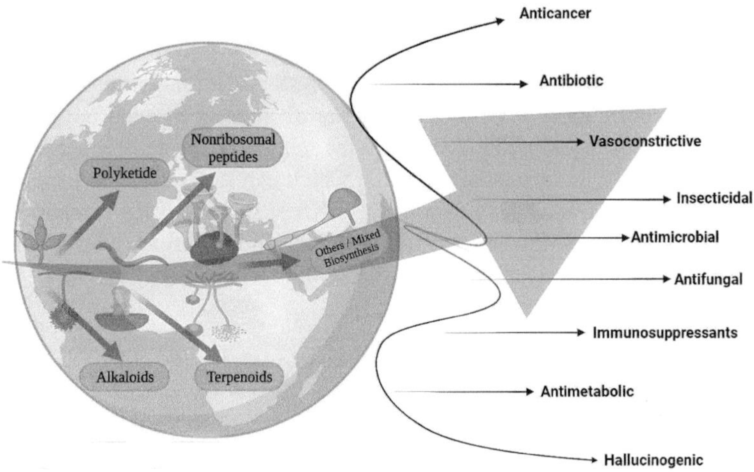

Fig. 2. Marine endophytic fungi and their therapeutic roles.

enabling fungi to transform basic building blocks into a variety of chemically diverse and bioactive compounds (Avalos and Limón, 2022). Enzymes linked to shikimate-derived pathways, terpenoid synthases, polyketide synthases (PKSs), and non-ribosomal peptide synthetases (NRPSs) are among the major biosynthetic pathways. To synthesize metabolites that are both ecologically powerful and structurally complex, many of these pathways work independently, while others work in hybrid configurations (Wang et al., 2014).

3.1 *Polyketide Synthases*

Polyketides rank among the most chemically varied classes of fungal secondary metabolites. Compounds such as aflatoxins, lovastatin, and tetracyclines illustrate the wide-ranging biological roles of this group. Synthesized by polyketide (PKSs) synthases through successive condensations of acyl-CoA precursors (typically acetyl-CoA besides malonyl-CoA), these molecules emerge from modular enzymatic assemblies. Fungal PKSs are generally grouped as three varieties: Type I PKSs are bulky, multifunctional enzymes often operating iteratively; Type II PKSs comprise discrete mono functional enzymes; and Type III PKSs are simpler, often dimeric enzymes that do not require carrier proteins (Singh et al., 2019). Enzymes such as methyltransferases, hydroxylases, and glycosyl transferases that diversify the core scaffold and fine-tune biological function increase the structural complexity of polyketides. PKS clusters show the dynamic evolution of secondary metabolism in fungi and can be anywhere from tightly compact units to scattered, multi-locus arrangements (Avalos and Limón, 2022).

3.2 *Non-Ribosomal Peptide Synthetases*

The NRPSs are an essential component of fungal secondary metabolism, in addition to PKSs. These large, modular enzymes put together peptides without the help of the ribosome, frequently adding non-proteinogenic building blocks, N-methylated residues, or D-amino acids. A thiolation (T) domain for intermediate shuttling, an adenylation (A) domain for selection of the substrate, and a condensation (C) domain involved in the formation of peptide bonds are all present in each NRPS module (Patel et al., 2023). Enzymes encoded within the same gene cluster frequently cyclize and further diversify the resultant products, which can range from siderophores to antibiotics (Matilla et al., 2012). Similar to PKS systems, NRPS clusters usually contain accessory genes for transport, regulation, and tailoring, which guarantee that metabolite biosynthesis is precisely synchronized with environmental cues and fungal development (Komaki and Tamura, 2023).

3.3 *Terpenoid and Shikimate Pathways*

Terpenoids (isoprenoids) are a common cluster of metabolites produced by fungi. Terpenoids, such as monoterpenes, sesquiterpenes, and diterpenes, are produced by the mevalonate (MVA) pathway from five-carbon isoprene units. Different carbon skeletons are produced by fungi using terpene synthases, and oxidases, methyltransferases, and other modifying enzymes elaborate these skeletons (Vavitsas et al., 2018). A variety of fungal terpenoids have strong bioactivities; fumonisins and trichothecenes, for instance, are well known for their cytotoxic, phytotoxic, and antifungal qualities. These substances aid in pathogenesis, environmental adaptation, and fungal competitiveness (Dai et al., 2021). The biosynthesis of aromatic amino acids, which are building blocks for a variety of secondary metabolites of alkaloids, depends heavily on the shikimate pathway. This pathway supports the synthesis of lead molecules in fungi, including benzylisoquinolines, phenylpropanoids, and indole alkaloids (Jan et al., 2021). Prenylation, methylation, or oxidative tailoring are common processes that greatly increase the pharmacological potential of these metabolites. Centuries of ethnopharmacological use and contemporary therapeutic development are supported by ergot alkaloids, which are derived from tryptophan in *Claviceps* species and exhibit neurotropic and vasoconstrictive properties (Li et al., 2012).

3.4 *Hybrid Biosynthetic Pathways*

Hybrid biosynthetic systems, which combine aspects of peptide and polyketide biosynthesis, are commonly found in fungal genomes. The PKS-NRPS hybrid system is one of the best-characterized; modules from both enzymatic families work together to create complex scaffolds that contain both amino acid and acyl components (Cox and Gulder, 2023). An example of this class is the neurotoxic metabolite cyclopsidonic acid, which is produced by *Aspergillus* species. The structural variety and functional adaptability of hybrid metabolites highlight their significance in ecology and pharmacology and indicate a high degree of metabolic innovation in fungi (Gallagher et al., 2010). The biosynthesis of prenylated indole alkaloids, a class of metabolites with improved chemical and biological characteristics, depends on dimethylallyl tryptophan synthases (DMATSs). Tryptophan-derived indoles undergo a significant change in solubility, stability, and bioactivity when DMATS enzymes add prenyl groups to them (Luk and Tanner, 2009). These substances are especially abundant in marine-derived fungi, which produce structurally distinct prenylated indole alkaloids with cytotoxic, neuroactive, and antimicrobial qualities. The adaptive reactions of marine fungi to extremely competitive and chemically dynamic environments are reflected in these metabolites (Meng et al., 2020).

3.5 *Meroterpenoid and Loline Alkaloid Pathways*

Meroterpenoids are natural products that are hybrids of terpenoid moieties and scaffolds based on polyketides or indoles. Their biosynthesis starts with the formation of a core structure, usually through an aromatic amino acid pathway or PKS. Terpene cyclases and modifying enzymes then prenylate and cyclize the core structure. The fusion of two structurally complex compounds, namely, Austinol and andrastin A, results in strong antimicrobial, anticancer, and anti-inflammatory properties. Meroterpenoids are appealing leads in drug discovery because of their structural flexibility (Nazir et al., 2021). Loline alkaloids are another unique class of chemicals that are mostly produced by grasses' symbiotic endophytic fungi. After cyclization and oxidation processes, these saturated pyrrolizidine compounds are created from amino acid precursors such as proline and homoserine. The co-regulated LOL gene cluster, which is up regulated during symbiotic interaction, contains the genes that produce loline. As an example of the useful functions of fungal metabolites in mutualistic relationships, loline alkaloids provide host plants with considerable ecological advantages by conferring resistance against herbivorous insects (Spiering et al., 2005).

4. Influence of the Marine Environment on Biosynthetic Gene Clusters

The distinct selective pressures imposed by the marine environment significantly influence the BGCs present in fungi and other microorganisms that are derived from the ocean. Marine environments, which are characterized by a variety of ecological niches, low temperatures, high salinity, hydrostatic pressure, and little light, are what propel the evolution of distinct metabolic strategies (Reen et al., 2015). Marine bacteria and fungi have developed specialized biosynthesis pathways in response to these stressors, producing secondary metabolites that are unique in structure and significant in pharmacological activity. As many studies highlight, the complex conditions of the ocean have produced natural products with distinct chemical scaffolds that are uncommon in terrestrial organisms (Petersen et al., 2020). This metabolic innovation is especially evident in deep-sea fungi. These organisms, which are isolated from extremely deep depths, flourish in environments of high pressure and low temperature. They frequently produce polyketides, alkaloids, and non-ribosomal peptides with structural characteristics that have never been seen before (Arifeen et al., 2019).

The rich chemical diversity present in marine biosynthetic systems is highlighted by the fact that several fungal strains derived from marine environments have already produced strong antimicrobial and anticancer agents (Hasan et al., 2015). The immense biosynthetic potential of marine microbial communities has been further revealed by recent developments in metagenomics and genome mining. Thanks to these instruments, thousands of hitherto unidentified BGCs have been found in ocean samples, especially among symbiotic or uncultivated microorganisms (Huang et al., 2023). An enormous repertoire of novel gene clusters, many from distinct and uncharted gene cluster families, were recovered by a historic global ocean sequencing project, underscoring the fact that the marine microbiome's biosynthetic landscape is still largely unexplored. Interestingly, some sponge-associated microbial consortia contain remarkably varied BGCs, such as distinct peptidyl brominases and halogenases, which reflect the chemistry of marine environments that are rich in halides (Paoli et al., 2022). Furthermore, the modulation of gene cluster expression is significantly influenced by ecological interactions in the sea. Previously silent gene clusters have often been activated by co-culture studies between marine fungi and their symbionts or microbial competitors. According to these results, ecological cues and chemical communication are the key players in controlling secondary metabolism in the oceanic environment. The emergence of unique BGC architectures leads to the dynamic interaction between microbial biosynthesis as well as marine environmental factors, highlighting the ocean's critical role in broadening the known repertoire of fungal secondary metabolites (Li et al., 2023).

4.1 *Silent Gene Clusters and Strategies for Activation*

In a typical laboratory condition, a major fraction of secondary metabolite multigene assemblies remains transcriptionally silent, despite the astounding diversity encoded within fungal genomes. The BGCs outnumber the known metabolites, as comparative genomics has repeatedly demonstrated (Brakhage and Schroeckh, 2011). This suggests that many fungal pathways are either poorly expressed or inactive *in vitro*. The goal of natural product discovery is to unlock these mysterious clusters, and different creative protocols have been developed to do so (Hur et al., 2023). Epigenetic reprogramming of fungal chromatin is one of the most popular methods. Many fungi have heterochromatic structures that restrict transcription, thereby suppressing BGCs. By interfering with these repressive markers, small-molecule inhibitors that target DNA methyltransferases or histone deacetylases (HDACs) can encourage the transcriptional activation of silent clusters (Pillay et al., 2022). As an example, it has been demonstrated that administering negative modulators of HDAC, like valproic acid or suberoylanilide hydroxamic acid (SAHA), can cause the synthesis of hitherto unidentified biological compounds, such as nygerone A and new sesquiterpenoids from *Aspergillus* species (Henrikson et al., 2009).

Likewise, by changing DNA methylation patterns, DNA demethylating agents such as 5-azacytidine can trigger the expression of dormant pathways (Xue et al., 2023). The benefit of epigenetic strategies is that they are accessible and widely applicable because they do not require genetic manipulation. Nevertheless, they frequently have pleiotropic effects, activating several gene clusters at once, which can make isolation and characterization more difficult later on (Brakhage and Schroeckh, 2011). Chromatin-based derepression is still an effective method for finding new secondary metabolites in a variety of fungal taxa, though (Al Aboud et al., 2025). Co-culturing a fungal strain alongside another microorganism to mimic ecological interactions like competition or symbiosis is another effective method for triggering dormant BGCs. The "One Strain, Many Compounds" (OSMAC) paradigm, responsible for metabolite profiles that can be significantly changed by changing growth conditions, is consistent with this method. Co-cultivation can kindle the synthesis of an entirely novel compound or a mimetic of known bioactive molecules, as numerous studies have shown (Arora et al., 2020; Giddings and Newman, 2022).

Co-culturing *Aspergillus nidulans* with *Epicoccum* species, for instance, altered global regulatory complexes such as LaeA/VelB, which in turn induced the biosynthesis of novel alkaloids. These interactions frequently entail signaling pathways that activate otherwise dormant genes and mimic natural environmental pressures (Wang et al., 2022). Researchers also alter abiotic culture conditions to extract novel chemistry, building on the OSMAC methodology. Different metabolic products can result from variations in temperature, pH, salinity, carbon or nitrogen sources, and even the physical condition of the culture (e.g., static vs. shaking) (Romano et al., 2018). To interfere with metabolic regulation, chemical elicitors like metal ions or enzyme inhibitors may also be added. Combinatorial methods are especially promising, such as combining modified culture conditions with epigenetic modifiers. For instance, combining OSMAC variables with HDAC inhibition has been successful for initial metabolite screening, increasing the possibility of finding cryptic compounds (Daniel-Ivad et al., 2017). When combined, these tactics highlight the latent biosynthetic potential that fungal genomes contain. Researchers can access a multitude of secondary metabolites that were previously unavailable by artificially disrupting regulatory mechanisms or by mimicking natural environmental cues. These initiatives are essential for improving our knowledge of fungal chemical ecology and gene regulation as well as for drug discovery (Singh et al., 2020).

5. Classes of Bioactive Metabolites

Marine endophytic fungi, which reside asymptomatically in the internal tissues of marine organisms such as algae, seagrasses, and invertebrates, are prolific producers of secondary metabolites with diverse chemical structures and remarkable bioactivities (El-Bondkly et al., 2021). These compounds are not directly involved in the organism's growth or reproduction but play essential roles in ecological interactions, such as defense against pathogens, competition, and adaptation to extreme marine conditions. Over recent years, marine-derived fungi have attracted significant attention in pharmaceutical research for their potential to yield novel therapeutic agents. Among the innumerable clusters of secondary bioactive molecules produced by marine endophytes, alkaloids, polyketides, terpenoids, peptides (and depsipeptides), and phenolic compounds stand out due to their broad-spectrum pharmacological activities (Saeed et al., 2021; Okechukwu et al., 2024). Their structural novelty, often shaped by unique BGCs and environmental pressures, makes them a valuable source for drug discovery.

5.1 *Alkaloids*

This group of chemicals (alkaloids) represents a key category of nitrogen-based bioactive molecules that exhibit an array of pharmacological activities, including cancer suppressive (Dhyani et al., 2022), antimicrobial (Othman et al., 2019), antiviral (Faisal et al., 2023), antiparasitic (Bessa et al., 2024), and neuropharmacological effects (Nigdelioglu Dolanbay et al., 2021). Marine endophytes have been reported to synthesize a variety of alkaloids that are capable of exhibiting potent bioactivities. *Penicillium citrinum* harvested from marine sponge was shown to form citrinin, a mycotoxin responsible for antibacterial and antioxidant properties (Zhang et al., 2020). Similarly, species of *Aspergillus* have been reported to synthesize fumiquinazoline alkaloids, which exhibit notable antiproliferative activity against cultured human cancer cells (Wang et al., 2024b). The biological efficacy of fungal alkaloids is often attributed to their mechanisms of action, such as DNA intercalation and enzyme inhibition, such as topoisomerases, acetylcholinesterase, and receptor modulation (Zorrilla and Evidente, 2022). The structural complexity of these alkaloids, including the presence of multiple fused rings, halogen substitutions, and stereo centers, contributes to their selectivity and potency (Seipp et al., 2021). The wide pharmacological spectrum and unique marine-derived scaffolds of alkaloids mark them as a potential scaffold for the drug development of lead molecules in medicine discovery programs.

5.2 *Polyketides*

Polyketides are one of the structurally diverse and medicinally significant classes of secondary metabolites produced by marine endophytic fungi (Bai et al., 2023). Synthesized via PKSs, these compounds are biosynthetically related to fatty acids and exhibit a wide range of pharmacological activities, such as antibacterial (Arasu et al., 2013), antifungal (Wang et al., 2023), and antineoplastic (Jiménez-Romero et al., 2025), properties. Another noteworthy example is monacolin K (lovastatin), a polyketide cholesterol-lowering agent isolated from *Aspergillus terreus*, a marine-associated endophyte (Na et al., 2024). The unique environmental pressures in marine ecosystems, including osmotic stress, nutrient limitation, and symbiotic interactions, likely contribute to the biosynthesis of structurally novel polyketides. These molecules often feature macrocyclic lactone rings, polyene chains, and various functional groups (Mayer et al., 2011; Imhoff, 2016; Blunt et al., 2016), making them ideal for chemical modification and optimization in medicinal chemistry). Due to their remarkable structural variety and therapeutic relevance, marine fungal polyketides remain at the forefront of the exploration of nature-based drug discovery processes.

5.3 *Terpenoids*

Terpenoids (isoprenoids) are another important group of metabolites of natural origin characterized by their derivation from isoprene units. Marine endophytes have been shown to produce an array of terpenoids, namely, mono-, sesqui-, di-, and tetraterpenes, many of which exhibit potent bioactivities (Núñez-Pons et al., 2020). *Trichoderma* species isolated from marine algae produce diterpenoids, which have demonstrated antiproliferative potential against a panel of cancer cells. Likewise, sesquiterpenoids isolated from *Aspergillus versicolor* have shown broad-spectrum antimicrobial effects. The complex ring structures, hydroxylations, and unusual side chains found in marine fungal terpenoids contribute to their chemical uniqueness and bioactivity. Terpenoids often function as signaling molecules or defense compounds in nature, and their ability to modulate inflammatory pathways, inhibit microbial growth, and interfere with tumor progression underlines their therapeutic potential (Carpena et al., 2024; Zhang et al., 2025b). The study of marine-derived terpenoids thus opens new avenues for the advance of antimicrobial, anti-inflammatory, and anticancer lead discovery.

5.4 *Peptides and Depsipeptides*

Oceanic fungi also have the capability to produce a diverse range of medicinally active peptides and depsipeptides, which are often cyclic in nature and feature unusual amino acid residues or non-proteinogenic components (Gupta et al., 2023; Hafez Ghoran et al., 2023). These compounds exhibit numerous therapeutic properties, such as cytotoxic (Suarez-Jimenez et al., 2012; Luan et al., 2021), antiviral (Agarwal and Gabrani, 2021), antifungal (Diehl et al., 2024), and immunosuppressive (Thell et al., 2014) properties. Beauvericin, a well-studied cyclodepsipeptide isolated from *Fusarium* species associated with marine organisms, exhibits strong antimicrobial activity and functions as an ionophore, disrupting ionic balance in microbial cells (Wang and Xu, 2012). Similarly, enniatins from *Emericella* species demonstrate antifungal properties and potential as anticancer agents (Eshboev et al., 2024). The structural novelty of marine fungal peptides is highlighted by their complex ring systems, lactone bridges, and the integration of D-amino acids or N-methylated residues, which enhance their stability and biological activity (Phyo et al., 2018; Hafez Ghoran et al., 2023). Due to their high specificity, membrane permeability, and resistance to enzymatic degradation, these peptides are increasingly being investigated as templates for designing novel peptide-based therapeutics, especially in the fields of oncology and infectious diseases (Hafez Ghoran et al., 2023).

5.5 *Phenolic Compounds*

In marine endophytic fungi, phenolic compounds are primarily identified on the basis of antioxidant, anti-inflammatory, and antimicrobial properties (Metsämuuronen and Sirén, 2019; Ecevit et al., 2022). Phenolic compounds are the key mediators in neutralizing free radicals and protecting cells and cell organelles from oxidative stress, which is implicated in many life-threatening diseases, including cancer, neurodegeneration (Arias-Sánchez et al., 2023), and cardiovascular disorders (Torres-Fuentes et al., 2022). Phomalactone, a phenolic compound isolated from marine-derived *Phoma* species, has shown significant anti-leishmanial and anti-inflammatory activity (Tracanna et al., 2015). Additionally, *Talaromyces* species have been reported to produce phenylpropanoid derivatives with notable antioxidant potential (Manganyi and Ateba, 2020). Phenolic compounds from marine fungi often feature glycosidic linkages, halogenations, or extended conjugation systems that contribute to their enhanced bioactivity. Their ability to modulate signaling pathways, inhibit lipid peroxidation, and suppress pro-inflammatory enzymes like COX and LOX (Vitale et al., 2020; Papikinou et al., 2024) underscores their utility in pharmaceutical development, particularly for chronic and degenerative diseases.

5.6 *Structural Novelty and Chemical Diversity*

One of the most compelling attributes of marine endophytic fungal metabolites is their extraordinary structural novelty and chemical diversity. These unique chemical entities are shaped by the fungi's adaptation to extreme marine environments and their symbiotic relationships with host organisms. The genomic repertoire of marine fungi includes a wealth of BGCs encoding PKS, NRPS, and hybrid PKS-NRPS systems, which collectively orchestrate the synthesis of diverse secondary metabolites (Debbab et al., 2012; Zhang et al., 2016; Naranjo-Ortiz and Gabaldón, 2019; El-Bondkly et al., 2021; Liao et al., 2025). Moreover, environmental cues such as salinity, pressure, and co-culture conditions can activate silent metabolic processes resulting in the biosynthesis of previously undetectable compounds. This has resulted in the unearthing of halogenated polyketides, sulfur-bridged alkaloids, and spirocyclic or fused ring systems with no terrestrial analogues. These unique structures often exhibit innovative pathways of biological effect and high potency, resulting in high desirability for drug development. Furthermore, progress in metabolomics, biodesign, and genome mining is accelerating the discovery and characterization of these novel metabolites (Wang et al., 2021; Liu et al., 2022; Gao and Zhang, 2023). The immense chemical space offered by marine endophytic fungi thus represents a largely untapped reservoir for future pharmaceuticals, especially for combating multidrug-resistant infections and complex diseases like cancer and neurodegeneration.

6. Pharmacological Applications

Marine endophytic fungi have emerged as a prolific source of chemically unique secondary metabolites with immense pharmacological promise. Living in extreme and dynamic marine environments, these fungi produce a wide array of bioactive molecules that enable them to survive under competitive and hostile conditions. Many of these metabolites possess potent activities against human diseases and have shown efficacy in preclinical models. The chemical novelty and broad-spectrum bioactivities of these compounds make marine endophytic fungi an exciting frontier for novel drug discovery (Alam et al., 2021) (Table 2).

Table 2. Recent advances in secondary metabolites and their pharmacological bioactivities.

Metabolite Class	**Metabolites**	**Pharmacological Activities**	**Key Reference(s)**
Alkaloids	Fumiquinazoline C, Amesto knoids A–C	Neuroprotective (AChE inhibition), antiviral (SARS-CoV-2 Mpro & spike–ACE2 blockade)	Netz and Opatz (2015); Li et al. (2025)
Flavonoids (flavonoid-like)	Quercetin-derivatives, chrysosporazines	Antioxidant (DPPH/ ABTS radical scavenging), anti-inflammatory (NF-κB suppression)	El-Bondkly et al. (2021)
Glycosides/ glycolipids	β-Glucans (fungal polysaccharides)	Immunomodulatory (macro-phage activation, cytokine modulation)	Batbayar et al. (2012); Suraiya et al. (2024)
Phenolics/ polyphenols	Resveratrol-like stilbenes, chaetoglobosins, penicoffrazins B-C	Antioxidant, anti-neuroinflammatory, anti-amyloid aggregation	Wu et al. (2021); Shao et al. (2022)
Polyketides	Versicolorin A, citrinin, chaetomugilin A/B, alterporriol Q, halorosellinic acid	Anti-bacterial (MRSA, VRE), anticancer (topoisomerase & microtubule inhibition), antiviral (SARS-CoV-2)	Liang et al. (2023); Subramani et al. (2013); Das et al. (2022)
Terpenoids	Brefeldin A, trichothecenes, penicilindoles, territrem analogues	Anticancer (ER stress & apoptosis), antifungal (membrane disruption), antibacterial (cell-wall inhibition)	Xia et al. (2014); Zheng et al. (2021b); Matsuda and Abe (2016); Carew et al. (2006)

The following subsections highlight the pharmacological relevance of their metabolites in major therapeutic areas, including antimicrobial, antiviral, antineoplastic, antioxidant, anti-inflammatory, neuroprotective, and immunomodulatory domains.

6.1 Antibacterial Activity

The global rise of antibiotic resistance, particularly among pathogens like methicillin-resistant *Staphylococcus aureus* (MRSA), vancomycin-resistant *Enterococci* (VRE), and multi drug resistant (MDR) *Pseudomonas aeruginosa*, has necessitated the urgent discovery of novel antibacterial agents. Oceanic endophytic microbes have shown remarkable potential in antimicrobial drug discovery. Several isolates have been shown to synthesize secondary metabolites that are active against many resistant bacterial species. For example, *Aspergillus versicolor*, derived from a marine sponge, produced versicolorin A, which showed significant inhibition of MRSA and VRE (Liang et al., 2023). Similarly, *Penicillium citrinum* from marine algae yielded citrinin derivatives with broad-spectrum antibacterial activity (Subramani et al., 2013). Anthraquinone compounds from *Aspergillus versicolor* (Hawas et al., 2012) and emodin analogues from marine *Aspergillus* strains demonstrated inhibitory effects against MRSA *Enterococcus faecalis* (Ismaiel et al., 2016). Polyketides such as chaetomugilin A and B from *Chaetomium globosum* and isocitreohybridone A from *Citreohybridum* species act by modulating efflux pumps (Yamada et al., 2009), while peptides including beauvericin and enniatins from *Fusarium* species, as well as cyclo-(L-Leu-L-Pro) from *Aspergillus terreus*, exhibit membrane-disrupting and quorum-sensing inhibitory activities (Martín-Rodríguez et al., 2014; Křížová et al., 2021).

Meroterpenoids such as penicilindoles from *Penicillium* species and territrem analogues from *Aspergillus terreus* interfere with cell wall biosynthesis in resistant strains (Matsuda and Abe, 2016). Halogenated metabolites like halorosellinic acid from *Halorosellinia oceanica* and halogenated chromones from marine *Aspergillus* species enhance membrane permeability and metabolic stability (Chinworrungsee et al., 2001; Chen et al., 2022). Additionally, pestalone from *Pestalotia* species displays low micromolar activity against MRSA and VRE (Augner et al., 2013), gliotoxin derivatives from marine *Aspergillus fumigatus* induce oxidative stress, and aspergillomarasmine A from *Aspergillus versicolor* inhibits metallo-β-lactamases and restores β-lactam antibiotic efficacy (Sychantha et al., 2021). These compounds often target bacterial cell walls, protein synthesis machinery, or efflux pump systems, making them valuable templates for overcoming resistance. Moreover, halogenated polyketides and alkaloids from marine fungi demonstrate improved membrane penetration and metabolic stability, enhancing their drug-likeness. The antibacterial activity of marine fungal compounds could serve as a novel arsenal against superbugs threatening global health (Rateb and Ebel, 2011; Imhoff, 2016).

6.2 Antifungal Activity

Systemic fungal infections, especially among immunocompromised individuals, represent a growing clinical challenge. Current antifungal drugs are limited by toxicity, resistance, and narrow spectra of activity. Marine endophytic fungi offer a promising solution by producing metabolites that exhibit antifungal activity against both superficial and invasive pathogenic fungi, such as *Candida albicans*, *Cryptococcus neoformans*, and *Aspergillus fumigatus* (Brown et al., 2012; Jin et al., 2016; Perfect, 2017; Hang et al., 2024). Compounds such as enniatins and beauvericin, isolated from marine-derived *Fusarium* and *Emericella* species, respectively, exhibit potent antifungal effects by disrupting fungal membrane integrity and ion homeostasis (Bharathi and Lee, 2024). Acremostrictin A (from marine *Acremonium* sp.) and diketopiperazines (from *Aspergillus* sp.) inhibit fungal growth through ROS-mediated pathways (Julianti et al., 2011); anthraquinone derivatives such as altersolanol B (from *Alternaria* sp.) that interfere with fungal cell wall biosynthesis; and halorosellinic acid (from *Halorosellinia* sp.) exhibiting fungistatic activity against multidrug-resistant *Candida* species (Song et al., 2021). Peptides and depsipeptides like scopularides A and B (from *Scopulariopsis* sp.) have also been reported to synergize with azoles and echinocandins, reducing effective doses and overcoming resistance (Kumar et al., 2024). Furthermore, phenolic metabolites from marine *Phoma* and *Talaromyces* species have shown inhibition of fungal biofilm formation, a major mechanism of resistance in clinical isolates (Caruso et al., 2022). Some marine peptides and depsipeptides have demonstrated synergy with conventional antifungals, allowing dose reduction and overcoming resistance (Buda De Cesare et al., 2020). Given the rising incidences of fungal infections in the backdrop of global warming and increased use of immunosuppressants, marine fungal metabolites are poised to offer novel antifungal therapies with unique mechanisms of action (Laverdière, 1994; Saxena et al., 2024).

6.3 Antiviral Activity

Viral diseases, including those caused by HIV, hepatitis viruses, influenza, and coronaviruses (e.g., SARS-CoV-2), exhibit a key risk to the health issues of society (Zhou et al., 2020). The need for effective antiviral agents with novel targets has led researchers to explore marine fungi as alternative sources. Phytochemicals produced by the endophytes have exhibited a prospective antiviral activity by targeting different phases of the viral lifecycle, namely, attachment, entry, replication, and assembly (Yasuhara-Bell and Lu, 2010). For instance, diketopiperazines from marine Aspergillus species have been reported to inhibit HIV-1 reverse transcriptase, while phenolic derivatives from Talaromyces have shown potent inhibition of influenza virus neuraminidase (Bojarska et al., 2021). Isochroman derivatives from *Penicillium* species that block hepatitis C virus (HCV) RNA polymerase (Nishikori et al., 2016), xanthone derivatives from *Emericella* species that interfere with viral envelope fusion, and halogenated azaphilones from *Chaetomium* species showing broad-spectrum activity against both influenza A and B viruses (Roy, 2017; Ghazawi et al., 2023). During the COVID-19 pandemic, marine-derived alkaloids, depsipeptides, and polyketides such as alterporriol Q (from *Alternaria* sp.) (Das et al., 2022) and Amesto knoids A–C (from *Talaromyces amestolkiae*) were identified via *in silico* and *in vitro* studies to inhibit SARS-CoV-2 main protease (Mpro) and disrupt spike–ACE2 interactions (Huang et al., 2021). These results emphasize the promising marine-derived metabolites as antiviral scaffolds, particularly in developing broad-spectrum antivirals or targeting emerging viral threats (Kálai et al., 2022; Arumugam et al., 2022).

6.4 Anticancer Activity

Cancer remains a major cause of deaths across the globe, and resistance to chemotherapy has resulted in the heightened health care interest in novel, selective anticancer agents (Garg et al., 2024). Marine endophytic fungi have yielded several cytotoxic compounds that target multiple cancer hallmarks, such as uncontrolled proliferation (Evidente, 2024), angiogenesis (Hulikere et al., 2016; Papikinou et al., 2024), metastasis (Liu et al., 2023), and evasion of apoptosis (Deshmukh et al., 2018). For example, the cyclic depsipeptide scopularide A from *Scopulariopsis* species has demonstrated significant cytotoxicity against lung and colon cancer cell lines through apoptosis induction. Likewise, marine-derived polyketides like mycophenolic acid (Chen et al., 2012) and chlorinated xanthones (Pontius et al.,

2008) show inhibition of topoisomerase and microtubule polymerization, causing the blockade of the cell cycle. Brefeldin A from *Penicillium* species, which interferes with the secretory pathway and induces apoptosis in leukemia cells (Huang et al., 2017), alterporriols from *Alternaria* species, which exhibit antiangiogenic activity by downregulating VEGF expression (Pompeng et al., 2013) and gliotoxin derivatives from *Aspergillus* species, which modulate NF-κB signaling and sensitize resistant ;cancer cells to apoptosis (Xia et al., 2014). Furthermore, epidithiodiketopiperazine DC1149B (ETPs) isolated from marine *Trichoderma* strains have demonstrated potent inhibition of endoplasmic reticulum (ER) signaling rather than PI3K/Akt and MAPK pathways (Tang et al., 2020). Importantly, marine fungal metabolites often exhibit selectivity toward cancer cells with minimal toxicity to normal cells, indicating their potential as lead candidates for targeted therapy. Combination studies have also revealed their ability to synergize with existing chemotherapeutics, paving the way for combination regimens that reduce side effects and delay resistance (Imhoff, 2016; Uzma et al., 2018; Alves and Diederich, 2021).

6.5 Antioxidant and Anti-Inflammatory Agents

Chronic oxidative stress and persistent inflammation are central to the pathogenesis of various degenerative diseases, including cardiovascular disorders (El-Agamy et al., 2019), diabetes (Bai et al., 2025), neurodegenerative conditions (Dimitrova et al., 2025), and certain cancers (Venkatachalam et al., 2024). Marine endophytic fungi have emerged as prolific producers of bioactive secondary metabolites, particularly phenolic compounds (Gribble, 2023), flavonoid-like molecules (El-Bondkly et al., 2021), Xanthones (Khattab and Farag, 2022), and quinones (Li et al., 2017), which exhibit robust antioxidant properties (Hashem et al., 2023). These metabolites act through multiple mechanisms, including free radical scavenging (e.g., DPPH, ABTS assays) (Kannan et al., 2024), metal ion chelation (Gunasekaran et al., 2015), and upregulation of endogenous antioxidant defense systems such as glutathione, catalase, and superoxide dismutase (Kumari et al., 2018). Phenylpropanoid derivatives like Chrysosporazines from *Talaromyces* spp. (Mohamed et al., 2020), xanthones from *Aspergillus* spp. (Wang et al., 2024a), and isocoumarins like penicoffrazins B and C from *Penicillium* spp. have demonstrated notable free radical scavenging potential (Cao et al., 2019). Mechanistically, these effects are often mediated via the inhibition of NF-κB and MAPK signaling pathways, thereby attenuating the expression of pro-inflammatory mediators. The combined antioxidant and anti-inflammatory potential of these fungal metabolites positions them as promising leads for the development of novel therapeutic agents against chronic inflammatory and oxidative stress-associated disorders (Xia et al., 2022; Dash et al., 2025).

6.6 Neuroprotective and Immunomodulatory Properties

Emerging evidence underscores the pivotal role of marine endophytic fungal metabolites in modulating neural and immune responses (Pant and Vasundhara,

2023). Neurodegenerative disorders such as Alzheimer's and Parkinson's diseases are driven by multifactorial processes, including oxidative stress (Zhang et al., 2018), protein aggregation, mitochondrial dysfunction (Girich et al., 2020), and chronic inflammation (Yang et al., 2018). Several marine-derived endophytic compounds, particularly indole alkaloids (e.g., fumiquinazoline C from *Aspergillus fumigatus*) (Netz and Opatz, 2015), terpenoids (e.g., trichothecenes from *Trichoderma* spp.) (Zheng et al., 2021b), and phenolic derivatives (e.g., chaetoglobosins from *Chaetomium* spp.) (Tian and Li, 2022), have shown neuroprotective activities by inhibiting acetylcholinesterase, mitigating neuroinflammation, and reducing β-amyloid aggregation. In addition, extracts from marine *Aspergillus* spp. have demonstrated significant neuroprotection in *in vitro* models of oxidative stress-induced neuronal injury (Kim et al., 2016). In parallel, marine fungal polysaccharides (e.g., β-glucans) exhibit potent immunomodulatory properties, including enhancement of macrophage phagocytic activity, modulation of cytokine secretion, and regulation of T-cell proliferation and differentiation (Papikinou et al., 2024). These dual actions suppressing pathological inflammation while supporting protective immune responses render marine endophytic metabolites attractive candidates for therapeutic intervention in autoimmune diseases, cancer immunotherapy, and vaccine adjuvant design (Catanesi et al., 2021; Pant and Vasundhara, 2023).

7. Approved Therapeutics and Metabolites

To date, no compound isolated directly from marine fungal endophytes has received approval from the U.S. Food and Drug Administration (FDA) for therapeutic use in humans. This absence persists despite the rapidly growing scientific and commercial interest in marine natural products as promising sources of novel bioactive molecules with antioxidant, anti-inflammatory, anticancer, neuroprotective, and antimicrobial properties (Gomes et al., 2015; Deshmukh et al., 2018a,b). Rather than originating from marine fungi, all FDA-approved "marine-derived" medications that are currently available on the market come from other marine organisms, including algae, sponges, tunicates, and mollusks (Gomes et al., 2015). Notably, four anticancer agents of marine origin—brentuximab vedotin (Adcetris), eribulin mesylate (Halaven) (Barreca et al., 2020), trabectedin (Yondelis), and cytarabine (Ara-C) (Pereira et al., 2019) were developed from marine sponges or tunicates have showed significant clinical impact. Likewise, marine sponges and cone snails have contributed to the development of other therapeutic categories, such as the analgesic peptide ziconotide (Prialt) derived from *Conus magus* venom and the antiviral nucleoside analogue vidarabine (Ara-A) (Sagar et al., 2010; Safavi-Hemami et al., 2019). Currently there is no approved therapeutic compound from marine fungal endophytes in the pharmaceutical market (Qiu et al., 2024). Most marine fungal metabolites remain at the preclinical or early clinical research stages, with promising activities demonstrated *in vitro* or in animal models, but they face challenges including low natural abundance, complex structural synthesis, variable yields in culture, and lengthy regulatory pathways (Li et al., 2025b). Continuous exploration of endophytic fungal biodiversity, genome mining, heterologous expression systems, and synthetic biology approaches is anticipated to accelerate the translation of marine fungal natural products into clinical candidates (Leal et al., 2024; Zhang et al., 2025a).

7.1 Compounds in Pre-Clinical Development

A number of notable compounds demonstrate the remarkable biosynthetic abilities of above listed endophytic fungi. In experiments on cells and animals, several of these secondary metabolites that were separated from marine fungal endophytes show strong biological activity. These substances are currently in preclinical stages, which include medicinal chemistry, animal testing, and discovery. They have not yet progressed to human clinical trials. Table 3 identifies representative examples with new structural frameworks and promising potency, as indicated by their IC_{50} or MIC values. These metabolites are still at least a generation away from receiving therapeutic approval, despite being investigated mainly for their antimicrobial or anticancer properties. When taken as a whole, they provide both great therapeutic potential and insightful information about the intricate chemistry of natural products.

Many structurally diverse antibacterial agents, some of which exhibit remarkable potency against multidrug-resistant pathogens, are produced by marine fungal endophytes (Fig. 3). Importantly, pestalone, a chlorinated benzophenone that was isolated from a species of *Pestalotia* that was co-cultured with *Streptomyces*, is only biosynthesized in the presence of microbial interaction and demonstrates nanomolar activity against MRSA (MIC = 37 ng/mL) and VRE (MIC = 78 ng/mL), underscoring the ecological complexity of marine secondary metabolism (Cueto et al., 2001). Aspergillomarasmine A, which is derived from *Aspergillus versicolour*, restores the effectiveness of carbapenem by blocking metallo-β-lactamases (NDM-1, VIM-2) (King et al., 2014). Other strong antibacterial agents include 14,16-epiascomylactam B, which exhibits broad-spectrum activity at sub-microgram concentrations (Yao et al., 2024), Spiromastixone (MIC = 0.5 μM MRSA) (Huang et al., 2024), Zopfiellamide C (MIC ≈ 12.5 μM) (Jiao et al., 2025), and Agonodepside C (IC_{50} ~2.5 μM) (Anh et al., 2022). Additionally, deep-sea Chaetoketoics B inhibits Gram-negative bacteria like *E. coli* and Vibrio with MICs of 4–8 μg/mL (Li et al., 2024; Pan et al., 2025).

The identification of cytotoxic substances with possible anticancer uses is a related area of emphasis. Because of its special spirobisnaphthalene scaffold, Rhytidenone H, which was isolated from *Rhytidhysteron rufulum*, is especially effective against lymphoma and lung cancer cells, displaying picomolar to low-nanomolar $IC_{50} \approx 0.018$ and 0.252 μM respectively (Siridechakorn et al., 2017). In the same model, *Penicillium brocae's* spirobrocazine G exhibits strong nanomolar activity ($IC_{50} \approx 0.66$ μM) against ovarian cancer cells, significantly outperforming cisplatin (Meng et al., 2016). Other noteworthy substances include Aspergisocoumarin A (IC_{50} = 5.08 μM) (Wu et al., 2019), 7-O-methylnigrosporolide (IC_{50} = 0.7 μM) (Liu et al., 2016; Gao et al., 2024), and Penochalasin K, a chaetoglobosin-class

Table 3. Notable bioactive metabolites from marine endophytic fungi.

Compound	Fungal Source	Bioactivity	Chemical Class	References
Agonodepside C	*Aspergillus unguis* (marine)	Anti-Gram-positive activity (MIC ≈ 5.3–22.1 μM); general cytotoxicity (IC_{50} down to 2.5 μM) on various cancer cell lines.	Phenolic polyketide (depside)	Anh et al. (2022)
Aspergillomarasmine A	*Aspergillus versicolor*	Metallo-β-lactamase inhibitor – rapid, potent inhibition of NDM-1 and VIM-2 (restores carbapenem activity).	Aminopolycarboxylate (polyamino acid)	King et al. (2014)
Aspergisocoumarin A	*Aspergillus* sp. HN15-5D	Cytotoxic to human breast cancer (MDA-MB-435) cells (IC_{50} = 5.08 μM).	Isocoumarin (polyketide)	Wu et al. (2019)
Chaetoketoic B	*Chaetomium globosum*	Antibacterial (*E. coli, E. tarda*) MIC = 4.0 μg/mL	Sesterterpenoid polyketide	Li et al. (2024); Pan et al. (2025)
14,16-Epiascomylactam B	*Phoma* sp. (deep-sea)	Inhibits multiple bacteria at MIC = 0.20–0.80 μg/mL, synthetic analogs explored	Tetracyclic lactam alkaloid	Yao et al. (2024)
7-O-Methylnigrosporolide	*Pestalotiopsis microspora*	Highly cytotoxic (L5178Y IC_{50} = 0.7 μM).	Phthalide lactone derivative	Liu et al. (2016); Gao et al. (2024)
Penochalasin K	*Penicillium chrysogenum*	Anticancer (breast/lung/gastric) IC_{50} ≈ 4.7–8.7 μM	Macrocyclic chaetoglobosin	Zhu et al. (2017)
Pestalone	*Pestalotia* sp. (marine)	Potent antibiotic – inhibits MRSA (MIC 37 ng/mL) and VRE (MIC 78 ng/mL); moderate cytotoxicity (GI_{50} ≈ 6.0 μM).	Chlorinated benzophenone	Cueto et al. (2001)
Pestalotiopsone F	*Pestalotiopsis* sp.	Cytotoxic to murine lymphoma L5178Y (EC_{50} = 8.93 μg/mL).	Chromone polyketide	Gao et al. (2024)
Rhytidenone H	*Rhytidhysteron rufulum*	Highly potent anticancer (0.018–17.98 μM) on lymphoma, NSCLC cell lines	Spirobisnaphthalene alkaloid	Siridechakorn et al. (2017)
Spirobrocazine G	*Penicillium brocae*	Anticancer (ovarian cancer) IC_{50} = 0.664 μM (A2780, cisplatinsensitive)	Spirocyclic alkaloid	Meng et al. (2016)
Sporothrin A	*Sporothrix* sp. (mangrove)	Acetylcholinesterase inhibitor (IC_{50} = 1.05 μM); modest cytotoxicity against several cancer lines.	Aryl-polyketide (benzofuranone)	Deshmukh et al. (2018a)
Tribrominated spiromastixone	*Spiromastix* sp. (deep-sea)	Very potent antibiotic – inhibits MRSA (MIC = 0.5 μM) and VRE (MIC = 1.0 μM).	Brominated depsidone (polyketide)	Huang et al. (2024)
Xyloketal B	*Xylaria* sp. (mangrove)	Broad cytoprotective/neuroprotective activity – antioxidant, anti-ischemic, anti-inflammatory, Ca^{2+}-channel blocking, etc.; reduces ischemic injury and glioblastoma growth. Xyloketal B is one of the most potent of this class.	Polycyclic ketal polyketide	Gong et al. (2022); Udayan et al. (2024)
Zopfiellamide C	*Aspergillus* sp. NF666	Potent antibacterial – inhibits Gram-positive bacteria (MIC ≈ 12.5 μM).	Decalin-type tetramic acid (hybrid)	Jiao et al. (2025)

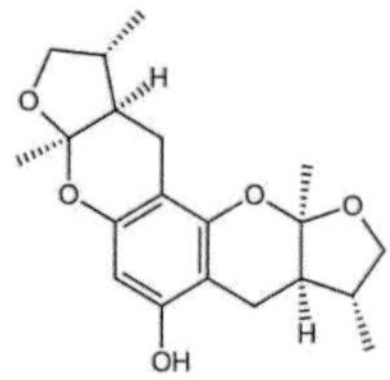

Xyloketal B

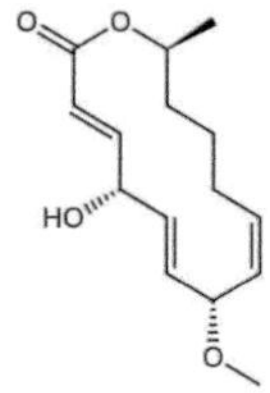

7-O-Methylnigrosporolide

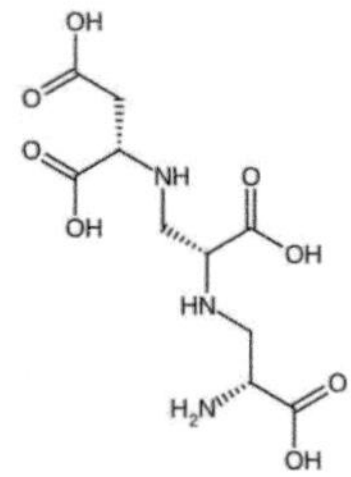

Aspergillomarasmine A

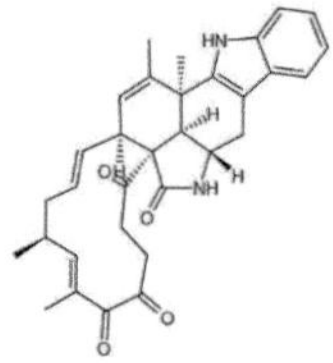

Penochalasin K

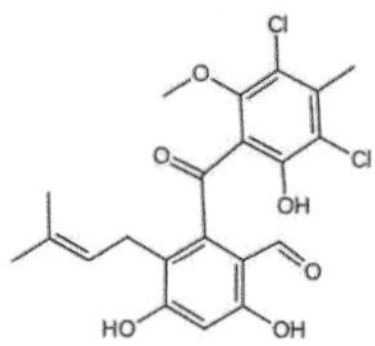

Pestalone

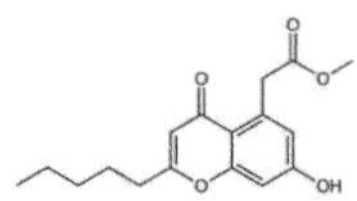

Pestalotiopsone F

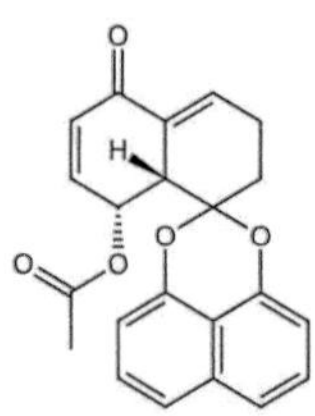

Rhytidenone H

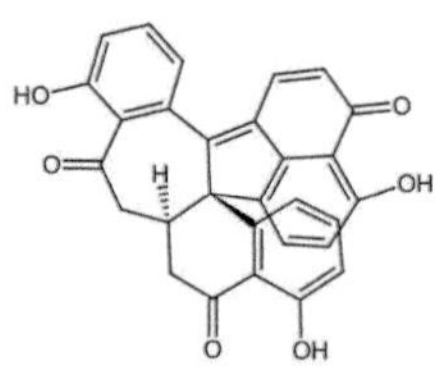

Sporothrin A

Fig. 3. Structural diversity of key metabolites of marine fungi.

compound (IC_{50} = 4.7–8.7 μM) that most likely functions by disrupting actin (Zhu et al., 2017). Additionally, Pestalotiopsone F (EC_{50} = 8.93 μg/mL) (Gao et al., 2024), Ascomylactam B (IC_{50} = 4–20 μM) (Chen et al., 2019), and Agonodepside C (Anh et al., 2022) (which exhibits dual antibacterial and anticancer activity) exhibit moderate cytotoxic activity (Anh et al., 2022). Collectively, these compounds offer structurally novel scaffolds such as spirocyclic alkaloids, macrocyclic peptides, and chromone-based polyketides, many of which are unique to marine biosynthesis.

Several fungal metabolites derived from marine sources have neuroactive and enzyme-inhibitory qualities in addition to their antimicrobial and anticancer effects. From the Sporothrix species, Sporothrin A has a rare spiroketal-fused benzofuran scaffold and is a strong acetylcholinesterase inhibitor (IC_{50} = 1.05 μM) (Deshmukh et al., 2018a). Extracted from mangrove-associated *Xylaria*, Xyloketal B exhibits protective properties against ischemia-reperfusion injury and oxidative stress (Gong et al., 2022; Udayan et al., 2024). It also inhibits calcium channels, indicating potential uses in neuroprotection and cardiovascular disease. The broad pharmacological potential of marine fungal endophytes, whose metabolites comprise an array of chemical classes and biological targets, is highlighted by these findings.

Their potency and novelty support the marine ecosystem as a vital source of next-generation therapeutic leads, even though the majority are still in the preclinical stage.

7.2 Compounds in Clinical Development

Plurabulin (NPI-2358) is the only known secondary metabolite that has progressed to human clinical trials among those derived from marine fungal endophytes. Phenylahistin, a cyclic dipeptide (diketopiperazine) that was first isolated from a marine *Aspergillus* species connected to a marine sponge, is the synthetic derivative of plinabulin. Plurabulin is considered a derivative of a marine fungal endophyte due to its origin. Plinabulin disrupts tumor cell microtubules by attaching itself to ®-tubulin, which functions as a microtubule destabilizer. This molecule upregulates the neutrophil counts and strengthens antitumor immune responses by inducing the release of guanine nucleotide exchange factor-H1 (GEF-H1). Plurabulin is currently in a more advanced clinical development stage (Ding et al., 2022). It is at this time in Phase III trials for treating chemotherapy-induced neutropenia as well as non-small-cell lung cancer (NSCLC). For the patients who are receiving docetaxel-based chemical therapeutic intervention, plinabulin (40 mg) was compared to pegfilgrastim in the pivotal PROTECTIVE-1 Phase III trial. The study showed that the effect of plinabulin was on par with pegfilgrastim in lowering the length of grade 3–4 neutropenia, thereby meeting its primary endpoint (Blayney et al, 2022).

In a different clinical trial (NCT02504489), subjects having metastatic non-small cell lung cancer had plinabulin compared to docetaxel alone. As of 2022, Beyond Spring Pharmaceuticals, the drug's sponsor, filed a New Drug Application (NDA) for NSCLC indications based on the findings of these studies. Plinabulin has shown strong anticancer activity *in vitro*; in NSCLC cell lines, its IC_{50} is about 26 nM (Ding et al,., 2022). Unlike traditional granulocyte colony-stimulating factor (G-CSF) agents like pegfilgrastim, it has a dual mechanism of action that combines immunomodulatory effects with microtubule destabilization (Ding et al., 2022). There are currently no other compounds from marine fungal endophytes that are undergoing clinical trials. Other than plinabulin, no Phase I–III candidates discovered through searches of ClinicalTrials.gov and other databases are specifically derived from marine endophytic fungi. Currently, no analgesics, antibiotics, or other therapeutic agents derived from marine fungal endophytes are being developed for clinical use. Plurabulin is still the only clinical-stage candidate in this group as a result. Its creation highlights the potential and difficulties of converting marine fungal metabolites from discovery into licensed treatments. Even though marine fungal endophytes' chemical diversity and bioactivity are widely known, it is still very uncommon for them to successfully advance to the clinical trial stage.

8. Biotechnological Applications

Traditionally, top-down, bioactivity-guided screening techniques have been used to find natural products from marine endophytic fungus (Kildgaard et al., 2017). Due to the high rate of rediscovering known compounds and their incapacity to identify metabolites that are only produced under particular physiological or environmental conditions, these methods frequently result in diminishing returns. However, by enabling researchers to directly examine the genomic potential of marine fungi, genome mining has transformed this field. Today, BGCs, including those encoding polyketides, nonribosomal peptides (NRPSs), terpenes, and hybrid molecules, can be predicted and prioritized without the need for cultivation-based screening by combining whole-genome and metagenomic sequencing with bioinformatic tools like antiSMASH (Blin et al., 2019).

More than 97% of the BGCs found in marine fungal isolates, including *Calcarisporium* sp. KF525 (36.8 Mb; 60 BGCs) and *Pestalotiopsis* sp. KF079 (47.5 Mb; 67 BGCs), are novel, according to recent genomic studies. Under typical laboratory conditions, however, fewer than 30% are actively expressed. Transcriptomic profiling can be used to find "silent" or low-expression BGCs, which can then be targeted for activation in order to get past this bottleneck. The recovery of BGCs from uncultured endophytes through sequence-guided library construction and subsequent heterologous expression is also made possible by metagenomic approaches, which broaden the scope of discovery to entire microbial communities. When taken as a whole, these tactics mark a paradigm shift from random culturing to a logical, gene-guided process that successfully unlocks the enormous, unexplored chemodiversity of marine fungal endophytes (Kennedy et al., 2010; Kumar et al., 2018).

8.1 CRISPR-Cas and Gene Editing for Metabolite Enhancement

Of late, CRISPR-Cas technologies have greatly accelerated the fungal genome modifications, allowing for the improvement of metabolite production as well as the functional characterization of BGCs. For example, Chen and colleagues successfully applied a single-plasmid CRISPR-Cas9 toolkit and a polyethylene glycol (PEG)-mediated transformation system to marine-derived novel bioactive metabolites. *Spiromastix* sp. SCSIO F190 isolated from South China Sea sediments was engineered to produce a novel series of diphenyl ethers (1–3), including atropisomers (3a/3b) and cyclopentenones (4a/4b) with remarkable antibacterial potency. These metabolites demonstrated minimum inhibitory concentrations (MICs) ranging from 0.5 to 4 μg/mL against multidrug-resistant strains such as MRSA, *Enterococcus faecalis*, and *Bacillus subtilis*, underscoring the value of gene editing in enhancing bioactive compound yields (Chen et al, 2022; Cai et al, 2023). *Aspergillus* sp. SCSIO SX7S7, a coral-associated marine endophyte, was subjected to CRISPR-Cas9-based rpd3 disruption attaining up to 40% gene-editing efficiency to activate silent metabolic pathways, resulting in the production of halogenated depsidones, including the novel asperdepside A and its congeners. Among these, triple-halogenated analogues displayed strong antibacterial activity with MICs ranging from 1 to 8 μg/mL, notably against vancomycin-resistant *Enterococcus faecium* (Chen et al.,, 2022). Particularly, *Aspergillus* sp. produced new secondary metabolic products that were not found in the wild-type strain when the histone deacetylase gene rpd3 was specifically deleted. This resulted in the activation of several silent BGCs (Chen et al., 2022). Beyond simple knockouts, catalytically inactive Cas proteins fused to transcriptional regulators enable the reversible modification of gene regulations through CRISPR interference (CRISPRi) and CRISPR activation (CRISPRa) techniques. Using dCas9–VP64, for instance, has made it possible to upregulate important polyketide synthase (PKS) genes in filamentous fungi, increasing target metabolite levels by up to 3.5 times without causing irreversible genetic alterations.

These adjustable systems enable the dynamic control of whole biosynthesis pathways and lessen cellular stress (Jiang et al,, 2021). Additionally, simultaneous edits at multiple genomic loci are being made possible by multiplexed CRISPR strategies. PEG-protoplast techniques for the efficient delivery of Cas9 ribonucleoprotein complexes in *Pestalotiopsis fici* produced dual-locus editing efficiencies of 44.4%, including accurate insertions of DNA fragments between 5–10 kb, all in less than a week. This is a significant improvement in speed and efficiency when compared to more conventional techniques like *Agrobacterium*-mediated transformation. When used in tandem, CRISPR-based systems provide accurate, scalable instruments for analyzing BGC function and creating strains with improved biosynthetic potential (Gil-Durán et al., 2023).

8.2 Heterologous Expression of Biosynthetic Gene Clusters

Although marine fungi frequently have rich repertoires of cryptic BGCs that can be found through genome mining, many of these organisms are either too slow to be exploited or genetically intractable. In these circumstances, functional analysis and scalable production depend on heterologous expression in genetically compatible host organisms, such as *Aspergillus oryzae*, *Aspergillus nidulans*, *Saccharomyces cerevisiae*, and *Streptomyces* species (Han et al,, 2024). Transformation-associated recombination (TAR) in yeast and Streptomyces artificial chromosomes (SACs) enable the capture and manipulation of clusters up to approximately 100 kb in size, which is necessary to accommodate large BGCs. A 65 kb meroterpenoid BGC from the mangrove-derived *Helotiales* sp. BL73, for example, was cloned using TAR, redesigned with yeast-optimized regulatory elements, and expressed in *A. oryzae*. This led to the biosynthesis of both novel and known terpenoids with strong anti-inflammatory effects (Burgaud et al., 2022).

The potential of heterologous expression is being further expanded by high-throughput workflows. The refactoring and expression of 39 marine fungal terpene BGCs were made possible by Zhang et al.'s (2022) automated pipeline in *A. oryzae*. Notably, 185 different terpenoids were found in 208 engineered strains, demonstrating the effectiveness and strength of standardized genetic components like terminators, promoters, and 5′-untranslated regions in pathway prototyping (Zhai et al., 2022; Zhang et al., 2022). By splitting a 28 kb hybrid PKS–NRPS cluster from *Aspergillus terreus* between two codon-optimized plasmids and co-expressing it with a phosphopantetheinyl transferase, a novel polyketide glycoside that was not seen in the native fungus was produced in *Escherichia coli*. Through intron splicing and precursor supply optimization, multi copy integration platforms in *S. cerevisiae* have increased the production of macrolides from marine fungal pathways by up to ten times. These developments highlight how adaptable and powerful heterologous systems are at bridging the gap between metabolite realization and genome mining (Luo et al., 2016). Natural product biosynthesis is starting to change as a result of the use of synthetic biology, which is based on the concepts of standardization, modularity, and abstraction. To fully realize their potential, however, marine fungal systems pose special biological and technical difficulties that need to be overcome (Liu et al., 2021). Although a number of microbial toolkits are available, including fungal promoter libraries and yeast Modular Cloning (MoClo), there are still no components that are especially tailored for marine fungi.

To better support workflows in synthetic biology, efforts are now being made to expand and systematically characterize collections of marine fungal terminators, ribosome binding sites, and promoters (Martins-Santana et al., 2018; Yuan et al., 2022). Marine endophytic fungi frequently produce metabolites in response to host-derived cues and have adapted to flourish in high-salinity, nutrient-limited environments. Unfortunately, the physiological machinery needed to replicate these conditions is often absent from standard laboratory strains used for heterologous expression (El-Bondkly et al., 2021). Marine-derived proteins may exhibit poor expression or misfolding due to incompatibilities in osmotic stress responses and cofactor availability, such as for halide-dependent enzymes like vanadium haloperoxidases. Although media optimization is still essential for robust expression, partial solutions include engineering host strains with marine-adapted enzymes (such as glycerol-3-phosphate dehydrogenases) or halide transporters (Driver et al., 2017). The use of synthetic biology with marine endophytes raises ethical and safety issues, just like with any other potent technology. When released into natural environments, synthetic "mirror" organisms engineered with non-canonical chirality may outcompete native species and evade immune responses, posing serious ecological risks (Rottinghaus et al., 2022). The possibility of horizontal gene transfer, unforeseen ecological effects, and bioterrorism abuse are further worries. Strict containment procedures, strong international governance frameworks, and fail-safe mechanisms (like kill-switch circuits) will all be necessary to address these risks. A fair distribution of the advantages of marine biodiversity must also be guaranteed by ethical considerations, particularly for the countries that provide these priceless organisms (Lei et al., 2024).

9. Cultivation of Endophytic Fungi

The enduring "microbial unculturability" of marine fungi remains a fundamental obstacle to the discovery of natural products, notwithstanding revolutionary developments in genome mining and synthetic biology (Kumar et al., 2021). According to estimates, 90–99% of marine fungal taxa are still resistant to common cultivation methods, thereby preventing the great majority of their BGCs from being accessed experimentally. These challenges are intricately linked to ecological, physiological, and regulatory complexities; they are not just technical (Gonçalves et al., 2022; Xu et al., 2022). The fact that marine endophytic fungi typically have low metabolite titers when cultivated in axenic cultures—often below milligram-per-liter thresholds—is one of the main obstacles to their use (Anjum and Xuewei 2022).

Furthermore, in laboratory settings, most BGCs are either transcriptionally silent or only weakly expressed (Kwon et al., 2021). More than 60 BGCs were found in each genome of *Pestalotiopsis* and *Calcarisporium* species, for example, but less than 30% of these genes exhibited detectable expression *in vitro*. Although techniques like transcription factor overexpression, promoter swapping, and epigenetic modifications have the potential to increase yield, they necessitate highly specialized genetic tools that are not yet generally accessible across a variety of marine fungal lineages (Kumar et al., 2018; Meng et al., 2022).

Several endophytic fungi have slow growth kinetics—doubling times frequently reach days or weeks—because they are adapted to oligotrophic, highly salinized marine environments. The risk of contamination increases with the length of the cultivation period, especially from faster-growing bacterial or fungal species that could outcompete or parasitize target strains (Gonçalves et al., 2022). Selective media with high salinity and specific carbon sources can help suppress undesirable microbial growth, but it frequently takes careful and expensive sterile procedures to strike a balance that promotes fungal viability (Kjer et al., 2010; Debbab et al., 2013). A unique set of difficulties arises when converting small-scale fermentations into commercially feasible procedures. Because of their sensitivity to osmotic stress, shear forces, and oxygen gradients, marine fungi are challenging to control in traditional bioreactors (Soerjawinata et al., 2022).

Furthermore, certain marine enzymes rely on signaling molecules from their natural host environment or need uncommon cofactors like vanadium or halides. Few marine fungal metabolites are produced at levels appropriate for commercialization without specially designed bioprocesses that mimic these conditions (Cochereau et al., 2023). Co-culture techniques, which mimic natural microbial interactions, have shown a great deal of promise in triggering dormant metabolic pathways. For instance, new antifungal polyketides not found in monocultures were produced when *Penicillium citrinum* and an actinomycete derived from mangroves were co-cultured. Unique sesquiterpene biosynthesis has been induced by similar chemical conversations between fungal partners, indicating that cross-kingdom signaling is a potent lever for metabolite induction (Yu et al., 2019; Rosero-Chasoy et al., 2021).

Marine fungal BGCs can now be refactored into tractable model hosts like *Aspergillus oryzae* or *Saccharomyces cerevisiae,* thanks to synthetic biology tools. Researchers have found that CRISPR-based promoter engineering and targeted transcription factor activation can increase yields for previously unavailable compounds by up to four times (Li, 2023; Hur et al., 2023). Rational prioritization of BGCs based on structural novelty and predicted bioactivity is becoming more and more possible thanks to genome mining pipelines like antiSMASH and DeepBGC. These technologies eliminate the need for thorough empirical screening by enabling the selective activation of promising gene clusters when combined with transcriptomic or epigenomic datasets. The discovery landscape is also expanding beyond cultivable strains as metagenomic techniques are starting to recover BGCs from uncultured endophytes (Lee et al., 2020; Zhu et al., 2025).

A coordinated, multidisciplinary approach is necessary to overcome the complex obstacles of marine fungal bioprospecting. By offering vital insights into fungal taxonomy and host ecology, mycologists play a fundamental role in guiding customized isolation and cultivation techniques (Amend et al., 2019). To unlock the potential of biosynthesis, molecular biologists develop genome editing tools and species-specific transformation systems. To better predict metabolite production, bioinformaticians simultaneously integrate multi-omics datasets and improve biosynthetic gene cluster (BGC) prediction algorithms (Hannigan et al., 2019). By creating bioreactors that are tailored for marine-specific factors like high salinity and the requirement for cofactor supplementation, chemical engineers aid in this endeavor. By examining host-symbiont interactions and environmental cues that control secondary metabolism in natural environments, marine ecologists advance scientific understanding. Marine ecologists further enhance discovery by studying host–symbiont interactions and environmental cues that regulate secondary metabolism in natural habitats. Collaborative initiatives, such as the Marine Fungal Bioprospecting Consortium, embody this integrative strategy—merging field sampling, genome sequencing, and high-throughput fermentation to expedite the identification and production of novel natural products (Gonçalves et al., 2022; Hyde et al., 2024).

10. Conclusion and Future Avenues

Marine endophytic fungi stand at the frontier of novel drug discovery, offering a vast reservoir of bioactive secondary metabolites with unparalleled chemical diversity and therapeutic relevance. Their ability to thrive in extreme and dynamic marine environments has led to the evolution of unique biosynthetic pathways, making them a critical yet underexploited resource. This chapter underscores the importance of exploring these fungal symbionts not only for their pharmacological potential but also for their ecological contributions to marine biodiversity. Although challenges such as low metabolite yields, cultivation difficulties, and sustainability concerns remain, emerging technologies in genome mining, co-culture strategies, and synthetic biology provide innovative solutions to bridge these gaps. The sustainable harnessing of marine fungal metabolites requires a multidisciplinary approach, integrating marine biology, pharmacology, biotechnology, and computational sciences. As the field advances, the incorporation of AI-driven analytics and comprehensive metabolomic databases will be pivotal in accelerating discovery pipelines. Ultimately, a deeper and more strategic exploration of marine endophytic fungi could transform the landscape of therapeutic development and contribute significantly to the future of personalized and precision medicine.

By making it possible to interpret complex data more quickly and accurately, artificial intelligence (AI) and machine learning platforms have the potential to revolutionize the discovery of marine metabolites. These technologies can help deconvolute datasets from mass spectrometry, predict the bioactivity of hypothetical scaffolds, and even use *in silico* modeling to design novel analogues. AI has the potential to significantly enhance the prioritization and dereplication of high-value BGCs when combined with genome mining (Hamdy et al., 2023; Chi et al., 2024). Databases like the Comprehensive Marine Natural Products Database (CMNPD), which already catalogued over 31,000 compounds, are vital resources for metabolite identification. However, the next generation of repositories must do more: integrating genomic annotations with spectral and bioactivity data will enable direct linkage between BGCs and their molecular outputs, accelerating functional characterization (Lyu et al., 2020).

Microfluidic devices that emulate marine microenvironments incorporating salinity gradients, host-derived extracts, and low-shear flow conditions hold promise for unlocking the biosynthetic potential of previously "unculturable" fungi. When coupled with single-cell omics, these tools may reveal novel triggers for cryptic metabolism (Gonçalves et al., 2022; Richter et al., 2022). Emerging bioreactor designs that employ continuous perfusion, closed-loop nutrient recycling, and microbial co-culture systems can significantly reduce environmental impact. By promoting synergy between fungi and beneficial bacteria, these systems may also enhance metabolite yields while suppressing contaminants—offering a scalable and sustainable solution for marine fungal fermentation (Dutta et al., 2024). As engineered marine fungal strains edge closer to industrial deployment, robust governance frameworks are essential. These should encompass containment protocols, biosafety regulations, and mechanisms for equitable benefit-sharing—particularly with coastal communities and biodiverse regions that provide source organisms. Ethical oversight must also address conservation priorities and prevent exploitative bioprospecting in ecologically sensitive marine ecosystems (Bohua et al., 2023; Lensch et al., 2024).

References

Abd El-Rahman, T.M.A., Tharwat, N.A., Abo El-Souad, S.M.S., El-Beih, A.A. and El-Diwany, A.I. (2020). Biological activities and variation of symbiotic fungi isolated from Coral reefs collected from Red Sea in Egypt. Mycology, 11: 243–255. https://doi.org/10.1080/21501203.2020.1741470.

Abdelrazek, H.M., Shams El-Din, N.G., Ghozlan, H.A., Sabry, S.A. and Abouelkheir, S.S. (2024). Distribution and functional perspective analysis of epiphytic and endophytic bacterial communities associated with marine seaweeds, Alexandria shores, Egypt. BMC Microbiol., 24: 293. https://doi.org/10.1186/s12866-024-03426-x.

Abeygunawardane, S., Thambugala, K.M., Kumara, W. and Daranagama, D. (2025). Seaweed-associated fungal endophytes from southern Sri Lanka and their biocontrol potential against selected fungal phytopathogens. Stud. Fungi, 10: e002. https://doi.org/10.48130/sif-0025-0002.

Agarwal, G. and Gabrani, R. (2021). Antiviral peptides: Identification and validation. Int. J. Pept. Res. Ther., 27: 149–168. https://doi.org/10.1007/s10989-020-10072-0.

Al Aboud, N.M., Tupper, C. and Jialal, I. (2025). Genetics, Epigenetic Mechanism. In: StatPearls. StatPearls Publishing, Treasure Island, FL, USA.

Alam, B., Lǐ, J., Gě, Q., Khan, M.A., Gōng, J. et al. (2021). Endophytic fungi: From symbiosis to secondary metabolite communications or vice versa? Front. Plant. Sci., 12: 791033. https://doi.org/10.3389/fpls.2021.791033.

Alves, C. and Diederich, M. (2021). Marine natural products as anticancer agents. Mar., Drugs, 19: 447. https://doi.org/10.3390/md19080447.

Amend, A., Burgaud, G., Cunliffe, M., Edgcomb, V.P., Ettinger, C.L. et al. (2019). Fungi in the marine environment: open questions and unsolved problems. mBio, 10: 10.1128/mbio.01189-18. https://doi.org/10.1128/mBio.01189-18.

Anh, C.V., Kwon, J.H., Kang, J.S., Lee, H.S., Heo, C.S. et al. (2022). Antibacterial and cytotoxic phenolic polyketides from two marine-derived fungal strains of *Aspergillus unguis*. Pharmaceuticals, 15: 74. https://doi.org/10.3390/ph15010074.

Anjugam, M., Bharathidasan, R., Rani, A.S.S. and Ambikapathy, V. (2019). Isolation and identification of endophytic fungi from *Avicennia marina* in mangrove environments. J. Emerg. Tech. Innov. Res., 6: 447–451.

Anjum, K. and Xuewei, Y. (2022). Epigenetic strategies to discover novel fungal secondary metabolites. J. Biomed. Res. Environ. Sci., 3: 246–263. https://doi.org/10.37871/jbres1430.

Arasu, M.V., Duraipandiyan, V. and Ignacimuthu, S. (2013). Antibacterial and antifungal activities of polyketide metabolite from marine *Streptomyces* sp. AP-123 and its cytotoxic effect. Chemosphere, 90: 479–487. https://doi.org/10.1016/j.chemosphere.2012.08.006.

Ariantari, N.P., Putra, I.P.Y.A., Wibowo, J.T., Putri, S.A.M.D.A., Nabila, C.I.B. et al. (2024). Evaluation of antibacterial and toxicity properties of marine endophytic fungi from red algae *Gracilaria salicornia* and *Chondrus* sp. BIO Web Conf., 92: 02019. https://doi.org/10.1051/bioconf/20249202019.

Arias-Sánchez, R.A., Torner, L. and Fenton Navarro, B. (2023). Polyphenols and neurodegenerative diseases: Potential effects and mechanisms of neuroprotection. Mol. Basel Switz., 28: 5415. https://doi.org/10.3390/molecules28145415.

Arifeen, M.Z.U., Xue, Y.R. and Liu, C.H. (2019). Deep-sea fungi: Diversity, enzymes, and bioactive metabolites. pp. 331–347. *In*: Tiquia-Arashiro, S.M. and Grube, M. (eds.). Fungi in Extreme Environments: Ecological Role and Biotechnological Significance. Springer International Publishing, Cham.

Arora, D., Gupta, P., Jaglan, S., Roullier, C., Grovel, O. et al. (2020). Expanding the chemical diversity through microorganisms co-culture: Current status and outlook. Biotechnol. Adv., 40: 107521. https://doi.org/10.1016/j.biotechadv.2020.107521.

Arumugam, G.S., Damodharan, K., Doble, M. and Thennarasu, S. (2022). Significant perspectives on various viral infections targeted antiviral drugs and vaccines including COVID-19 pandemicity. Mol. Biomed., 3: 21. https://doi.org/10.1186/s43556-022-00078-z.

Augner, D., Krut, O., Slavov, N., Gerbino, D.C., Sahl, H.G. et al. (2013). On the antibiotic and antifungal activity of pestalone, pestalachloride A, and structurally related compounds. J. Nat. Prod., 76: 1519–1522. https://doi.org/10.1021/np400301d.

Avalos, J. and Limón, M.C. (2022). Fungal secondary metabolism. Encyclopedia, 2: 1–13. https://doi.org/10.3390/encyclopedia2010001.

Balai, X., Sheng, Y., Tang, Z., Pan, J., Wang, S. et al. (2023). Polyketides as secondary metabolites from the genus *Aspergillus*. J. Fungi, 9: 261. https://doi.org/10.3390/jof9020261.

Bai, Y., Guo, F, Hu, J. and Pan, H. (2025). Discovery of antioxidants and α-glycosidase inhibitors from an endophytic fungus *Penicillium brefeldianum* F4a. Fitoterapia, 182: 106479. https://doi.org/10.1016/j.fitote.2025.106479.

Bakhtra, D., Yanwirasti, Y., Wahyuni, F.S., Aminah, I. and Handayani, D. (2022). Antimicrobial and cytotoxic activities screening of marine invertebrate-derived fungi extract from West Sumatera, Indonesia. Open Access Maced. J. Med. Sci., 10: 1427–1432. https://doi.org/10.3889/oamjms.2022.10374.

Barreca, M., Spanò, V., Montalbano, A., Cueto, M., Díaz Marrero, A.R. et al. (2020). Marine anticancer agents: an overview with a particular focus on their chemical classes. Mar. Drugs, 18: 619. https://doi.org/10.3390/md18120619.

Batbayar, S., Lee, D.H. and Kim, H.W. (2012). Immunomodulation of Fungal β-Glucan in host defense signaling by Dectin-1. Biomol. Ther., 20: 433–445. https://doi.org/10.4062/biomolther.2012.20.5.433.

Bessa, C.D.P.B., Feu, A.E., de Menezes, R.P.B., Scotti, M.T., Lima, J.M.G. et al. (2024). Multitarget anti-parasitic activities of isoquinoline alkaloids isolated from *Hippeastrum aulicum* (Amaryllidaceae). Phytomedicine, 128: 155414. https://doi.org/10.1016/j.phymed.2024.155414.

Bharathi, D. and Lee, J. (2024). Recent advances in marine-derived compounds as potent antibacterial and antifungal agents: A comprehensive review. Mar. Drugs, 22: 348. https://doi.org/10.3390/md22080348.

Blayney, D.W., Mohanlal, R., Adamchuk, H., Kirtbaya, D.V., Chen, M. et al. (2022). Efficacy of plinabulin vs pegfilgrastim for prevention of docetaxel-induced neutropenia in patients with solid tumors: A randomized clinical trial. JAMA Netw Open, 5: e2145446. https://doi.org/10.1001/jamanetworkopen.2021.45446.

Blin, K., Kim, H.U., Medema, M.H. and Weber, T. (2019). Recent development of antiSMASH and other computational approaches to mine secondary metabolite biosynthetic gene clusters. Brief. Bioinform., 20: 1103–1113. https://doi.org/10.1093/bib/bbx146.

Blunt, J.W., Copp, B.R., Keyzers, R.A., Munro, M.H.G. and Prinsep, M.R. (2016). Marine natural products. Nat. Prod. Rep., 33: 382–431. https://doi.org/10.1039/C5NP00156K.

Bohua, L., Yuexin, W., Yakun, O., Kunlan, Z., Huan, L. et al. (2023). Ethical framework on risk governance of synthetic biology. J. Biosaf. Biosecurity., 5: 45–56. https://doi.org/10.1016/j.jobb.2023.03.002.

Bojarska, J., Mieczkowski, A., Ziora, Z.M., Skwarczynski, M., Toth, I. et al. (2021). Cyclic dipeptides: The biological and structural landscape with special focus on the anti-cancer proline-based scaffold. Biomolecules, 11: 1515. https://doi.org/10.3390/biom11101515.

Brakhage, A.A. and Schroeckh, V. (2011). Fungal secondary metabolites: Strategies to activate silent gene clusters. Fungal Genet. Biol., 48: 15–22. https://doi.org/10.1016/j.fgb.2010.04.004.

Brown, G.D., Denning, D.W. and Levitz, S.M. (2012). Tackling Human Fungal Infections. Science, 336: 647–647. https://doi.org/10.1126/science.1222236.

Buda De Cesare, G., Cristy, S.A., Garsin, D.A. and Lorenz, M.C. (2020). Antimicrobial peptides: A new frontier in antifungal therapy. mBio, 11: 10.1128/mbio.02123-20. https://doi.org/10.1128/mbio.02123-20.

Burgaud, G., Edgcomb, V., Hassett, B.T., Kumar, A., Li, W. et al. (2022). Marine fungi. pp. 243–295. *In*: Stal, L.J., and Cretoiu, M.S. (eds.). The Marine Microbiome. Springer International Publishing, Cham.

Cadamuro, R.D., da Silveira Bastos, I.M.A., Silva, I.T., da Cruz, A.C.C., Robl, D. et al. (2021). Bioactive compounds from mangrove endophytic fungus and their uses for microorganism control. J. Fungi Basel Switz., 7: 455. https://doi.org/10.3390/jof7060455.

Cai, C., Chen, Y., Zhou, L., Gong, N., Zhang, H. et al. (2023). Antimicrobial polyketides from the marine-derived fungus *Spiromastix* sp. SCSIO F190. J. Nat. Prod., 86: 589–595. https://doi.org/10.1021/acs.jnatprod.2c00900.

Cao, J., Li, X.M., Li, X., Li, H.L., Meng, L.H. et al. (2019). New lactone and isocoumarin derivatives from the marine mangrove-derived endophytic fungus *Penicillium coffeae* MA-314. Phytochem. Lett., 32: 1–5. https://doi.org/10.1016/j.phytol.2019.04.018.

Carew, J.S., Nawrocki, S.T., Krupnik, Y.V., Dunner, K., McConkey, D.J. and Huang, M.K.P. (2006). Targeting endoplasmic reticulum protein transport: A novel strategy to kill malignant B cells and overcome fludarabine resistance in CLL. Blood, 107: 222–231. https://doi.org/10.1182/blood-2005-05-1923.

Carpena, M., Pereira, C.S., Silva, A., Barciela, P., Jorge, A.O.S. et al. (2024). Metabolite profiling of macroalgae: Biosynthesis and beneficial biological properties of active compounds. Mar. Drugs, 22: 478. https://doi.org/10.3390/md22100478.

Caruso, D.J., Palombo, E.A., Moulton, S.E. and Zaferanloo, B. (2022). Exploring the promise of endophytic fungi: A review of novel antimicrobial compounds. Microorganisms, 10: 1990. https://doi.org/10.3390/microorganisms10101990.,

Catanesi, M., Caioni, G., Castelli, V., Benedetti, E., d'Angelo, M. et al. (2021). Benefits under the Sea: The role of marine compounds in neurodegenerative disorders. Mar. Drugs, 19: 24. https://doi.org/10.3390/md19010024.

Chaudhary, P., Agri, U., Chaudhary, A., Kumar, A. and Kumar, G. (2022). Endophytes and their potential in biotic stress management and crop production. Front. Microbiol., 13: 933017. https://doi.org/10.3389/fmicb.2022.933017.

Chen, D., Zhang, P., Liu, T., Wang, X.F., Li, Z.X. et al. (2018). Insecticidal activities of chloramphenicol derivatives isolated from a marine alga-derived endophytic fungus, *Acremonium vitellinum*, against the Cotton Bollworm, *Helicoverpa armigera* (Hübner) (Lepidoptera: Noctuidae). Mol. Basel Switz., 23: 2995. https://doi.org/10.3390/molecules23112995.

Chen, J., Zhang, P., Ye, X., Wei, B., Emam, M. et al. (2020). The Structural diversity of marine microbial secondary metabolites based on co-culture strategy: 2009–2019. Mar. Drugs, 18: 449. https://doi.org/10.3390/md18090449.

Chen, Y., Cai, C., Yang, J., Shi, J., Song, Y. et al. (2022a). Development of the CRISPR-Cas9 system for the marine-derived fungi *Spiromastix* sp. SCSIO F190 and *Aspergillus* sp. SCSIO SX7S7. J. Fungi Basel Switz., 8: 715. https://doi.org/10.3390/jof8070715.

Chen, Y., Liu, Z., Huang, Y., Liu, L., He, J. et al. (2019). Ascomylactams A-C, cytotoxic 12- or 13-membered-ring macrocyclic alkaloids isolated from the mangrove endophytic fungus *Didymella* sp. CYSK-4, and structure revisions of phomapyrrolidones A and C. J. Nat. Prod., 82: 1752–1758. https://doi.org/10.1021/acs.jnatprod.8b00918.

Chen, Y., Xu, L.C., Liu, S., Zhang, Z.X. and Cao, G.Y. (2022c). Halometabolites isolated from the marine-derived fungi with potent pharmacological activities. Front. Microbiol., 13: 1038487. https://doi.org/10.3389/fmicb.2022.1038487.

Chen, Z., Zheng, Z., Huang, H., Song, Y., Zhang, X. et al. (2012). Penicacids A–C, three new mycophenolic acid derivatives and immunosuppressive activities from the marine-derived fungus *Penicillium* sp. SOF07. Bioorg. Med. Chem. Lett., 22: 3332–3335. https://doi.org/10.1016/j.bmcl.2012.02.106.

Chi, J., Shu, J., Li, M., Mudappathi, R., Jin, Y. et al. (2024). Artificial intelligence in metabolomics: A current review. TrAC Trends Anal. Chem., 178: 117852. https://doi.org/10.1016/j.trac.2024.117852.

Chinworrungsee, M., Kittakoop, P., Isaka, M., Rungrod, A., Tanticharoen, M. et al. (2001). Antimalarial halorosellinic acid from the marine fungus *Halorosellinia oceanica*. Bioorg. Med. Chem. Lett., 11: 1965–1969. https://doi.org/10.1016/s0960-894x(01)00327-4.

Cochereau, B., Le Strat, Y., Ji, Q., Pawtowski, A., Delage, L. et al. (2023). Heterologous expression and biochemical characterization of a new chloroperoxidase isolated from the deep-sea hydrothermal vent black yeast *Hortaea werneckii* UBOCC-A-208029. Mar. Biotechnol., 25: 519–536. https://doi.org/10.1007/s10126-023-10222-7.

Corral, P., Amoozegar, M.A. and Ventosa, A. (2019). Halophiles and their biomolecules: Recent advances and future applications in biomedicine. Mar. Drugs, 18: 33. https://doi.org/10.3390/md18010033.

Cox, J. and Gulder, T.A.M. (2023). Introduction to engineering the biosynthesis of fungal natural products. Nat. Prod. Rep., 40: 7–8. https://doi.org/10.1039/D2NP90047E.

Cueto, M., Jensen, P.R., Kauffman, C., Fenical, W., Lobkovsky, E. et al. (2001). Pestalone, a new antibiotic produced by a marine fungus in response to bacterial challenge. J. Nat. Prod., 64: 1444–1446. https://doi.org/10.1021/np0102713.

Dai, Q., Zhang, F.L. and Feng, T. (2021). Sesquiterpenoids specially produced by fungi: Structures, biological activities, chemical, and biosynthesis (2015–2020). J. Fungi, 7: 1026. https://doi.org/10.3390/jof7121026.

Daniel-Ivad, M., Hameed, N., Tan, S., Dhanjal, R., Socko, D. et al. (2017). An engineered allele of *afsQ1* facilitates the discovery and investigation of cryptic natural products. ACS Chem. Biol., 12: 628–634. https://doi.org/10.1021/acschembio.6b01002.

Das, S., Singh, A., Samanta, S.K. and Singha Roy, A. (2022). Naturally occurring anthraquinones as potential inhibitors of SARS-CoV-2 main protease: An integrated computational study. Biologia (Bratisl), 77: 1121–1134. https://doi.org/10.1007/s11756-021-01004-4.

Dash, U.C., Bhol, N.K., Swain, S.K., Samal, R.R., Nayak, P.K. et al. (2025). Oxidative stress and inflammation in the pathogenesis of neurological disorders: Mechanisms and implications. Acta Pharm. Sin. B, 15: 15–34. https://doi.org/10.1016/j.apsb.2024.10.004.

Debbab, A., Aly, A.H. and Proksch, P. (2012). Endophytes and associated marine-derived fungi—ecological and chemical perspectives. Fungal Divers., 57: 45–83. https://doi.org/10.1007/s13225-012-0191-8.

Debbab, A., Aly, A.H. and Proksch, P. (2013). Mangrove derived fungal endophytes: A chemical and biological perception. Fungal Divers., 61: 1–27. https://doi.org/10.1007/s13225-013-0243-8.

Dembitsky, V.M., Ermolenko, E., Savidov, N., Gloriozova, T.A. and Poroikov, V.V. (2021). Antiprotozoal and antitumor activity of natural polycyclic endoperoxides: Origin, structures, and biological activity. Mol. Basel Switz., 26: 686. https://doi.org/10.3390/molecules26030686.

Deng, L., Zhong, M., Li, Y., Hu, G., Zhang, C. et al. (2023). High hydrostatic pressure harnesses the biosynthesis of secondary metabolites via the regulation of polyketide synthesis genes of hadal sediment-derived fungi. Front. Microbiol., 14: 1207252. https://doi.org/10.3389/fmicb.2023.1207252.

Deshmukh, S.K., Gupta, M.K., Prakash, V. and Reddy, M.S. (2018a). Mangrove-associated fungi: A novel source of potential anticancer compounds. J. Fungi, 4: 101. https://doi.org/10.3390/jof4030101.

Deshmukh, S.K., Prakash, V. and Ranjan, N. (2018b). Marine Fungi: A source of potential anticancer compounds. Front. Microbiol., 8: 2536. https://doi.org/10.3389/fmicb.2017.02536.

Dhyani, P., Quispe, C., Sharma, E., Bahukhandi, A., Sati, P. et al. (2022). Anticancer potential of alkaloids: A key emphasis to colchicine, vinblastine, vincristine, vindesine, vinorelbine, and vincamine. Cancer Cell Int., 22: 206. https://doi.org/10.1186/s12935-022-02624-9.

Diehl, C., Pinzan, C.F., de Castro, P.A., Delbaje, E., García Carnero, L.C. et al. (2024). Brilacidin, a novel antifungal agent against *Cryptococcus neoformans*. mBio, 15: e01031-24. https://doi.org/10.1128/mbio.01031-24.

Dimitrova, D., Dimitrova, S., Kehayova, G. and Dragomanova, S. (2025). Meroterpenoids from terrestrial and marine fungi: Promising agents for neurodegenerative disorders: An updated review. Curr. Issues Mol. Biol., 47: 96. https://doi.org/10.3390/cimb47020096.

Ding, Z., Li, F., Xie, L., Gu, M., Li, C. et al. (2022). Design and synthesis of novel phenylahistin derivatives based on co-crystal structures as potent microtubule inhibitors for anti-cancer therapy. Mar. Drugs, 20: 752. https://doi.org/10.3390/md20120752.

Dolashki, A., Abrashev, R., Kaynarov, D., Krumova, E., Velkova, L. et al. (2024). Structural and functional characterization of cold-active sialidase isolated from Antarctic fungus *Penicillium griseofulvum* P29. Biochem. Biophys Rep., 37: 101610. https://doi.org/10.1016/j.bbrep.2023.101610.

dos Santos, T.B., Ribas, A.F., de Souza, S.G.H., Budzinski, I.G.F. and Domingues, D.S. (2022). Physiological responses to drought, salinity, and heat stress in plants: A review. Stresses, 2: 113–135. https://doi.org/10.3390/stresses2010009.

Driver, T., Trivedi, D.K., McIntosh, O.A., Dean, A.P., Goodacre, R. et al. (2017). Two Glycerol-3-phosphate dehydrogenases from chlamydomonas have distinct roles in lipid metabolism. Plant Physiol., 174: 2083–2097. https://doi.org/10.1104/pp.17.00491.

Dutta, D., Kumar, P., Singh, A. and Khade, S. (2024). Bioprocess strategies for enhanced performance in single-use bioreactors for biomolecule synthesis: A biokinetic approach. Food Bioeng., 3: 337–351. https://doi.org/10.1002/fbe2.12104.

Ecevit, K., Barros, A.A., Silva, J.M. and Reis, R.L. (2022). Preventing microbial infections with natural phenolic compounds. Future Pharmacol., 2: 460–498. https://doi.org/10.3390/futurepharmacol2040030.

El-Agamy, D.S., Ibrahim, S.R., Ahmed, N., Khoshhal, S., Abo-Haded, H.M. et al. (2019). Aspernolide F, as a new cardioprotective butyrolactone against doxorubicin-induced cardiotoxicity. Int. Immunopharmacol., 72: 429–436. https://doi.org/10.1016/j.intimp.2019.04.045.

El-Bondkly, E.A.M., El-Bondkly, A.A.M. and El-Bondkly, A.A.M. (2021a). Marine endophytic fungal metabolites: A whole new world of pharmaceutical therapy exploration. Heliyon, 7: e06362. https://doi.org/10.1016/j.heliyon.2021.e06362.

Elhamouly, N.A., Hewedy, O.A., Zaitoon, A., Miraples, A., Elshorbagy, O.T. et al. (2022). The hidden power of secondary metabolites in plant-fungi interactions and sustainable phytoremediation. Front. Plant. Sci., 13: 1044896. https://doi.org/10.3389/fpls.2022.1044896.

Elnaggar, M.S., Ebada, S.S., Ashour, M.L., Ebrahim, W., Müller, W.E. et al. (2016). Xanthones and sesquiterpene derivatives from a marine-derived fungus *Scopulariopsis* sp. Tetrahedron, 72: 2411–2419. https://doi.org/10.1016/j.tet.2016.03.073.

Eshboev, F., Mamadalieva, N., Nazarov, P.A., Hussain, H., Katanaev, V. et al. (2024). Antimicrobial action mechanisms of natural compounds isolated from endophytic microorganisms. Antibiotics, 13: 271. https://doi.org/10.3390/antibiotics13030271.

Ettinger, C.L., Vann, L.E. and Eisen, J.A. (2021). Global diversity and biogeography of the *Zostera marina* mycobiome. Appl. Environ. Microbiol., 87: e02795-20. https://doi.org/10.1128/AEM.02795-20.

Evidente, A. (2024). Advances on anticancer fungal metabolites: Sources, chemical, and biological activities in the last decade (2012–2023). Nat. Prod. Bioprospecting, 14: 31. https://doi.org/10.1007/s13659-024-00452-0.

Fagorzi, C. and Mengoni, A. (2022). Endophytes: Improving plant performance. Microorganisms, 10: 1777. https://doi.org/10.3390/microorganisms10091777.

Faisal, S., Badshah, S.L., Kubra, B., Emwas, A.H. and Jaremko, M. (2023). Alkaloids as potential antivirals. A comprehensive review. Nat. Prod. Bioprospecting, 13: 4. https://doi.org/10.1007/s13659-022-00366-9.

Gallagher, K.A., Fenical, W. and Jensen, P.R. (2010). Hybrid isoprenoid secondary metabolite production in terrestrial and marine actinomycetes. Curr. Opin. Biotechnol., 21: 794–800. https://doi.org/10.1016/j.copbio.2010.09.010.

Gao, L.W. and Zhang, P. (2023). An update on chemistry and bioactivities of secondary metabolites from the marine algal-derived endophytic fungi. Phytochem. Rev., 22: 587–614. https://doi.org/10.1007/s11101-022-09852-x.

Gao, Y., Wang, J., Meesakul, P., Zhou, J., Liu, J. et al. (2024). Cytotoxic compounds from marine fungi: Sources, Structures, and Bioactivity. Mar. Drugs, 22: 70. https://doi.org/10.3390/md22020070.

Garg, P., Malhotra, J., Kulkarni, P., Horne, D., Salgia, R. et al. (2024). Emerging therapeutic strategies to overcome drug resistance in cancer cells. Cancers, 16: 2478. https://doi.org/10.3390/cancers16132478.

Ghazawi, K.F., Fatani, S.A., Mohamed, S.G., Mohamed, G.A. and Ibrahim, S.R. (2023). *Aspergillus nidulans*—Natural metabolites powerhouse: Structures, biosynthesis, bioactivities, and biotechnological potential. Fermentation, 9: 325. https://doi.org/10.3390/fermentation9040325.

Giddings, L.A. and Newman, D.J. (2022). Extremophilic fungi from marine environments: Underexplored sources of antitumor, anti-infective, and other biologically active agents. Mar. Drugs, 20: 62. https://doi.org/10.3390/md20010062.

Gil-Durán, C., Palma, D., Marcano, Y., Palacios, J.L., Martínez, C. et al. (2023). CRISPR/Cas9-mediated disruption of the pcz1 gene and its impact on growth, development, and penicillin production in *Penicillium rubens*. J. Fungi Basel Switz., 9: 1010. https://doi.org/10.3390/jof9101010.

Girich, E.V., Yurchenko, A.N., Smetanina, O.F., Trinh, P.T.H., Ngoc, N.T.D. et al. (2020). Neuroprotective metabolites from Vietnamese marine derived fungi of *Aspergillus* and *Penicillium* Genera. Mar Drugs, 18: 608. https://doi.org/10.3390/md18120608.

Gomes, N.G.M., Lefranc, F., Kijjoa, A. and Kiss, R. (2015). Can some marine-derived fungal metabolites become actual anticancer agents? Mar. Drugs, 13: 3950–3991. https://doi.org/10.3390/md13063950.

Gonçalves, M.F.M., Esteves, A.C. and Alves, A. (2022c). Marine fungi: Opportunities and challenges. Encyclopedia, 2: 559–577. https://doi.org/10.3390/encyclopedia2010037.

Gong, H., Bandura, J., Wang, G.L., Feng, Z.P. and Sun, H.S. (2022). Xyloketal B: A marine compound with medicinal potential. Pharmacol. Ther., 230: 107963. https://doi.org/10.1016/j.pharmthera.2021.107963.

Gostinčar, C., Zalar, P. and Gunde-Cimerman, N. (2022). No need for speed: Slow development of fungi in extreme environments. Fungal Biol. Rev., 39: 1–14. https://doi.org/10.1016/j.fbr.2021.11.002.

Gouda, S., Das, G., Sen, S.K., Shin, H.S. and Patra, J.K. (2016). Endophytes: A treasure house of bioactive compounds of medicinal importance. Front. Microbiol., 7. https://doi.org/10.3389/fmicb.2016.01538.

Grabka, R., d'Entremont, T.W., Adams, S.J., Walker, A.K., Tanney, J.B. et al. (2022). Fungal endophytes and their role in agricultural plant protection against pests and pathogens. Plants, 11: 384. https://doi.org/10.3390/plants11030384.

Gribble, G.W. (2023). Naturally occurring organohalogen compounds: A comprehensive review. pp. 1–546. *In*: Kinghorn, A.D., Falk, H., Gibbons, S., Asakawa, Y., Liu, J.K. and Dirsch, V.M. (eds.). Naturally Occurring Organohalogen Compounds. Springer Nature Switzerland, Cham.

Gunasekaran, S., Sundaramoorthy, S., Anitha, U., Sathiavelu, M. and Arunachalam, S. (2015). Endophytic fungi with antioxidant activity: A review. Res. J. Pharm. Technol., 8: 731–737. https://doi.org/10.5958/0974-360X.2015.00116.X.

Gupta, A., Meshram, V., Gupta, M., Goyal, S., Qureshi, K.A. et al. (2023). Fungal endophytes: Microfactories of novel bioactive compounds with therapeutic interventions; A comprehensive review on the biotechnological developments in the field of fungal endophytic biology over the last decade. Biomolecules, 13: 1038. https://doi.org/10.3390/biom13071038.

Hafez Ghoran, S., Taktaz, F., Sousa, E., Fernandes, C. and Kijjoa, A. (2023). Peptides from marine-derived fungi: Chemistry and biological activities. Mar. Drugs, 21: 510. https://doi.org/10.3390/md21100510.

Hamdy, R., Altaie, A.M., El-Labbad, E.M., Ashmawy, N.S., Ozsahin, D.U. et al. (2023). Artificial intelligence and discovery of microbial natural products. pp. 37–78. *In*: Cruz, J.N. (ed.). Drug Discovery and Design Using Natural Products. Springer Nature Switzerland, Cham.

Hamzah, T.N.T., Lee, S.Y., Hidayat, A., Terhem, R., Faridah-Hanum, I. et al. (2018). Diversity and characterization of endophytic fungi isolated from the tropical mangrove species, *Rhizophora mucronata*, and Identification of potential antagonists against the soil-borne fungus, *Fusarium solani*. Front. Microbiol., 9: 1707. https://doi.org/10.3389/fmicb.2018.01707.

Han, C., Song, A., He, Y., Yang, L., Chen, L. et al. (2024). Genome mining and biosynthetic pathways of marine-derived fungal bioactive natural products. Front. Microbiol., 15: 1520446. https://doi.org/10.3389/fmicb.2024.1520446.

Hang, S., Lu, H. and Jiang, Y. (2024). Marine-derived metabolites act as promising antifungal agents. Mar. Drugs, 22: 180. https://doi.org/10.3390/md22040180.

Hannigan, G.D., Prihoda, D., Palicka, A., Soukup, J., Klempir, O. et al. (2019). A deep learning genome-mining strategy for biosynthetic gene cluster prediction. Nucleic Acids Res., 47: e110. https://doi.org/10.1093/nar/gkz654.

Hasan, S., Ansari, M.I., Ahmad, A. and Mishra, M. (2015). Major bioactive metabolites from marine fungi: A review. Bioinformation, 11: 176–181. https://doi.org/10.6026/97320630011176.

Hashem, A.H., Attia, M.S., Kandil, E.K., Fawzi, M.M., Abdelrahman, A.S. et al. (2023). Bioactive compounds and biomedical applications of endophytic fungi: A recent review. Microb. Cell Factories, 22: 107. https://doi.org/10.1186/s12934-023-02118-x.

Hawas, U.W., Al-Farawati, R., Abou El-Kassem, L.T. and Turki, A.J. (2016). Different culture metabolites of the Red Sea fungus *Fusarium equiseti* optimize the inhibition of Hepatitis C Virus NS3/4A Protease (HCV PR). Mar. Drugs, 14: 190. https://doi.org/10.3390/md14100190.

Hawas, U.W., El-Beih, A.A. and El-Halawany, A.M. (2012). Bioactive anthraquinones from endophytic fungus *Aspergillus versicolor* isolated from red sea algae. Arch. Pharm. Res., 35: 1749–1756. https://doi.org/10.1007/s12272-012-1006-x.

Henrikson, J.C., Hoover, A.R., Joyner, P.M. and Cichewicz, R.H. (2009). A chemical epigenetics approach for engineering the *in situ* biosynthesis of a cryptic natural product from *Aspergillus niger*. Org. Biomol. Chem., 7: 435–438. https://doi.org/10.1039/B819208A.

Höller, U., Wright, A.D., Matthee, G.F., Konig, G.M., Draeger, S. et al. (2000). Fungi from marine sponges: Diversity, biological activity, and secondary metabolites. Mycol. Res., 104: 1354–1365. https://doi.org/10.1017/S0953756200003117.

Huang, H., Liu, T., Guo, J., Yu, L., Wu, X. et al. (2017). Brefeldin A enhances docetaxel-induced growth inhibition and apoptosis in prostate cancer cells in monolayer and 3D cultures. Bioorg. Med. Chem. Lett., 27: 2286–2291. https://doi.org/10.1016/j.bmcl.2017.04.047.

Huang, R., Wang, Y., Liu, D., Wang, S., Lv, H. et al. (2023). Long-read metagenomics of marine microbes reveals diversely expressed secondary metabolites. Microbiol., 11: e01501-23. https://doi.org/10.1128/spectrum.01501-23.

Huang, Z., Liu, D., Chen, S., Ren, J., Gao, C. et al. (2024) Brominated depsidones with antibacterial effects from a deep-sea-derived fungus *Spiromastix* sp. Mar. Drugs, 22: 78. https://doi.org/10.3390/md22020078.

Huang, Z.H., Liang, X., Li, C.J., Gu, Q., Ma, X. et al. (2021). Talaromynoids A–I, highly oxygenated meroterpenoids from the marine-derived fungus Talaromyces purpureogenus SCSIO 41517 and their lipid accumulation inhibitory activities. J. Nat. Prod., 84: 2727–2737. https://doi.org/10.1021/acs.jnatprod.1c00681.

Hulikere, M.M., Joshi, C.G., Ananda, D., Poyya, J. and Nivya, T., (2016). Antiangiogenic, wound healing and antioxidant activity of *Cladosporium cladosporioides* (Endophytic Fungus) isolated from seaweed (*Sargassum wightii*). Mycology, 7: 203. https://doi.org/10.1080/21501203.2016.1263688.

Hur, J.Y., Jeong, E., Kim, Y.C. and Lee, S.R. (2023b). Strategies for natural product discovery by unlocking cryptic biosynthetic gene clusters in fungi. Separations, 10: 333. https://doi.org/10.3390/separations10060333.

Hyde, K.D., Baldrian, P., Chen, Y., Thilini Chethana, K.W., De Hoog, S. et al. (2024). Current trends, limitations, and future research in the fungi? Fungal Divers., 125: 1–71. https://doi.org/10.1007/s13225-023-00532-5.

Imhoff, J.F. (2016). Natural products from marine fungi: Still an underrepresented resource. Mar. Drugs, 14: 19. https://doi.org/10.3390/md14010019.

Ismaiel, A.A., Rabie, G.H. and Abd El-Aal, M.A. (2016). Antimicrobial and morphogenic effects of emodin produced by Aspergillus awamori WAIR120. Biologia (Bratisl), 71: 464–474. https://doi.org/10.1515/biolog-2016-0067.

Jan, R., Asaf, S., Numan, M., Lubna and Kim, K.M. (2021). Plant secondary metabolite biosynthesis and transcriptional regulation in response to biotic and abiotic stress conditions. Agronomy, 11: 968. https://doi.org/10.3390/agronomy11050968.

Jha, P., Kaur, T., Chhabra, I., Panja, A., Paul, S. et al. (2023). Endophytic fungi: Hidden treasure chest of antimicrobial metabolites interrelationship of endophytes and metabolites. Front. Microbiol., 14: 1227830. https://doi.org/10.3389/fmicb.2023.1227830.

Jiang, C., Lv, G., Tu, Y., Cheng, X., Duan, Y. et al. (2021). Applications of CRISPR/Cas9 in the synthesis of secondary metabolites in filamentous fungi. Front. Microbiol., 12: 638096. https://doi.org/10.3389/fmicb.2021.638096.

Jiao, F., Liu, T., Wang, K., Li, S., Jiao, R. et al. (2025a). Zopfiellamides C and D, new decalin-type tetramic acid derivatives from the marine-derived fungus *Aspergillus* sp. NF666. Molecules, 30: 1502. https://doi.org/10.3390/molecules30071502.

Jiménez-Romero, C., Amador, L.A., Castro-Falcón, G. and Rodríguez, A.D. (2025). Antimicrobial and anticancer potential of polyketides isolated from the Caribbean marine sponge plakortis halichondrioides. Drugs Drug Candidates, 4(1): 6. https://doi.org/10.3390/ddc4010006.

Jin, L., Quan, C., Hou, X. and Fan, S. (2016). Potential pharmacological resources: Natural bioactive compounds from marine-derived fungi. Mar. Drugs, 14: 76. https://doi.org/10.3390/md14040076.

Joe Xavier Sneha, M., Thangavel, M., Mani, I., Rajapriya, P., Ponnuraj, N. et al. (2024). Endophytic fungal diversity in *Hardwickia binata*: Bridging the gap between traditional and modern techniques. Microbiol. Res., 15: 823–840. https://doi.org/10.3390/microbiolres15020053.

Julianti, E., Oh, H., Jang, K.H., Lee, J.K., Lee, S.K. et al. (2011). Acremostrictin, a highly oxygenated metabolite from the marine fungus *Acremonium strictum*. J. Nat. Prod., 74: 2592–2594. https://doi.org/10.1021/np200707y.

Kálai, T., Pongrácz, J.E. and Mátyus, P. (2022). Recent advances in influenza, HIV and SARS-CoV-2 infection prevention and drug treatment: The need for precision medicine. Chemistry, 4: 216–258. https://doi.org/10.3390/chemistry4020019.

Kalaiselvam, M. (2015). Marine fungal diversity and bioprospecting. pp. 13–25. *In*: Kim, S.K. (ed.). Springer Handbook of Marine Biotechnology. Springer, Berlin, Heidelberg.

Kannan, J., Pang, K.L., Ho, Y.N., Hsu, P.H. and Chen, L.L. (2024). A comparison of the antioxidant potential and metabolite analysis of marine fungi associated with the red algae *Pterocladiella capillacea* from Northern Taiwan. Antioxidants, 13: 336. https://doi.org/10.3390/antiox13030336.

Kaushik, N.K., Murali, T.S., Sahal, D. and Suryanarayanan, T.S. (2014). A search for antiplasmodial metabolites among fungal endophytes of terrestrial and marine plants of southern India. Acta Parasitol., 59: 745–757. https://doi.org/10.2478/s11686-014-0307-2.

Kennedy, J., Flemer, B., Jackson, S.A., Lejon, D.P., Morrissey, J.P. et al. (2010). Marine metagenomics: New tools for the study and exploitation of marine microbial metabolism. Mar. Drugs, 8: 608–628. https://doi.org/10.3390/md8030608.

Khalmuratova, I., Choi, D.H., Woo, J.R., Jeong, M.J., Oh, Y. et al. (2020). Diversity and plant growth-promoting effects of fungal endophytes isolated from salt-tolerant plants. J. Microbiol. Biotechnol., 30: 1680–1687. https://doi.org/10.4014/jmb.2006.06050.

Khare, E., Mishra, J. and Arora, N.K. (2018). Multifaceted interactions between endophytes and plants: Developments and prospects. Front. Microbiol., 9: 2732. https://doi.org/10.3389/fmicb.2018.02732.

Khattab, A.R. and Farag, M.A. (2022). Marine and terrestrial endophytic fungi: A mine of bioactive xanthone compounds, recent progress, limitations, and novel applications. Crit. Rev. Biotechnol., 42: 403–430. https://doi.org/10.1080/07388551.2021.1940087.

Kildgaard, S., Subko, K., Phillips, E., Goidts, V., De la Cruz, M. et al. (2017). A dereplication and bioguided discovery approach to reveal new compounds from a marine-derived fungus Stilbella fimetaria. Mar. Drugs, 15: 253. https://doi.org/10.3390/md15080253.

Kim, D.C., Cho, K.H., Ko, W., Yoon, C.S., Sohn, J.H. et al (2016). Anti-Inflammatory and cytoprotective effects of tmc-256c1 from marine-derived fungus Aspergillus sp. SF-6354 via up-regulation of heme oxygenase-1 in murine hippocampal and microglial cell lines. Int. J. Mol. Sci., 17: 529. https://doi.org/10.3390/ijms17040529.

King, A.M., Reid-Yu, S.A., Wang, W., King, D.T., De Pascale, G. et al. (2014). Aspergillomarasmine A overcomes metallo-β-lactamase antibiotic resistance. Nature, 510: 503–506. https://doi.org/10.1038/nature13445.

Kjer, J., Debbab, A., Aly, A.H. and Proksch, P. (2010a). Methods for isolation of marine-derived endophytic fungi and their bioactive secondary products. Nat. Protoc., 5: 479–490. https://doi.org/10.1038/nprot.2009.233.

Klinges, J.G., Patel, S.H., Duke, W.C., Muller, E.M. and Vega Thurber, R.L. (2023). Microbiomes of a disease-resistant genotype of *Acropora cervicornis* are resistant to acute, but not chronic, nutrient enrichment. Sci. Rep., 13: 3617. https://doi.org/10.1038/s41598-023-30615-x.

Komaki, H. and Tamura, T. (2023). Profile of PKS and NRPS gene clusters in the genome of *Streptomyces cellostaticus* NBRC 12849T. Fermentation, 9: 924. https://doi.org/10.3390/fermentation9110924.

Křížová, L., Dadáková, K., Dvořáčková, M. and Kašparovský, T. (2021). Feedborne mycotoxins beauvericin and enniatins and livestock animals. Toxins, 13: 32. https://doi.org/10.3390/toxins13010032.

Kumar, A., Sørensen, J.L., Hansen, F.T., Arvas, M., Syed, M.F. et al. (2018c). Genome sequencing and analyses of two marine fungi from the North Sea unraveled a plethora of novel biosynthetic gene clusters. Sci. Rep., 8: 10187. https://doi.org/10.1038/s41598-018-28473-z.

Kumar, M., Kajal, Taranum, N., Khyati, Balan, B., Kumar, P. and Dhaulaniya, A.S. (2024). Antifungal potential of bioactive compounds derived from microbes and other natural sources: Challenges and future scope. pp. 591–609. *In*: Manzoor, N. (ed.). Advances in Antifungal Drug Development: Natural Products with Antifungal Potential. Springer Nature, Singapore.

Kumar, V., Sarma, V.V., Thambugala, K.M., Huang, J.J., Li, X.Y. et al. (2021). Ecology and evolution of marine fungi with their adaptation to climate change. Front. Microbiol., 12: 719000. https://doi.org/10.3389/fmicb.2021.719000.

Kumari, M., Taritla, S., Sharma, A. and Jayabaskaran, C. (2018). Antiproliferative and antioxidative bioactive compounds in extracts of marine-derived endophytic fungus *Talaromyces purpureogenus*. Front. Microbiol., 9: 1777. https://doi.org/10.3389/fmicb.2018.01777.

Kwon, M.J., Steiniger, C., Cairns, T.C., Wisecaver, J.H., Lind, A.L. et al. (2021). Beyond the biosynthetic gene cluster paradigm: Genome-wide coexpression networks connect clustered and unclustered transcription factors to secondary metabolic pathways. Microbiol. Spectr., 9: e0089821. https://doi.org/10.1128/Spectrum.00898-21.

Laverdière, M. (1994). Systemic antifungal drugs: Are we making any progress? Can. J. Infect. Dis., 5: 59. https://doi.org/10.1155/1994/828061.

Leal, K., Rojas, E., Madariaga, D., Contreras, M.J., Nuñez-Montero, K. et al. (2024). Unlocking fungal potential: The CRISPR-Cas system as a strategy for secondary metabolite discovery. J. Fungi, 10: 748. https://doi.org/10.3390/jof10110748.

Lee, N., Hwang, S., Kim, J., Cho, S., Palsson, B. et al. (2020). Mini review: Genome mining approaches for the identification of secondary metabolite biosynthetic gene clusters in *Streptomyces*. Comput. Struct. Biotechnol. J., 18: 1548–1556. https://doi.org/10.1016/j.csbj.2020.06.024.

Lei, R., Peng, Y., He, Y. and Li, J. (2024). Some remarks on the argument appealing to nature against synthetic biology. Front. Bioeng. Biotechnol., 12: 1428832. https://doi.org/10.3389/fbioe.2024.1428832.

Lengrand, S., Pesenti, L., Bragard, C. and Legrève, A. (2024) Bacterial endophytome sources, profile and dynamics: A conceptual framework. Front. Sustain. Food. Syst., 8: 1378436. https://doi.org/10.3389/fsufs.2024.1378436.

Lensch, A., Lindfors, H.A., Duwenig, E., Fleischmann, T., Hjort, C. et al. (2024). Safety aspects of microorganisms deliberately released into the environment. EFB Bioeconomy J., 4: 100061. https://doi.org/10.1016/j.bioeco.2023.100061.

Li, H.L., Li, X.M., Li, X., Wang, C.Y., Liu, H. et al. (2017). Antioxidant hydroanthraquinones from the marine algal-derived endophytic fungus *Talaromyces islandicus* EN-501. J. Nat. Prod., 80: 162–168. https://doi.org/10.1021/acs.jnatprod.6b00797.

Li, J., Jia, H., Chen, C., An, S., Yu, J. et al. (2025a). Congested heptacyclic meroterpenoids with anti-SARS-CoV-2 and anti-inflammatory activities from mangrove endophytic fungus *Talaromyces amestolkiae* SCNU-F0041. Phytochem., 231: 114344. https://doi.org/10.1016/j.phytochem.2024.114344.

Li, L. (2023). Next-generation synthetic biology approaches for the accelerated discovery of microbial natural products. Eng. Microbiol., 3: 100060. https://doi.org/10.1016/j.engmic.2022.100060.

Li, L., Zhang, W., Cheng, S., Cao, D. and Parent, M. (2012). Isoprenoids and related pharmacological interventions: Potential application in Alzheimer's disease. Mol. Neurobiol., 46: 64–77. https://doi.org/10.1007/s12035-012-8253-1.

Li, S., Xiao, Y., Li, Q., Su, M., Guo, Y. et al. (2025b). Recent advances in natural products derived from marine echinoderms and endophytic microbes: Chemical insights and therapeutic potential. Mar. Drugs, 23: 33. https://doi.org/10.3390/md23010033.

Li, X., Xu, H., Li, Y., Liao, S. and Liu, Y. (2023). Exploring diverse bioactive secondary metabolites from marine microorganisms using co-culture strategy. Molecules, 28: 6371. https://doi.org/10.3390/molecules28176371.

Li, X.D., Li, X.M., Wang, B.G. and Li, X. (2024). Antimicrobial sesterterpenoids with a unique 5/8/6/5 tetracyclic carbon-ring-system and diepoxide polyketides from a deep sea-sediment-sourced fungus *Chaetomium globosum* SD-347. Org. Biomol. Chem., 22: 3979–3985. https://doi.org/10.1039/d4ob00449c.

Liang, J., She, J., Fu, J., Wang, J., Ye, Y. et al. (2023). Advances in natural products from the marine-sponge-associated microorganisms with antimicrobial activity in the last decade. Mar. Drugs, 21: 236. https://doi.org/10.3390/md21040236.

Liao, C., Doilom, M., Jeewon, R., Hyde, K.D., Manawasinghe, I.S. et al. (2025). Challenges and update on fungal endophytes: Classification, definition, diversity, ecology, evolution, and functions. Fungal Divers., 131: 301–367. https://doi.org/10.1007/s13225-025-00550-5.

Liu, S., Dai, H., Makhloufi, G., Heering, C., Janiak, C. et al. (2016). Cytotoxic 14-membered macrolides from a mangrove-derived endophytic fungus, *Pestalotiopsis microspora*. J. Nat. Prod., 79: 2332–2340. https://doi.org/10.1021/acs.jnatprod.6b00473.

Liu, Y., Lin, L., Zheng, H., He, Y.L., Li, Y. et al. (2023). Mechanisms of antitumor invasion and metastasis of the marine fungal derivative Epi-Aszonalenin A in HT1080 cells. Mar. Drugs, 21: 156. https://doi.org/10.3390/md21030156.

Liu, Z., Li, M., Wang, S., Huang, H. and Zhang, W., (2022). Sulfur-containing metabolites from marine and terrestrial fungal sources: Origin, structures, and bioactivities. Mar. Drugs, 20: 765. https://doi.org/10.3390/md20120765.

Liu, Z., Lin, Z. and Nielsen, J. (2021). Expression of fungal biosynthetic gene clusters in *S. cerevisiae* for natural product discovery. Synth. Syst. Biotechnol., 6: 20–22. https://doi.org/10.1016/j.synbio.2021.01.003.

Luan, X., Wu, Y., Shen, Y.W., Zhang, H., Zhou, Y.D. et al. (2021). Cytotoxic and antitumor peptides as novel chemotherapeutics. Nat. Prod. Rep., 38: 7–17. https://doi.org/10.1039/D0NP00019A.

Lücking, R., Aime, M.C., Robbertse, B., Miller, A.N., Ariyawansa, H.A. et al. (2020). Unambiguous identification of fungi: Where do we stand and how accurate and precise is fungal DNA barcoding? IMA Fungus, 11: 14. https://doi.org/10.1186/s43008-020-00033-z.

Luk, L.Y.P. and Tanner, M.E. (2009). Mechanism of dimethylallyltryptophan synthase: Evidence for a dimethylallyl cation intermediate in an aromatic prenyltransferase reaction. J. Am. Chem. Soc., 131: 13932–13933. https://doi.org/10.1021/ja906485u.

Luo, Y., Enghiad, B. and Zhao, H. (2016). New tools for reconstruction and heterologous expression of natural product biosynthetic gene clusters. Nat. Prod. Rep., 33: 174–182. https://doi.org/10.1039/c5np00085h.

Lyu, C., Chen, T., Qiang, B., Liu, N., Wang, H. et al. (2020). CMNPD: A comprehensive marine natural products database towards facilitating drug discovery from the ocean. Nucleic Acids Res., 49: D509–D515. https://doi.org/10.1093/nar/gkaa763.

Manganyi, M.C. and Ateba, C.N. (2020). Untapped potentials of endophytic fungi: A review of novel bioactive compounds with biological applications. Microorganisms, 8: 1934. https://doi.org/10.3390/microorganisms8121934.

Martín-Rodríguez, A.J., Reyes, F., Martín, J., Pérez-Yépez, J., León-Barrios, M. et al. (2014). Inhibition of Bacterial quorum sensing by extracts from aquatic fungi: First report from marine endophytes. Mar. Drugs, 12: 5503–5526. https://doi.org/10.3390/md12115503.

Martins-Santana, L., Nora, L.C., Sanches-Medeiros, A., Lovate, G.L., Cassiano, M.H. et al. (2018). Systems and synthetic biology approaches to engineer fungi for fine chemical production. Front. Bioeng. Biotechnol., 6: 117. https://doi.org/10.3389/fbioe.2018.00117.

Matilla, M.A., Stöckmann, H., Leeper, F.J. and Salmond, G.P.C. (2012). Bacterial biosynthetic gene clusters encoding the anti-cancer haterumalide class of molecules: Biogenesis of the broad spectrum antifungal and anti-oomycete compound, oocydin A. J. Biol. Chem., 287: 39125–39138. https://doi.org/10.1074/jbc.M112.401026.

Matsuda, Y. and Abe, I. (2016). Biosynthesis of fungal meroterpenoids. Nat. Prod. Rep., 33: 26–53. https://doi.org/10.1039/C5NP00090D.

Mayer, A.M.S., Rodríguez, A.D., Berlinck, R.G.S. and Fusetani, N. (2011). Marine pharmacology in 2007–8: Marine compounds with antibacterial, anticoagulant, antifungal, anti-inflammatory, antimalarial, antiprotozoal, antituberculosis, and antiviral activities; Affecting the immune and nervous system, and other miscellaneous mechanisms of action. Comp. Biochem. Physiol. C. Toxicol. Pharmacol., 153: 191–222. https://doi.org/10.1016/j.cbpc.2010.08.008.

Meng, L.H., Wang, C.Y., Mándi, A., Li, X.M., Hu, X.Y. et al. (2016). Three diketopiperazine alkaloids with spirocyclic skeletons and one bisthiodiketopiperazine derivative from the mangrove-derived endophytic fungus *Penicillium brocae* MA-231. Org. Lett., 18: 5304–5307. https://doi.org/10.1021/acs.orglett.6b02620.

Meng, X., Fang, Y., Ding, M., Zhang, Y., Jia, K. et al. (2022). Developing fungal heterologous expression platforms to explore and improve the production of natural products from fungal biodiversity. Biotechnol. Adv., 54: 107866. https: //doi.org/10.1016/j.biotechadv.2021.107866.

Meng, Z.H., Sun, T.T., Zhao, G.Z., Yue, Y.F., Chang, Q.H. et al. (2020). Marine-derived fungi as a source of bioactive indole alkaloids with diversified structures. Mar. Life. Sci. Technol., 3: 44–61. https://doi.org/10.1007/s42995-020-00072-w.

Metsämuuronen, S. and Sirén, H. (2019). Bioactive phenolic compounds, metabolism, and properties: A review on valuable chemical compounds in Scots pine and Norway spruce. Phytochem, Rev., 18: 623–664. https://doi.org/10.1007/s11101-019-09630-2.

Mohamed, O.G., Salim, A.A., Khalil, Z.G., Elbanna, A.H., Bernhardt, P.V. et al. (2020). Chrysosporazines F–M: P-Glycoprotein inhibitory phenylpropanoid piperazines from an Australian marine fish derived fungus, *Chrysosporium* sp. CMB-F294. J. Nat. Prod., 83: 497–504. https://doi.org/10.1021/acs.jnatprod.9b01181.

Mukhopadhyay, S., Bhunjun, C.S., Phukhamsakda, C., Apurillo, C.C.S., Al-Otibi, F. et al. (2024). Introducing a new species, Vaginatispora acrostichi (Lophiostomataceae), based on morphology and multigene phylogeny. Bot. Mar., 67: 401–410. https://doi.org/10.1515/bot-2024-0008.

Na, H., Zheng, Y.Y., Jia, Y., Feng, J., Huang, J. et al. (2024). Screening and genetic engineering of marine-derived *Aspergillus terreus* for high-efficient production of lovastatin. Microb. Cell. Fac., 23: 134. https://doi.org/10.1186/s12934-024-02396-z.

Naranjo-Ortiz, M.A. and Gabaldón, T. (2019). Fungal evolution: Major ecological adaptations and evolutionary transitions. Biol. Rev. Camb. Philos. Soc., 94: 1443–1476. https://doi.org/10.1111/brv.12510.

Nazir, M., Saleem, M., Tousif, M.I., Anwar, M.A., Surup, F. et al. (2021). Meroterpenoids: A comprehensive update insight on structural diversity and biology. Biomolecules, 11: 957. https://doi.org/10.3390/biom11070957.

Netz, N. and Opatz, T. (2015). Marine indole alkaloids. Mar. Drugs, 13: 4814–4914. https://doi.org/10.3390/md13084814.

Nicoletti, R. and Trincone, A. (2016). Bioactive compounds produced by strains of *Penicillium* and *Talaromyces* of marine origin. Mar. Drugs, 14: 37. https://doi.org/10.3390/md14020037.

Nigdelioglu Dolanbay, S., Kocanci, F.G. and Aslim, B. (2021). Neuroprotective effects of allocryptopine-rich alkaloid extracts against oxidative stress-induced neuronal damage. Biomed. Pharmacother., 140: 111690. https://doi.org/10.1016/j.biopha.2021.111690.

Nishikori, S., Takemoto, K., Kamisuki, S., Nakajima, S., Kuramochi, K. et al. (2016). Anti-hepatitis C virus natural product from a fungus, *Penicillium herquei*. J. Nat. Prod., 79: 442–446. https://doi.org/10.1021/acs.jnatprod.5b00555.

Noor, S., Begum, M.N., Rony, S.R., Chowdhury, A.A., Sohrab, M.H. et al. (2025). Bioactivity and chemical screening of endophytic fungi associated with seaweeds *Gracilaria* sp. and *Sargassum* sp. of the Bay of Bengal, Bangladesh. Sci. Rep., 15: 16121. https://doi.org/10.1038/s41598-025-00099-y.

Núñez-Pons, L., Shilling, A., Verde, C., Baker, B.J. and Giordano, D. (2020) Marine terpenoids from polar latitudes and their potential applications in biotechnology. Mar. Drugs, 18: 401. https://doi.org/10.3390/md18080401.

Okechukwu, Q.N., Adepoju, F.O., Kanwugu, O.N., Adadi, P., Serrano-Aroca, Á. et al. (2024). Marine-derived bioactive metabolites as a potential therapeutic intervention in managing viral diseases: Insights from the SARS-CoV-2 *in silico* and pre-clinical studies. Pharmaceuticals, 17: 328. https://doi.org/10.3390/ph17030328.

Ortega, H.E., Torres-Mendoza, D., Caballero, E. Z. and Cubilla-Rios, L. (2021). Structurally uncommon secondary metabolites derived from endophytic fungi. J. Fungi, 7: 570. https://doi.org/10.3390/jof7070570.

Othman, L., Sleiman, A. and Abdel-Massih, R.M. (2019). Antimicrobial activity of polyphenols and alkaloids in middle eastern plants. Front. Microbiol., 10: 911 https://doi.org/10.3389/fmicb.2019.00911.

Pan, C., Hassan, S.S.U., Muhammad, I. and Jin, H. (2025). Marine fungi as a goldmine for novel antibiotics: A 2024 perspective. Front. Mar. Sci., 11. https://doi.org/10.3389/fmars.2024.1538136.

Pang, K.L., Overy, D.P., Jones, E.G., da Luz Calado, M., Burgaud, G. et al. (2016). Marine fungi and 'marine-derived fungi' in natural product chemistry research: Toward a new consensual definition. Fungal. Biol. Rev., 30: 163–175. https://doi.org/10.1016/j.fbr.2016.08.001.

Pant, A. and Vasundhara, M. (2023). Endophytic fungi: A potential source for drugs against central nervous system disorders. Braz. J. Microbiol., 54: 1479. https://doi.org/10.1007/s42770-023-00997-1.

Paoli, L., Ruscheweyh, H.J., Forneris, C.C., Hubrich, F., Kautsar, S. et al. (2022). Biosynthetic potential of the global ocean microbiome. Nature, 607: 111–118. https://doi.org/10.1038/s41586-022-04862-3.

Papikinou, M.A., Pavlidis, K., Cholidis, P., Kranas, D., Adamantidi, T. et al. (2024). Marine fungi bioactives with anti-inflammatory, antithrombotic, and antioxidant health-promoting properties against inflammation-related chronic diseases. Mar. Drugs, 22: 520. https://doi.org/10.3390/md22110520.

Patel, K.D., Ahmed, S.F., MacDonald, M.R. and Gulick. A.M. (2023). Structural studies of modular nonribosomal peptide synthetases. pp. 17–46. *In*: Burkart, M. and Ishikawa, F. (eds.). Non-Ribosomal Peptide Biosynthesis and Engineering: Methods and Protocols. Springer US, New York, NY.

Pereira, R.B., Evdokimov, N.M., Lefranc, F., Valentão, P., Kornienko, A. et al. (2018). Butyrolactone-I from coral-derived fungus Aspergillus terreus attenuates neuro-inflammatory response via suppression of NF-κB pathway in BV-2 cells. Mar. Drugs, 16: 202. https://doi.org/10.3390/md16060202.

Pereira, R.B., Evdokimov, N.M., Lefranc, F., Valentão, P., Kornienko, A. et al. (2019). Marine-derived anticancer agents: Clinical benefits, innovative mechanisms, and new targets. Mar. Drugs, 17: 329. https://doi.org/10.3390/md17060329.

Perfect, J.R. (2017). The antifungal pipeline: A reality check. Nat. Rev. Drug. Discov., 16: 603–616. https://doi.org/10.1038/nrd.2017.46.

Petersen, L.E., Kellermann, M.Y. and Schupp, P.J. (2020). Secondary metabolites of marine microbes: From natural products chemistry to chemical ecology. *In*: YOUMARES 9—The Oceans: Our Research, Our Future. Springer, Cham, pp. 159–180.

Phyo, Y.Z., Ribeiro, J., Fernandes, C., Kijjoa, A. and Pinto, M.M. (2018). Marine natural peptides: Determination of absolute configuration using liquid chromatography methods and evaluation of bioactivities. Mol. Basel Switz., 23: 306. https://doi.org/10.3390/molecules23020306.

Pillay, L.C., Nekati, L., Makhwitine, P.J. and Ndlovu, S.I. (2022). Epigenetic activation of silent biosynthetic gene clusters in endophytic fungi using small molecular modifiers. Front. Microbiol., 13: 815008. https://doi.org/10.3389/fmicb.2022.815008.

Pivkin, M. (2000). Filamentous fungi associated with holothurians from the sea of Japan, off the primary coast of Russia. Biol. Bull., 198: 101–109. https://doi.org/10.2307/1542808.

Pompeng, P., Sommit, D., Sriubolmas, N., Ngamrojanavanich, N., Matsubara, K. et al. (2013). Antiangiogenetic effects of anthranoids from Alternaria sp., an endophytic fungus in a Thai medicinal plant Erythrina variegata. Phytomedicine, 20: 918–922. https://doi.org/10.1016/j.phymed.2013.03.019.

Pontius, A., Krick, A., Kehraus, S., Brun, R. and König, G.M. (2008) Antiprotozoal activities of heterocyclic-substituted xanthones from the marine-derived fungus *Chaetomium* sp. J. Nat. Prod., 71: 1579–1584. https://doi.org/10.1021/np800294q.

Proksch, P., Edrada, R. and Ebel, R. (2002). Drugs from the seas: Current status and microbiological implications. Appl. Microbiol. Biotechnol., 59: 125–134. https://doi.org/10.1007/s00253-002-1006-8.

Pushparaj, K., Meyyazhagan, A., Bhotla, H.K., Arumugam, V.A., Pappuswamy, M. et al. (2023). The crux of bioactive metabolites in endophytic and thermophilic fungi and their proximal prospects in biotechnological and industrial domains. Toxicon, 223: 107007. https://doi.org/10.1016/j.toxicon.2022.107007.

Qader, M.M., Hamed, A.A., Soldatou, S., Abdelraof, M., Elawady, M.E. et al. (2021). Antimicrobial and antibiofilm activities of the fungal metabolites isolated from the marine endophytes *Epicoccum nigrum* M13 and *Alternaria alternata* 13A. Mar. Drugs, 19: 232. https://doi.org/10.3390/md19040232.

Qiu, Y., Chen, S., Yu, M., Shi, J., Liu, J. et al. (2024). Natural products from marine-derived fungi with anti-inflammatory activity. Mar. Drugs, 22: 433. https://doi.org/10.3390/md22100433.

Quang, T.H., Phong, N.V., Hanh, T.T.H., Cuong, N.X., Ngan, N.T.T. et al. (2021). Cytotoxic and immunomodulatory phenol derivatives from a marine sponge-derived fungus Ascomycota sp. VK12. Nat. Prod. Res., 35: 5153–5159. https://doi.org/10.1080/14786419.2020.1786829.

Rajulu, M.G., Rajamani, T., Murali, T.S., Suryanarayanan, T.S. and Minj, D. (2022). The fungal endobiome of seaweeds of the Andaman Islands, India. Curr. Sci., 123: 1508. https://doi.org/10.18520/cs/v123/i12/1508-1514.

Ramesh, T., Yamunadevi, R., Sundaramanickam, A., Thangaraj, M., Kumaran, R. et al. (2021). Biodiversity of the fungi in extreme marine environments. pp. 75–100. *In*: Sharma, V.K., Shah, M.P., Parmar, S. and Kumar, A. (eds.). Fungi Bio-Prospects in Sustainable Agriculture, Environment, and Nano-technology. Academic Press.

Rateb, M.E. and Ebel, R. (2011a). Secondary metabolites of fungi from marine habitats. Nat. Prod. Rep., 28: 290–344. https://doi.org/10.1039/C0NP00061B.

Reen, F.J., Romano, S., Dobson, A.D.W. and O'Gara, F. (2015). The sound of silence: Activating silent biosynthetic gene clusters in marine microorganisms. Mar. Drugs, 13: 4754–4783. https://doi.org/10.3390/md13084754.

Ren, C., Liu, Z., Wang, X. and Qin, S. (2022). The seaweed holobiont: From microecology to biotechnological applications. Microb. Biotechnol., 15: 738–754. https://doi.org/10.1111/1751-7915.14014.

Richter, F., Bindschedler, S., Calonne-Salmon, M., Declerck, S., Junier, P. et al. (2022). Fungi-on-a-Chip: microfluidic platforms for single-cell studies on fungi. FEMS Microbiol. Rev., 46: fuac039. https://doi.org/10.1093/femsre/fuac039.

Risdian, C., Mozef, T. and Wink, J. (2019). Biosynthesis of polyketides in *Streptomyces*. Microorganisms, 7: 124. https://doi.org/10.3390/microorganisms7050124.

Roik, A., Reverter, M. and Pogoreutz, C. (2022). A roadmap to understanding diversity and function of coral reef-associated fungi. FEMS Microbiol. Rev., 46(6): p.fuac028. DOI: 10.1093/femsre/fuac028.

Romano, S., Jackson, S.A., Patry, S. and Dobson, A.D.W. (2018). Extending the "One Strain Many Compounds" (OSMAC) principle to marine microorganisms. Mar. Drugs, 16: 244. https://doi.org/10.3390/md16070244.

Rosero-Chasoy, G., Rodríguez-Jasso, R.M., Aguilar, C.N., Buitrón, G., Chairez, I. et al. (2021). Microbial co-culturing strategies for the production of high value compounds, a reliable framework towards sustainable biorefinery implementation: An overview. Bioresour. Technol., 321: 124458. https://doi.org/10.1016/j.biortech.2020.124458.

Rottinghaus, A.G., Ferreiro, A., Fishbein, S.R., Dantas, G. and Moon, T.S. (2022). Genetically stable CRISPR-based kill switches for engineered microbes. Nat. Commun., 13: 672. https://doi.org/10.1038/s41467-022-28163-5.

Roy, B.G. (2017). Potential of small-molecule fungal metabolites in antiviral chemotherapy. Antivir. Chem. Chemother., 25: 20–52. https://doi.org/10.1177/2040206617705500.

Saeed, A.F.U.H., Su, J. and Ouyang, S. (2021). Marine-derived drugs: Recent advances in cancer therapy and immune signaling. Biomed. Pharmacother, 134: 111091. https://doi.org/10.1016/j.biopha.2020.111091.

Safavi-Hemami, H., Brogan, S.E. and Olivera, B.M. (2019). Pain therapeutics from cone snail venoms: From ziconotide to novel non-opioid pathways. J. Proteomics, 190: 12–20. https://doi.org/10.1016/j.jprot.2018.05.009.

Sagar, S., Kaur, M. and Minneman, K.P. (2010). Antiviral lead compounds from marine sponges. Mar. Drugs, 8: 2619–2638. https://doi.org/10.3390/md8102619.

Sahoo, S., Subban, K. and Chelliah, J. (2021). Diversity of marine macro-algicolous endophytic fungi and cytotoxic potential of biscogniauxia petrensis metabolites against cancer cell lines. Front. Microbiol., 12: 650177. https://doi.org/10.3389/fmicb.2021.650177.

Sakayaroj, J., Preedanon, S., Supaphon, O., Jones, E.B.G. and Phongpaichit, S. (2010). Phylogenetic diversity of endophyte assemblages associated with the tropical seagrass Enhalus acoroides in Thailand. Fungal Divers., 42: 27–45. https://doi.org/10.1007/s13225-009-0013-9.

Saxena, S., Dufossé, L., Deshmukh, S.K., Chhipa, H. and Gupta, M.K. (2024). Endophytic fungi: A treasure trove of antifungal metabolites. Microorganisms, 12: 1903. https://doi.org/10.3390/microorganisms12091903.

Seipp, K., Geske, L. and Opatz, T. (2021). Marine pyrrole alkaloids. Mar. Drugs, 19: 514. https://doi.org/10.3390/md19090514.

Sen, K., Sen, B. and Wang, G. (2022). Diversity, abundance, and ecological roles of planktonic fungi in marine environments. J. Fungi, 8: 491. https://doi.org/10.3390/jof8050491.

Shao, S., Wang, X., She, J., Zhang, H., Pang, X. et al. (2022). Diversified chaetoglobosins from the marine-derived fungus *Emericellopsis* sp. SCSIO41202. Molecules, 27: 1823. https://doi.org/10.3390/molecules27061823.

Singh, A., Chaudhary, S., Shankar, A. and Prasad, V. (2019). Polyketide and its derivatives. pp. 219–227. *In*: Singh, H.B., Gupta, V.K. and Jogaiah, S. (eds.). New and Future Developments in Microbial Biotechnology and Bioengineering. Elsevier, Amsterdam.

Singh, B., Mohanty, D. and Upadhyay, A.K. (2020). Computational approach to understand fungal secondary metabolites and responsible biological molecules. pp. 105–114. *In*: Singh, J. and Gehlot, P. (eds). New and Future Developments in Microbial Biotechnology and Bioengineering. Elsevier.

Singh, V.K. and Kumar, A. (2023). Secondary metabolites from endophytic fungi: Production, methods of analysis, and diverse pharmaceutical potential. Symbiosis, 90: 111–125. https://doi.org/10.1007/s13199-023-00925-9.

Siridechakorn, I., Yue, Z., Mittraphab, Y., Lei, X. and Pudhom, K. (2017). Identification of spirobisnaphthalene derivatives with anti-tumor activities from the endophytic fungus *Rhytidhysteron rufulum* AS21B. Bioorg. Med. Chem., 25: 2878–2882. https://doi.org/10.1016/j.bmc.2017.02.054.

Soerjawinata, W., Kockler, I., Wommer, L., Frank, R., Schüffler, A. et al. (2022). Novel bioreactor internals for the cultivation of spore-forming fungi in pellet form. Eng. Life Sci., 22: 474–483. https://doi.org/10.1002/elsc.202100094.

Song, Z., Hou, Y., Yang, Q., Li, X. and Wu, S. (2021). Structures and biological activities of diketopiperazines from marine organisms: A review. Mar. Drugs, 19: 403. https://doi.org/10.3390/md19080403.

Spiering, M.J., Moon, C.D., Wilkinson, H.H. and Schardl, C.L. (2005). Gene clusters for insecticidal loline alkaloids in the grass-endophytic fungus *Neotyphodium uncinatum*. Genetics, 169: 1403–1414. https://doi.org/10.1534/genetics.104.035972.

Sridhar, K.R. (2018). Diversity, ecology, and significance of fungal endophytes. pp. 1–40. *In*: Jha, S. (ed.). Endophytes and Secondary Metabolites. Springer International Publishing, Cham.

Suarez-Jimenez, G.M., Burgos-Hernandez, A. and Ezquerra-Brauer, J.M. (2012). Bioactive peptides and depsipeptides with anticancer potential: Sources from marine animals. Mar. Drugs, 10: 963–986. https://doi.org/10.3390/md10050963.

Subramani, R., Kumar, R., Prasad, P. and Aalbersberg, W. (2013). Cytotoxic and antibacterial substances against multi drug resistant pathogens from marine sponge symbiont: Citrinin, a secondary metabolite of *Penicillium* sp. Asian Pac. J. Trop. Biomed., 3: 291–296. https://doi.org/10.1016/S2221-1691(13)60065-9.

Supaphon, P., Phongpaichit, S., Rukachaisirikul, V. and Sakayaroj, J. (2013). Antimicrobial potential of endophytic fungi derived from three seagrass species: *Cymodocea serrulata, Halophila ovalis*, and *Thalassia hemprichii*. PLoS One, 8: e72520. https://doi.org/10.1371/journal.pone.0072520.

Suraiya, S., Jang, W.J., Haq, M. and Kong, I.S. (2024). Isolation and characterization of β-glucan containing polysaccharides from *Monascus* spp. using *Saccharina japonica* as submerged fermented substrate. Polysaccharides, 5: 435–449. https://doi.org/10.3390/polysaccharides5030027.

Suryanarayanan, T.S. (2012). Fungal endosymbionts of seaweeds. pp. 53–69. *In*: Raghukumar, C. (ed.). Biology of Marine Fungi. Springer, Berlin, Heidelberg.

Sychantha, D., Rotondo, C.M., Tehrani, K.H., Martin, N.I. and Wright, G.D. (2021). Aspergillomarasmine A inhibits metallo-β-lactamases by selectively sequestering Zn2+. J. Biol. Chem., 297: 100918. https://doi.org/10.1016/j.jbc.2021.100918.

Tacconelli, E. (2017). Global priority list of antibiotic-resistant bacteria to guide research, discovery, and development. chrome-extension://efaidnbmnnnibpcajpcglclefindmkaj/https://www.quotidianosanita.it/allegati/allegato4135670.pdf.

Tang, R., Kimishima, A., Ishida, R., Setiawan, A. and Arai, M. (2020). Selective cytotoxicity of epidithiodiketopiperazine DC1149B, produced by marine-derived Trichoderma lixii on the cancer cells adapted to glucose starvation. J. Nat. Med., 74: 153–158. https://doi.org/10.1007/s11418-019-01357-w.

Tapfuma, K.I., Nyambo, K., Adu-Amankwaah, F., Baatjies, L., Smith, L. et al. (2022). Antimycobacterial activity and molecular docking of methanolic extracts and compounds of marine fungi from Saldanha and False Bays, South Africa. Heliyon, 8: e12406. https://doi.org/10.1016/j.heliyon.2022.e12406.

Thell, K., Hellinger, R., Schabbauer, G. and Gruber, C.W. (2014). Immunosuppressive peptides and their therapeutic applications. Drug Discov. Today, 19: 645–653. https://doi.org/10.1016/j.drudis.2013.12.002.

Thomas, T.R.A., Kavlekar, D.P. and LokaBharathi, P.A. (2010). Marine drugs from sponge-microbe association: A Review. Mar. Drugs, 8: 1417–1468. https://doi.org/10.3390/md8041417.

Tian, Y. and Li, Y. (2022). A review on bioactive compounds from marine-derived *Chaetomium* Species. J. Microbiol. Biotechnol., 32(5): 541–550. https://doi.org/10.4014/jmb.2201.01007.

Torres-Fuentes, C., Suarez, M., Aragones, G., Mulero, M., Ávila-Román, J. et al. (2022). Cardioprotective properties of phenolic compounds: A role for biological rhythms. Mol. Nutr. Food Res., 66: e2100990. https://doi.org/10.1002/mnfr.202100990.

Tracanna, M.I., Fortuna, A.M., Contreras Cárdenas, A.V., Marr, A.K., McMaster, W.R. et al. (2015). Anti-leishmanial, anti-inflammatory, and antimicrobial activities of phenolic derivatives from *Tibouchina paratropica*. Phytother. Res., 29: 393–397. https://doi.org/10.1002/ptr.5263.

Udayan, S.P., Hariharan, S. and Nevin, K.G. (2024). Multifaceted bioactivity of marine fungal derived secondary metabolite, xyloketal B: A review. Toxicol. Res., 13: tfae156. https://doi.org/10.1093/toxres/tfae156.

Uzma, F., Mohan, C.D., Hashem, A., Konappa, N.M., Rangappa, S. et al. (2018). Endophytic fungi-alternative sources of cytotoxic compounds: A review. Front. Pharmacol., 9: 309. https://doi.org/10.3389/fphar.2018.00309.

Valli, J., Kanimozhi, K., Ambikapathy, V. and Panneerselvam, A. (2024). Isolation and identification with different enzyme production from marine associated plants. Int. J. Biosci., 25: 12–20. https://doi.org/10.12692/ijb/25.3.12-20.

Vavitsas, K., Fabris, M. and Vickers, C.E. (2018). Terpenoid metabolic engineering in photosynthetic microorganisms. Genes, 9: 520. https://doi.org/10.3390/genes9110520.

Venkatachalam, J., Jeyadoss, V.S., Bose, K.S.C. and Subramanian, R. (2024). Marine seaweed endophytic fungi-derived active metabolites promote reactive oxygen species-induced cell cycle arrest and apoptosis in human breast cancer cells. Mol. Biol. Rep., 51: 611. https://doi.org/10.1007/s11033-024-09511-8.

Vitale, G.A., Coppola, D., Palma Esposito, F., Buonocore, C., Ausuri, J. et al. (2020). Antioxidant molecules from marine fungi: Methodologies and perspectives. Antioxidants, 9: 1183. https://doi.org/10.3390/antiox9121183.

Wang, B., Cai, J., Huang, L., Chen, Y., Wang, R. et al. (2024a). Significance of research on natural products from marine-derived Aspergillus species as a source against pathogenic bacteria. Front. Microbiol., 15: 1464135. https://doi.org/10.3389/fmicb.2024.1464135.

Wang, C., Lu, H., Lan, J., Zaman, K.A. and Cao, S. (2021) A review: Halogenated compounds from marine fungi. Molecules, 26: 458. https://doi.org/10.3390/molecules26020458.

Wang, G., Ran, H., Fan, J., Keller, N.P., Liu, Z. et al. (2022). Fungal-fungal cocultivation leads to widespread secondary metabolite alteration requiring the partial loss-of-function VeA1 protein. Sci. Adv., 8: eabo6094. https://doi.org/10.1126/sciadv.abo6094.

Wang, H., Fewer, D.P., Holm, L., Rouhiainen, L. and Sivonen, K. (2014). Atlas of nonribosomal peptide and polyketide biosynthetic pathways reveals common occurrence of nonmodular enzymes. Proc. Natl. Acad. Sci., 111: 9259–9264. https://doi.org/10.1073/pnas.1401734111.

Wang, H., Sun, L., Ma, X., Jin, S., Xi, Y. et al. (2024b). Polyketides and alkaloids from the fungus Aspergillus Fumigatus YB4-17 and ent-Fumiquinazoline J induce apoptosis, paraptosis in human hepatoma HepG2 cells. Front. Pharmacol., 15: 1487977. https://doi.org/10.3389/fphar.2024.1487977.

Wang, L., Lu, H. and Jiang, Y. (2023). Natural polyketides act as promising antifungal agents. Biomolecules, 13: 1572. https://doi.org/10.3390/biom13111572.

Wang, Q. and Xu, L. (2012). Beauvericin, a bioactive compound produced by fungi: A short review. Mol. Basel Switz., 17: 2367–2377. https://doi.org/10.3390/molecules17032367.

Waqar, S., Bhat, A.A. and Khan, A.A. (2024). Endophytic fungi: Unravelling plant-endophyte interaction and the multifaceted role of fungal endophytes in stress amelioration. Plant Physiol. Biochem., 206: 108174. https://doi.org/10.1016/j.plaphy.2023.108174.

Wu, X.Z., Huang, W.J., Liu, W., Mandi, A., Zhang, Q. et al. (2021). Penicisteckins A–F, Isochroman-Derived Atropisomeric Dimers from *Penicillium steckii* HNNU-5B18. J. Nat. Prod., 84: 2953–2960. https://doi.org/10.1021/acs.jnatprod.1c00787.

Wu, Y., Chen, S., Liu, H., Huang, X., Liu, Y. et al. (2019). Cytotoxic isocoumarin derivatives from the mangrove endophytic fungus *Aspergillus* sp. HN15-5D. Arch. Pharm. Res., 42: 326–331. https://doi.org/10.1007/s12272-018-1019-1.

Wu, Y., Chen, Y., Huang, X., Pan, Y., Liu, Z. et al. (2018). α-Glucosidase inhibitors: diphenyl ethers and phenolic bisabolane sesquiterpenoids from the mangrove endophytic fungus *Aspergillus flavus* QQSG-3. Mar. Drugs, 16: 307. https://doi.org/10.3390/md16090307.

Xia, G., Li, J., Li, H., Long, Y., Lin, S.E. et al. (2014). Alterporriol-type dimers from the mangrove endophytic fungus, *Alternaria* sp. (SK11), and their MptpB inhibitions. Mar. Drugs, 12: 2953–2969. https://doi.org/10.3390/md12052953.

Xia, Y., Wang, D., Li, J., Chen, M., Wang, D. et al. (2022). Compounds purified from edible fungi fight against chronic inflammation through oxidative stress regulation. Front. Pharmacol., 13: 974794. https://doi.org/10.3389/fphar.2022.974794.

Xu, W., Gong, L., Pang, K.L. and Luo, Z.H. (2018). Fungal diversity in deep-sea sediments of a hydrothermal vent system in the Southwest Indian Ridge. Deep Sea Res., I: Oceanogr. Res. Pap. 131: 16–26. https://doi.org/10.1016/j.dsr.2017.11.001.

Xu, Y., Du, X., Yu, X., Jiang, Q., Zheng, K. et al. (2022). Recent Advances in the heterologous expression of biosynthetic gene clusters for marine natural products. Mar. Drugs, 20: 341. https://doi.org/10.3390/md20060341.

Xue, M., Hou, X., Fu, J., Zhang, J., Wang, J. et al. (2023). Recent advances in search of bioactive secondary metabolites from fungi triggered by chemical epigenetic modifiers. J. Fungi, 9: 172. https://doi.org/10.3390/jof9020172.

Yamada, T., Muroga, Y. and Tanaka, R. (2009). New Azaphilones, seco-chaetomugilins a and d, produced by a marine-fish-derived *Chaetomium globosum*. Mar Drugs, 7: 249–257. https://doi.org/10.3390/md7020249.

Yang, L.H., Ou-Yang, H., Yan, X., Tang, B.W., Fang, M.J. et al. (2018). Open-ring butenolides from a marine-derived anti-neuroinflammatory fungus *Aspergillus terreus* Y10. Mar. Drugs, 16: 428. https://doi.org/10.3390/md16110428.

Yao, F.H., Liang, X., Shen, W.B., Lu, X.H., Li, G.C. et al. (2024). Microascones, decahydrofluorene-class alkaloids from the marine-derived fungus *Microascus* sp. SCSIO 41821. J. Nat. Prod., 87: 810–819. https://doi.org/10.1021/acs.jnatprod.3c0098.

Yasuhara-Bell, J. and Lu, Y. (2010). Marine compounds and their antiviral activities. Antiviral. Res., 86: 231–240. https://doi.org/10.1016/j.antiviral.2010.03.009.

Yu, M., Li, Y., Banakar, S.P., Liu, L., Shao, C. et al. (2019). New metabolites from the co-culture of marine-derived actinomycete *Streptomyces rochei* MB037 and fungus *Rhinocladiella similis* 35. Front. Microbiol., 10: 915. https://doi.org/10.3389/fmicb.2019.00915.

Yuan, Y., Cheng, S., Bian, G., Yan, P., Ma, Z. et al. (2022). Efficient exploration of terpenoid biosynthetic gene clusters in filamentous fungi. Nat. Catal., 5: 277–287. https://doi.org/10.1038/s41929-022-00762-x.

Zhai, X., Yao, L. and Zhou, Y.J. (2022). Construction of microbial chassis for terpenoid discovery. Synth. Syst. Biotechnol., 7: 1181–1182. https://doi.org/10.1016/j.synbio.2022.09.002.

Zhang, B., Zhang, T., Xu, J., Lu, J., Qiu, P. et al. (2020). Marine sponge-associated fungi as potential novel bioactive natural product sources for drug discovery: A review. Mini. Rev. Med. Chem., 20: 1966–2010. https://doi.org/10.2174/1389557520666200826123248.

Zhang, D., Xu, F., Wang, F., Le, L. and Pu, L. (2025a). Synthetic biology and artificial intelligence in crop improvement. Plant Commun., 6: 101220. https://doi.org/10.1016/j.xplc.2024.101220.

Zhang, L., Li, D., Chen, X. and Zhao, F. (2025b). Marine-derived diterpenes from 2019 to 2024: Structures, biological activities, synthesis, and potential applications. Mar. Drugs, 23: 72. https://doi.org/10.3390/md23020072.

Zhang, P., Li, X. and Wang, B.G. (2016). Secondary metabolites from the marine algal-derived endophytic fungi: Chemical diversity and biological activity. Planta Med., 82: 832–842. https://doi.org/10.1055/s-0042-103496.

Zhang, P., Wu, G., Heard, S.C., Niu, C., Bell, S.A. et al. (2022). Identification and characterization of a cryptic bifunctional type i diterpene synthase involved in talaronoid biosynthesis from a marine-derived fungus. Org. Lett., 24: 7037–7041. https://doi.org/10.1021/acs.orglett.2c02904.

Zheng, H., Qiao, M., Xu, J. and Yu, Z. (2021a). Culture-based and culture-independent assessments of endophytic fungal diversity in aquatic plants in southwest China. Front. Fungal. Biol., 2: 692549. https://doi.org/10.3389/ffunb.2021.692549.

Zheng, R., Li, S., Zhang, X. and Zhao, C. (2021b). Biological activities of some new secondary metabolites isolated from endophytic fungi: A review study. Int. J. Mol., Sci., 22: 959. https://doi.org/10.3390/ijms22020959.

Zhong, M., Li, Y., Deng, L., Fang, J. and Yu, X. (2024). Insight into the adaptation mechanisms of high hydrostatic pressure in physiology and metabolism of hadal fungi from the deepest ocean sediment. mSystems, 9(1): e01085-23. https://doi.org/10.1128/msystems.01085-23.

Zhou, J., Diao, X., Wang, T., Chen, G., Lin, Q. et al. (2018). Phylogenetic diversity and antioxidant activities of culturable fungal endophytes associated with the mangrove species *Rhizophora stylosa* and *R. mucronata* in the South China Sea. PLoS One, 13: e0197359. https://doi.org/10.1371/journal.pone.0197359.

Zhou, J., Feng, Z., Zhang, W. and Xu, J. (2022). Evaluation of the antimicrobial and cytotoxic potential of endophytic fungi extracts from mangrove plants *Rhizophora stylosa* and *R. mucronata*. Sci. Rep., 12: 2733. https://doi.org/10.1038/s41598-022-06711-9.

Zhou, P., Yang, X.L., Wang, X.G., Hu, B., Zhang, L. et al. (2020). A pneumonia outbreak associated with a new coronavirus of probable bat origin. Nature, 579: 270–273. https://doi.org/10.1038/s41586-020-2012-7.

Zhu, H., Zeng, W., Chen, M., He, D., Cheng, X. et al. (2024). Endophytic fungal diversity of mangrove ferns *Acrostichum speciosum* and *A. aureum* in China. Plants, 13: 685. https://doi.org/10.3390/plants13050685.

Zhu, S., Xu, H., Liu, Y., Hong, Y., Yang, H. et al. (2025). Computational advances in biosynthetic gene cluster discovery and prediction. Biotechnol. Adv., 79: 108532. https://doi.org/10.1016/j.biotechadv.2025.108532.

Zhu, X., Zhou, D., Liang, F., Wu, Z., She, Z. et al. (2017). Penochalasin K, a new unusual chaetoglobosin from the mangrove endophytic fungus *Penicillium chrysogenum* V11 and its effective semi-synthesis. Fitoterapia, 123: 23–28. https://doi.org/10.1016/j.fitote.2017.09.016.

Zorrilla, J.G. and Evidente, A. (2022). Structures and biological activities of alkaloids produced by mushrooms, a fungal subgroup. Biomolecules, 12: 1025. https://doi.org/10.3390/biom12081025.

Index

About the Editors

Kandikere R. Sridhar is an Adjunct Professor (Department of Biosciences, Mangalore University, India) and a Visiting Professor at NITTE (Deemed to be University), Mangalore, India. His research areas are the Diversity and Ecology of Fungi of the Western Ghats and maritime habitats. He was an NSERC fellow and visiting professor at Mount Allison University, Canada; the Helmholtz Centre for Environmental Research at UFZ and Martin Luther University, Germany; and the Centre of Biology, University of Minho, Portugal. He is a recipient of the Dr. Shome Memorial Award (2004), Vice President (2013), President (2018) of the Mycological Society of India (IMSI). He is a Fellow of the Indian Mycological Society, Kolkata (2014), Distinguished Asian Mycologist (2015), Outstanding Leader in Education and Research, Association of Agricultural Technology, Bangkok (2016), and Fellow of the Karnataka Science and Technology Academy, India (2023). He received the Lifetime Achievement Award from MSI (2019) and the Association of Fungal Biologists (AFB), India (2025). He has mentored 25 Ph.D. theses, 5 M.Phil. theses, and 25 MSc projects (mycology, food science, and myriapodology). He was a UGC-BSR Faculty Fellow, India (2014–2017) and in the top 2% of scientists in the field of mycology (2020–2025). He has published over 600 research papers, edited 22 books and reviewed over 200 research articles, 25 project proposals, 15 book proposals, and has evaluated 25 Ph.D. theses (Australia, India, Mauritius, and Thailand).

Sunil K. Deshmukh is a Scientific Advisor to Greenvention Biotech, Uruli Kanchan, Pune, India, and AGPharm Bioinnovations LLP, Patiala, Punjab, India. He is a veteran industrial mycologist who spent a substantial part of his career in drug discovery at Hoechst Marion Roussel Limited [now Sanofi India Ltd.], Mumbai, and Piramal Enterprises Limited, Mumbai. He has also served at the TERI-Deakin Nanobiotechnology Centre, New Delhi, and as an Adjunct Associate Professor at Deakin University, Australia. He has to his credit 8 patents, 158 publications, and 27 books on various aspects of fungi and natural products of microbial origin. He is the president of the Association of Fungal Biologists (AFB) and a past president of the Mycological Society of India (MSI). Dr. Deshmukh is Managing Editor of *Mycological Spectrum*, the official journal of the Association of Fungal Biologists. He has approximately four decades of research experience dealing with bioactives from fungi, and various aspects of keratinophilic fungi.

Patricia Velez serves as a faculty member (Research Professor) at the Institute of Biology, National Autonomous University of Mexico (UNAM) where her work aims to explore the diversity, ecology, and potential utilization of fungal communities, particularly in marine and threatened ecosystems. She teaches courses on microbial molecular ecology and taxonomy in the Graduate School of Biological Sciences at the UNAM, and has mentored 31 graduate and undergraduate students in diverse topics of mycology. She has published over 40 peer-reviewed publications in international journals, being recognized with numerous awards and fellowships, including the prestigious Alfonso Caso Medal for her doctoral research, the City U–Molitoris Scholarship for Marine Mycology, and the Early-Career Scientists Award (17th Deepsea Biology Symposium) for her research on marine and deepsea fungal diversity. Presently, she serves as a member of the IUCN Specialist Group on Aquatic Fungi and as editor of the *Revista Mexicana de Biodiversidad* and *MycoAsia* journals.

For Product Safety Concerns and Information please contact our
EU representative GPSR@taylorandfrancis.com Taylor & Francis
Verlag GmbH, Kaufingerstraße 24, 80331 München, Germany